abelian groups and modules

PURE AND APPLIED MATHEMATICS

A Program of Monographs, Textbooks, and Lecture Notes

LECTURE NOTES IN PURE AND APPLIED MATHEMATICS

1. *N. Jacobson,* Exceptional Lie Algebras
2. *L.-Å. Lindahl and F. Poulsen,* Thin Sets in Harmonic Analysis
3. *I. Satake,* Classification Theory of Semi-Simple Algebraic Groups
4. *F. Hirzebruch, W. D. Newmann, and S. S. Koh, Differentiable Manifolds and Quadratic Forms*
5. *I. Chavel,* Riemannian Symmetric Spaces of Rank One
6. *R. B. Burckel,* Characterization of C(X) Among Its Subalgebras
7. *B. R. McDonald, A. R. Magid, and K. C. Smith,* Ring Theory: Proceedings of the Oklahoma Conference
8. *Y.-T. Siu,* Techniques of Extension on Analytic Objects
9. *S. R. Caradus, W. E. Pfaffenberger, and B. Yood,* Calkin Algebras and Algebras of Operators on Banach Spaces
10. *E. O. Roxin, P.-T. Liu, and R. L. Sternberg,* Differential Games and Control Theory
11. *M. Orzech and C. Small,* The Brauer Group of Commutative Rings
12. *S. Thomier,* Topology and Its Applications
13. *J. M. Lopez and K. A. Ross,* Sidon Sets
14. *W. W. Comfort and S. Negrepontis,* Continuous Pseudometrics
15. *K. McKennon and J. M. Robertson,* Locally Convex Spaces
16. *M. Carmeli and S. Malin,* Representations of the Rotation and Lorentz Groups: An Introduction
17. *G. B. Seligman,* Rational Methods in Lie Algebras
18. *D. G. de Figueiredo,* Functional Analysis: Proceedings of the Brazilian Mathematical Society Symposium
19. *L. Cesari, R. Kannan, and J. D. Schuur,* Nonlinear Functional Analysis and Differential Equations: Proceedings of the Michigan State University Conference
20. *J. J. Schäffer,* Geometry of Spheres in Normed Spaces
21. *K. Yano and M. Kon,* Anti-Invariant Submanifolds
22. *W. V. Vasconcelos,* The Rings of Dimension Two
23. *R. E. Chandler,* Hausdorff Compactifications
24. *S. P. Franklin and B. V. S. Thomas,* Topology: Proceedings of the Memphis State University Conference
25. *S. K. Jain,* Ring Theory: Proceedings of the Ohio University Conference
26. *B. R. McDonald and R. A. Morris,* Ring Theory II: Proceedings of the Second Oklahoma Conference
27. *R. B. Mura and A. Rhemtulla,* Orderable Groups
28. *J. R. Graef,* Stability of Dynamical Systems: Theory and Applications
29. *H.-C. Wang,* Homogeneous Branch Algebras
30. *E. O. Roxin, P.-T. Liu, and R. L. Sternberg,* Differential Games and Control Theory II
31. *R. D. Porter,* Introduction to Fibre Bundles
32. *M. Altman,* Contractors and Contractor Directions Theory and Applications
33. *J. S. Golan,* Decomposition and Dimension in Module Categories
34. *G. Fairweather,* Finite Element Galerkin Methods for Differential Equations
35. *J. D. Sally,* Numbers of Generators of Ideals in Local Rings
36. *S. S. Miller,* Complex Analysis: Proceedings of the S.U.N.Y. Brockport Conference
37. *R. Gordon,* Representation Theory of Algebras: Proceedings of the Philadelphia Conference
38. *M. Goto and F. D. Grosshans,* Semisimple Lie Algebras
39. *A. I. Arruda, N. C. A. da Costa, and R. Chuaqui,* Mathematical Logic: Proceedings of the First Brazilian Conference
40. *F. Van Oystaeyen,* Ring Theory: Proceedings of the 1977 Antwerp Conference
41. *F. Van Oystaeyen and A. Verschoren,* Reflectors and Localization: Application to Sheaf Theory
42. *M. Satyanarayana,* Positively Ordered Semigroups
43. *D. L Russell,* Mathematics of Finite-Dimensional Control Systems
44. *P.-T. Liu and E. Roxin,* Differential Games and Control Theory III: Proceedings of the Third Kingston Conference, Part A
45. *A. Geramita and J. Seberry,* Orthogonal Designs: Quadratic Forms and Hadamard Matrices
46. *J. Cigler, V. Losert, and P. Michor,* Banach Modules and Functors on Categories of Banach Spaces
47. *P.-T. Liu and J. G. Sutinen,* Control Theory in Mathematical Economics: Proceedings of the Third Kingston Conference, Part B
48. *C. Byrnes,* Partial Differential Equations and Geometry

49. *G. Klambauer,* Problems and Propositions in Analysis
50. *J. Knopfmacher,* Analytic Arithmetic of Algebraic Function Fields
51. *F. Van Oystaeyen,* Ring Theory: Proceedings of the 1978 Antwerp Conference
52. *B. Kadem,* Binary Time Series
53. *J. Barros-Neto and R. A. Artino,* Hypoelliptic Boundary-Value Problems
54. *R. L. Sternberg, A. J. Kalinowski, and J. S. Papadakis,* Nonlinear Partial Differential Equations in Engineering and Applied Science
55. *B. R. McDonald,* Ring Theory and Algebra III: Proceedngs of the Third Oklahoma Conference
56. *J. S. Golan,* Structure Sheaves Over a Noncommutative Ring
57. *T. V. Narayana, J. G. Williams, and R. M. Mathsen,* Combinatorics, Representation Theory and Statistical Methods in Groups: YOUNG DAY Proceedings
58. *T. A. Burton,* Modeling and Differential Equations in Biology
59. *K. H. Kim and F. W. Roush,* Introduction to Mathematical Consensus Theory
60. *J. Banas and K. Goebel,* Measures of Noncompactness in Banach Spaces
61. *O. A. Nielson,* Direct Integral Theory
62. *J. E. Smith, G. O. Kenny, and R. N. Ball,* Ordered Groups: Proceedings of the Boise State Conference
63. *J. Cronin,* Mathematics of Cell Electrophysiology
64. *J. W. Brewer,* Power Series Over Commutative Rings
65. *P. K. Kamthan and M. Gupta,* Sequence Spaces and Series
66. *T. G. McLaughlin,* Regressive Sets and the Theory of Isols
67. *T. L. Herdman, S. M. Rankin III, and H. W. Stech,* Integral and Functional Differential Equations
68. *R. Draper,* Commutative Algebra: Analytic Methods
69. *W. G. McKay and J. Patera,* Tables of Dimensions, Indices, and Branching Rules for Representations of Simple Lie Algebras
70. *R. L. Devaney and Z. H. Nitecki,* Classical Mechanics and Dynamical Systems
71. *J. Van Geel,* Places and Valuations in Noncommutative Ring Theory
72. *C. Faith,* Injective Modules and Injective Quotient Rings
73. *A. Fiacco,* Mathematical Programming with Data Perturbations I
74. *P. Schultz, C. Praeger, and R. Sullivan,* Algebraic Structures and Applications: Proceedings of the First Western Australian Conference on Algebra
75. *L Bican, T. Kepka, and P. Nemec,* Rings, Modules, and Preradicals
76. *D. C. Kay and M. Breen,* Convexity and Related Combinatorial Geometry: Proceedings of the Second University of Oklahoma Conference
77. *P. Fletcher and W. F. Lindgren,* Quasi-Uniform Spaces
78. *C.-C. Yang,* Factorization Theory of Meromorphic Functions
79. *O. Taussky,* Ternary Quadratic Forms and Norms
80. *S. P. Singh and J. H. Burry,* Nonlinear Analysis and Applications
81. *K. B. Hannsgen, T. L. Herdman, H. W. Stech, and R. L. Wheeler,* Volterra and Functional Differential Equations
82. *N. L. Johnson, M. J. Kallaher, and C. T. Long,* Finite Geometries: Proceedings of a Conference in Honor of T. G. Ostrom
83. *G. I. Zapata,* Functional Analysis, Holomorphy, and Approximation Theory
84. *S. Greco and G. Valla,* Commutative Algebra: Proceedings of the Trento Conference
85. *A. V. Fiacco,* Mathematical Programming with Data Perturbations II
86. *J.-B. Hiriart-Urruty, W. Oettli, and J. Stoer,* Optimization: Theory and Algorithms
87. *A. Figa Talamanca and M. A. Picardello,* Harmonic Analysis on Free Groups
88. *M. Harada,* Factor Categories with Applications to Direct Decomposition of Modules
89. *V. I. Istrăţescu,* Strict Convexity and Complex Strict Convexity
90. *V. Lakshmikantham,* Trends in Theory and Practice of Nonlinear Differential Equations
91. *H. L. Manocha and J. B. Srivastava,* Algebra and Its Applications
92. *D. V. Chudnovsky and G. V. Chudnovsky,* Classical and Quantum Models and Arithmetic Problems
93. *J. W. Longley,* Least Squares Computations Using Orthogonalization Methods
94. *L. P. de Alcantara,* Mathematical Logic and Formal Systems
95. *C. E. Aull,* Rings of Continuous Functions
96. *R. Chuaqui,* Analysis, Geometry, and Probability
97. *L. Fuchs and L. Salce,* Modules Over Valuation Domains
98. *P. Fischer and W. R. Smith,* Chaos, Fractals, and Dynamics
99. *W. B. Powell and C. Tsinakis,* Ordered Algebraic Structures
100. *G. M. Rassias and T. M. Rassias,* Differential Geometry, Calculus of Variations, and Their Applications

101. *R.-E. Hoffmann and K. H. Hofmann,* Continuous Lattices and Their Applications
102. *J. H. Lightbourne III and S. M. Rankin III,* Physical Mathematics and Nonlinear Partial Differential Equations
103. *C. A. Baker and L. M. Batten,* Finite Geometrics
104. *J. W. Brewer, J. W. Bunce, and F. S. Van Vleck,* Linear Systems Over Commutative Rings
105. *C. McCrory and T. Shifrin,* Geometry and Topology: Manifolds, Varieties, and Knots
106. *D. W. Kueker, E. G. K. Lopez-Escobar, and C. H. Smith,* Mathematical Logic and Theoretical Computer Science
107. *B.-L. Lin and S. Simons,* Nonlinear and Convex Analysis: Proceedings in Honor of Ky Fan
108. *S. J. Lee,* Operator Methods for Optimal Control Problems
109. *V. Lakshmikantham,* Nonlinear Analysis and Applications
110. *S. F. McCormick,* Multigrid Methods: Theory, Applications, and Supercomputing
111. *M. C. Tangora,* Computers in Algebra
112. *D. V. Chudnovsky and G. V. Chudnovsky,* Search Theory: Some Recent Developments
113. *D. V. Chudnovsky and R. D. Jenks,* Computer Algebra
114. *M. C. Tangora,* Computers in Geometry and Topology
115. *P. Nelson, V. Faber, T. A. Manteuffel, D. L. Seth, and A. B. White, Jr.,* Transport Theory, Invariant Imbedding, and Integral Equations: Proceedings in Honor of G. M. Wing's 65th Birthday
116. *P. Clément, S. Invernizzi, E. Mitidieri, and I. I. Vrabie,* Semigroup Theory and Applications
117. *J. Vinuesa,* Orthogonal Polynomials and Their Applications: Proceedings of the International Congress
118. *C. M. Dafermos, G. Ladas, and G. Papanicolaou,* Differential Equations: Proceedings of the EQUADIFF Conference
119. *E. O. Roxin,* Modern Optimal Control: A Conference in Honor of Solomon Lefschetz and Joseph P. Lasalle
120. *J. C. Díaz,* Mathematics for Large Scale Computing
121. *P. S. Milojevič,* Nonlinear Functional Analysis
122. *C. Sadosky,* Analysis and Partial Differential Equations: A Collection of Papers Dedicated to Mischa Cotlar
123. *R. M. Shortt,* General Topology and Applications: Proceedings of the 1988 Northeast Conference
124. *R. Wong,* Asymptotic and Computational Analysis: Conference in Honor of Frank W. J. Olver's 65th Birthday
125. *D. V. Chudnovsky and R. D. Jenks,* Computers in Mathematics
126. *W. D. Wallis, H. Shen, W. Wei, and L. Zhu,* Combinatorial Designs and Applications
127. *S. Elaydi,* Differential Equations: Stability and Control
128. *G. Chen, E. B. Lee, W. Littman, and L. Markus,* Distributed Parameter Control Systems: New Trends and Applications
129. *W. N. Everitt,* Inequalities: Fifty Years On from Hardy, Littlewood and Pólya
130. *H. G. Kaper and M. Garbey,* Asymptotic Analysis and the Numerical Solution of Partial Differential Equations
131. *O. Arino, D. E. Axelrod, and M. Kimmel,* Mathematical Population Dynamics: Proceedings of the Second International Conference
132. *S. Coen,* Geometry and Complex Variables
133. *J. A. Goldstein, F. Kappel, and W. Schappacher,* Differential Equations with Applications in Biology, Physics, and Engineering
134. *S. J. Andima, R. Kopperman, P. R. Misra, J. Z. Reichman, and A. R. Todd,* General Topology and Applications
135. *P Clément, E. Mitidieri, B. de Pagter,* Semigroup Theory and Evolution Equations: The Second International Conference
136. *K. Jarosz,* Function Spaces
137. *J. M. Bayod, N. De Grande-De Kimpe, and J. Martínez-Maurica,* p-adic Functional Analysis
138. *G. A. Anastassiou,* Approximation Theory: Proceedings of the Sixth Southeastern Approximation Theorists Annual Conference
139. *R. S. Rees,* Graphs, Matrices, and Designs: Festschrift in Honor of Norman J. Pullman
140. *G. Abrams, J. Haefner, and K. M. Rangaswamy,* Methods in Module Theory
141. *G. L. Mullen and P. J.-S. Shiue,* Finite Fields, Coding Theory, and Advances in Communications and Computing
142. *M. C. Joshi and A. V. Balakrishnan,* Mathematical Theory of Control: Proceedings of the International Conference

143. *G. Komatsu and Y. Sakane*, Complex Geometry: Proceedings of the Osaka International Conference
144. *I. J. Bakelman*, Geometric Analysis and Nonlinear Partial Differential Equations
145. *T. Mabuchi and S. Mukai*, Einstein Metrics and Yang–Mills Connections: Proceedings of the 27th Taniguchi International Symposium
146. *L. Fuchs and R. Göbel*, Abelian Groups: Proceedings of the 1991 Curaçao Conference
147. *A. D. Pollington and W. Moran*, Number Theory with an Emphasis on the Markoff Spectrum
148. *G. Dore, A. Favini, E. Obrecht, and A. Venni*, Differential Equations in Banach Spaces
149. *T. West*, Continuum Theory and Dynamical Systems
150. *K. D. Bierstedt, A. Pietsch, W. Ruess, and D. Vogt*, Functional Analysis
151. *K. G. Fischer, P. Loustaunau, J. Shapiro, E. L. Green, and D. Farkas*, Computational Algebra
152. *K. D. Elworthy, W. N. Everitt, and E. B. Lee*, Differential Equations, Dynamical Systems, and Control Science
153. *P.-J. Cahen, D. L. Costa, M. Fontana, and S.-E. Kabbaj*, Commutative Ring Theory
154. *S. C. Cooper and W. J. Thron*, Continued Fractions and Orthogonal Functions: Theory and Applications
155. *P. Clément and G. Lumer*, Evolution Equations, Control Theory, and Biomathematics
156. *M. Gyllenberg and L. Persson*, Analysis, Algebra, and Computers in Mathematical Research: Proceedings of the Twenty-First Nordic Congress of Mathematicians
157. *W. O. Bray, P. S. Milojević, and Č. V. Stanojević*, Fourier Analysis: Analytic and Geometric Aspects
158. *J. Bergen and S. Montgomery*, Advances in Hopf Algebras
159. *A. R. Magid*, Rings, Extensions, and Cohomology
160. *N. H. Pavel*, Optimal Control of Differential Equations
161. *M. Ikawa*, Spectral and Scattering Theory: Proceedings of the Taniguchi International Workshop
162. *X. Liu and D. Siegel*, Comparison Methods and Stability Theory
163. *J.-P. Zolésio*, Boundary Control and Variation
164. *M. Křížek, P. Neittaanmäki, and R. Stenberg*, Finite Element Methods: Fifty Years of the Courant Element
165. *G. Da Prato and L. Tubaro*, Control of Partial Differential Equations
166. *E. Ballico*, Projective Geometry with Applications
167. *M. Costabel, M. Dauge, and S. Nicaise*, Boundary Value Problems and Integral Equations in Nonsmooth Domains
168. *G. Ferreyra, G. R. Goldstein, and F. Neubrander*, Evolution Equations
169. *S. Huggett*, Twistor Theory
170. *H. Cook, W. T. Ingram, K. T. Kuperberg, A. Lelek, and P. Minc*, Continua: With the Houston Problem Book
171. *D. F. Anderson and D. E. Dobbs*, Zero-Dimensional Commutative Rings
172. *K. Jarosz,* Function Spaces: The Second Conference
173. *V. Ancona, E. Ballico, and A. Silva*, Complex Analysis and Geometry
174. *E. Casas*, Control of Partial Differential Equations and Applications
175. *N. Kalton, E. Saab, and S. Montgomery-Smith*, Interaction Between Functional Analysis, Harmonic Analysis, and Probability
176. *Z. Deng, Z. Liang, G. Lu, and S. Ruan*, Differential Equations and Control Theory
177. *P. Marcellini, G. Talenti, and E. Vesentini*, Partial Differential Equations and Applications: Collected Papers in Honor of Carlo Pucci
178. *A. Kartsatos*, Theory and Applications of Nonlinear Operators of Accretive and Monotone Type
179. *M. Maruyama*, Moduli of Vector Bundles
180. *A. Ursini and P. Aglianò*, Logic and Algebra
181. *X. H. Cao, S. X. Liu, K. P. Shum, and C. C. Yang,* Rings, Groups, and Algebras
182. *D. Arnold and R. M. Rangaswamy*, Abelian Groups and Modules

Additional Volumes in Preparation

abelian groups and modules

proceedings of the international conference at Colorado Springs

edited by

David M. Arnold
Baylor University
Waco, Texas

Kulumani M. Rangaswamy
University of Colorado
Colorado Springs, Colorado

Marcel Dekker, Inc. **New York • Basel • Hong Kong**

Library of Congress Cataloging-in-Publication Data

Abelian groups and modules : proceedings of the international conference at Colorado Springs / edited by David M. Arnold, Kulumani M. Rangaswamy.
p. cm. — (Lecture notes in pure and applied mathematics ; v. 182)
Papers from the International Conference on Abelian Groups and Modules held Aug. 7–12, 1995.
ISBN 0-8247-9789-2 (alk. paper)
1. Abelian groups—Congresses. 2. Modules (Algebra) —Congresses. I. Arnold, David M. II. Rangaswamy, Kulumani M. III. International Conference on Abelian Groups and Modules (1995: Colorado Springs, Colo.) IV. Series.
QA180.A227 1996
512'.2—dc20 96-23964
CIP

The publisher offers discounts on this book when ordered in bulk quantities. For more information, write to Special Sales/Professional Marketing at the address below.

This book is printed on acid-free paper.

MARCEL DEKKER, INC.
270 Madison Avenue, New York, New York 10016

Current printing (last digit):
10 9 8 7 6 5 4 3 2 1

PRINTED IN THE UNITED STATES OF AMERICA

Preface

The International Conference on Abelian Groups and Modules was organized by the University of Colorado at Colorado Springs. It was held on the campus of The Colorado College. A total of 72 researchers and graduate students from 15 countries came together to exchange ideas and to discuss their current research. Seven one-hour survey lectures and 42 twenty-minute talks were given.

The papers collected herein reflect the ideas discussed at the conference, as well as the current state of research in several areas of abelian group theory and related subjects.

As in the recent past, torsion-free groups are drawing much attention. In particular, Butler groups are the subject of several papers. In the finite rank case, questions about classification and representation type are addressed and partially answered. There is a flurry of activity in the area of almost completely decomposable groups, a special class of finite rank Butler groups previously thought to be too complicated for serious consideration. Several papers extend ideas from torsion-free groups of finite rank to those of infinite rank. As for torsion-free groups of "large" infinite rank, new sophisticated set-theoretic techniques are employed to reexamine nonfree Whitehead groups and to realize rings with free additive groups as endomorphism rings of almost free groups.

The survey article on the structure of Tor and a paper on totally projective groups aptly demonstrate that the theory of p-groups is alive and well. As for mixed groups, one article presents criteria for a mixed group to be (fully) transitive arising from known p-group criteria. In addition, the reader will find extensions of results on finite rank torsion-free groups to mixed groups with finite torsion-free rank, via functorial equivalences.

Noteworthy is the extent of cross-fertilization of concepts between different areas. For example, representations of partially ordered sets, set theory, and lattices over orders are used as tools for the investigation of abelian groups. On the other hand, classical results from abelian group theory are extended to modules over valuation domains, used for a description of units in group algebras, and applied to representations of partially ordered sets.

Included in this collection are several papers on modules over von Neumann regular rings, serial rings, discrete valuation rings, and classes of modules over more general rings. Some of the results are intimately and immediately connected with abelian groups; other papers contain more general results and techniques which could yield new areas of interest for abelian group theorists.

Scattered throughout the papers, including the survey articles, are a number of open questions and problems which we hope will stimulate further interest and research in abelian groups and modules.

It is our privilege to extend our sincere thanks to all who contributed to the success of the conference and to the publication of the proceedings. Special thanks go to the three coorganizers of this conference, Gene Abrams, Jeremy Haefner, and Mike Siddoway, for their valuable help. We are indebted to Laszlo Fuchs for his insightful ideas and constant support.

We are grateful to the Mathematics Department and the College of Engineering and Applied Science at the University of Colorado, Colorado Springs, and to The Colorado College, for their financial and in-kind support. Administrative and organizational help was provided by many people; specifically, we are grateful to Sarah Ayer, Debra Maresh, Steve Narvet, Scott Quillesh Nelson, Siew Nylund, Ken Rebman, and Kyle Van Buskirk, for their hours of tireless hard work in making this an enjoyable and successful conference.

All the papers appearing in this proceedings were carefully refereed according to international standards and they will not be published anywhere else. Many referees spent many hours reading and rereading articles for consideration for inclusion in these proceedings; we are deeply appreciative of their behind-the-scenes work.

David M. Arnold
Kulumani M. Rangaswamy

Contents

Contributors

Ulrich Albrecht, Auburn University, Auburn, AL
David M. Arnold, Baylor University, Waco, TX
Robert El Bashir, Charles University, Prague, Czech Republic
Khalid Benabdallah, University de Montreal, Montreal, Quebec, Canada
Ekaterina Blagoveshchenskaya, St. Petersburg State Tech, St. Petersburg, Russia
Ladislav Bican, Charles University, Prague, Czech Republic
Andreas R. Blass, University of Michigan, Ann Arbor, MI
H. H. Brungs, University of Alberta, Edmonton, Alberta, Canada
John Dauns, Tulane University, New Orleans, LA
Clorinda De Vivo, Universita Federico II di Napoli, Napoli, Italy
Manfred Dugas, Baylor University, Waco, TX
Paul C. Eklof, University of California at Irvine, Irvine, CA
Theodore G. Faticoni, Fordham University, Bronx, NY
Steve Files, University of Essen, Essen, Germany
Laszlo Fuchs, Tulane University, New Orleans, LA
Rüdiger Göbel, Universität Essen, Essen, Germany
H. Pat Goeters, Auburn University, Auburn, AL
Jutta Hausen, University of Houston, Houston, TX
Paul Hill, Auburn University, Auburn, AL
Dinh Van Huynh, Institute of Mathematics, Hanoi, Vietnam
John Irwin, Wayne State University, Detroit, MI
S.K. Jain, Dept of Mathematics, Ohio University, Athens, OH
Ralf Jarisch, Universität Würzburg, Würzburg, Germany
Patrick W. Keef, Whitman College, Walla Walla, WA
Menaka Lelwala, McMaster University, Hamilton, Ontario, Canada, and University of Sri Jayewardenepura, Gangowila, Nugegoda, Sri Lanka
Sergio R. Lopez-Permouth, Ohio University, Athens, OH
Ar. Meenakshi, Annamalai University, Annamalainagar, India
Claudia Metelli, Universita Federico II di Napoli, Napoli, Italy
Bruno Müller, McMaster University, Hamilton, Ontario, Canada
Otto Mutzbauer, Universität Würzburg, Würzburg, Germany
Loyiso Nongxa, University of the Western Cape, Bellville, S. Africa
Mohamed A. Ouldbeddi, University of Montreal, Montreal, Canada
John D. O'Neill, University of Detroit Mercy, Detroit, MI
K.M. Rangaswamy, University of Colorado at Colorado Springs, CO
J.D. Reid, Wesleyan University, Middletown, CT
Fred Richman, Florida Atlantic University, Boca Raton, FL
Helmut Schäfer, Universität Würzburg, Würzburg, Germany
Phillip Schultz, University of Western Australia, Nedlands, Australia
Saharon Shelah, The Hebrew University, Jerusalem, Israel
Daniel Simson, Nicholas Copernicus University, Torun, Poland
Elias Toubassi, University of Arizona, Tucson, AZ

Günter Törner, Gerhard-Mercator-Universität, Duisburg, Germany
Jan Trlifaj, Charles University, Prague, Czech Republic
William Ullery, Auburn University, Auburn, AL
Charles Vinsonhaler, University of Connecticut, Storrs, CT
Jane Kirchner West, Auburn University, Auburn, AL
William Wickless, University of Connecticut, Storrs, CT

Conference Participants

Gene Abrams, University of Colorado at Colorado Springs, CO
Ulrich Albrecht, Auburn University, Auburn, AL
Giuseppe Baccella, Universite de L'Aquila, Italy
Robert El Bashir, Charles University, Prague, Czech Republic
Silvana Bazzoni, University of Padova, Padova, Italy
Khalid Benabdallah, University de Montreal, Montreal, Quebec, Canada
Ladislav Bican, Charles University, Prague, Czech Republic
Andreas R. Blass, University of Michigan, Ann Arbor, MI
John Clark, Otago University, Dunedin, New Zealand
Rad Dimitric, University of California at Berkeley, Berkeley, CA
Manfred Dugas, Baylor University, Waco, TX
Paul C. Eklof, University of California at Irvine, Irvine, CA
Alberto Facchini, University of Udine, Italy
Theodore Faticoni, Fordham University, Bronx, NY
Steve Files, University of Essen, Essen, Germany
Michael Friger, Ben-Gurion University of the Negev, Beer-Sheva, Israel
Laszlo Fuchs, Tulane University, New Orleans, LA
Anthony Giovannitti, University of Southern Mississippi, Hattiesburg, MS
Pat Goeters, Auburn University, Auburn, AL
Brendan Goldsmith, Dublin Institute of Technology, Dublin, Ireland
Jeremy A. Haefner, University of Colorado at Colorado Springs, CO
Alfred W. Hales, UCLA, Los Angeles, CA
Jutta Hausen, University of Houston, Houston, TX
Kin-ya Honda, Meisei University, Japan
Edgar J. Howard, San Diego State University, San Diego, CA
John Irwin, Wayne State University, Detroit, MI
Patrick W. Keef, Whitman College, Walla Walla, WA
Toshiko Koyama, Ochanomizu University, Tokyo, Japan
Peter Krog, University of Connecticut, CT
Franz-Viktor Kuhlmann, Universität Heidelberg, Heidelberg, Germany
Salma Kuhlmann, Universität Heidelberg, Heidelberg, Germany
Lawrence Levy, University of Wisconsin, Madison, WI
Wolfgang Liebert, Technische Universität München, München, Germany
Sergio Lopez-Permouth, Ohio University, Athens, OH
Adolf Mader, University of Hawaii,, Honolulu, HI
Warren May, University of Arizona, Tucson AZ
AR Meenakshi, Annamalai University, Annamalainagar, India
Claudia Metelli, Universita Federico II di Napoli, Napoli, Italy
Ray Mines, New Mexico State University, Las Cruces NM
Otto Mutzbauer, Universitat Würzburg, Würzburg, Germany
Loyiso Nongxa, University of the Western Cape, Bellville, S. Africa
Kevin O'Meara, University of Canterbury, Christchurch, New Zealand

John D. O'Neill, University of Detroit, Detroit, MI
Takashi Okuyama, Universite de Montreal, Canada
Barbara Osofsky, Rutgers University, New Brunswick, NJ
Edwin P. Oxford, Baylor University, Waco, TX
K.M. Rangaswamy, University of Colorado at Colorado Springs, CO
Kenneth R. Rebman, University of Colorado at Colorado Springs, CO
James Reid, Wesleyan University, Middletown, CT
Fred Richman, Florida Atlantic University, Boca Raton, FL
Laurel Rogers, University of Colorado at Colorado Springs, CO
Mitsushi Sakamoto, Toba National College of Maritime Technology, Osaka, Japan
Luigi Salce, University of Padova, Padova, Italy
Phillip Schultz, University of Western Australia, Nedlands, Australia
Anatoly Sebeldin, Pedagogical State University, Russia
Michael Siddoway, Colorado College, Colorado Springs, CO
Daniel Simson, Nicholas Copernicus University, Torun, Poland
Carlo Toffalori, Universita di Camerino, Italy
Noriko Tone, Tokyo Denki University, Tokyo, Japan
Jan Trlifaj, Charles University, Prague, Czech Republic
William Ullery, Auburn University, Auburn, AL
Gert Viljoen, University of Orange Free State, Bloemfontein, South Africa
Charles Vinsonhaler, University of Connecticut, Storrs, CT
Carol Walker, New Mexico State University, Las Cruces, NM
Elbert A. Walker, New Mexico State University, Las Cruces, NM
Jane West, Auburn University, Auburn, AL
William Wickless, University of Connecticut, Storrs, CT
Takeshi Yasuda, Tokyo Metropolitan Sumidagawa High School, Tokyo, Japan

A Survey of Butler Groups and the Role of Representations

DAVID M. ARNOLD and MANFRED H. DUGAS Department of Mathematics, Baylor University, Waco, Texas

1 INTRODUCTION

Recent years have seen a flurry of activity in the theory of Butler groups of both finite and infinite rank, as demonstrated in several survey articles in the forthcoming Proceedings of the 1994 Padua Conference on Abelian Groups and Modules. In this paper, we will zoom in on one particular aspect of the theory, namely the role of representations and associated constructions. Here, a *representation* over a commutative ring R is an R-module with a list of distinguished submodules, usually with some additional conditions imposed. This survey is meant to be a companion to (Arnold 1995). In order to minimize duplication, the reader is referred to that paper and its references for additional historical comments and definitions of terms.

2 BUTLER GROUPS OF FINITE RANK

The classical example of a functorial connection between Butler groups and representations over Q is given in Theorem 1. Let T be a finite lattice of types and B(T) the class of all finite rank Butler groups whose typeset is contained in T. An element $\sigma \in T$ is called *join-irreducible* if whenever $\alpha, \beta \in T$ with $\sigma = \max\{\alpha, \beta\}$, then $\alpha = \sigma$ or $\beta = \sigma$. Define $JI(T)^{op}$ to be the subposet of T of all join-irreducible elements with inverse order, i.e. $\sigma \leq^{op} \tau$ if and only if $\tau \leq \sigma$ in T. Let $Rep(Q,JI(T)^{op})$ denote the class of all representations over Q of the form

$$W = (V;\ V(\sigma),\ \sigma \in JI(T))$$

such that V is a finite dimensional Q-vector space, the $V(\sigma)$'s are subspaces of V and whenever $\sigma, \tau \in JI(T)$ with $\sigma \leq^{op} \tau$, then $V(\sigma) \leq V(\tau)$. Then $Rep(Q,JI(T)^{op})$ is an additive category with morphisms Q-linear maps $\varphi: V \rightarrow V'$ such that $\varphi(V(\sigma)) \leq V'(\sigma)$ for all σ. Define a functor $D_T: B(T) \rightarrow Rep(Q,JI(T)^{op})$ by $D_T(G) = (G \otimes Q;\ G(\sigma) \otimes Q,\ \sigma \in JI(T))$ for $G \in B(T)$ and $G(\sigma) = \{g \in G \mid \text{type}(g) \geq \sigma\}$. The *quasi-isomorphism category* of B(T) is the category with objects those of B(T) but morphism sets $Q \otimes Hom(G,H)$ for G, H in B(T). Indecomposables in this category are called strongly indecomposables.

THEOREM 1 (Butler 1968, 1987) The functor D_T is an isomorphism between the quasi-isomorphism category of B(T) and $Rep(Q,JI(T)^{op})$.

This elementary, yet striking, result links Butler groups—up to quasi-isomorphism—to the rich and deep theory of representations of posets over Q. The bijective functor D_T associates to each strongly indecomposable group in B(T) an indecomposable representation in $Rep(Q,JI(T)^{op})$. Each finite poset S has either *finite* representation type (there are only finitely many indecomposable objects in Rep(Q,S)), or *tame* representation type, or is *wild.* The case of tame representation type refers to infinite representation type, but a classification of the indecomposable objects in Rep(Q,S) is feasible. In the case of wild representation type such a classification is deemed hopeless. Detailed definitions, further results, and references can be fopund in (Arnold 1989). For example, it is known what posets have finite representation type and that strongly indecomposable Butler groups with these type sets have rank ≤ 6.

With the huge body of knowledge on representations of posets and associated algebras over fields having translations via the category isomorphism mentioned above, why couldn't the theory of Butler groups of finite rank get a free "piggy back ride" from the theory of representations? First of all, both theories ask different questions. For example, given a Butler group, the first thing you want to do is determine its type set—often a non-trivial task— while in representation theory the poset is usually already fixed. Secondly, classifications arising from representations of posets of tame type are just too limited from the point of view of abelian group theory. In particular, the typeset must be relatively "narrow", see (Arnold 1989). Even more importantly, most of the time studying a Butler group up to quasi-isomorphism is simply not good enough. For example, *almost completely decomposable (acd) groups,* finite extensions of *completely decomposable (cd) groups*, are known to be a terrifically complicated class of groups, but quasi-isomorphic to the well understood class of cd groups, direct sums of subgroups of Q. Many pathological examples are actually acd groups (Fuchs 1973) or (Arnold 1982). Acd groups provide striking examples that indecomposable (Butler) groups usually are not strongly indecomposable.

On the other hand, the isomorphism category of finite rank Butler groups is just too complicated to expect any reasonable classification results, see (Dugas, Oxford 1993) for an illustration in the class of rigid acd groups. What is needed at this point is a notion of isomorphism that is located somewhere between isomorphism and quasi-isomorphism' yet preserves indecomposables. There is a growing body of evidence that near-isomorphism, locally called isomorphism at p, is the right notion. E. L. Lady (1975) defined two groups A, B to be *nearly isomorphic* if for any given $m \in N$, there are maps $f_m: A \to B$ and $g_m: B \to A$ such that $B/f_m(A)$ and $A/g_m(B)$ are finite of order relatively prime to m. Equivalently, there is $k \in N$ with $A^k \approx B^k$, see [Arnold 1982]. It is known that if A is indecomposable and B is nearly isomorphic to A, then B is indecomposable, i.e. near-isomorphism preserves indecomposability as desired. If, in the above definition, one fixes m = p, for a prime p, then one obtains the notion of *isomorphism at p*. It is not difficult to see that isomorphism at p is isomorphism in the category of finite rank torsion-free abelian groups with morphism sets $Z_p \otimes Hom(G,H)$. Thus, A and B are nearly isomorphic if and only if A and B are isomorphic at p for each prime p. Near isomorphism also arises naturally by considering acd groups as extensions of finite groups by cd groups (Mader, Vinsonhaler 1994).

A special case of a result due to Richman extends Butler's classical functor to the isomorphism at p setting. Here, $B(T)_p$ denotes the isomorphism at p category of B(T). A lattice T of types is *p-locally free* for a prime p if $pX \neq X$ for each rank-1 group X with type $X \in T$. For a finite poset S, define $Rep(Z_p, S)$ to be the class of objects $W = (V; V(\sigma), \sigma \in S)$ such that V is a finitely generated free Z_p-module, the $V(\sigma)$'s are submodules of V, and whenever $\sigma, \tau \in \Sigma$ with $\sigma \leq^{op} \tau$, then $V(\sigma) \leq V(\tau)$.

THEOREM 2 (Richman, 1994) Let T be a p-locally free lattice of types. There is a full embedding F_p of the category $B(T)_p$ into $Rep(Z_p, JI(T)^{op})$ given by

$$F_p(H) = (H_p; H(\tau)_p, \tau \in JI(T)).$$

The range of F_p consists of all $(U;U_\tau) \in Rep(Z_p, JI(T)^{op})$ such that each U_τ is pure in U.

In fact, as a consequence of the following theorem, one might as well work with Z/p^mZ-representations. Note that if S is a poset with n elements, then $Rep(Z_p,S)$ is naturally embedded in $Rep_n(Z_p) = Rep(Z_p,S_n)$, S_n an anti-chain of n elements. Assume $W = (U; U_1, \ldots, U_n)$ and W' $= (U'; U_1', \ldots U_n')$ are in $Rep_n(Z_p)$. Then *W is isomorphic to W' mod p^m* provided that $W/p^mW \cong W'/p^mW'$ in $Rep_n(Z/p^mZ)$ where $W/p^mW = (U/p^mU;(U_1+p^mU)/p^mU, \ldots, (U_n + p^m U/p^mU)$.

THEOREM 3 (Arnold, Dugas 1995A) Assume $W = (U; U_1, \ldots, U_n)$ and $W' = (U'; U'_1, \ldots U'_n)$ are in $Rep_n(Z_p)$. There is $m \geq 0$ such that W isomorphic to W' modulo p^m implies $W \cong W'$.

We are now ready for the main thrust of this article: look for suitable classes C of Butler groups such that there is a functor from the isomorphism at p category of C into representations over some rings. In view of Theorem 2, fields will no longer do. Here is the trade off: the functors will preserve indecomposability but will have restricted domains and images in a class of representations that are not that well understood. In order to avoid extraneous complications, we restrict attention to a special set of types. It remains unclear as to just how much of this can be generalized.

DEFINITION 4 For $n \in N$, let $T_n = \{\tau_0, \tau_1, \ldots, \tau_n\}$ denote a set of distinct types with $\tau_i \wedge \tau_j = \tau_0$ for all $1 \leq i, j \leq n$.

Define $B(T_n)$ to be the class of finite rank Butler groups with typeset contained in T_n. For $H \in B(T_n)$, let $C_H = H(\tau_1) + \ldots + H(\tau_n)$. Observe that, due to the special nature of T_n, C_H is the unique regulator subgroup of H (in the terminology of (Mader, Vinsonhaler, 1994)). It suffices to assume that H has no rank one summands of type τ_0, as explained in (Arnold, Dugas 1995A). In this case, H/C_H is finite. To reduce to an isomorphism at p setting, let p be a prime, $m \in N$, and define $B(T_n, p^m) = \{H \in B(T_n) \mid p^m H \leq C_H\}$. Let $Rep_{n+1}(Z/p^mZ)$ be the class of representations over Z/p^mZ with n+1 distinguished submodules; equivalently representations of an anti-chain S_{n+1}. Let $B(T_n,p^m)/p^m$ be the category with objects in $B(T_n,p^m)$, but morphisms in $Hom(H,H')/Hom(C_H,p^mC_{H'})$.

THEOREM 5 (Arnold, Dugas 1995B) Let p be a prime such that T_n is p-locally free. There is a functorial embedding $F: B(T_n,p^m)/p^m \to Rep_{n+1}(Z/p^mZ)$ defined by

$$F(H) = (C_H/p^mC_H; (H(\tau_i) + p^mC_H) / p^mC_H: 1 \leq i \leq n, p^mH/p^mC_H)$$

that is full for acd groups.

(a) $(U; U_1, \ldots, U_{n+1}) \in Rep_{n+1}(Z/p^mZ)$ is in the range of F iff and only if

(i) each U_i is pure in $U = U_1 + \ldots + U_n$;

(ii) U and all U_i, $1 \leq i \leq n$, are free Z/p^mZ-modules,

(iii) $U_i \cap U_{n+1} = 0$ for $1 \leq i \leq n$, and

(iv) rank U_i + rank $U_j \leq$ rank U for each $i \neq j \leq n$.

(b) If H and H' are acd groups, then H and H' are nearly isomorphic if and only if F(H) is isomorphic to F(H') and H is indecomposable if and only if F(H) is indecomposable.

We now slightly modify the functor F: For $m \geq e$, let F_m: $B(T, p^e) \to Rep_{n+1}(Z/p^mZ)$, $n = |JI(T)|$ be defined by $F_m(H) = (C_H/p^mC_H; p^mH/p^mC_H, (H(\tau)+p^mC_H)/p^mC_H, \tau \in JI(T))$. It follows from Theorems 2 and 3 that:

COROLLARY 6 (Arnold, Dugas 1995A) Given H, H' $\in B(T, p^e)$, there is some $m = m(H, H') \in N$, such that $F_m(H) \cong F_m(H')$ implies H and H' are isomorphic at p.

As an application of the functor F, one can now use Theorem 5 and some fairly easily constructed representations to construct examples of indecomposable Butler groups. These examples can be used to determine representation type, as given in Corollary 11.

COROLLARY 7 (Arnold, Dugas 1995B) Let T_n be p-locally free and $n \geq 3$.
(a) For each j, $m \geq 1$, there is an indecomposable H in $B(T_n, p^m)$ such that H is *not* an acd group and rank $H = j(n-1)$.
(b) There is an indecomposable acd group H in $B(T_n, p^{m+1})$ with rank $H = mn$.
(c) If $n \geq 4$, then there are indecomposable acd groups of arbitrarily large rank in $B(T_n, p)$.

There is a bound on indecomposables in case $n = 2$ and for a special case of $n = 3$.

COROLLARY 8 (Arnold, Dugas 1995B) Let T_n be p-locally free with $n \leq 3$.
(a) For $n = 2$, the isomorphism class of H/C_H, rank H, rank $H(\tau_1)$, rank $H(\tau_2)$ is a set of complete near isomorphism invariants for groups $H \in B(T_2)$. Moreover, each group in $B(T_2)$ is the direct sum of acd groups of rank ≤ 2.
(b) If $\tau_1 \vee \tau_2 \vee \tau_3$ = type of Q, then
  (i) The isomorphism class of H/C_H, rank H, rank $H(\tau_1)$, rank $H(\tau_2)$, rank $H(\tau_3)$ is a set of complete near isomorphism invariants for acd groups $H \in B(T_3)$.
  (ii) If $H \in B(T_3)$ is acd and indecomposable, then rank $H \leq 3$. If rank $H = 3$, then H/C_H is torsion cyclic.

In (Brenner 1974) one finds a list of all the indecomposable elements in $Rep_3(k)$, k a field. We can use that list to give a partial list of indecomposables in $B(T_3,p)$. The proof is done by inspecting Brenner's list and checking which satisfy our conditions in Theorem 5 on $F(H) \in Rep_4(k)$. The fact that F need not be full gives rise to our assumption that F(H), rather than H, be indecomposable. Also, groups H with $H = C_H$ remain mysterious, a topic that we will briefly address following Theorem 10 and in more detail in (Arnold, 1995). In particular, it is demonstrated that classification of acd groups plays a large role for classification of groups with $H = C_H$.

COROLLARY 9 (Arnold, Dugas 1995B) Let $T_3 = \{\tau_0, \tau_1, \tau_2, \tau_3\}$ be p-locally free, $H \in B(T_3, p)$ with rank $H(\tau_i) = r + r_i$, rank $H = 2r + r_1 + r_2 + r_3$, and $k = Z/pZ$. The only indecomposable H's in $B(T_3, p)$ with $H \neq C_H$ and F(H) indecomposable are:
(a) $r_i = 0$ for $1 \leq i \leq 3$, $H/C_H \approx k^r$, $r \geq 1$ and $H/C_H = (I + M)k^r$ for some $r \times r$ indecomposable matrix M such that 0, 1 are not eigenvalues of M.
(b) $r_i = 0$ for some i, $r_j = 1$ for $j \neq i$ and $H/C_H \cong k^{r+1}$ for some $r \geq 0$.
(c) $r_i = 1$ for $1 \leq i \leq 3$ and $H/C_H \cong k^{r+2}$ for some $r \geq 0$ or else $r_i = 1$ for some i and $r_j = 0$ for $j \neq i$ and $H/C_H \cong k^r$, $r \geq 1$.
(d) $r_i = 1$ for $1 \leq i \leq 3$ and $H/C_H \cong k^{r+1}$ for some $r \geq 0$.

The next result demonstrates how group theoretically motivated restrictions can change the behavior of representations.

THEOREM 10 [Arnold, Dugas 1993A) Let R be a proper subring of Q.
(a) If $V = (U; U_1, U_2) \in Rep_2(R)$ with each U_i pure in U, then V is a direct sum of representations of rank ≤ 2.
(b) For each $n \geq 1$ there is an indecomposable $V = (U; U_1, U_2)$ in $Rep_2(R)$ with rank $U = 2n$.
(c) Let $V = (U; U_1, U_2, U_3) \in Rep_3(R)$ such that $U = U_1 \oplus U_2$ is finitely generated free, $U_i \cap U_3 = 0$ for $i = 1, 2$ and U_3 is pure in U. Then U is a direct sum of indecomposable representations of rank ≤ 2.
(d) Let $k \in N$ be an even integer and p a prime. There exists indecomposable $W = (V; V_1, V_2, V_3)$ in $Rep_3(Z)$ of rank $\geq k$ such that $V = (V_1 \oplus V_2) + V_3$, $V/(V_1 \oplus V_2)$ is a p-group, and each V_i is pure in V.

REMARK As a consequence of Theorem 10.d, there exists $H = C_H \in B(T_3)$, indecomposable of rank $\geq k$. If, on the other hand, $H = C_H = (A_1U_1 \oplus A_2U_2) + A_3U_3 \in B(T_3)$ such that U_3 is pure in $U_1 \oplus U_2$, then H is a direct sum of indecomposables of rank ≤ 2.

We will now relax our conditions on T_n and fix $1 \in A_i < Q$ of type τ_i, $1 \leq i \leq n$, such that $\{A_1, \ldots, A_n\}$ is *rigid*, i.e. Hom $(A_i, A_j) = 0$ for $i \neq j$. Fix a prime p such that $1/p \notin A_i$, $1 \leq i \leq n$ and define the class C(n,j) to consist of all acd groups H with $C_H \approx A_1^{r(1)} \oplus \ldots \oplus A_n^{r(n)}$ and $p^jH < C_H$. We call groups $H \in C(n,j)$ *block-rigid*. Moreover, $H \in C(n,j)$ is *uniform*, if H/C_H is a free Z/p^jZ-module.

In (Dugas, Oxford 1993) one finds a complete list of near isomorphism invariants for uniform groups $H \in C(n,j)$ that are rigid. In light of our functor F, this translates into a complete set of isomorphism invariants for the corresponding F(H)'s. Representations don't show up in this paper, but the work done with the representing matrices can immediately be translated into representation language. Let UC(n, j) denote the class of uniform groups in C(n, j).

COROLLARY 11 (Arnold, Dugas, 1993B)
(a) C(n, j) has finite representation type iff $n = 2$ or else $n = 3$ and $j = 1$. In each case indecomposables have rank $\leq n$.
(b) If $n \geq 4$, then UC(n, j) has infinite representation type.
(c) UC(3, j) has finite representation type provided $j \leq 2$ and infinite representation type if $j \geq 4$.

REMARK The case $j = 3$ in Corollary 11.c remains open, as discussed at the end of the next example.

EXAMPLE 12 As a concrete illustration of a part of the proof of Corollary 11, we'll construct large indecomposables in C(3, j), $j \geq 2$. Write $H = C_H + (1/p^j)V \in C(3,j)$, where $C_H = A_1^r \oplus A_2^r \oplus A_3^r$ and V is the row space of

$$M = \begin{pmatrix} I & I & A \\ 0 & pI & pI \end{pmatrix},$$

I is the $r \times r$ identity matrix, and A is an indecomposable $r \times r$ matrix with entries in Z.

Indecomposablity of H can be determined from its endomorphism ring. To this end, let $\varphi \in \text{End } H$. Then φ can be represented by $\varphi = (X_1, X_2, X_3)$ where the X_i's are $r \times r$ matrices operating on A_i^r from the right and $M(X_1, X_2, X_3)^T$ is - modulo p^jC_H - contained in the row space of M. This leads to equations

$$(X_1, X_2, AX_3) \equiv Y_{11}(I, I, A) + Y_{12}(O, pI, pI) \bmod p^j$$
$$(0, pX_2, pX_3) \equiv Y_{21}(I, I, A) + Y_{22}(O, pI, pI) \bmod p^j$$

where the $Y_{i,j}$'s are $r \times r$ integer matrices. This implies

$$Y_{21} \equiv 0 \bmod p^j$$
$$X_2 \equiv Y_{22} \equiv X_3 \bmod p^{j-1}.$$

Moreover,

$$X_1 \equiv Y_{11} \bmod p^j$$
$$X_2 \equiv Y_{11} + pY_{12} \bmod p^j$$
$$AX_3 \equiv Y_{11}A + pY_{12} \bmod p^{j-1}.$$

Since $j \geq 2$, we conclude

$$X_1 \equiv X_2 \equiv X_3 \bmod p \text{ and}$$

$$AX_1 \equiv X_1 A \bmod p.$$

To guarantee that H is indecomposable, we just have to choose A such that A mod p has only trivial idempotents in its centralizer and is invertible (to get that $A_1{}^r$ is pure in H). This can easily be done by choosing A mod p to be the companion matrix of a power of an irreducible polynomial mod p. Note that H is *not uniform:*, in fact H/C_H is isomorphic to $(Z/p^jZ)^r \oplus (Z/p^{j-1}Z)^r$.

The groups $H \in UC(3, j)$ can be described in terms of $M = (M_1 \mid M_2 \mid M_3)$ as above such that each row of M contains at least two units mod p. For $j \geq 4$, the matrix

$$M = \left[\begin{array}{cccc|cccc|cccc} I & 0 & 0 & 0 & I & 0 & 0 & 0 & 0 & 0 & A & pI \\ 0 & I & 0 & 0 & 0 & p^2I & 0 & 0 & I & 0 & pI & p^2I \\ 0 & 0 & I & 0 & 0 & 0 & 0 & 0 & 0 & I & p^2I & 0 \\ 0 & 0 & 0 & p^3I & 0 & 0 & I & 0 & I & 0 & 0 & 0 \\ 0 & 0 & 0 & 0 & 0 & 0 & 0 & I & 0 & I & 0 & 0 \end{array}\right]$$

defines an $H \in UC(3, p^j)$ as well as any matrix obtained from M by performing mod p invertible row operations on M and mod p invertible column operations inside any of the blocks M_1, M_2, M_3. This represents a matrix problem typical for representation theory except that our entries are in Z/p^jZ and *not* in a field.

In order to completely determine the indecomposable H's in UC(3, 2) we're required to find a normal form of matrices $M = (M_1 \mid M_2 \mid M_3)$ with entries in Z/p^jZ using row/column operations as above. This is done in detail in [Mouser 1993]. The result is that all indecomposables in UC(3, 2) have rank ≤ 9 and a complete, finite list of all indecomposables in UC (3, 2) is given.

All attempts to repeat that task for the case $j = 3$ have failed. In the absence of sophisticated methods, brute force approaches lead to complicated matrices and one gets lost in a messy maze. It's not even known if the ranks of indecomposables in UC(3, 3) are bounded! It seems that representations over Z/p^3Z are much more complicated than over Z/p^2Z, let alone Z/pZ. A theory of representations over rings like Z/p^jZ, $j \geq 2$, is sorely missed and would definitely *not* be a trivial generalization of the case $j = 1$. Some results are in the literature, e.g. (Plahotnik 1976). Unfortunately, these results don't help us, since we have to put extra conditions on the representation to be in a range of F restricted to some class of groups as illustrated in the case of C(3, 2) versus UC(3, 2).

So far we have exploited the functor F to study Butler groups of finite rank. This functor works best for acd groups. We will now pursue another approach that will also yield some results for Butler groups of infinite rank. The main idea is to strip the group from its types to find its "skeleton" that is to be a representation over Z or some other subring of Q as in (Arnold, Dugas 1993A).

Let E be a finite set of primes and $B(T_n, E)$ be the subclass of $B(T_n)$ of groups H with H/C_H a direct sum of p-groups, $p \in E$, $C_H = H(\tau_1) + \ldots + H(\tau_n)$.

Very elementary but crucial is:

LEMMA 13 Let H be a τ-homogeneous cd group, $1 \in X < Q$, type $X = \tau$. Then there is a *uniquely* determined free End(X)-module $U \leq H$ with $H = XU$.

Henceforth, fix $A_i < Q$ of type $\tau_i \in T_n$ with $1/p \notin A_i$ for all $p \in E$ and $1 \in A_i$.

DEFINITION 14 An (n + 1)-tuple $V = (U; U_1, \ldots, U_n)$ is called an *(E, A)-adjusted representation*, $A = (A_1, \ldots, A_n)$, if

(a) U is a finite rank Butler group, $U_i < U$ and $U/(U_1 + \ldots + U_n)$ an E-group.

(b) Each U_i is a finitely generated *free* End(A_i)-module and *E-pure* in U, i.e. U_i is p-pure in U for each $p \in E$.

(c) $U_i \cap U_j = 0$ for $1 \leq i \neq j \leq n$.

(d) Each A_jU_j is pure in $A_1U_1 + \ldots + A_nU_n$, a subgroup of QU.

Note that (d) can be dropped if $\Sigma U_i = \oplus U$. Let $C_n(E, A)$ denote the category of all (E, A)-adjusted representations with morphisms $f: U \to U'$ such that $f(U_i) \leq U'_i$ $1 \leq i \leq n$.

THEOREM 15 (Arnold, Dugas 1993A) There is a category equivalence $F_A: B(T_n, E) \to C_n(E,A)$ given by $F_A(H) = (U; U_1, \ldots, U_n)$ where $H(\tau_i) = A_iU_i$ as in Lemma 13 and U is the E-purification of $U_1 + \ldots + U_n$ in H. The inverse of F_A in D_A, given by $D_A(V) = U + A_1U_1 + \ldots + A_nU_n$.

Recall that $1 \in A_i < Q$ of type τ_i and $1/p \notin A_i$ for all $p \in E$. Let R = End A_0. Note that for R_i = End A_i, we have $R_i \cap R_j = R$ for $i \neq j$. Moreover, R is a p.i.d. and we have:

COROLLARY 16 Let $R = R_i$ for $1 \leq i \leq n$. Then $F_A: B(T_n, E) \to Rep_n(R)$ is a full embedding with $C_n(E, A)$ the image of F_A.

3 BUTLER GROUPS OF INFINITE RANK

Let $B(T_n, \omega)$ be the class of countable torsion-free groups G such that typeset $G < T_n$ and each finite rank pure subgroup H of G is a Butler group, i.e. G is a Butler group of at most countable rank. Moreover, $Rep_n(R, \omega)$ is the class of all representations $V = (U; U_1, \ldots, U_n)$ where U is a free R-module of at most countable rank and the U_i's are (free) R-submodules of U. Corollary 16 generalizes to:

THEOREM 17 (Arnold, Dugas 1993A) Let $R = R_i$ for $1 \le i \le n$ and E a set of primes as above.
(a) $F_A: B(T_n, E, \omega) \to Rep_n(R, \omega)$ is a full embedding where $B(T_n, E, \omega)$ is the class of all $G \in B(T_n, \omega)$ such that G/C_G is an E-group.
(b) Image F_A is the class of $V = (U; U_1, \ldots, U_n) \in Rep_n(R, \omega)$ such that $U/(U_1 + \ldots + U_n)$ is an E-group; each U_i is pure in U with $U_i \cap U_j = 0$ for $1 \le i \ne j \le n$ and each A_jU_j is pure in $A_1U_1 + \ldots + A_nU_n$, a subgroup of QU.

It is well known by now how to use combinatorical tools like Shelah's Black Box to construct complicated algebraic structures. Unfortunately, these methods, by their very nature, won't yield Butler groups, i.e. B_2-groups. A detour using representations now comes in handy. Recall that Albrecht and Hill (1987) characterized B_2-groups as groups admitting an axiom 3-family of decent subgroups. We will see that many representations naturally admit axiom 3-families:

LEMMA 18 Let E be a set of primes and $V = (F_0; F_1, \ldots, F_n) \in Rep_n(Z)$, F_0 free of infinite rank, F_i pure in F_0, $F_1 \oplus F_2 \oplus \ldots \oplus F_n < F_0$, and $F_0 / (F_1 \oplus \ldots \oplus F_n)$ an E-group. Then there is an axiom 3-family F of subrepresentations of V such that for all $X = (X_0; X_1, \ldots, X_n) \in F$, $X_i = X_0 \cap F_i$ is a summand of F_i for $0 \le i \le n$ and $X_0 / (X_1 \oplus \ldots \oplus X_n)$ is an E-group.

Now let V be as in Lemma 18, $G = F_0 + A_1F_1 + \ldots + A_nF_n$ and, for $X \in F$, let $H = X_0 + A_1X_1 + \ldots + A_nX_n$. Then it can be shown that H is a pure and decent subgroup in G, and consequently G is a B_2-group with typeset $G = T_n$. To summarize, a $V \in Rep_n(Z)$, i.e. a free abelian group with distinct subgroups, gives rise—quite naturally—to B_2-groups. This is another example of some strange connection between free abelian groups and B_2-groups. In (Fuchs, Rangaswamy 1995) a class is defined for p a prime:

$$T_{B,p} = \{T \text{ a p-group: if } T \cong G/B, B \text{ a } B_2\text{-group, then } G \text{ is a } B_2\text{-group}\}.$$

In (Dugas, Irwin 1992) the class

$$T_{F,p} = \{T \text{ a p-group: if } T \cong G/B, B \text{ free abelian, then } G \text{ is free abelian}\}$$

was investigated.

One of the results in (Fuchs, Rangaswamy 1995) is that always $T_{F,p} = T_{B,p}$. This is somewhat surprising and might be somehow explained with the connection between B_2-groups and $Rep_n(Z)$. Lemma 18 can now be used to prove:

THEOREM 19 (Arnold, Dugas 1993A) Let $R < Q$ be a proper subring, p a prime of $R = \text{End } A_0$, $1 \in A_i$, and $1/p \notin A_i$ for $i = 1, 2$. For each cardinal $\kappa = \kappa^{\omega}$, there is a B_2-group H of cardinality κ with typeset $H < T_2$ and End H $= R$.

The same methods of proof are used to generalize Theorem 19:

THEOREM 20 (Dugas, Goebel 1994A) Let R be a ring such that $1 \in R$ and R^+, the additive group of R, is a B_2-group, and let $\lambda > |R|$ be a cardinal with $\lambda = \lambda^{\omega}$. If R^+ is p-reduced for at least three distinct primes p, then there is a B_2-group H such that $|H| = \lambda$ and End H $\cong$ R. If R^+ is free abelian, then typeset $H = T_2$.

There is also a version of Theorem 20 providing a topological isomorphism between End H with the finite topology and R, a complete topological ring with some additional conditions (Dugas, Goebel 1994A) Recall that most rings R used to produce pathological examples in a setting like Theorem 20, have free additive groups. Thus we have justification to say that infinite rank B_2-groups G with typeset $G \leq T_2$ are as 'wild' as the whole class of torsion-free abelian groups of infinite rank. By way of contrast, recall from Corollary 7 that groups of finite rank in $B(T_2)$ are direct sums of indecomposables of rank ≤ 2.

There have been recent studies of representations of posets over rings with an emphasis on producing examples of representations with prescribed endomorphism algebras. In (Dugas, Goebel, May 1995), there is a construction of $V = (F_0; F_1, F_2)$ where F_i, $0 \leq i \leq 2$, are free R-modules, $F_1 < F_0$, and End V is a prescribed R-algebra A. As we have seen, this has applications in the construction of large B_2-groups with type set T_2. The same representations have also been used to construct rigid fields [Dugas, Goebel 1994B]. One has to wonder what other applications might be found in the future.

REFERENCES

Albrecht U. and Hill, P. Butler groups of infinite rank and axiom 3, *Czech. Math. J.* 37 (1987), 293-309.

Arnold, D.M. Finite Rank Torsion-Free Abelian Groups and Rings, *Lect. Notes in Math.* 931, Springer-Verlag, New York, 1982.

___________ Representations of partially ordered sets and abelian groups, *Contemporary Math.* 87 (1989), 91-109.

___________ (1995) Near isomorphism of Butler groups and representations of finite posets, Proc. of 1994 Padua Conf. on Abelian Groups and Modules, to appear.

Arnold, D.M. and Dugas, M. (1993A) Butler groups with finite typesets and free groups with distinguished subgroups, *Comm. in Alg.* 21 (1993), 1947-1982.

______________________ (1993B) Block rigid almost completely decomposable groups and lattices over multiple pullback rings, *J. Pure and Appl. Alg.*, 87(1993), 105-121.

______________________ (1995A) Locally free finite rank Butler groups and near isomorphism, Proc. of 1994 Padua Conf. on Abelian Groups and Modules, to appear,.

______________________ (1995B) Representations of finite posets and near-isomorphism of finite rank Butler groups, Rocky Mountain J. Math. 25 (1995), 590-609

Arnold D.M. and Vinsonhaler, C. Finite rank Butler groups: A survey of recent results, Abelian Groups, *Lect. Notes in Pure and Applied Math.*, Marcel Dekker, New York, 1993, 17-41.

Brenner, S. On four subspaces of a vector space, *J. Alg.*, 6 (1967), 100-114.

Butler, M.C.R. A class of torsion-free abelian groups of finite rank, *Proc. London Math. Soc.* (3) 15 (1965), 680-698.

___________ Torsion-free modules and diagrams of vector spaces, *Proc. London Math. Soc.* (3) 18 (1968), 635-652.

___________ Some almost split sequences in torsion-free abelian group theory, *Abelian Group Theory,* Gordon and Breach, New York, 1987, 291-302.

Dugas, M. and Goebel, R. (1994A) Endomorphism rings of B_2-groups of infinite rank, preprint.

___________________ (1994B) Automorphism groups of fields, *Manuscripta Mathematica* 85 (1994), 227-242.

Dugas, M., Goebel, R. and May, W. (1995) Free submodules with two distinguished submodules, preprint.

Dugas, M. and Irwin, J. On a class of abelian p-groups, Forum Math. 4 (1992), 147-158.

Dugas, M. and Oxford, E. Near isomorphism invariants for a class of almost completely decomposable groups, Abelian Groups, *Lect. Notes in Pure and Applied Math.*, Marcel Dekker, New York, 1993, 129-150.

Fuchs, L. *Infinite Abelian Groups*, Vol. I I, Academic Press, New York, 1973.

Fuchs, L. and Rangaswamy, K.M. (1995) Torsion-free extensions of Butler groups by primary groups, preprint.

Lady, E.L. Nearly isomorphic torsion-free abelian groups, *J. Alg.* 35 (1975), 235-238.

Mader, A. and Vinsonhaler, C. Classifying almost completely decomposable groups, J. Alg. 170 (1994), 754-780.

Mouser, M.A. On Z/p^2Z-modules with distinguished submodules, Master's thesis, Baylor University, 1993.

Plahotnik, V.V. Representations of partially ordered sets over commutaive rings, *Math. USSR Isvestija* 10 (1976), 497-514.

Richman, F. (1994) Isomorphism of Butler groups at a prime, Proc. of 1993 Oberwolfach Conference on Abelian Groups, to appear.

New Nonfree Whitehead Groups by Coloring

Paul C. Eklof*
University of California, Irvine

Saharon Shelah†
Hebrew University and Rutgers University

Abstract

We extend the method of uniformization by considering colorings which take values in a group. We use this to construct new non-free Whitehead groups, in particular answering an open problem in [1] by showing that it is consistent that there is a strongly $\aleph_1$-free $\aleph_1$-coseparable group of cardinality $\aleph_1$ which is not $\aleph_1$-separable.

0 INTRODUCTION

An abelian group A is called a Whitehead group, or W-group for short, if $\mathrm{Ext}(A, \mathbb{Z}) = 0$. For historical reasons, A is called an $\aleph_1$-coseparable group if $\mathrm{Ext}(A, \mathbb{Z}^{(\omega)}) = 0$, but for convenience we shall use non-standard terminology and say A is a W_ω-group when A is $\aleph_1$-coseparable. Obviously a W_ω-group is a W-group. In 1973–75, the second author proved that it is consistent with ZFC + GCH that every W-group is free and consistent with ZFC that there are non-free W_ω-groups of cardinality $\aleph_1$ ([8], [9]); he later showed that it is consistent with ZFC + GCH that there are non-free W_ω-groups of cardinality $\aleph_1$ ([10], [11]). Before 1973 it was known (in ZFC) that every W-group is $\aleph_1$-free, separable, and slender, and assuming CH, every W-group is strongly $\aleph_1$-free. (See, for example, [3, pp. 178–180].) These turned out, by the results of the second author, to be essentially all that could be proved without additional set-theoretic hypotheses.

However, new questions of what could be proved in ZFC arose, inspired by the consistency results and their proofs. One of the most intriguing was:

(0.1) Does every strongly $\aleph_1$-free W_ω-group of cardinality $\aleph_1$ satisfy the stronger property that it is $\aleph_1$-separable?

*Thanks to Rutgers University for its support of this research through its funding of the authors' visits to Rutgers.

†Partially supported by Basic Research Fund, Israeli Academy of Sciences. Pub. No. 559

(See [1, p. 454, Problem 5]). As we shall explain below, not only was the answer to this question affirmative in every known model of ZFC, but the nature of the known constructions of non-free Whitehead groups was such as to lead to the suspicion that the answer might be affirmative (provably in ZFC). However, in this paper we show that it is consistent that the answer is negative.

First we recall the key definitions. An abelian group A is *$\aleph_1$-free* if every countable subgroup of A is free; A is *strongly $\aleph_1$-free* if every countable subset is contained in a countable free subgroup B such that A/B is $\aleph_1$-free. A is *$\aleph_1$-separable* if every countable subset is contained in a countable free subgroup B which is a direct summand of A; so an $\aleph_1$-separable group is strongly $\aleph_1$-free. It is a consequence of CH (or even of $2^{\aleph_0} < 2^{\aleph_1}$) that there are strongly $\aleph_1$-free groups of cardinality $\aleph_1$ which are not $\aleph_1$-separable (see [12]). However, the existence of such groups is not settled by the hypothesis $2^{\aleph_0} = 2^{\aleph_1}$; specifically, in a model of MA + $\neg$CH every strongly $\aleph_1$-free group of cardinality $\aleph_1$ *is* $\aleph_1$-separable; but the methods of [6] show that it is consistent with $2^{\aleph_0} = 2^{\aleph_1}$ that there are strongly $\aleph_1$-free groups of cardinality $\aleph_1$ which are *not* $\aleph_1$-separable.

Now suppose A is strongly $\aleph_1$-free and is a W_ω-group. Consider a countable subgroup B of A such that A/B is $\aleph_1$-free. We have a short exact sequence

$$0 \to B \to A \to A/B \to 0$$

where the map of B into A is inclusion. Since B is a free group of countable rank, if we knew that A/B were a W_ω-group, then we would have $\mathrm{Ext}(A/B, \mathbb{Z}^{(\omega)}) = \mathrm{Ext}(A/B, B) = 0$ and we could conclude that this sequence splits and hence B is a direct summand of A. In every previously known model where there are non-free W_ω-groups, the construction of a W_ω-group A is such that A/B shares the properties of A closely enough that A/B is also a W_ω-group — when B is a countable subgroup such that A/B is $\aleph_1$-free; cf. the Remark after Theorem 2 below. Thus in these models the answer to (0.1) is affirmative. This motivates question (0.1) as well as the related question

> (0.2) If a group A of cardinality $\aleph_1$ is strongly $\aleph_1$-free and a W_ω-group, and B is a countable subgroup of A such that A/B is $\aleph_1$-free, is A/B a W_ω-group?

By what we have just remarked, a positive answer to (0.2) implies a positive answer to (0.1). The converse holds as well: if A and B are as in the hypotheses of (0.2) and A is $\aleph_1$-separable, $A = F \oplus A'$ where F is countable and contains B; then A/B is a W_ω-group because $A/B = F/B \oplus A'$ and F/B is free by hypothesis on B.

We shall prove that there is a model of ZFC in which not only the answer to (0.2) but also the answer to the following weaker question is "no":

> (0.3) If a group A of cardinality $\aleph_1$ is strongly $\aleph_1$-free and a W_ω-group, and B is a countable subgroup of A such that A/B is $\aleph_1$-free, is A/B necessarily a W-group?

The proof involves an extension of the method of uniformization which was first used by the second author in [10] and [11] to construct non-free Whitehead groups in a model of CH. We recall the method and describe the extended method in the next section. That section and the following one do not require a knowledge of forcing. In the last section we construct a model of ZFC with the desired properties; the construction is by means of an iterated forcing, with finite support, of c.c.c. posets.

It remains open whether negative answers to (0.1), (0.2) and (0.3) are consistent with CH.

1 COLORING METHODS

Throughout, E will be a stationary subset of ω_1 consisting of limit ordinals, with (for technical reasons) $\omega \notin E$. We begin with a general construction of a group, related to that in [1, XII.3.4].

Definition 1 *For each $\delta \in E$ let η_δ be a ladder on δ, that is, a strictly increasing function $\eta_\delta : \omega \to \delta$ whose range approaches δ. Let φ be a function from $E \times \omega$ to $\{0\} \cup \{x_k : k \in \omega\}$. Let F be the free abelian group with basis $\{x_\nu : \nu \in \omega_1\} \cup \{z_{\delta,n} : \delta \in E, n \in \omega\}$ and let K be the subgroup of F generated by $\{w_{\delta,n} : \delta \in E, n \in \omega\}$ where*

$$w_{\delta,n} = 2z_{\delta,n+1} - z_{\delta,n} - x_{\eta_\delta(n)} - \varphi(\delta, n). \tag{1}$$

Let $A = F/K$.

Clearly A is an abelian group of cardinality $\aleph_1$. Notice that because the right-hand side of (1) is 0 in A, we have for each $\delta \in E$ and $n \in \omega$ the following relation in A:

$$2^{n+1} z_{\delta,n+1} = z_{\delta,0} + \sum_{k=0}^{n} 2^k (x_{\eta_\delta(k)} + \varphi(\delta, k)) \tag{2}$$

where, in an abuse of notation, we write, for example, $z_{\delta,n+1}$ instead of $z_{\delta,n+1} + K$. If we let

$$A_\alpha = \langle \{x_\nu : \nu < \alpha\} \cup \{z_{\delta,n} : \delta \in E \cap \alpha,\, n \in \omega\} \rangle . \tag{3}$$

for each $\alpha < \omega_1$, then for each $\delta \in E$ $z_{\delta,0} + A_\delta$ is non-zero and divisible in $A_{\delta+1}/A_\delta$ by 2^n for all $n \in \omega$. Thus $A_{\delta+1}/A_\delta$ is not free and hence A is not free. (In fact $\Gamma(A) \supseteq \tilde{E}$; see [1, pp. 85f].) Moreover, A is strongly $\aleph_1$-free; in fact, for every $\alpha < \omega_1$, using Pontryagin's Criterion we can show that A/A_α is $\aleph_1$-free whenever $\alpha \notin E$.

The following definitions are, by now, standard. A *ladder system* on a stationary subset E of $\lim(\omega_1)$ is an indexed family of functions $\eta = \langle \eta_\delta : \delta \in E \rangle$ such that each $\eta_\delta : \omega \to \delta$ is a ladder on δ. The ladder system η is called *tree-like* if for all $\delta_1, \delta_2 \in E$ and all $n, m \in \omega$, $\eta_{\delta_1}(n) = \eta_{\delta_2}(m)$ implies $n = m$ and $\eta_{\delta_1}(k) = \eta_{\delta_2}(k)$ for all $k \leq n$. If λ is a cardinal, a *λ-coloring* of η is an indexed family, $c = \langle c_\delta : \delta \in E \rangle$, of functions $c_\delta : \omega \to \lambda$. (We are particularly interested in the cases when λ is 2 or ω.) We say that η has the *λ-uniformization property* provided that for every λ-coloring c of η, there exists a pair of functions (f, f^*) such that $f : \omega_1 \to \lambda$, $f^* : E \to \omega$, and for all $\delta \in E$, $f(\eta_\delta(n)) = c_\delta(n)$ for all $n \geq f^*(\delta)$.

The following theorem is due to the second author; a proof can be found, for example, in [1, pp. 371–374].

Theorem 2 *Suppose A is constructed as in Definition 1 with $\varphi(\delta, n) = 0$ for all δ, n.*

(1) If η has the 2-uniformization (respectively, ω-uniformization) property, then A is a W-group (respectively, a W_ω-group).

(2) If η is tree-like and A is a W-group (respectively, a W_ω-group), then η has the 2-uniformization (respectively, ω-uniformization) property. □

Since it is consistent with ZFC + GCH that there is a ladder system η with the ω-uniformization property, it is consistent with ZFC + GCH that there are non-free W_ω-groups (cf. [10]).

It is also true, though not needed for our present purposes, that if there exists any non-free W-group (respectively, W_ω-group) of cardinality $\aleph_1$, then there is a ladder system with the 2-uniformization (respectively, ω-uniformization) property; see [2, pp. 92–98] for a proof.

Remark. Note that if A and η are as in Theorem 2 (1) and $B = A_\alpha$ for $\alpha \notin E$, then A/B is $\aleph_1$-free and A/B is also a W-group (resp. W_ω-group), because Theorem 2 (1) applies to A/B since it is constructed as in Definition 1, using essentially the same ladder system η (with a countable initial segment deleted).

For our new construction, we shall need an extension of the notions of coloring and uniformization. Let M be a countable group. (We are particularly interested in the cases when M is $\mathbb{Z}$ or $\mathbb{Z}^{(\omega)}$.) Let η and φ be as above but such that φ takes values in $\{x_k : k \in \omega\}$. Denote by $\bar{\varphi}(\delta, n)$ the unique $k \in \omega$ such that $\varphi(\delta, n) = x_k$. Define an *M-coloring* of (η, φ) to be an indexed family $c = \langle c_\delta : \delta \in E \rangle$ of functions $c_\delta : \omega \to M$. Say that the pair (f, f^*) *uniformizes* c if $f : \omega_1 \to M$ and $f^* : E \to \omega$ and for all $\delta \in E$ and all $n \geq f^*(\delta)$,

$$f(\eta_\delta(n)) + f(\bar{\varphi}(\delta, n)) = c_\delta(n). \tag{4}$$

(Addition in the group M.) Say that (η, φ) has the *M-uniformization property* iff for every M-coloring c there is a pair (f, f^*) which uniformizes c.

Theorem 3 *Let A be constructed as in Definition 1, where φ takes values in $\{x_k : k \in \omega\}$. If (η, φ) has the M-uniformization property, then $Ext(A, M) = 0$.*

PROOF. It suffices to prove that every homomorphism $\psi : K \to M$ extends to a homomorphism $\theta : F \to M$. Given ψ, define $c_\delta(n) = -\psi(w_{\delta,n})$ for each $\delta \in E$, $n \in \omega$. Suppose that (f, f^*) uniformizes this coloring. Define $\theta(x_\nu) = f(\nu)$ and $\theta(z_{\delta,n}) = 0$ for $n \geq f^*(\delta)$. It then follows from equations (1) and (4) that $\theta(w_{\delta,n}) = \psi(w_{\delta,n})$ for all $n \geq f^*(\delta)$. Define $\theta(z_{\delta,n})$ for $n < f^*(\delta)$ by downward induction, using the equation

$$\theta(z_{\delta,n}) = 2\theta(z_{\delta,n+1}) - \theta(x_{\eta_\delta(n)}) - \theta(\varphi(\delta, n)) - \psi(w_{\delta,n}).$$

so that θ extends ψ. □

Corollary 4 *Suppose that there are a tree-like ladder system η and a function $\varphi : E \times \omega \to \{x_k : k \in \omega\}$ such that (η, φ) has the $\mathbb{Z}^{(\omega)}$-uniformization property but such that η does not have the 2-uniformization property. Then there is a strongly $\aleph_1$-free W_ω-group A of cardinality $\aleph_1$ with a countable subgroup B of A such that A/B is $\aleph_1$-free but is not a W-group.*

PROOF. Let A be constructed as in Definition 1, using the given η and φ. Then by Theorem 3, with $M = \mathbb{Z}^{(\omega)}$, A is a W_ω-group. Let B be A_ω. Then A/B is $\aleph_1$-free since $\omega \notin E$; moreover A/B is isomorphic to a group constructed as in Definition 1 but with $\varphi(\delta, n) = 0$ for all δ, n. Thus by the hypothesis and Theorem 2 (2), A/B is not a W-group. □

In what follows we prove that there is a model of ZFC in which the hypotheses of Corollary 4 are true. In the next section we describe the basic partial ordering used in the iterated forcing and prove that it is c.c.c. In the final section we describe the iteration and prove its properties.

2 A c.c.c PARTIAL ORDERING

Let M, η, and φ be as in the previous section where η is tree-like. Let c be a fixed M-coloring of (η, φ). We shall define a partial order which consists of finite approximations to a uniformization of c.

Definition 5 *Let Q_c be the set of all pairs $(p,\ p^*)$ such that for some finite subset S of E, and some $r \in \omega$, $p^* : S \to \omega$ and p is the restriction to $\bigcup\{\{\eta_\delta(m), \bar{\varphi}(\delta, m)\} : \delta \in S,\ m < r\}$ of a function $f_p : \bigcup\{\{\eta_\delta(n), \bar{\varphi}(\delta, n)\} : \delta \in S,\ n \in \omega\} \to M$ such that for every $\delta \in S$, $f_p(\eta_\delta(n)) + f_p(\bar{\varphi}(\delta, n)) = c_\delta(n)$ if $n \geq p^*(\delta)$. Partially order Q_c by: $(p,\ p^*) \leq (q,\ q^*)$ if and only if q is an extension of p and q^* is an extension of p^*. Define $cont((p, p^*)) = S$ and $num((p, p^*) = r$.*

Lemma 6 *Q_c is c.c.c., that is, every antichain in Q_c is countable.*

PROOF. The proof is similar to that in [1, Proof of VI.4.6]. Let κ be large enough for Q_c. Identify M with ω. We shall show that 0 is N-generic whenever $N = \bigcup_{i\in\omega} N_i$ where $N_i \prec N_{i+1} \prec H(\kappa)$ and $N_i \cap \omega_1 < N_{i+1} \cap \omega_1$. Given such an N, let $\alpha_i = N_i \cap \omega_1$ and $\alpha = N \cap \omega_1$. Suppose that $(p, p^*) \in Q_c$ and $D \in N$ is a dense subset of Q_c. We need to show that there is an element of $D \cap N$ which is compatible with (p, p^*). Let f_p be as in the definition of Q_c, so that p is a restriction of f_p. Choose i so that $D \in N_i$, $\mathrm{dom}(p) \cap \alpha \subseteq \alpha_i$, and $\mathrm{dom}(p^*) \cap \alpha \subseteq \alpha_i$. Let

$$\begin{aligned} Y = \bigcup\{\{\eta_\delta(\ell), \bar{\varphi}(\delta, \ell)\} : \delta \in \mathrm{dom}(p^*) \setminus \alpha_i, \eta_\delta(\ell) < \alpha_i\} \cup \\ \{\bar{\varphi}(\delta, k) : \eta_\delta(k) = \eta_\gamma(k) \text{ for some } \delta \neq \gamma \in \mathrm{dom}(p^*)\}. \end{aligned}$$

Then Y is finite since $\delta > \alpha_i$ if $\delta \in \mathrm{dom}(p^*) \setminus \alpha_i$ and since ladders on different ordinals have different limits and hence their ranges must have finite intersection. Since N_i is a model of set theory, there exists $q \in N_i$ such that $q = p \restriction (\mathrm{dom}(p) \cap \alpha_i) \cup f_p \restriction Y$. Let $q^* = p^* \restriction (\mathrm{dom}(p) \cap \alpha_i)$. Then $(q, q^*) \in N_i \cap Q_c$ so since $N_i \models$ "D is dense in Q_c", there exists $(\tilde{q},\ \tilde{q}^*) \in D \cap N_i$ such that $(q, q^*) \leq (\tilde{q}, \tilde{q}^*)$. By choice of q and q^* and Y, we can modify f_p to agree with $f_{\tilde{q}}$ where necessary and show that for some extension q' of $p \cup \tilde{q}$, $(q', p^* \cup \tilde{q}^*) \in Q_c$. Therefore $(\tilde{q}, \tilde{q}^*) \in D \cap N$ is compatible with (p, p^*). □

We shall also need some results about denseness of certain subsets of Q_c. For $\delta \in E$, let $D_\delta = \{(p,\ p^*) \in Q_c : \delta \in \mathrm{dom}(p^*)\}$. We claim that D_δ is dense in Q_c. Indeed, given any $(p,\ p^*) \in Q_c$ and $\delta \notin \mathrm{dom}(p^*)$ we can find k such that if $n \geq k$, then

$$\eta_\delta(n) \notin \{\eta_\gamma(m) : \gamma \in \mathrm{dom}(p^*),\ m \in \omega\} \cup \mathrm{dom}(p).$$

Let $\tilde{p}^* = p^* \cup \{(\delta,\ k)\}$; we can easily extend f_p to show that $(\tilde{p}, \tilde{p}^*)$ belongs to Q_c for some $\tilde{p}$.

For any $\alpha \in \{\eta_\delta(n) : \delta \in E,\ n \in \omega\}$, let $D'_\alpha = \{(p, p^*) \in Q_c : \alpha \in \mathrm{dom}(p)\}$. Then it is easy to see that D'_α is dense in Q_c.

Simply as an exercise, let us see how we can use these notions to show that non-free W_ω-groups exist in a model of MA + ¬CH.

Theorem 7 *Assume MA + ¬CH. Let A be constructed as in Definition 1. Then for any countable group M, $Ext(A, M) = 0$.*

PROOF. By Theorem 3, we need only prove that (η, φ) has the M-uniformization property. So let c be any M-coloring of (η, φ). Let Q_c be as in Definition 5. By Lemma 6, Q_c is c.c.c. Hence, by MA, there is a directed subset G of Q_c which meets each D_δ and each D'_α. Let $f^* = \cup\{p^* : (p, p^*) \in G$ for some $p\}$. Then f^* is a function — because G is directed — with domain E — because G meets every D_δ. Define $f' = \cup\{p : (p, p^*) \in G$ for some $p^*\}$. Then f' is a function with domain $\{\eta_\delta(n) : \delta \in E,\, n \in \omega\}$ — because G meets each D'_α — and we can arbitrarily extend f' to f so that (f, f^*) is a uniformization of c. □

Now η has the 2-uniformization property in any model of MA $+\neg$CH, so this is not the model we seek (satisfying the hypotheses of Corollary 4). However, in the next section we shall construct the desired model by imitating the proof of the consistency of MA: employing an iterated forcing with finite support, but instead of iterating over all c.c.c. posets, iterating over just those of the form Q_c defined in Definition 5.

3 THE FORCING CONSTRUCTION

In this section we prove the following theorem.

Theorem 8 *It is consistent with ZFC that there exists a tree-like ladder system η on a stationary set E and a function $\varphi : E \times \omega \to \{x_k : k \in \omega\}$ such that (η, φ) has the $Z^{(\omega)}$-uniformization property, but such that η does not have the 2-uniformization property.*

PROOF. We begin with a ground model where GCH holds and fix a stationary set E of limit ordinals such that $\omega \notin E$. We shall first do a forcing to create a generic family of ladders $\langle \eta_\delta : \delta \in E\rangle$ and a generic function φ. Let Q_0 be the set of all finite functions p such that dom(p) is a finite subset of E and for some $r^p \in \omega$, for all $\delta \in$ dom(p), $p(\delta)$ is a pair $(\eta^p_\delta, \varphi^p_\delta)$ where η^p_δ is a strictly increasing function: $r^p \to \delta$ and $\varphi^p_\delta : \{\delta\} \times r^p \to \{x_k : k \in \omega\}$; moreover, we require that if $\delta, \gamma \in$ dom(p), $n, m < r^p$, and $\eta^p_\delta(n) = \eta^p_\gamma(m)$, then $n = m$ and $\eta^p_\delta(\ell) = \eta^p_\gamma(\ell)$ for all $\ell < n$.

The group A is defined in V^{Q_0} as in Definition 1 using the generic ladders η_δ and the generic $\varphi : E \times \omega \to \{x_k : k \in \omega\}$. Then

$$P = \left\langle P_\alpha, \dot{Q}_\alpha : 0 \le \alpha < \omega_2 \right\rangle$$

will be constructed to be a finite support iteration of length ω_2 so that for each $\alpha \ge 1$ $\Vdash_{P_\alpha} \dot{Q}_\alpha = Q_{\dot{c}^{(\alpha)}}$ where

$$\Vdash_{P_\alpha} \dot{c}^{(\alpha)} \text{is a } Z^{(\omega)}\text{-coloring of } (\eta, \varphi).$$

We choose our enumeration of names $\{\dot{c}^{(\alpha)} : 1 \le \alpha < \omega_2\}$ so that if G is P-generic and $c \in V[G]$ is a $Z^{(\omega)}$-coloring of (η, φ), then for some $\alpha < \omega_2$, $\dot{c}^{(\alpha)}$ is a name for c in V^{P_α}. Then P is c.c.c. and it is clear from our construction that if G is P-generic, in $V[G]$ (η, φ) has the $Z^{(\omega)}$-uniformization property.

It remains to prove that in $V[G]$ η does not have the 2-uniformization property. In fact, we claim that in $V[G]$ the coloring c defined by

$$c_\delta(n) = \begin{cases} 0 & \text{if } \bar{\varphi}(\delta, n) \text{ is even} \\ 1 & \text{if } \bar{\varphi}(\delta, n) \text{ is odd} \end{cases}$$

cannot be uniformized.

Suppose, to the contrary, that there are names $\dot{f}$ and $\dot{f}^*$ and a $p \in G$ such that

$$p \Vdash (\dot{f}, \dot{f}^*) \text{ uniformizes } c.$$

For each $\delta \in E$, there exists $p_\delta \geq p$ and $m_\delta \in \omega$ such that

$$p_\delta \Vdash \dot{f}^*(\delta) = m_\delta.$$

Since ω is countable, there is a stationary $E' \subseteq E$ and $m \in \omega$ such that for all $\delta \in E'$, $m_\delta = m$. We claim

> ($\star$) *there exist* $\delta_1 < \delta_2$ *in* E', $r, k \in \omega$, *and* $q \geq p_{\delta_i}$ $(i = 1, 2)$ *such that* $r \geq m$ and q *forces the following:* $\eta_{\delta_1}(r) = \eta_{\delta_2}(r)$; $\varphi(\delta_1, r) = x_{2k}$; *and* $\varphi(\delta_2, r) = x_{2k+1}$.

Suppose that ($\star$) holds. We obtain a contradiction by considering what happens in a generic extension $V[\tilde{G}]$ where $q \in \tilde{G}$. In $V[\tilde{G}]$ we have: $f^*(\delta_i) = m$ for $i = 1, 2$, so $f(\eta_{\delta_i}(r)) = c_{\delta_i}(r)$ for $i = 1, 2$. But this is impossible by definition of c, since $\bar{\varphi}(\delta_1, r) = 2k$ and $\bar{\varphi}(\delta_2, r) = 2k + 1$.

So it remains to prove ($\star$). Without loss of generality, we can assume that for every $\delta \in E'$ and every $\beta > 0$, $p_\delta(\beta)$ is a finite function (in the ground model) and not just a name, and that $\delta \in \text{dom}(p_\delta(0))$. Moreover, we can assume that for all $\delta \in E'$, $p_\delta(0)$ determines $\eta_\sigma(n)$ for all $\sigma \in \text{cont}(p_\delta(\beta))$ and $n < \text{num}(p_\delta(\beta))$ for all $\beta \in \text{supp}(p_\delta) \setminus \{0\}$. We can also assume that there is an $r \geq m$ such that for all $\delta \in E'$ $r^{p_\delta(0)} = r$ and that for all $\beta \in \text{supp}(p_\delta) \setminus \{0\}$, $\text{num}(p_\delta(\beta)) = r$. Furthermore, by Fodor's Lemma, we can assume that $\langle \eta_\delta^{p_\delta}(n) : n < r \rangle$ and $\langle \varphi^{p_\delta}(\delta, n) : n < r \rangle$ are independent of δ. Finally, by the Δ-system lemma we can find $\delta_1 < \delta_2$ such that for all $\beta < \omega_2$, $p_{\delta_1}(\beta) \cup p_{\delta_2}(\beta)$ is a function.

Now we define $q \geq p_{\delta_i}$ $(i = 1, 2)$ by defining $q(\beta)$ for each $\beta < \omega_2$. In order to define $q(0)$, choose γ such that $\eta_{\delta_1}^{p_{\delta_1}}(r - 1) < \gamma < \delta_1$, and $k \in \omega$ such that x_{2k} and x_{2k+1} do not "occur in" any $p_{\delta_i}(\beta)$ $(i = 1, 2)$. Then define $q(0)$ extending $p_{\delta_1}(0)$ and $p_{\delta_2}(0)$ so that it forces $\eta_{\delta_1}(r) = \gamma = \eta_{\delta_2}(r)$, $\varphi(\delta_1, r) = x_{2k}$, and $\varphi(\delta_2, r) = x_{2k+1}$. We choose γ_1 and γ_2 so that $\gamma < \gamma_1 < \delta_1 < \gamma_2 < \delta_2$ and let $q(0)$ force $\eta_{\delta_i}(r + 1) = \gamma_i$ $(i = 1, 2)$; then we will have guaranteed that $q(0)$ forces $\eta_{\delta_1}(k) \neq \eta_{\delta_2}(\ell)$ and $\eta_{\delta_1}(\ell) \neq \eta_{\delta_2}(k)$ for all $k \geq r + 1$ and $\ell \in \omega$. Similarly we can define $q(0)$ to force values for $\eta_\sigma(j)$ for $j = r, r + 1$ for all other $\sigma \in \text{dom}(p_{\delta_1}(0)) \cup \text{dom}(p_{\delta_2}(0))$ so that $q(0)$ forces $\eta_{\sigma_1}(\ell) \neq \eta_{\sigma_2}(k)$ whenever $k \geq r$, $\ell \in \omega$, $\sigma_1 \neq \sigma_2 \in \text{dom}(p_{\delta_1}(0)) \cup \text{dom}(p_{\delta_2}(0))$, and either $k > r$ or $\sigma_2 \notin \{\delta_1, \delta_2\}$.

For any $\beta > 0$, if $\beta \in \text{supp}(p_{\delta_i}) \setminus \text{supp}(p_{\delta_j})$ for $i \neq j \in \{1, 2\}$, let $q(\beta) = p_{\delta_i}(\beta)$. If $\beta \in \text{supp}(p_{\delta_1}) \cap \text{supp}(p_{\delta_2})$, let $q(\beta)$ be a name $(\dot{q}, \dot{q}^*)$ for a condition extending $p_{\delta_1}(\beta) \cup p_{\delta_2}(\beta)$ so that (in V^{P_β}): $\text{num}(q(\beta)) = r + 2$; $\text{cont}(q(\beta)) = \text{cont}(p_{\delta_1}(\beta)) \cup \text{cont}(p_{\delta_2}(\beta))$; $\dot{q}(\gamma) + \dot{q}(2k) = \dot{c}_{\delta_1}^{(\beta)}(r)$; and $\dot{q}(\gamma) + \dot{q}(2k + 1) = \dot{c}_{\delta_2}^{(\beta)}(r)$. This is certainly possible because we have freedom to choose $\dot{q}(2k)$ and $\dot{q}(2k + 1)$ independently and because we have forced the ranges of the relevant ladders to be disjoint above the rth rung (except for $\gamma = \eta_{\delta_1}(r) = \eta_{\delta_2}(r)$). $\square$

References

[1] P. C. Eklof and A. H. Mekler, **Almost Free Modules: Set-theoretic Methods**, North-Holland (1990).

[2] P.C. Eklof and S. Shelah, *A combinatorial principle equivalent to the existence of non-free Whitehead groups*, in **Abelian Group Theory and Related Topics** (R. Göbel et al, eds), Contemporary Math. **171** (1994), 79–98.

[3] L. Fuchs, **Infinite Abelian Groups**, vol II, Academic Press (1973).

[4] T. Jech, **Multiple Forcing**, Cambridge University Press (1986).

[5] K. Kunen, **Set Theory: An Introduction to Independence Proofs**, North-Holland (1980).

[6] A. Mekler and S. Shelah, *Diamond and λ-systems*, Fund. Math. **131** (1988), 45–51.

[7] S. Shelah, **Proper Forcing**, Lecture Notes in Mathematics No. 940, Springer-Verlag (1982).

[8] S. Shelah, *Infinite abelian groups, Whitehead problem and some constructions*, Israel J. Math **18** (1974), 243–25.

[9] S. Shelah, *A compactness theorem for singular cardinals, free algebras, Whitehead problem and transversals*, Israel J. Math. **21** (1975), 319–349.

[10] S. Shelah, *Whitehead groups may not be free even assuming CH, I*, Israel J. Math. **28** (1977), 193–203.

[11] S. Shelah, *Whitehead groups may not be free even assuming CH, II*, Israel J. Math. **35** (1980), 257–285.

[12] S. Shelah, *On endo-rigid strongly $\aleph_1$-free abelian groups in $\aleph_1$*, Israel J. Math. **40** (1981), 291–295.

Simply Presented Torsion Modules Over Valuation Domains

LASZLO FUCHS
Department of Mathematics
Tulane University
New Orleans, LA 70118, USA

Recall that an abelian p-group (p a prime) is said to be simply presented if it can be generated by a set X of symbols such that all the defining relations are of the form $px = 0$ or $px = y$ with $x, y \in X$. One of the great achievements of the theory of abelian groups is the classification of simply presented (i.e. totally projective) p-groups, due to Nunke [10], Hill [9], Crawley-Hales [3], and Walker [12], generalizing the classical Ulm-Zippin theory of countable p-groups. This theory has been extended by Warfield [13] to mixed groups. So far no attempt has been made to carry it over to modules beyond the obvious generalization to modules over discrete rank one valuation domains.

At first glance, such an attempt looks futile, since in most proofs several properties of abelian groups are needed which no longer hold even for modules over general valuation domains. In fact, the naive approach of imitating the theory of simply presented p-groups breaks down before getting off the ground. However, in view of the success in introducing the divisible module ∂ and the S-divisible modules ∂_S for multiplicative subsemigroups S of a domain R in [5] and [8] (which are generators of the category of divisible and S-divisible R-modules, respectively), an attempt of developing a theory for simple presentation seems after all not entirely hopeless, provided one is very careful with the definitions.

Actually, the first difficulty one is confronted with is the very definition of simple presentation: it is no longer enough to stipulate that all the defining relations be of the form $rx = 0$ or $rx = y$ (with $r \in R$ and x, y in the set X of generators). In fact, this hypothesis alone does not provide enough control over the module. However, it turns out that the Crawley-Hales idea of the "tree structure" of the generating system can remedy the situation. Accordingly, two additional conditions have to be added (which are basically always satisfied in the case of abelian groups and hold for the presentations of the modules ∂ and ∂_S as well): first, r cannot be a unit of R, and secondly, every $x \in X$ occurs at most once with a coefficient $r \neq 1$ in the defining relations. As we shall see, these restrictive conditions enable us to carry over several useful features from abelian groups to modules, though - needless to say - more delicate proofs are required. In order to restrict our present attempt to more tractable cases, where we have a good chance for accomplishing our primary goal of finding a structure theorem along with a satisfactory classification in terms of cardinal invariants, we will deal with modules over valuation domains R only. Even in this special case, several difficult hurdles have to be passed.

In the focus of our attention are those simply presented torsion modules M for which the first Ulm submodule $M^1 = \bigcap_{r \neq 0} rM$ vanishes. It is hard to imagine that the general case can be handled successfully without a thorough knowledge of this special case which itself turns out to be quite challenging (in fact, the Ulm factors of simply presented torsion modules are again simply presented). Besides, without the indicated restriction, we would have needed a more general notion of height than the one available in the literature, and therefore we would have had to develop a general theory of heights, by calculating heights in the higher Ulm submodules of M, before entering into the discussion of the main issue. However, the complications involved would have distracted our attention from the genuine structural problems, so we started to assume $M^1 = 0$. Since the divisible case could be handled by making use of the results in [6] and [7], it turned out that without much extra effort we could manage with the assumption that M^1 was divisible.

For discrete rank one valuation domains our assumptions on the module M guarantee that M is a direct sum of cyclic modules and modules isomorphic to Q/R. In the general case, the situation is much more complicated. Our main results are inspired to a certain extent by the remarkable equivalent conditions for an abelian p-group to be simply presentable. Even the ideas of some proofs bear resemblance to their group-theoretical analogues, though of course much more elaborate techniques are required in our present setting.

We will establish an equivalence theorem on simply presented torsion modules M with divisible M^1. The equivalent conditions are analogs of the well-known conditions on an abelian p-group to be simply presented with the following exceptions. First, total projectivity is ignored: it is automatic for bounded p-groups, but for bounded modules this is not the case; we plan to return to this question at another time. Secondly, we could not decide whether or not a summand of a simply presented module is again simply presented. We expect to learn more about this question when we investigate the invariants of simply presented modules.

1. Simple presentation

Throughout, R will denote a valuation domain, i.e. a commutative domain with 1 whose ideals form a chain under inclusion. We let Q denote its field of quotients, and P its maximal ideal.

An R-module M is said to be *simply presented* if it is generated by a set X of elements subject to a set Ξ of defining relations such that

(a) all relations in Ξ are of the form

$$rx = 0 \quad \text{or} \quad rx = y \qquad (x, y \in X;\ r \in P)$$

subject to the condition that no $x \in X$ occurs more than once in Ξ with coefficient $\neq 1$.

Notice that relations like $x = y$ or $tx = y$ with t a unit of R are excluded, and so are relations involving more than two generators from X.

A presentation of the indicated sort is called a *simple presentation*, and we will write $M = \langle X; \Xi \rangle$. A module is *simply presentable* if it admits a simple presentation.

It can very well happen that in a simply presented R-module M, different elements of the generating set X become equal or some $x \in X$ becomes 0. As it is more convenient to work with presentations where such situations do not occur, we proceed to show that from any given simple presentation it is always possible to obtain another one, $M = \langle X; \Xi \rangle$, which is *faithful* in the following sense:

(b) no $x \in X$ is equal to 0 in M;

(c) $x \neq y$ in X implies $x \neq y$ in M.

Condition (b) can at once be satisfied by omitting those x's from X which become 0 in M, replacing them by 0 in the relations in Ξ and dropping the meaningless relations like $r0 = 0$. By doing so, (a) is not violated. (b) implies that $x \neq y$ must hold if $rx = y$ is a relation in Ξ (otherwise $(1-r)x = 0$ and so $x = 0, 1-r$ being a unit in R).

We now concentrate on (c) and show that (a) and (b) already imply (c).

To this end, we introduce a partial order in the set $X^* = X \cup \{0\}$ as the transitive extension of the covering relation:

$y \mathrel{-\!\!<} x$ in X if and only if there is a relation $rx = y$ in Ξ; and

$0 \mathrel{-\!\!<} x$ $(x \in X)$ if and only if there is a relation $rx = 0$ in Ξ.

To show that this is a partial order, only its antisymmetry has to be checked. Observe that if we can find a sequence of relations like $r_1x_1 = x_2$, $r_2x_2 = x_3, \ldots, r_nx_n = x_1$ $(x_i \in X, r_i \in P)$, then $(1 - r_1 \ldots r_n)x_1 = 0$. The last coefficient being a unit in R, we obtain $x_1 = 0$, and hence $x_2 = \ldots = x_n = 0$ in M. This means that all of $x_1, \ldots, x_n$ have been dropped from the generating system and replaced by 0 in the relations, in accordance with (b). Consequently, the indicated binary relation in X^* is in fact a partial order. Henceforth X^* will be considered to be equipped with this partial order.

Using this partial order, we can verify that (c) holds true whenever (a) and (b) are satisfied. Assuming (a) and (b), let u and v be two different elements of X which supposedly become equal in M. This means that in the free R-module on X there is a relation of the form

$$(1) \qquad u - v = \sum_{i=1}^{n} t_i(y_i - r_ix_i) + \sum_{i=n+1}^{n+k} t_i(r_ix_i) \qquad (t_i \in R, r_i \in P)$$

for some $n + k \geq 1$ where $r_ix_i = y_i$ $(i \leq n)$ and $r_ix_i = 0$ $(i \geq n+1)$ are relations in Ξ. In the partial order introduced above there is a maximal x_j on the right member of (1). By (a), it can not occur more than once, so it must occur also on the left. This is impossible, since $t_ir_i \neq \pm 1$. We conclude that (b) assures the faithfulness of the simple presentation.

Let us record what we have proved.

THEOREM 1.1. *A simple presentation is faithful if and only if it satisfies* (b). *Every simply presentable R-module has a faithful simple presentation.* □

The existence of a faithful simple presentation can be interpreted as an evidence that the generating system has a tree structure indeed.

An argument similar to the one above applies to show:

LEMMA 1.2. *Let* $M = \langle X;\Xi\rangle$ *be a faithful simple presentation. If* $tu = v$ *(or* $tu = 0$*) holds in* M *where* $u, v \in X$ *and* $t \in R$*, then there must exist relations*

$$r_0u = x_1, r_1x_1 = x_2, \dots, r_nx_n = v \qquad (resp.\ r_nx_n = 0)$$

in Ξ *such that* $t = r_0r_1\dots r_n$ *(resp.* $r_0r_1\dots r_n$ *divides* t*). In particular,* $tu = v$ *with* t *a unit of* R *implies* $t = 1$. □

Next we wish to exhibit examples for simply presented R-modules. The following examples in the literature are the four prototypes we will need.

EXAMPLE 1.3. A cyclic R-module is simply presented if and only if it is cyclically presented (i.e. is of the form R/Rr with $r \in R$). The presentation is $\langle x; rx = 0\rangle$. □

EXAMPLE 1.4. A uniserial module U of projective dimension 1 is simply presentable (for the converse see (3.5)). Note that p.d. $U \le 1$ amounts to saying that (i) U is countably generated, and (ii) the annihilator ideals of the elements of U are principal ideals; see [5]. In other words, U is the union of a properly ascending chain $Ru_1 < Ru_2 < \dots < Ru_n < \dots (n < \omega)$ of cyclically presented R-modules. If we set $\operatorname{Ann} u_1 = Rr_1$ and $r_nu_n = u_{n-1}$ $(n \ge 2)$ with suitable $r_n \in R$, then $U \cong \langle X;\Xi\rangle$ where

$$X = \{x_n \mid n < \omega\} \quad \text{and} \quad \Xi = \{r_1x_1 = 0, r_nx_n = x_{n-1}\ (n \ge 2)\}.$$

In fact, the module $M = \langle X;\Xi\rangle$ is uniserial with $\operatorname{Ann} x_1 = Rr_1$, so the epimorphism $M \to U$ induced by $x_n \mapsto u_n$ has trivial kernel. □

EXAMPLE 1.5. It is readily seen from the definition that the modules $V = V(U^-, Rr)$ see [7, p.158] are simply presented for any limit height U^- and any $r \in P$. Recall that if $U = J/R$ with $J \le Q$ is generated by s_i^{-1} $(s_i \in R, i \in I)$ (for convenience we assume that $Rs_i \ne Rs_j$ if $i \ne j$ in I), then V is generated by

$$X = \{a, x_i (i \in I)\} \quad \text{subject to the relations} \quad \Xi = \{ra = 0, s_ix_i = a\}.$$

It is clear that $V(U^-, Rr)$ is not uniquely determined by U^- and r, since it also depends on the way the generators of U are chosen. (Thus the symbol $V(U^-, Rr)$ denotes a certain kind of module rather than a definite one.) □

EXAMPLE 1.6. The divisible module ∂ defined in [7, p.123] is likewise simply presented (and so are the S-divisible modules ∂_S introduced in [8]). Recall that ∂ is generated by all k-tuples $(r_1, \dots, r_k)$ of non-zero and non-unit elements r_i of R $(k \ge 0)$ subject to the relations

$$r_k(r_1, \dots, r_k) = (r_1, \dots, r_{k-1}) \qquad (k \ge 1).$$

We shall need the torsion version ∂^* which is the quotient of ∂ modulo the cyclic submodule generated by the generator $(\emptyset)$. Thus ∂^* is generated by all the k-tuples $(r_1, \ldots, r_k)$ with $k \geq 1$ subject to the indicated relations with $(\emptyset)$ replaced by 0. It will be helpful to note that in the definition it suffices to use only those r_i whose reciprocals generate Q as an R-module.

If Q/R is countably generated, then ∂^* as defined in the preceding paragraph is a direct sum of (countably many) copies of Q/R. In this case, it is more convenient to set $\partial^* = Q/R$. □

It is worth while observing that the modules in (1.3)-(1.5) can also be defined to be generated by k-tuples with the same kind of defining relations as in (1.6). Indeed, for (1.3), we have a single generator (r). For the module in (1.4), the generators are $(r_1, \ldots, r_n)$ for all $n \geq 1$ with fixed r_i, while for (1.5) the generators are (r, s_i) for all $i \in I$.

Further examples for non-torsion simply presented modules are as follows:

EXAMPLE 1.7. The R-module R, countably generated ideals J of R, as well as the mixed version of (1.5). For the latter one, let $X = \{x_n, y_n \mid n < \omega\}$ and $\Xi = \{r_n x_n = x_{n+1}, s_n y_n = x_n \mid n < \omega\}$ where r_n, s_n are nonzero elements of P. The module $M = \langle X; \Xi \rangle$ has torsion-free rank 1 and all the generators in X have 0 annihilator ideals. □

It is evident that direct sums of simply presented modules are likewise simply presented.

Let us have a closer look at the partial order on $X^* = X \cup \{0\}$. The last example shows that in the partial order of X there may exist descending chains of length ω. (It is not difficult to see that no such chain of length $> \omega$ can exist.) However, if M is torsion, then all descending chains in X are finite. In fact, this follows at once from (1.2). In this case an element x in X is either minimal in the partial order of X or else the set $\{y \in X \mid y < x\}$ contains a unique maximal element. Hence it follows that X is a union of trees (of lengths $\leq \omega$), and X^* is itself a tree. It is clear that there are no relations in X which involve elements from different connected components of X.

THEOREM 1.8. *Let $\{X_j \mid j \in J\}$ be the set of connected components of X where $M = \langle X; \Xi \rangle$ is a faithful simple presentation. Then*

$$M = \bigoplus_{j \in J} M_j$$

where $M_j = \langle X_j; \Xi_j \rangle$ are simply presented summands; here Ξ_j stands for the subset of Ξ involving the elements in X_j and 0 only. □

Evidently, each M_j has torsion-free rank at most 1. From now on we assume M is a torsion module.

We introduce a notation. Let $M = \langle X; \Xi \rangle$ be a faithful simple presentation of the torsion module M, and $x \in X$. By condition (a), there is exactly one relation in Ξ of the form $rx = y$ which involves x for some $y \in X^*$. Using this $r \in P$, we set $A(x) = Rr$. The ideal $A(x)$ associated with x is uniquely determined by the given faithful simple presentation.

The next result shows that the elements of $M = \langle X; \Xi \rangle$ admit canonical forms.

LEMMA 1.9. *A non-zero element a of a simply presented torsion R-module $M = \langle X; \Xi \rangle$ (faithful) can be written in the form*

$$a = s_1 x_1 + \ldots + s_k x_k \qquad (k \geq 1) \tag{2}$$

where

(i) $x_1, x_2, \ldots, x_k \in X$ *are pairwise incomparable;*

(ii) $s_i \in R \backslash A(x_i)$ *for each* i.

Moreover, the x_i are unique and the coefficients s_i are uniquely determined $\operatorname{mod} A(x_i)$.

PROOF. We can write $a = t_1 x_1 + \ldots + t_m x_m$ $(t_j \in R, x_j \in X)$. There are two kinds of steps to be done to obtain (2) for a.

1° If $r_j \mid t_j$ where $r_j x_j = y_j$ is in Ξ $(A(x_j) = Rr_j)$, then we replace $t_j x_j$ by $t_j r_j^{-1} y_j$.

2° If, in the above partial order, we have $x_j < x_i$, say $x_j = s x_i$ $(s \in R)$ for some $i, j \in \{1, \ldots, m\}$, then replace $t_j x_j + t_i x_i$ by $(t_j s + t_i) x_i$.

From the tree structure of X it is clear that the set $\{x \in X \mid x \leq x_j$ for some $j = 1, \ldots, m\}$ is finite. Hence after a finite number of applications of the above procedures, we must arrive at a stage where neither of the hypotheses of 1° or 2° occurs, i.e. we have form (2). The proof of uniqueness is similar to the argument above in (1), using again the free R-module on X. □

We shall call (2) satisfying (i) and (ii) a *canonical form* for $a \in M = \langle X; \Xi \rangle$.

2. Stable submodules

Suppose that $M = \langle X; \Xi \rangle$ is a fixed faithful simple presentation of the torsion module M.

Submodules generated by subsets of X are called *stable*. (Actually, this notion depends on the given presentation, but the reference to the fixed presentation will usually be suppressed whenever there is no danger of confusion.) Evidently, N is a stable submodule of M if and only if $N = \langle N \cap X \rangle$. For a finite subset Y of X, $\langle Y \rangle \cap X$ is again finite, while if $Y \subset X$ is of infinite cardinality, then $\langle Y \rangle \cap X$ has the same cardinality as Y.

LEMMA 2.1. *Let N be a stable submodule of a simply presented R-module $M = \langle X; \Xi \rangle$. Then both N and M/N are simply presentable.*

PROOF. We can generate N by $N \cap X$. Collect all the defining relations in X which involve exclusively elements of $N \cap X$ and 0. Then $N \cap X$ together with these relations yield a faithful simple presentation for N. In fact, from (1.9) it is easy to conclude that N does not admit any relation other than the consequences of the indicated relations.

If we omit from X the elements of $N \cap X$, we replace them in X by 0, and drop the meaningless relations, then we get a simple presentation for M/N. □

In an R-module M of projective dimension ≤ 1, a submodule N satisfying p.d. $M/N \leq 1$ is said to be *tight*. (Then necessarily p.d. $N \leq 1$ as well.) The following result is fundamental.

THEOREM 2.2. *Simply presentable torsion R-modules have projective dimension ≤ 1. Stable submodules in any simple presentation are tight.*

PROOF. Because of (2.1), the second assertion is a consequence of the first. To justify the first claim, let $M = \langle X; \Xi \rangle$ be a faithful simple presentation. We well-order the elements of X such that in this well-ordering $X = \{x_\alpha \mid \alpha < \tau\}$, we have 1) x_0 is a minimal element in X, and 2) for each $\alpha > 0$, either x_α is minimal in X or in the unique relation $r_\alpha x_\alpha = y$ in X involving x_α with a coefficient $\neq 1, y$ equals some x_β with $\beta < \alpha$. In this way, we obtain a continuous well-ordered ascending chain

$$0 = M_0 < M_1 < \ldots < M_\alpha < \ldots \qquad (\alpha < \tau)$$

of submodules where M_α is generated by all x_β with $\beta < \alpha$ and $\cup M_\alpha = M$. As $M_{\alpha+1}/M_\alpha$ is cyclically presented for each α, p.d. $M_{\alpha+1}/M_\alpha \leq 1$. From a well-known lemma of Auslander [2] we infer that p.d. $M \leq 1$ as well. □

Example (3.11) below will show that a torsion module of projective dimension 1 need not be simply presentable. However, divisible torsion modules of projective dimension 1 have to be simply presentable (cf. (7.2)).

By a *tight system* of an R-module M of projective dimension ≤ 1 is meant a collection $\mathcal{T}$ of submodules of M such that (i) $0, M \in \mathcal{T}$; (ii) $\mathcal{T}$ is closed under unions of chains; (iii) if $M_i, M_j \in \mathcal{T}$ and $M_i < M_j$ then p.d. $M_j/M_i \leq 1$; (iv) given $M_i \in \mathcal{T}$ and a countable subset Δ of M, there is an $M_j \in \mathcal{T}$ such that $\langle M_i, \Delta \rangle \leq M_j$ and M_j/M_i is countably generated.

The existence of tight systems in modules with p.d. $M \leq 1$ has been established in [7, p.87]. In the present case, this can be sharpened:

THEOREM 2.3. *Let $M = \langle X; \Xi \rangle$ be a simply presented R-module. The set $\mathcal{S}$ of all stable submodules in M is a tight system. If $\mathcal{T}$ is any tight system in M then*

$$\mathcal{S} \cap \mathcal{T} = \{N \in \mathcal{T} \mid N \textit{ is stable in } M\}$$

is likewise a tight system in M.

PROOF. The first assertion is an immediate consequence of (2.2), since manifestly $\mathcal{S}$ satisfies conditions (i), (ii) and (iv) above. It is well known and easy to verify (by a standard back-and-forth argument) that the intersection of two families of submodules both satisfying (i), (ii) and (iv) will also satisfy these conditions.
□

3. Heights and nice submodules

Heights of elements in modules over valuation domains have been introduced in Salce [11] and more fully developed in Fuchs-Salce [7]. These heights turned out to be quite useful in dealing with elements in modules, though they have the shortcoming of not distinguishing between elements of the first Ulm submodule $M^1 = \cap_{r \neq 0} rM$. If one wishes to develop a general theory for simply presented torsion modules, then this notion of height is no longer satisfactory, and one has to look for a more delicate way to measure heights. As pointed out in the introduction, we shall limit ourselves to the discussion of simply presented torsion modules M with M^1 a divisible module in which case the heights in the old sense will suffice.

Recall that the *height ideal* $H(a)$ of an element $a \in M$ is the submodule J of Q generated by all r^{-1} $(r \in R)$ such that $rx = a$ has a solution for x in M. Setting $U = J/R$, the *height* $h_M(a)$ of a in M (or simply $h(a)$) is then defined as U (*non-limit height*) or U^- (*limit height*) according as there is or there is not a homomorphism $J \to M$ mapping 1 onto a. If J is singly generated, we say J/R is a principal height. The height R/R is referred to as 0 height, and to the elements in the divisible submodule the height ∞ is assigned. The heights of elements in M^1 are either $(Q/R)^-$ or ∞.

We define division of heights by elements $0 \neq t \in R$ as follows: $t^{-1}U = t^{-1}J/R$ and $t^{-1}U^- = (t^{-1}J/R)^-$ where again $U = J/R$. Note that $h(tx) \geq t^{-1}h(x)$ for all $0 \neq t \in R$ and $x \in M$.

LEMMA 3.1. *Suppose $M = \langle X; \Xi \rangle$ (faithful) is a torsion R-module with M^1 divisible. Let $a \in M$ be given in its canonical form (2). Then*

(i) $0 \neq t \in R$ *divides a exactly if t divides each $s_i x_i$ $(i = 1, \ldots, k)$.*

(ii) t *divides $s_i x_i$ if and only if ts_i^{-1} divides x_i.*

PROOF. The first assertion is trivial in one way, so assume $a = t(t_1 y_1 + \ldots + t_m y_m)$ with $y_j \in X, t, t_j \in R$. If we perform steps 1° and 2° of (1.9) on the expression $a = tt_1 y_1 + \ldots + tt_m y_m$, then after a finite number of applications this process terminates at a canonical form of a. The property of having all terms divisible by t is preserved under these two steps, so each $s_i x_i$ must be divisible by t. The second claim follows from the uniqueness statement of (1.9). □

Using the last lemma, we can conclude at once that the first Ulm submodule $M^1 = \cap_{r \neq 0} rM$ is stable. Moreover, from (2.1) it is readily derived:

LEMMA 3.2. *In a simply presented module, all Ulm submodules are stable. Hence all Ulm submodules, all Ulm factors and the maximal divisible submodule are simply presented.* □

For a characterization of those heights which can occur in simply presented torsion R-modules, we require a general lemma. *Maximal* uniserial submodule obviously means a uniserial submodule not properly contained in any uniserial submodule.

LEMMA 3.3. *In an R-module of projective dimension ≤ 1, maximal uniserial submodules are countably generated and have projective dimension ≤ 1.*

PROOF. Let p.d.$M \leq 1$ and U a maximal uniserial submodule of M. From the existence of tight systems (cf. [7, p.87]) we conclude that there is a continuous well-ordered ascending chain of tight submodules in M,

$$0 = M_0 < M_1 < \ldots < M_\alpha < \ldots < M_\delta$$

for some ordinal δ such that $M_{\alpha+1}/M_\alpha$ is countably generated for each $\alpha < \delta$ and

$$0 = U_0 < U_1 < \ldots < U_\alpha < \ldots < U_\delta = U$$

where $U_\alpha = M_\alpha \cap U$. Then $U_{\alpha+1}/U_\alpha = U_{\alpha+1}/(M_\alpha \cap U_{\alpha+1}) \cong (U_{\alpha+1} + M_\alpha)/M_\alpha \leq M_{\alpha+1}/M_\alpha$. The annihilators of elements in modules of projective dimension 1 are principal ideals (see [7]), so U_α must be cyclic for each α. This can happen only if $\delta \leq \omega$. Thus U is contained in a countably generated tight submodule of M. The proof is thus reduced to countably generated modules M.

A countably generated module M of projective dimension ≤ 1 is the union of a countable ascending chain of finitely presented submodules: $0 = F_0 < F_1 < \ldots < F_n < \ldots < F_\omega = M$; see [7]. An argument similar to the one given above shows that either $U < F_n$ for some n or U is the countable union of cyclically presented modules, and thus countably generated. In the first alternative, F_n is a direct sum of cyclically presented R-modules and U is isomorphic to a submodule of one of them. U being maximal in M must be maximal in F_n, so it has to be cyclic. As U is countably generated and the annihilator ideals of its elements are principal ideals, we have p.d.$U \leq 1$. □

We can now easily derive:

PROPOSITION 3.4. *In an R-module M of projective dimension ≤ 1 (in particular, in a simply presented R-module) every element $a \notin M^1$ has a height which is either a principal height, or a limit height, or a countably generated non-limit height. All these heights do occur in simply presented modules.*

PROOF. Suppose $a \in M \backslash M^1$ has non-limit height $U = J/R$ in M where J is an uncountably generated submodule of Q. By definition, there is a homomorphism $\phi : J \to M$ with $\phi 1 = a$. Here ϕJ must be a maximal uniserial submodule of M. A simple appeal to (3.3) completes the proof of the first claim. The second claim is a consequence of (1.3), (1.4) and (1.5). □

PROPOSITION 3.5. *A uniserial R-module U is simply presentable exactly if p.d.$U \leq 1$.*

PROOF. The 'only if' part follows from (3.3) at once, while (1.5) implies the 'if' part. □

The heights mentioned in (3.4) will be called *admissible*. We now verify an important formula for the height of an element.

THEOREM 3.6. *Let $M = \langle X; \Xi \rangle$ be a faithful simple presentation of a torsion R-module M with $M^1 = 0$, and (2) a canonical form of $a \in M$. Then the height of a satisfies*

$$h(a) = \inf\{h(s_1x_1), \ldots, h(s_kx_k)\}$$

where $h(s_ix_i) = s_i^{-1}h(x_i)$.

PROOF. Let $h(a) = \sigma$. In view of (3.1), the only non-trivial case is when σ is a countably generated non-limit height, so assume σ is such. Let a_n $(n < \omega)$ be elements of M such that $a = a_0, a_n = r_n a_{n+1}$ for $n < \omega$ where $\sigma = \cup(a_n^{-1}R)/R$. The proof of (3.1) shows that if $a_1 = t_1y_1 + \ldots + t_my_m$, then $t_1y_1 + \ldots + t_my_m$ can be decomposed into the sum of $k+1$ pieces such that r_0 times the i-th piece gives $s_ix_i \bmod A(x_i)x_i$ for $i \leq k$, while t times the $k+1$st piece vanishes. Repeating this argument for a_1 in place of a, and continuing, it is readily seen that each s_ix_i must have a height $\geq \sigma$. □

Thus the heights of elements can be computed easily once the heights of the generators have been determined.

Given a height σ, it will be convenient to have a unified notation and denote by $V(\sigma, r)$ the cyclically presented module in (1.3) if σ is a principal height, and the uniserial module in (1.4) if σ is a non-limit height. In view of (3.6), if σ is a principal height, then every non-zero element in $V(\sigma, r)$ has principal height, and if σ is a non-limit height, then every non-zero element in the uniserial module $V(\sigma, r)$ has a non-limit height. Furthermore, from (3.6) we conclude at once:

LEMMA 3.7. *If σ is a limit height, then precisely the elements of $V(\sigma, r) = V$ contained in Rr ($= V^{\sigma}$) have limit heights, while the rest have principal heights.* □

The so-called nice submodules (introduced by Hill [9]) play a relevant role in the theory of simply presented p-groups. Their analogues for modules are equally important in our case. A submodule N of an R-module M is said to be *nice* if every coset $a + N$ ($a \in M$) has a representative $a + x$ ($x \in N$) such that

$$h_M(a + x) = h_{M/N}(a + N).$$

Such a representative $a + x$ has the largest height in its coset and is called *proper with respect to* N. Let us point out that if $a \in M$ is proper with respect to N, then $h(a + y) = \inf\{h(a), h(y)\}$ holds for all $y \in N$.

An easy but useful consequence of (3.6) is:

THEOREM 3.8. *In a simply presented module M with $M^1 = 0$, a stable submodule N is nice and satisfies : $(M/N)^1 = 0$.*

PROOF. Write $a \in M = \langle X; \Xi \rangle$ in the canonical form (2). Let N be a stable submodule of M, and suppose that $x_1, \ldots, x_j \notin N$, but $x_{j+1}, \ldots, x_k \in N$. Then (3.6) implies that $a^* = a - s_{j+1}x_{j+1} - \ldots - s_k x_k = s_1 x_1 + \ldots + s_j x_j$ has height $\inf\{h(s_1 x_1), \ldots, h(s_k x_k)\}$ which is evidently $\geq$ the height of any element in the coset $a + N$. This is sufficient to ensure that a^* is proper with respect to N. The second assertion follows from the first. □

The information about the heights can be utilized to verify the next lemma which improves on (3.3) in the simply presented case.

LEMMA 3.9. *Let $M = \langle X; \Xi \rangle$ be a faithful simple presentation of a torsion R-module M with $M^1 = 0$, and suppose $x \in X$ has a countably generated non-limit height σ. Then x can be embedded in a stable uniserial submodule U of M in which x has the same height σ.*

PROOF. Let $r \in R$ divide x in M. Then there is a $b = s_1 x_1 + \ldots + s_k x_k$ (canonical form (2)) in M such that $rb = x$ and $h(b) = r\sigma$. If we assume that b is chosen so that k is as small as possible, then all the x_i have to be $> x$ in the partial order of X, and one of the terms $s_j x_j$ has to have the same height as b. Changing the element b and the coefficient s_j if necessary, we can see that the choice $k = 1$ possible: $x = rsx_j$. Repeating this argument for x_j, the elements $x, x_j, \ldots$ generate in M a uniserial module U as desired. □

The *indicator* $i_M(a)$ of an element a in M is a function from the positivity domain Γ^+ of the value group Γ of R into the set of all heights, $i_M(a)$:

$\gamma \mapsto r_\gamma h_M(r_\gamma a)$ where $\gamma \in \Gamma^+$. We say that it has a *gap* at an ideal L of R if $\sup\{rh_M(ra) \mid r \in R\backslash L\} < \inf\{rh_M(ra) \mid r \in L\}$. E.g. a gap at Rr signals a jump of the height of ra. (For details, we refer to [7].)

THEOREM 3.10. *In a simply presentable torsion R-module M, the indicator of an element contains only a finite number of gaps; they can occur only at principal ideals.*

PROOF. Write the element $a \in M$ in a canonical form. The first gap of the indicator $i(a)$ can occur at the principal ideal Rt ($t \in R$) only if, for some $j, Rtr_j = A(x_j)$. Consequently, there are at most as many gaps in $i(a)$ as the number of elements of X which are $\leq$ some of $x_1, \ldots, x_k$ occurring in (2). □

EXAMPLE 3.11. The following countably generated torsion module M is of projective dimension 1, but is not simply presentable. Let R be a valuation domain whose value group is the ordered group $\mathbb{Q}$ of the rational numbers. For $n < \omega$ let $r_n \in R$ have value 2^{-n}, and $s \in R$ value 1. Define the module M to be generated by the set $X = \{x_n, y_{n+1} \mid n < \omega\}$ subject to the following relations:

$$r_{n+1}x_n = x_{n+1}, \; s^{n+1}y_{n+1} = x_{n+1} \quad (n < \omega) \qquad \text{and} \qquad s^2 x_0 = 0.$$

Then the indicator of x_0 has gaps at $Rr_0r_1 \ldots r_{n+1}$ for each $n < \omega$, so by (3.10) M cannot be simply presented. To show that M has projective dimension 1, observe that M is the factor module of the free module $F = \bigoplus_{n<\omega} Rx_n \oplus \bigoplus_{n<\omega} Ry_{n+1}$ modulo the free submodule

$$H = Rs^2x_0 \oplus \bigoplus_{n<\omega} R(r_{n+1}x_n - x_{n+1}) \oplus \bigoplus_{n<\omega} R(s^{n+1}y_{n+1} - x_{n+1}).$$

4. Balanced submodules

We continue with the discussion of a class of submodules which are needed in our study of simply presented modules. As we shall see, the simply presented modules possess the projective property relative to the class of exact sequences defined by these submodules.

For every height σ and for every R-module M, the set

$$M^\sigma = \{a \in M \mid h(a) \geq \sigma\}$$

is a fully invariant submodule of M. It is useful to point out that $M^{r\sigma} = rM^\sigma$ for each $0 \neq r \in R$. An exact sequence

$$0 \to A \to B \xrightarrow{\phi} C \to 0 \tag{3}$$

induces a sequence

$$0 \to A^\sigma \to B^\sigma \to C^\sigma \to 0 \tag{4}$$

which is in general not exact. It is exact at B^σ if and only if A is *equiheight* in B, i.e. the heights of elements of A are the same in A as in B, in symbols: $A^\sigma = A \cap B^\sigma$ for each height σ. Furthermore, (4) is exact at C^σ if and only if each coset of B mod A can be represented by an element which has the same height in B as the coset has in C, i.e. if A is *nice* in B. A submodule which is both equiheight and nice will be called *balanced*. Thus A is balanced in B if and only if (4) is exact for all heights σ. In this case (3) is called a *balanced-exact* sequence.

A weakened version of balancedness is needed in our study. (3) will be called ω*-balanced* if (4) is exact for all admissible heights (the symbol ω should be a reminder that a countability restriction is involved).

LEMMA 4.1. *The exact sequence (3) is ω-balanced-exact if and only if, for each coset $b + A$ $(b \in B)$ which belongs to C^σ (σ an admissible height) and for each $r \in \operatorname{Ann}(b + A)$, there exists an $a \in A$ such that*

$$h_B(b - a) \geq \sigma \quad \textit{and} \quad r(b-a) = 0.$$

PROOF. Suppose A is ω-balanced in B, and let $c \in C^\sigma[r] = \{x \in C^\sigma \mid rx = 0\}$, σ an admissible height. By hypothesis, some $b \in B^\sigma$ satisfies $\phi b = c$. Evidently, $rb \in rB^\sigma \cap A = B^{r\sigma} \cap A = A^{r\sigma} = rA^\sigma$ whence $ra = rb$ for some $a \in A^\sigma$. Therefore, $b - a \in B^\sigma[r]$; this element is mapped by ϕ onto c, and so the stated condition holds.

Conversely, assume the stated condition. The exactness of (4) at A^σ and at C^σ for admissible σ is clear. It only remains to show that $a \in A \cap B^\sigma$ implies $a \in A^\sigma$.

First, let σ be a limit height, say, $\sigma = (J/R)^-$. For every $r \in R$ with $r^{-1} \in J$ there is a $b \in B$ such that $rb = a$. Since $r \in \operatorname{Ann}(b + A)$, there is an $a' \in A$ satisfying $r(b - a') = 0$. Therefore, $ra' = a$ and $a \in A^\sigma$.

If σ is a countably generated non-limit height, and if $rb = a$ for some $b \in B$, then manifestly, $h(b) = rh(a)$ can be assumed. This means $b \in B^{r\sigma}$; thus by hypothesis, some $a_1 \in A$ satisfies $b - a_1 \in B^{r\sigma}$ and $r(b - a_1) = 0$. As before, we obtain $ra_1 = a$ with $a_1 \in A \cap B^{r\sigma}$. We can repeat this argument with a replaced by a_1, and continue to obtain a countable sequence $a, a_1, a_2, \ldots$ in A assuring that the height of a in A is likewise $\geq \sigma$. $\square$

An R-module M is said to be ω*-balanced-projective* if for every ω-balanced-exact sequence (3), the induced map $\operatorname{Hom}_R(M, B) \to \operatorname{Hom}_R(M, C)$ is surjective. In order to survey the ω- balanced-projectives, the following lemma is required.

LEMMA 4.2. *Let M be an R-module and u an element of M such that*

$$h(u) \geq \sigma \quad \textit{and} \quad \operatorname{Ann} u \geq Rr$$

for some admissible height σ and some $0 \neq r \in R$. Then there exists a homomorphism

$$\eta : V(\sigma, r) \to M \quad \textit{with} \quad \eta a = u$$

where $a \in V(\sigma, r)$ has height σ.

PROOF. The correspondence $a \mapsto u$ gives rise to a homomorphism $Ra \to Ru$ which does not decrease heights. $V(\sigma, r)$ is a simply presented module in which a is the only minimal generator. For each generator x of V with $x >- a$ (say, coming from a relation $sx = a$ in Ξ), choose for the image of x any element $v \in M$ such that $h(v) \geq h(x)$ and $sv = u$; it is easy to convince ourselves that such a choice is possible. If σ is non-limit, then we continue, repeating this step a countable number of times. In this way, we assign images to all the generators of $V(\sigma, r)$ in its simple presentation such that the defining relations are respected and heights are never decreased. Obviously, this yields a map η as desired. □

An immediate conclusion is:

COROLLARY 4.3. *An exact sequence is ω-balanced if and only if all the modules $V(\sigma, r)$ (where σ ranges over the admissible heights and r runs over R) enjoy the projective property relative to it.* □

We now apply routine arguments to establish the existence of 'enough' ω-balanced-projectives.

PROPOSITION 4.4. *Every torsion R-module M with $M^1 = 0$ can be embedded in an ω-balanced-exact sequence $0 \to A \to B \to M \to 0$ where B is a direct sum of modules each of which is isomorphic to some $V(\sigma, r)$ for an admissible height σ and $r \in R$.*

PROOF. For an $0 \neq u \in M$ consider all $V(\sigma, r)$ for all admissible heights σ with $h(u) \geq \sigma$ and let $0 \neq r \in R$ range over a generating set of $\operatorname{Ann} u$. By (4.2), the correspondence $a \mapsto u$ extends to a homomorphism $\eta_{\sigma,r} : V(\sigma, r) \to M$. Define B as the direct sum of these $V(\sigma, r)$, taken for all $0 \neq u \in M$, all admissible σ, and all the generators r of $\operatorname{Ann} u$, and let $\eta = \bigoplus \eta_{\sigma,r} : B \to M$. Visibly, η is an epimorphism whose kernel A has to be ω-balanced in B, because every $0 \neq u \in M$ has the required preimage in B; see (4.1). □

Hence we are led to the following conclusion.

THEOREM 4.5. *An R-module is ω-balanced-projective if and only if it is a summand of a direct sum of modules each of which is isomorphic to a module $V(\sigma, r)$ for some admissible height σ and some $0 \neq r \in R$.* □

As an application, we show:

LEMMA 4.6. *Let $V = V(\sigma, r)$ for some limit height σ and some $0 \neq r \in R$ as generated in (1.5). If J is a subset of I such that for each $i \in I$ there is a $j \in J$ satisfying $Rs_j < Rs_i$, then $V^* = \langle Rx_j \mid j \in J \rangle$ is a summand of V.*

PROOF. V^* is (a stable, so by (3.8)) a nice submodule of V. It is readily checked that it is equiheight, and hence balanced in V. V/V^* is a direct sum of cyclically presented modules, so it is ω-balanced projective. Thus V^* is a summand of V. □

5. Perfect elements

In abelian group theory, elements which are proper with respect to a subgroup are instrumental in extending homomorphisms between subgroups of simply presented p-groups. In the present case proper elements are not good enough to deal with the delicate problem of extending homomorphisms from submodules. A refined version of properness is indispensible, and to this end we introduce a more powerful concept.

Let N be a submodule of the R-module M. Call $g \in M$ *perfect with respect to* N if

(a) g is proper with respect to N;

(b) $\operatorname{Ann}(g+N)$ is a principal ideal Rr of R;

(c) the indicator of the coset $g+N$ in M/N is constant up to Rr.

If g is perfect with respect to 0, we simply say: g is a *perfect element*.

This definition is to be complemented with the following observation.

LEMMA 5.1. *If the element $g \in M$ is perfect with respect to the submodule N, then for each $s \notin \operatorname{Ann}(g+N), sg$ is perfect with respect to N.*

PROOF. If we had $h(sg+x) > h(sg)$ for some $s \notin \operatorname{Ann}(g+N)$ and $x \in N$, then $h(g) = h(g+N) = s^{-1}h(sg+N) > s^{-1}h(sg) \geq h(g)$ would be a contradiction. Thus (a) holds for sg. The equality $\operatorname{Ann}(sg+N) = Rrs^{-1}$ is obvious. From $sh(g) = h(sg+N) = h(sg)$ we derive property (c) for sg. □

It is easy to verify:

LEMMA 5.2. *Let $M = \langle X; \Xi \rangle$ be a simply presented torsion module with $M^1 = 0$, and N a stable submodule of M. If $x+N$ $(x \in X)$ is a minimal generator in M/N, then x is perfect with respect to N.*

PROOF. Property (a) follows from (3.8), (b) from (2.2), while property (c) is an immediate consequence of (3.10). □

We have come to the technical lemma which provides an important tool in the proof of (5.6) infra. This lemma imitates Lemma 80.3 in [4].

LEMMA 5.3. *Let*

$$\begin{array}{ccccccccc} & & 0 & \longrightarrow & N & \longrightarrow & M & & \\ & & & & \downarrow \psi & & \downarrow \phi & & \\ 0 & \longrightarrow & A & \longrightarrow & B & \xrightarrow{\eta} & C & \longrightarrow & 0 \end{array}$$

be a commutative diagram of torsion R-modules with exact top row and ω-balanced bottom row. Assume that ψ does not decrease heights (computed in M and B, respectively) and $g \in M$ is perfect with respect to N. If g has admissible height, then ψ can be extended to a map

$$\psi_1 : N + Rg \to B$$

such that $\eta\psi_1 g = \phi g$ and ψ_1 does not decrease heights either.

PROOF. If $h(g) = \sigma$ (admissible), then choose a $b \in B$ such that $h(b) \geq \sigma, \operatorname{Ann} b \geq \operatorname{Ann}(g + N) = Rr$, and $\eta b = \phi g$; owing to (4.1), this is possible. We then have $rb - \psi(rg) \in A$ of height $\geq r^{-1}\sigma$. Therefore, in view of (4.1) some $a \in A$ of height $\geq r^{-1}\sigma$ satisfies $ra = rb - \psi rg$. A map ψ_1 extending ψ can be defined by sending g onto $b - a$. Clearly, $h(b - a) \geq \sigma = h(g)$. By virtue of (5.1), for each $s \notin Rr$ and each $x \in N$ we have $h(sg + x) = \min\{h(sg), h(x)\} = \min\{s^{-1}\sigma, h(x)\}$. On the other hand,

$$h(s(a - b) + \psi x) \geq \min\{h(s(a - b)), h(\psi x)\} \geq \min\{s^{-1}\sigma, h(\psi x)\},$$

establishing the height-non-decreasing character of ψ_1. □

A continuous well-ordered ascending chain of submodules

$$0 = N_0 < N_1 < \ldots < N_\nu < \ldots < N_\lambda = M \tag{5}$$

of the module M is said to be a *nice composition chain for* M if each $N_{\nu+1}$ is of the form $N_{\nu+1} = N_\nu + Ra_\nu$ for some $a_\nu \in N_{\nu+1}$ which is perfect with respect to N_ν.

LEMMA 5.4. *All simply presented torsion modules M with $M^1 = 0$, in particular, all the modules $V(\sigma, Rr)$ with σ an admissible height and $0 \neq r \in R$, have nice composition chains.*

PROOF. It is evident from the proof of (3.8) that a simply presented R-module M admits a chain (5) where the N_ν are stable submodules such that, for each $\nu, N_{\nu+1} = N_\nu + Ra_\nu$ holds for some $a_\nu \in X$ which represents a minimal generator in M/N_ν. Owing to (5.2), a_ν is perfect with respect to N_ν. □

It is straightforward to verify:

LEMMA 5.5. *A direct sum of modules each of which has a nice composition chain also has such a chain.* □

We continue with the following lemma.

LEMMA 5.6. *A torsion R-module M with $M^1 = 0$ that admits a nice composition chain is ω-balanced-projective. In particular, simply presented torsion modules M with $M^1 = 0$ are ω- balanced-projective.*

PROOF. The proof of this relies heavily on (5.3). Suppose (5) is a nice composition chain for M, and assume $0 \to A \to B \xrightarrow{\eta} C \to 0$ is an ω-balanced-exact sequence of torsion R-modules with a map $\phi : M \to C$. We use transfinite induction on $\nu < \lambda$, starting at the zero map $\psi_0 : N_0 \to B$. At each non-limit step ν we apply (5.3) with $N = N_\nu$ and $g = a_\nu$ in order to extend the previously defined height-non-decreasing map $\psi_n : N_\nu \to B$ (satisfying $\eta\psi_n = \phi \mid N_\nu$) to a height-non-decreasing map $\psi_{\nu+1} : N_{\nu+1} \to B$ such that $\eta\psi_{\nu+1} = \phi \mid N_{\nu+1}$. At limit ordinals ν, we take unions. When we reach $N_\lambda = M$, we obtain a map $\psi_\lambda = \psi : N_\lambda \to B$ which will be the desired lifting of ϕ; indeed, it satisfies $\eta\psi = \phi$.

The last claim is a simple consequence of (5.4), (5.5) and what has been proved.

□

6. The α- and β-invariants

In the classification theory for simply presented modules, it is crucial to find sufficiently many cardinal invariants for the modules. We shall require the α-invariants discussed in [7] and the β-invariants which we shall introduce here. For the sake of convenience, we will work with the more suggestive vector spaces rather than the cardinals themselves.

The *α-invariant* $\alpha_M(\sigma, I)$ of M for the height σ and for the ideal I is defined by the formula (see [7]):

$$\alpha_M(\sigma, I) = M^\sigma[I]/(M^\sigma[I^+] + M^{\sigma+}[I])$$

where

$$M^{\sigma+} = \{x \in M \mid h(x) > \sigma\},\ M^\sigma[I] = \{x \in M^\sigma \mid \operatorname{Ann} x \geq I\},$$
$$M^\sigma[I^+] = \{x \in M^\sigma \mid \operatorname{Ann} x > I\}.$$

This definition is functorial: a homomorphism $\phi : M \to M'$ between modules induces a map $\phi' : \alpha_M(\sigma, I) \to \alpha_{M'}(\sigma, I)$ which is an isomorphism whenever ϕ is an isomorphism. The α-invariants are additive: they commute with direct sums.

Since the annihilators of elements in simply presented torsion modules are principal ideals, here only invariants of the form $\alpha_M(\sigma, Rr)$ need to be considered. (In this case $\alpha_M(\sigma, Rr)$ is a vector space over R/P.) In view of (3.4), the heights can also be restricted to admissible heights.

The non-zero elements of $\alpha_M(\sigma, Rr)$ can easily be characterized.

LEMMA 6.1. *An element $a \in M$ represents a non-zero element of $\alpha_M(\sigma, Rr)$ exactly if*

(i) $h(a) = \sigma$;
(ii) $\operatorname{Ann} a = Rr$;
(iii) $h(sa) = s^{-1}h(a)$ *for all* $s \in R \backslash Rr$.

PROOF. See Lemma 1.1 on p.198 in [7]. $\square$

LEMMA 6.2. *Let M be a simply presented module. For $r, t \in R$ with $Rr < Rt$, the map $a \mapsto ta$ induces a monomorphism*

$$\mu_t : \alpha_M(\sigma, Rr) \to \alpha_M(t^{-1}\sigma, Rt^{-1}r).$$

PROOF. The claim follows at once from (6.1). $\square$

It should be pointed out that the maps μ_t commute with the homomorphisms $\alpha_M(\sigma, I) \to \alpha_{M'}(\sigma, I)$ induced by a map $\phi : M \to M'$.

From the α-invariants we now derive new invariants which we shall need later on. Let

$$\beta_M(\sigma, Rr) = \alpha_M(\sigma, Rr) / \bigcup_{t \in P} \operatorname{Im} \mu_t$$

where $\mu_t : \alpha_M(t\sigma, Rtr) \to \alpha_M(\sigma, Rr)$.

As an illustration, consider a cyclically presented module $M = R/Rs$. Then $\beta(\sigma, Rr) = 0$ except when $\sigma = 0$ and $Rr = Rs$ in which case $\beta(\sigma, Rr) \cong R/P$ (i.e.

its dimension is 1). Next consider $V(\sigma, Rr)$ of (1.5) where σ is a limit ordinal. The only non-zero β-invariants are $\beta(\sigma, Rr)$ and $\beta(0, Rs_i)$, all 1-dimensional provided that the Rs_i are all different (notation as in (1.5)). A non-cyclic uniserial module has no non-zero β-invariants.

The analogue of (6.1) holds true:

LEMMA 6.3. *$a \in M$ represents a non-zero element of $\beta_M(\sigma, Rr)$ if and only if*

(i) $h(a) = \sigma$,
(ii) $\operatorname{Ann} a = Rr$,
(iii) *a is a perfect element, and*
(iv) *if $b \in M, t \in P$ satisfy $tb = a$, then $h(b) < t\sigma$.* □

7. Equivalence theorem

We have come to the discussion of more substantial results on simply presented torsion modules; we keep assuming that their first Ulm submodules are divisible. Our first goal is to obtain a direct decomposition, separating the divisible part, so that the two summands can be treated separately.

The next lemma shows that the divisible parts are always summands.

LEMMA 7.1. *Let M be a simply presented torsion R-module with $D = M^1$ divisible. Then D is a stable submodule of M and $M = D \oplus T$ where T is simply presented with $T^1 = 0$.*

PROOF. We have seen in (3.2) that D is a stable submodule of M. Divisible modules are evidently equiheight, and as stable submodules are nice, we conclude that D is a balanced submodule of M. From (3.2) it also follows that M/D is simply presented, so (5.6) implies that D is a summand of M, i.e. $M = D \oplus T$ where $T \cong M/D$. □

The following lemma is phrased under a milder hypothesis than we need here.

LEMMA 7.2. *A divisible torsion module of projective dimension one, in particular, a summand of a simply presented divisible torsion module is simply presented.*

PROOF. A divisible torsion module D of projective dimension 1 is completely characterized by a single invariant: the cardinality δ of the number of summands in a direct decomposition of $D[r]$ into a direct sum of cyclics (all $\cong R/Rr$), for any non-zero $r \in R$; see [7, p. 128]. In [6, Thm 3] it is shown that if Q/R is countably generated (as an R-module), then the cardinal number δ can be arbitrary; while if the minimal cardinality κ of generating sets of Q/R is $> \aleph_0$, then δ has to be $\geq \kappa$.

In the first alternative, D is isomorphic to the direct sum of δ copies of Q/R, so it is simply presented. In the second alternative, the direct sum E of δ copies of ∂^* is simply presented and $E[r]$ has the same number of summands in its decomposition into cyclic modules. Hence $D \cong E$, and we are done. □

The theorems we have proved so far show that the simply presented R-modules under consideration can be characterized in various ways each of which corresponds to a characterization of simply presented abelian p-groups. Because of (7.1) and (7.2), in the next theorem it suffices to deal with modules whose first Ulm submodules vanish.

THEOREM 7.3. *For a torsion R-module M with $M^1 = 0$ the following are equivalent:*

(i) *M is a summand of a simply presented module;*

(ii) *M is a summand of a module that admits a nice composition chain;*

(iii) *M is ω-balanced-projective;*

(iv) *M a summand of a direct sum of modules of types $V(\sigma, Rr)$ for various admissible heights σ and various $0 \neq r \in R$.*

PROOF. (i) $\Rightarrow$ (ii) This is an immediate consequence of (5.4).

(ii) $\Rightarrow$ (iii) is the content of our (5.6).

(iii) $\Rightarrow$ (iv) follows from (4.5).

(iv) $\Rightarrow$ (i) is trivial, since all the summands in (iv) are simply presented. □

It is still an open question if a summand of a simply presented torsion module is again simply presented. Our result (8.3) below will show that the term "summand of" can not be omitted from (iv).

8. Direct summands

It is evident that more infomation is needed about simply presented modules with trivial divisible submodules. Special attention will be given the two extreme cases in which no elements are of limit height and in which every element has limit height, respectively.

LEMMA 8.1. *Let T be a summand of a simply presented torsion R-module such that $T^1 = 0$. The following are equivalent :*

(i) *T is a direct sum of uniserials ;*

(ii) *T contains no elements of limit height;*

(iii) *all the β-invariants of T belonging to limit heights σ are 0.*

PROOF. The implication (i) $\Rightarrow$ (ii) is trivial. In order to verify (ii) $\Rightarrow$ (i), assume that T contains no elements of limit heights. By the proof of (4.4) T is an epimorphic image of a direct sum of simply presented uniserial modules with ω-balanced kernel (i.e. modules $V(\sigma, Rr)$ with limit σ are not needed in the resolution). As a summand of a simply presented torsion R-module, T is by (5.6) ω-balanced-projective, and so by (4.5) it is a summand of a direct sum of simply presented uniserial modules. Here the summands have local endomorphism rings [7], whence the well-known Krull-Schmidt-Azumaya theorem (see e.g. [1, p. 141]) leads us to the conclusion that T itself is a direct sum of (simply presented) uniserial modules.

Again, (ii) $\Rightarrow$ (iii) is clear in view of (6.3), so we prove (iii) $\Rightarrow$ (ii). By (5.4) and (5.6), T is a summand of a direct sum M of modules of types $V(\sigma, Rr)$ with admissible σ; thus M is simply presented, $M = \langle X; \Xi \rangle$. Evidently, if M contains elements of limit height, then at least one of its β-invariants for limit σ is $\neq 0$. What we want to show is that the same holds for T. Pick an $a \neq 0$ in T of limit height σ. If its indicator $i(a)$ has a gap, then let the first gap be at Rr (cf.(3.10)). Hence there is a $c \in T$ of height $> \sigma$ such that $rc = ra$, and if we replace a by $a-c$, then this new element will have the same height as a and will have a constant indicator. Thus we can without loss of generality assume that $\operatorname{Ann} a = Rr$ and $i(a)$ has no gaps at all up to Rr.

If $h(a) = \sigma$ is a limit height and $a = s_1x_1 + \ldots + s_kx_k$ is a canonical form (2) for a in M, then by (3.6) one of the terms, say s_kx_k, has the same height σ, and all other terms have heights $\geq \sigma$. It follows that there is an element b in T of height $s_k\sigma$ such that $s_kb = a$; it is necessarily of the form $b = c + x_k$ with $s_kc = s_1x_1 + \ldots + s_{k-1}x_{k-1}$ where $h(x_k) = h(b) = s_k\sigma$, still a limit height. But in our special case any generator $y \in X$ of M which is larger than a generator $x_k \in X$ of limit height must have a non-limit height. Thus if $td = b$ $(t \in P, d \in T)$ then necessarily $h(d) < th(b)$. It is easy to see that the indicator $i(b)$ can not have any gaps either, whence it readily follows from (6.3) that b represents an element of $\beta(s_k\sigma, \operatorname{Ann} b)$, so this β-invariant of T is not 0. □

The next example plays an important role in (8.3) which will exhibit simply presented modules with different direct decompositions.

EXAMPLE 8.2. Let U_i $(i \in I)$ be a set of pairwise non-isomorphic countably generated uniserial modules, and write $U_i \cong J_i/R$ where J_i are submodules of Q such that $J = \cup J_i$ $(< Q)$ is not principal.

Using the limit height $\sigma = (J/R)^-$, define the module $W = W(\sigma, Rr)$ to be generated by symbols b, x_{in} $(i \in I, n \geq 1)$ subject to the relations

$$rb = 0,\ s_{i1}x_{i1} = b,\ s_{in+1}x_{in+1} = x_{in} \qquad (s_{in} \text{ non-units } \in R)$$

where, for every fixed i, the inverses of s_{in} $(n \geq 1)$ generate J_i. The module W/Rb is isomorphic to the direct sum of the U_i. It is readily checked that no element in W has principal height. □

LEMMA 8.3. *$W(\sigma, Rr)$ is isomorphic to a summand of $M = V(\sigma, Rr) \oplus (\bigoplus U_i)$.*

PROOF. As in (1.6), let the module $V(\sigma, Rr)$ be generated by a, x_i $(i \in I)$ subject to the relations $ra = 0$, $s_ix_i = a$ (we keep assuming that $Rs_i \neq Rs_j$ if $i \neq j$ in I), where the elements s_i^{-1} $(s_i \in R, i \in I)$ generate the R-submodule J of Q. For each $i \in I$, choose an index $k \in I$ such that $J_i \leq s_k^{-1}R$; this is possible in view of our hypotheses. Let U_i be generated by the elements u_{in} subject to the relations $s_{i1}u_{i1} = 0$, $s_{in+1}u_{in+1} = u_{in}$, and let $\eta : U_i \to s_k^{-1}R/R$ be a monomorphism. Throughout in M, replace the generators u_{in} by $v_{in} = u_{in} - \eta u_{in}$. Then b and the new generators generate a submodule W' of M that is isomorphic to W. We can argue as in (4.6) (or prove directly) that W' is a summand of M. □

It is worth while adding to (8.3) the remark that by suitable choices of the modules involved, the complementary summand of $W(\sigma, Rr)$ in M will be a direct sum of cyclically presented modules.

Let $M = \langle X; \Xi \rangle$ (faithful) be a simply presented module and $x \in X$. We shall denote by V_x the stable submodule in M generated by $Y = \{y \in X \mid y \geq x\}$.

Set $A(x) = Rr$ and $rx = x^* \in X$. Assume that $h(x^*) > r^{-1}h(x)$, i.e. $i(x)$ has a gap (its first) at Rr. Then there are a $b \in M$ and $s \in R$ such that $sb = x^*$ and $s^{-1}R > r^{-1}h(x)$. Evidently, Rb is a uniserial submodule of M which intersects V_x in Rx^*. There is a homomorphism $\eta : V_x \longrightarrow M$ such that $\eta x = sr^{-1}b$; this η will fix x^*. For all $y \in Y$ let $y' = y - \eta y$ and $Y' = \{y' \mid y \in Y\}$. As in the proof of (8.3), it is straightforward to check:

LEMMA 8.4. *We have* $M = \langle X \backslash Y \rangle \oplus \langle Y' \rangle$ *where* V_x / Rx^*. □

Note that if Y is replaced by Y' in X, say $X' = X \backslash Y \cup Y'$, then in the new presentation of $M = \langle X'; \Xi' \rangle$ the set Ξ' of relations differs from Ξ – besides replacing every $y \in Y$ by the corresponding y' – only in that $rx = x^*$ is replaced by $rx' = 0$. Clearly, x' is a minimal generator in the presentation $M = \langle X'; \Xi' \rangle$. It is of importance to emphasize that $h(y) = h(y')$ for all $y \in Y$.

We shall refer to the process of replacing Y by Y' in X as *pruning the branch of x at x^**. Keep in mind that such a pruning is possible if $x^* = A(x)x$ and $i(x)$ has a gap at $A(x)$. (If $i(x)$ has no gap at $A(x)$, then pruning is still possible whenever $h(x^*) = A(x)^{-1}h(x)$ and the height of x^* stays the same after removing the set Y from X.)

The inverse process of pruning can be described as follows. Let x be a minimal generator in a presentation $M = \langle X; \Xi \rangle$, and $\operatorname{Ann} x = Rr$. Suppose $x^* \in X$ such that $h(x^*) > r^{-1}h(x)$, so that there exist a $b \in M$ and an $s \in R$ satisfying $sb = x^*$ and $s^{-1}R > r^{-1}h(x)$. As above, we denote by V_x the submodule of M generated by $Y = \{y \in X \mid y \geq x\}$. If $\operatorname{Ann} x^* = Rt$, then define A_x as the simply presented module with the same generators and relations as V_x with a sole exception: $rx = 0$ is replaced by $rtx = 0$. Then we can find a homomorphism $\xi : A_x \longrightarrow M$ such that $\xi x = sr^{-1}b$. Finally, we replace every $y \in Y$ by $y' = \xi y - y$ and form $X' = X \backslash Y \cup Y'$ with $Y' = \{y' \mid y \in Y\}$; the relations between the elements of Y' are the same as between the corresponding elements of Y, only $rx = 0$ is replaced by $rx' = x^*$. Clearly, $i(x')$ will have a gap at $A(x') = Rr$. This process may be called *copruning*. (It is easy to see that copruning is still possible if $h(x^*) = r^{-1}h(x)$.)

EXAMPLE 8.5. Suppose $J_1, \ldots, J_k, \ldots$ is a sequence of non-principal fractional ideals $> R$ such that the union of all the fractional ideals $J_1 \ldots J_n$ $(1 \leq n < \omega)$ is not all of Q. Let the module M be defined as follows: it is generated by all k-tuples $(r, r_1, \ldots, r_k)$ such that r is any fixed non-zero element of P, and $r_i \in R$ runs over all elements with $r_i^{-1} \in J_i$ $(i \geq 1)$, subject to the relations

$$r(r) = 0 \quad \text{and} \quad r_k(r, r_1, \ldots, r_k) = (r, r_1, \ldots, r_{k-1}) \qquad (k \geq 1).$$

It is easily checked that the height of $(r, r_1, \ldots, r_k)$ is $(K/R)^-$ where K is the union of the fractional ideals $J_{k+1} \ldots J_n$ for all $n \geq k+1$. This M is a simply presented module in which every element has limit height. (This example behaves like a "fractal".) □

Recall that a module is *superdecomposable* if it has no indecomposable summands $\neq 0$.

LEMMA 8.6. *For a simply presented R-module $M = \langle X; \Xi \rangle$ the following conditions are equivalent:*

(i) *every element of M has limit height;*

(ii) *M has no uniserial summand $\neq 0$.*

If M is superdecomposable, then these conditions are satisfied.

PROOF. The implication (i) $\Rightarrow$ (ii) is obvious.

To verify (ii) $\Rightarrow$ (i), suppose M has no uniserial summand. Consider the following subset:

$$C = \{x \in X \mid h(x) \text{ is a limit height, and } i(x) \text{ has a gap at } A(x)\}.$$

If C contains a maximal element y, then we prune the branch of y at $y^* = A(y)y$. This yields a direct decomposition of M where one of the summands is the cut branch B with a generator y' replacing y. Let N be the stable submodule of M generated by all $x \in X$ which are not $> y$. Then by the choice of $y, M/N \cong B/Ry'$ contains no elements of limit height, so it is by (8.1) a direct sum of uniserials. Hence B is the extension of Ry' by a direct sum of uniserials; and at y' we can cut either a branch which is cyclic or a branch which corresponds to a countably generated uniserial. Next suppose that C does not contain a maximal element, but X contains an element x of non-limit height. If $i(x)$ has no gaps before $\operatorname{Ann} x$, then x is contained in a stable uniserial submodule of M which must be equiheight, and hence balanced in M. Thus it is a summand of M. If $i(x)$ has a gap before $\operatorname{Ann} x$, then we can prune a uniserial branch at the element where this gap occurs. In either way, we get a uniserial summand. Thus (i) is satisfied in the only case left when (ii) can hold.

It is evident that a superdecomposable module must satisfy (ii). $\square$

References

[1] F.W. Anderson and K.R. Fuller, *Rings and Categories of Modules*, Graduate Texts in Math. (Springer, New York, 1974).

[2] L. Auslander, *On the dimension of modules and algebras. III*, Nagoya Math. J. 9 (1955), 67–77.

[3] P. Crawley and A. Hales, *The structure of torsion abelian groups given by presentations*, Bull. Amer. Math. Soc. **74** (1968), 954–956.

[4] L. Fuchs, *Infinite Abelian Groups*, vol. II (Academic Press, 1973).

[5] L. Fuchs, *On divisible modules over domains*, Proc. Conf. on Abelian Groups and Modules, CISM Lecture Notes, vol. **287** (Springer, 1985), 341–356.

[6] L. Fuchs, *On divisible modules over valuation domains*, J. Algebra **110** (1987), 498–506.

[7] L. Fuchs and L. Salce, *Modules over Valuation Domains*, Lecture Notes in Pure and Applied Math., vol. **97** (Marcel Dekker, 1985).

[8] L. Fuchs and L. Salce, *S-divisible modules over domains*, Forum Math. **4** (1992), 383–394.

[9] P. Hill, *On the classification of abelian groups* (unpublished manuscript, 1968).

[10] R. Nunke, *Homology and direct sums of countable abelian groups*, Math. Z. **101** (1967), 182–212.

[11] L. Salce, *Moduli di torsione su domini di valutazione* (manuscript), Trento Conference, 1980.

[12] E.A. Walker, *Ulm's theorem for totally projective groups*, Proc. Amer. Math. Soc. **37** (1973), 387–392.

[13] R.B. Warfield, Jr., *The structure of mixed abelian groups*, Abelian Group Theory, Lecture Notes of Mathematics **616** (Springer, 1977), 1–38.

Abelian Groups and the Torsion Product

PATRICK KEEF Department of Mathematics, Whitman College, Walla Walla, WA, 99362

1 INTRODUCTION

This survey is an attempt to describe some of the work of the last twenty or so years on the torsion product. In particular, the nearly pervasive influence of set theory in the study of Abelian groups during that period can be seen at several points in the exposition. Though most of the material is already in the literature, a few new results are included, as well as some different approaches to old results.

We begin with a few remarks regarding terminology and notation. We will denote the torsion product of the groups A and B by $A \circledT B$, instead of the more usual notation $\mathrm{Tor}(A, B)$. There are several reasons for this convention (originally suggested by Claudia Metelli during a seminar I gave at the University of Padova in 1992; see [19]). Not only is the notation simpler, using only three symbols as opposed to six, but it also parallels the usual notation for the tensor product, from which the torsion product is defined. In addition, we denote the iterated torsion product of the groups $A_1, \ldots, A_k$ by $\circledT_{j=1}^{k} A_j$. This convention also makes it easy to interpret the torsion product of a single group to be just the group, itself.

Since torsion products only depend upon torsion subgroups, and since these, in turn, split into primary components, unless specifically noted, when we use the term "group" we will actually mean "abelian p-group." By a Σ-*cyclic* we mean a group which is isomorphic to a direct sum of cyclic groups. Other notation and terminology will be standard and can

Research supported in part by NSF (RUI) Grant DMS-9405118

be found in [5] or [6]. In particular, $A \textcircled{T} B$ can be defined to be the group generated by all expressions of the form (a, n, b), where $a \in A$, $b \in B$, $n \in \mathbf{Z}$ and $na = nb = 0$, subject to the relations:

$$\begin{aligned} (a + a, n, b) &= (a, n, b) + (a', n, b) \\ (a, n, b + b') &= (a, n, b) + (a, n, b') \\ (a, nm, b) &= (ma, n, b) = (a, n, mb). \end{aligned}$$

What is perhaps unexpected is how infrequently it is necessary to utilize this description. This is less remarkable, however, when one recalls that the bifunctor's behavior on finitely generated groups is easy to determine, and that it commutes with direct limits.

Another fact which is somewhat surprising is that the study of torsion products can be conducted with next to no mention of tensor products. This is less remarkable, however, when one recalls that the tensor product of two groups is always Σ-cyclic. In addition, one might suppose that the tensor product is needed to complete some exact sequences. The torsion product is left-exact, in other words, if $E : 0 \to A_1 \to A_2 \to A_3 \to 0$ is a short exact sequence and B is a group, then

$$E \textcircled{T} B : 0 \to A_1 \textcircled{T} B \to A_2 \textcircled{T} B \to A_3 \textcircled{T} B$$

is exact, and the next term in this sequence is $A_1 \otimes B$. In practice, however, we usually are working with sequences E which are pure, and in this case $E \textcircled{T} B$ is also right exact (in fact, it is also a pure sequence).

In the 1960s, R. Nunke wrote a sequence of papers which are fundamental to our understanding of the torsion product (see [26], [27], [28], [29]). Not only were the socles, Ulm subgroups and Ulm functions of torsion products described, but the torsion product was utilized in a strong way in his study of generalizations of purity. Subsequent work on this bifunctor has been heavily influenced by the following questions:

Fuchs' Problem 50 ([5]): Relate the groups A and B if $A \textcircled{T} G \cong B \textcircled{T} G$ for every reduced group G.

Warfield's Problem ([31]: Is there a structure theory for groups of the form $A \textcircled{T} B$, where A and B are totally projective groups?

Nunke's Problem ([28]): When is $A \textcircled{T} B$ a dsc (i.e., a direct sum of countable groups)?

Oxford's Problem ([30]): If H is a group of infinite length α, and $0 \to M \to H \to \mathbf{Z}_{p^\infty} \to 0$ is a pure sequence such that

$$0 \to M \textcircled{T} H \to H \textcircled{T} H \to \mathbf{Z}_{p^\infty} \textcircled{T} H \to 0$$

splits, does it follow that H is p^α-projective (i.e., $p^\alpha \mathrm{Ext}(H, X) \equiv 0$)?

The theory of abelian p-groups naturally divides into two parts: the study of separable groups, for which the class of Σ-cyclic groups is fundamental, and the study of groups with elements of infinite height, for which the class of totally projective groups is fundamental. Although this is not apparent at first blush, Fuchs' problem 50 is really a question involving separable groups. It was addressed first in by P. Hill in [7], where it was shown that groups A and B satisfying this condition have a similar structure (for example, their Ulm functions agree), but whether they must actually be isomorphic was left unresolved. In [1], however, Cutler showed that this need not be the case.

Cutler's example is based upon the observation that when A and B are reduced groups, then $A \circledT B$ necessarily has large Σ-cyclic summands (a phenomenon which was initially noted by Irwin in [11]). This means that if A is, say, a large torsion-complete group, C is a small Σ-cyclic and $B = A \oplus C$, then A and B satisfy Fuchs' Problem 50, but are not isomorphic. The natural question is whether this example is general, i.e., if A and B satisfy this condition, do they necessarily differ by summands which are Σ-cyclic. In [20] it was shown that this is, indeed, the case, at least in any model of ZFC in which a certain mild set-theoretic hypothesis is valid. We include this result in a general discussion of the behavior of torsion products in the context of separable groups in section 2.

Turning to Warfield's Problem, it was shown in [16] that any such product is necessarily an A-group (this class having been defined and classified in [9]). The following, though, is a logical extension of this problem:

Warfield's Problem$^+$: Is there a structure theory for groups of the form $\circledT_{j=1}^{k} A_j$, where $A_1, \ldots, A_k$ are totally projective?

Half an answer to this question is given; a class containing the A-groups is described which contains these iterated torsion products. Unfortunately, I am not aware of a set of cardinal invariants which can be used to classify this general class of groups, as can be done with the subclass of A-groups. The result is part of a general discussion of the behavior of torsion products in the context of groups of infinite height contained in section 3. In particular, machinery is developed which makes it possible to approach the problem by a double induction on the number of groups involved, together with their lengths.

I presented work on Nunke's Problem at our meeting in Perth, Australia in 1985, and I discussed Oxford's Problem at our meeting in Padova, Italy last year. We will, therefore, be content with an over-view of results on these two problems, which is the content of the fourth section of these notes. Even the answer to certain cases of Nunke's problem strongly depends upon what set-theoretic environment is being employed. In particular, the question of when $A \circledT B$ is a dsc of length ω_1 is contingent upon a set-theoretic statement known as *Kurepa's Hypothesis.* The answer to Oxford's problem is *yes* if $\alpha \leq \omega$ and *no* for larger ordinals. This question is equivalent to asking if the only examples of representable p-coprimary radicals with enough projectives are of the form p^α, where α is an ordinal or the symbol ∞, are (see section 3 for an explanation of this terminology).

2 SOME PROPERTIES RELATED TO DIRECT SUMS OF CYCLICS

We begin by reviewing some basic facts contained in [5] (dispensing with individual citations). The torsion product is naturally commutative and associative. In addition, if X and Y are subgroups of A and B, then left-exactness implies that there is an injection $X \textcircled{T} Y \to A \textcircled{T} B$, which we will view as an inclusion. If G is a group, there is a natural isomorphism $\mathbf{Z}_{p^\infty} \textcircled{T} G \cong G$, and it follows that any group can be expressed in the form $A \textcircled{T} B$. One purpose of this section, however, is to show that if A and B are both reduced, then $A \textcircled{T} B$ resembles a Σ-cyclic group in several respects. Recall that for $n < \omega$ there is a natural isomorphism $\mathbf{Z}_{p^n} \textcircled{T} G \cong G[p^n]$. Since a bounded group is Σ-cyclic and torsion products preserve direct sums, if either A or B is Σ-cyclic, then so is $A \textcircled{T} B$.

Many important structural properties of groups are preserved under the torsion product. A simple example is the following useful identity:

(A) If A and B are groups and $n < \omega$, then $(A \textcircled{T} B)[p^n] = (A[p^n]) \textcircled{T} (B[p^n])$. In particular, the rank of $A \textcircled{T} B$ equals the product of the ranks of A and B individually.

The next result is proved by a transfinite induction using the representation of $A \textcircled{T} B$ in terms of generators and relators.

(B) If A and B are groups and α is an ordinal, then $p^\alpha(A \textcircled{T} B) = (p^\alpha A) \textcircled{T} (p^\alpha B)$. As a result, the length of $A \textcircled{T} B$ equals the minimum of the lengths of A and B individually.

We know that the torsion product is left exact, but if $0 \to B_1 \to B_2 \to B_3 \to 0$ is pure, then for every $n < \omega$, $0 \to B_1[p^n] \to B_2[p^n] \to B_3[p^n] \to 0$ splits. Together with (A), this implies that

$$0 \to (A \textcircled{T} B_1)[p^n] \to (A \textcircled{T} B_2)[p^n] \to (A \textcircled{T} B_2)[p^n] \to 0$$

also splits. We summarize this in the following:

(C) The torsion product preserves pure exact sequences.

The next two observations will also be required:

(D) If X is a subgroup of A and Y is a subgroup of B, then

$$X \textcircled{T} Y = (A \textcircled{T} Y) \cap (X \textcircled{T} B),$$

and if $K = A/X$ and $L = B/Y$, then there is a left exact sequence

$$0 \to X \textcircled{T} Y \to A \textcircled{T} B \to (A \textcircled{T} L) \oplus (K \textcircled{T} B).$$

To verify (D), consider the commutative diagram:

$$\begin{array}{ccccccc} & & 0 & & 0 & & 0 \\ & & \downarrow & & \downarrow & & \downarrow \\ 0 & \to & X \circledT Y & \to & X \circledT B & \to & X \circledT L \\ & & \downarrow & & \downarrow & & \downarrow \\ 0 & \to & A \circledT Y & \to & A \circledT B & \to & A \circledT L \\ & & \downarrow & & \downarrow & & \downarrow \\ 0 & \to & K \circledT Y & \to & K \circledT B & \to & K \circledT L \end{array}$$

from which the result follows by some elementary diagram chasing.

A similar argument can be used to verify the following:

(E) If X_1 and X_2 are subgroups of A and Y_1 and Y_2 are subgroups of B, then

$$(X_1 \cap X_2) \circledT (Y_1 \cap Y_2) = (X_1 \circledT Y_1) \cap (X_2 \circledT Y_2).$$

Another useful observation is the following, which will allow us to construct well behaved subgroups of torsion products.

(F) If C is an infinite subgroup of $A \circledT B$, then there are subgroups $X \subseteq A$ and $Y \subseteq B$ such that C such that $|C| = |X| = |Y|$ and $G \subseteq X \circledT Y$.

To verify (F), recall that an element of $A \circledT B$ is a a sum of expressions of the form (x, n, y). Let X and Y be generated by all elements $x \in A$ and $y \in B$ which appear in one of these expressions corresponding to an element of C.

We remark in passing that there are analogues of (A)-(F) for iterated torsion products which we will assume as needed. A group M is *fully starred* if for all subgroups $N \subseteq M$, N has the same cardinality as one of its basic subgroups. In other words, in a fully starred group, no subgroup is terribly far removed from a Σ-cyclic.

PROPOSITION 1 ([27]) *If A and B are reduced, then $A \circledT B$ is fully starred.*
Proof: Suppose G is an arbitrary infinite subgroup of $A \circledT B$ and C is a basic subgroup of G. Let $X \subseteq A$ and $Y \subseteq B$ be subgroups with $|C| = |X| = |Y|$ and $C \subseteq X \circledT Y$. If $K = A/X$ and $L = B/Y$, then $(A \circledT B)/(X \circledT Y)$ embeds in $(A \circledT L) \oplus (K \circledT B)$, and since the latter is reduced and G/C is divisible, we conclude that $G \subseteq X \circledT Y$. Therefore, $|G| \leq |X \circledT Y| = |C| \leq |G|$, and the result follows.

The last result says that a group which embeds in a torsion product of reduced groups is not too far from being Σ-cyclic. It seems reasonable, then, that a group which embeds in an iterated torsion product would come even closer to being Σ-cyclic. We will say a group G is in $\mathcal{T}_n$ if there are separable groups $X_1, \ldots, X_n$ and an embedding of G into

$\textcircled{T}_{j=1}^{n} X_j$. We say G is in $\mathcal{T}_\infty$ if it is in $\mathcal{T}_n$ for every $n < \omega$. The following easily checked properties will be left to the reader:

(G) If $n \in \omega \cup \{\infty\}$, then $\mathcal{T}_n$ is closed under direct sums and absorbs torsion products, in the sense that if $G \in \mathcal{T}_n$ and A is any group, then $G \textcircled{T} A \in \mathcal{T}_n$.

If G is a group, then a collection of subgroups $\{W_j\}_{j<\alpha}$ whose union is G is a *filtration* if for every $j < \alpha$, $W_j = \cup_{i<j} W_{i+1}$ and $|W_j| < |G|$. We will always assume our filtrations begin with $W_0 = 0$.

THEOREM 2 ([16]) A group G is in $\mathcal{T}_\infty$ iff it is Σ-cyclic.
Proof: It is straightforward to verify that a Σ-cyclic group is in $\mathcal{T}_\infty$, so assume $G \in \mathcal{T}_\infty$. We will induct on the cardinality of G, so assume that all groups in $\mathcal{T}_\infty$ of smaller cardinality are Σ-cyclic. If G is countable, the result follows from the observation that any countable separable group is Σ-cyclic. Suppose now that the result holds for all groups of smaller cardinality than G. For every n there exist separable groups $A_{n,1},\ldots,A_{n,n}$ and an embedding of G into $\textcircled{T}_{j=1}^{n} A_{n,j}$. Identifying G with its image, we may assume that this embedding is simply an inclusion. By (F) there is no loss of generality in assuming that each $A_{n,j}$ has the same cardinality as G. Using a standard "back-and-forth" argument, there is a pure filtration $\{G_i\}_{i<\alpha}$ of G, and pure filtrations $\{X_{n,j,i}\}_{i<\alpha}$ of $A_{n,j}$, respectively, such that for all $i \in I$ and $n \leq \omega$,

$$G_i = G \cap (\textcircled{T}_{j=1}^{n} X_{n,j,i}).$$

This implies that for every $i \in I$ and $n \leq \omega$ there are embeddings,

$$G_{i+1}/G_i \to (\textcircled{T}_{j=1}^{n} X_{n,j,i+1})/(\textcircled{T}_{j=1}^{n} X_{n,j,i}).$$

If $K_{n,j,i} = X_{n,j,i+1}/X_{n,j,i}$, then by the obvious generalization of (D) to iterated products, there is an embedding

$$(\textcircled{T}_{j=1}^{n} X_{n,j,i+1})/(\textcircled{T}_{j=1}^{n} X_{n,j,i}) \to \oplus_{k=1}^{n} [K_{n,k,i+1} \textcircled{T} (\textcircled{T}_{j\neq k} X_{n,j,i+1})] \stackrel{\text{def}}{=} Y_{n,i}$$

By (G), each $Y_{n,i} \in \mathcal{T}_{n-1}$, and it follows that $G_{i+1}/G_i \in \mathcal{T}_\infty$. By induction, therefore, G_{i+1}/G_i is Σ-cyclic. Since G_i is pure in G, it follows that $G_{i+1} \cong G_i \oplus (G_{i+1}/G_i)$ and so $G \cong \oplus_{i\in I}(G_{i+1}/G_i)$ is Σ-cyclic, as required.

We now wish to consider what type Σ-cyclic summands a torsion product will admit. Putting together (A) and (B), we obtain the following useful formula:

(H) If A and B are groups and α is an ordinal, then,

$$f_{A\textcircled{T}B}(\alpha) = f_A(\alpha) f_B(\alpha) + f_A(\alpha) r(p^{\alpha+1}B) + r(p^{\alpha+1}A) f_B(\alpha).$$

The next result shows that a torsion product has Σ-cyclic summands which are as large as possible.

PROPOSITION 3 ([11]) *If A and G are reduced groups, then $T \stackrel{\text{def}}{=} A \textcircled{T} G$ has a summand C which is Σ-cyclic and has the same finite Ulm invariants as T. In other words, C is isomorphic to a basic subgroup of T.*

Proof: Note that $A \cong A_1 \oplus A_2$, where A_1 is bounded and the rank of A_2 agrees with the final rank of A. Since $A \textcircled{T} G \cong (A_1 \textcircled{T} G) \oplus (A_2 \textcircled{T} G)$ and the first term is Σ-cyclic, it follows that we may assume $A = A_2$ has its rank and final rank the same. By similar reasoning, we may assume that the rank and final rank of G also agree.

Let A' be a lower basic subgroup of A (so $r(A/A') = r(A)$). There is a pure short exact sequence

$$0 \to A' \textcircled{T} G \to A \textcircled{T} G \to K \to 0, \qquad (*)$$

where

$$K = \oplus_{r(A)} G \cong \oplus_{n<\omega}(\oplus_{r(A)} G).$$

For $n < \omega$,

$$\oplus_{r(A)} G \cong (\oplus_{r(A)} G) \oplus (\oplus_{r(A) f_G(n)} Z_{p^{n+1}}),$$

so if $L = \oplus_{n<\omega}(\oplus_{r(A) f_G(n)} Z_{p^{n+1}})$, it follows that K has a summand isomorphic to L. The purity of $(*)$ easily implies that we can "pull this back" to a summand of T isomorphic to L. Since $L \cong L \oplus L$,

$$T \cong T \oplus (\oplus_{n<\omega}(\oplus_{r(A) f_G(n)} Z_{p^{n+1}})).$$

Similarly,

$$T \cong T \oplus (\oplus_{n<\omega}(\oplus_{f_A(n) r(G)} Z_{p^{n+1}})),$$

from which it follows that T has a summand isomorphic to,

$$C \stackrel{\text{def}}{=} \oplus_{n<\omega} (\oplus_{f_A(n) r(G) + r(A) f_G(n)} Z_{p^{n+1}}).$$

Since $r(G) = r(p^{n+1}G) \geq f_G(n)$ and $r(A) = r(p^{n+1}A) \geq f_A(n)$,

$$f_A(n) r(G) + r(A) f_G(n) = f_A(n) f_G(n) + f_A(n) r(p^{n+1}G) + r(p^{n+1}A) f_G(n)$$

and by (F), C and T have equal finite Ulm invariants.

COROLLARY 4 ([1]) *There are non-isomorphic groups A and B such that $A \textcircled{T} G \cong B \textcircled{T} G$ for all reduced groups G.*

Proof: Let X be a Σ-cyclic group such that $f_X(n) = \omega$ for each $n < \omega$, and let A be the torsion completion of X. If $B = A \oplus X$ and G is a reduced group, we claim that $T \stackrel{\text{def}}{=} A \textcircled{T} G$ is isomorphic to $T' \stackrel{\text{def}}{=} B \textcircled{T} G$. Since $r(p^n A) = c = r(p^n B)$ and $f_A(n) = \omega = f_B(n)$ for all $n < \omega$, we can conclude that $f_T(n) = f_{T'}(n)$. It follows that $T \cong T'$ whenever these are Σ-cyclic, for example, whenever G is bounded. If G is unbounded, then $f_T(n)$ is infinite for all n and is at least as large as $f_{X \textcircled{T} G}(n)$. Since $X \textcircled{T} G$ is Σ-cyclic, it follows from Proposition 3 that $T \cong T \oplus (X \textcircled{T} G) = (A \textcircled{T} G) \oplus (X \textcircled{T} G) \cong T'$.

Is the converse to the last Corollary valid, i.e., if A and B are groups such that $A \circledT G \cong B \circledT G$ for every reduced group G, does it follow that A and B differ by summands which are Σ-cyclic? Our purpose now is to verify that this holds, at least in any model of ZFC satisfying a mild set theoretic condition.

Suppose α is a limit ordinal. A subset $C \subseteq \alpha$ is a *cub* if it is closed (in the order topology) and unbounded. A subset $S \subseteq \alpha$ is *stationary* if for every cub $C \subseteq \alpha$, $S \cap C \neq \emptyset$. The *cofinality* of α is the minimal cardinality of its unbounded subsets. A cardinal κ is *regular* if it has cofinality κ. If κ is a regular cardinal, then $\mathrm{E}(\kappa)$ is the following statement:

There is a stationary subset $S \subseteq \kappa$ such that
(1) every $\alpha \in S$ is a limit ordinal of countable cofinality, and
(2) for every limit ordinal $\alpha < \kappa$, $\alpha \cap S$ is not a stationary subset of α.

LEMMA 5 *Suppose κ is a regular cardinal for which $\mathrm{E}(\kappa)$ is valid and $S \subseteq \kappa$ is as in the definition of $\mathrm{E}(\kappa)$. Then there is a group G of cardinality κ with a filtration $\{W_j\}_{j<\kappa}$ consisting of pure subgroups, such that whenever $j < \ell < \kappa$ and $j \in S$, then $W_\ell/W_j \cong \mathbf{Z}_{p^\infty} \oplus M$, where M is Σ-cyclic.*

Proof: For every $i < \kappa$ and $m < \omega$, let $C^i_m \cong \mathbf{Z}_{p^m}$ and $B^i = \oplus_{m<\omega} C^i_m$. For a moment, fix $j \in S$ and let $x_1, x_2, x_3, \ldots$ be a strictly increasing sequence of isolated ordinals with limit j. Let $\oplus_{m<\omega} C^{x_m}_m \subseteq Z_j \subseteq \prod_{m<\omega} C^{x_m}_m$ satisfy $Z_j / \oplus_{m<\omega} C^{x_m}_m \cong \mathbf{Z}_{p^\infty}$. Let $G \subseteq \prod_{i<\kappa} B^i$ be generated by $\oplus_{i<\kappa} B^i$ and the Z_j's for $j \in S$. If $j < \kappa$, let $G_j = G \cap \prod_{i<j} B^i$ and $G^j = G \cap \prod_{j \le i < \kappa} B^i$. Observe that if $j < \kappa$ and $\pi_j : \prod_{i<\kappa} B^i \to \prod_{i<j} B^i$ is the obvious projection, then whenever $\ell > j$ is in S, we have $\pi_j(Z_\ell) \subseteq \oplus_{i<j} B^i$. Therefore, $\pi_j(G) \subseteq G$, so that $G \cong G_j \oplus G^j$. In addition G_j will be generated by $\oplus_{i<j} B^i$ and the Z_i for $i \le j$ with $i \in S$. In particular $|G_j| = |j| \cdot \omega_0$. Finally, let $W_j = \cup_{i<j} G_i$. Clearly these form a filtration of G by pure subgroups.

We claim G_j is Σ-cyclic for all $j < \kappa$. We prove this by induction. Since $G_{j+1} \cong G_j \oplus B^j$, the interesting case is where j is a limit ordinal. Let $g : \mu \to j - S$ be a strictly increasing function whose image is a cub. For each $\beta < \mu$, $G_{g(\beta)}$ is a Σ-cyclic group which is a summand of G, hence of $G_{g(\beta+1)}$. Note that if $\sigma < \mu$ is a limit, then $g(\sigma)$ is not in S, so that $G_{g(\sigma)} = W_{g(\sigma)}$. Therefore, $W_j \cong \oplus_{\beta<\mu} G_{g(\beta+1)}/G_{g(\beta)}$ is Σ-cyclic. Note that if $j \notin S$, then $G_j = W_j$ and the claim follows, and if $j \in S$, then G_j is generated by W_j, and Z_j, hence $G_j/W_j \cong \mathbf{Z}_{p^\infty}$. There is a decomposition $W_j = K \oplus L$, where L is a countable subgroup containing $W_j \cap Z_j$. Clearly K is Σ-cyclic and since $L + Z_j$ is countable, it is Σ-cyclic, also. Therefore $G_j = K \oplus (L + Z_j)$ is Σ-cyclic, as required.

Note that from the last paragraph, if $j \in S$ and $j < \ell < \kappa$, then

$$\mathbf{Z}_{p^\infty} \cong G_j/W_j \subseteq W_\ell/W_j \subseteq G_\ell/W_j,$$

and therefore $W_\ell/W_j \cong \mathbf{Z}_{p^\infty} \oplus M$, where $M \subseteq G_\ell/G_j$ is Σ-cyclic.

Let λ be the supremum of the values of κ for which $\mathrm{E}(\kappa)$ is valid. So λ is either a cardinal, or the symbol ∞ if $\mathrm{E}(\kappa)$ is valid for a proper class of regular cardinals. It is known that $\lambda = \infty$ in the constructible universe (see [4]).

THEOREM 6 ([20]) *Suppose* $A \circledT G \cong B \circledT G$ *for every reduced* G, *and the rank of* A *and* B *is less than* λ. *Then there are* Σ*-cyclic groups* X *and* Y *such that* $A \oplus X \cong B \oplus Y$.
Proof: Let κ be a regular cardinal greater than the ranks of A and B for which $\mathrm{E}(\kappa)$ holds and let $S \subseteq \kappa$, G and $\{W_j\}_{j<\kappa}$ be as in Theorem 3. Suppose $g : A \circledT G \to B \circledT G$ is an isomorphism. Note that $\{A \circledT W_j\}_{j<\kappa}$ and $\{B \circledT W_j\}_{j<\kappa}$ are filtrations of $A \circledT G$ and $B \circledT G$, respectively. By a standard "back-and-forth" argument, there is a cub $C \subseteq \kappa$ such that for all $j \in C$, $g(A \circledT W_j) = B \circledT W_j$. Since S is stationary, there is a $j \in S \cap C$. Let $\ell > j$ with $\ell \in C$. Since W_j is a pure subgroup of W_ℓ and the torsion product is exact on pure sequences, there is a commutative diagram,

$$\begin{array}{ccccccccc} 0 & \to & A \circledT W_j & \to & A \circledT W_\ell & \to & A \circledT W_\ell/W_j & \to & 0 \\ & & \downarrow \cong & & \downarrow \cong & & \downarrow \exists\cong & & \\ 0 & \to & B \circledT W_j & \to & B \circledT W_\ell & \to & B \circledT W_\ell/W_j & \to & 0 \end{array}$$

If $W_\ell/W_j \cong \mathbf{Z}_{p^\infty} \oplus M$, then $X = A \circledT M$ and $Y = B \circledT M$ are Σ-cyclic, and

$$A \oplus X \cong (A \circledT \mathbf{Z}_{p^\infty}) \oplus (A \circledT M) \cong A \circledT W_\ell/W_j \cong$$
$$B \circledT W_\ell/W_j \cong (B \circledT \mathbf{Z}_{p^\infty}) \oplus (B \circledT M) \cong B \oplus Y,$$

which proves the result.

In fact, we could be even more specific: If λ is greater than the ranks of A and B, then $A \circledT G \cong B \circledT G$ for all reduced G iff $f_A(n) \cong f_B(n)$ for all $n < \omega$ and $A \oplus X \cong B \oplus Y$ for some Σ-cyclic groups X and Y.

We end this section with the following striking example of the relation between set theory, the torsion product and Σ-cyclic groups. Although this result is mentioned in [18], Ray Mines and others pointed out to me that it had circulated much earlier as an unpublished note.

THEOREM 7 *If* A *is an unbounded torsion complete group with countable basic subgroup* B, *then* $A \circledT A$ *is* Σ*-cyclic iff the continuum hypothesis is valid.*
Proof: Suppose first that the continuum hypothesis is valid. Construct a pure filtration $\{B_i\}_{i<c}$ of A, so that each B is countable. For each $i < c$ there are pure exact sequences

$$0 \to B_i \circledT B_i \to B_i \circledT B_{i+1} \to B_i \circledT B_{i+1}/B_i \to 0$$

and

$$0 \to B_i \circledT B_{i+1} \to B_{i+1} \circledT B_{i+1} \to B_{i+1}/B_i \circledT B_{i+1} \to 0.$$

Since B_i and B_{i+1} are countable and separable, they must be Σ-cyclic. Therefore, the right hand groups in these sequences are also Σ-cyclic, which means that the sequences both split. We can conclude that

$$A \circledT A \cong \oplus_{i<c}((B_i \circledT B_{i+1}/B_i) \oplus (B_{i+1}/B_i \circledT B_{i+1}))$$

is Σ-cyclic.

Conversely, suppose $A \textcircled{T} A$ is Σ-cyclic and $\omega_1 < c$. By a standard "back-and-forth" argument, we can find a pure subgroup $C \subseteq A$ of cardinality ω_1 containing B such that $C \textcircled{T} C$ is a summand of $A \textcircled{T} A$. Observe that C is not Σ-cyclic and A/C will be a divisible group of cardinality c. Since $A \textcircled{T} C$ is a subgroup of $A \textcircled{T} A$, it, too, is Σ-cyclic. Since $C \textcircled{T} C$ is a summand of $A \textcircled{T} A$, it is also a summand of $A \textcircled{T} C$. In other words, the sequence

$$0 \to C \textcircled{T} C \to A \textcircled{T} C \to A/C \textcircled{T} C \to 0$$

splits. We conclude that $A/C \textcircled{T} C \cong \oplus_c C$ is Σ-cyclic, and this contradiction proves the result.

3 PROPERTIES RELATED TO GROUPS OF INFINITE HEIGHT

We begin our discussion with the following somewhat surprising observation:

PROPOSITION 8 ([13]) *Suppose A and B are groups containing subgroups $X \subseteq A$ and $Y \subseteq B$. Then $X \textcircled{T} Y$ is an isotype subgroup of $(A \textcircled{T} Y) \oplus (X \textcircled{T} B)$*

Proof: Note that using the diagonal map we can identify $X \textcircled{T} Y$ with a subgroup of $(A \textcircled{T} Y) \oplus (X \textcircled{T} B)$, and if α is an ordinal, we have

$$\begin{aligned} p^\alpha((A \textcircled{T} Y) \oplus (X \textcircled{T} B)) \cap (X \textcircled{T} Y) &= ((p^\alpha A \textcircled{T} p^\alpha Y) \oplus (p^\alpha X \textcircled{T} p^\alpha B)) \cap (X \textcircled{T} Y) \\ &= (p^\alpha A \textcircled{T} p^\alpha Y) \cap (p^\alpha X \textcircled{T} p^\alpha B) \cap (X \textcircled{T} Y) \\ &= (p^\alpha A \cap p^\alpha X \cap X) \textcircled{T} (p^\alpha Y \cap p^\alpha B \cap Y) \\ &= p^\alpha X \textcircled{T} p^\alpha Y = p^\alpha (X \textcircled{T} Y), \end{aligned}$$

proving the result.

COROLLARY 9 *If A and B are any groups, then $A \textcircled{T} B$ is an isotype subgroup of a direct sum of copies of A and B.*

Proof: If D and E are divisible hulls of A and B, then $A \textcircled{T} B$ is an isotype subgroup of $(A \textcircled{T} E) \oplus (D \textcircled{T} B) \cong (\oplus A) \oplus (\oplus B)$.

By an *IT-group* we mean a group which can be embedded as an isotype subgroup of a totally projective group. The following is an immediate consequences of the last result.

COROLLARY 10 *The class of IT-groups is closed under the torsion product.*

One consequence of this result is that if $A_1, \ldots, A_k$ are all totally projective, then the iterated torsion product $\textcircled{T}_{j=1}^k A_j$ will be an IT-group. It is our main purpose in this section to describe the possible IT-groups which emerge as iterated torsion products of totally projective groups. One somewhat surprising consequence of our investigation is that such an iterated torsion product will only be totally projective if it is p^{ω_1}-bounded, in other words, if it a dsc.

We start our discussion by reviewing some terminology. A *preradical* is a function which assigns to every group G a subgroup RG such that for any group homomorphism $f : A \to B$, we have $f(RA) \subseteq RB$. If, in addition, we have $R(G/RG) = 0$ for every group G, then R is said to be a *radical*. An extension of the integers, $0 \to \mathbf{Z} \to T \to H \to 0$, represents a preradical by letting RG be the image of $\mathrm{Hom}(T, G) \to \mathrm{Hom}(\mathbf{Z}, G) \cong G$, in other words, the kernel of

$$\delta_G : G \cong \mathrm{Hom}(\mathbf{Z}, G) \to \mathrm{Ext}(H, G).$$

Such a representable preradical is *p-coprimary* if H can be chosen to be a p-group. The most important example of a p-coprimary representable radical is p^α, where α is an ordinal, or the symbol ∞. This is represented by the sequence $0 \to \mathbf{Z} \to T_\alpha \to H_\alpha \to 0$, where H_α is the generalized Prüfer group of length α and $p^\alpha T_\alpha = \mathbf{Z}$.

For any group G and representable preradical R there is a homomorphism

$$\partial_G : H \circledT G \to \mathbf{Z} \otimes G \cong G.$$

By a result of Nunke, there is an isomorphism $\mu : \mathrm{Ext}(H \circledT G, A) \to \mathrm{Ext}(H, G \circledT A)$ such that

$$\delta_{\mathrm{Ext}(G,A)} = \mu \circ \mathrm{Ext}(\partial_G, A) : \mathrm{Ext}(G, A) \to \mathrm{Ext}(H \circledT G, A) \xrightarrow{\cong} \mathrm{Ext}(H, \mathrm{Ext}(G, A)).$$

This implies that the elements of $R\mathrm{Ext}(G, A)$ can be identified with the set of sequences $0 \to A \to X \to G \to 0$ for which ∂_G factors through X. These sequences are called *R-pure*.

It is easily seen that the height of $1 \in \mathbf{Z} \subseteq T$ is independent of the particular sequence chosen to represent R, and we call this the *height* of the preradical. Alternately, the height of R is the largest ordinal such that $R \subseteq p^\alpha$. Since regular purity is equivalent to p^ω-purity, if R has infinite height, then any R-pure sequence is pure in the regular sense. In this case, when Y is a group, and $0 \to A \to X \to G \to 0$ is R-pure, then

$$0 \to A \circledT Y \to X \circledT Y \to G \circledT Y \to 0$$

is exact and since $\partial_G : H \circledT G \to G$ factors through X, $\partial_{G \circledT Y} = \partial_G \circledT Y : H \circledT G \circledT Y \to G \circledT Y$ factors $X \circledT Y$. This means that R-purity is preserved under the torsion product.

The notion of R-purity allows us to define the concepts of R-projectivity and R-injectivity. If P is R-projective and G is any group, then

$$\mathrm{Ext}(P \circledT G, A) \cong \mathrm{Ext}(P, \mathrm{Ext}(G, A)) = 0,$$

so that $P \circledT G$ is R-projective.

If R is a representable p-coprimary preradical of infinite height, and M is the torsion subgroup of T, then there is an exact sequence $0 \to M \to H \to T/(\mathbf{Z} + M) \to 0$. It can be checked that the last group in this sequence is isomorphic to $\mathbf{Z}_{p^\infty}$ and the last map in this sequence can be identified with $\partial_{\mathbf{Z}_{p^\infty}}$, so that the sequence is R-pure; it is sometimes referred to as a *corepresenting* sequence for R. If H is R-projective and G is an arbitrary group, then there is a sequence

$$0 \to M \circledT G \to H \circledT G \xrightarrow{\partial_G} G \to 0,$$

which is an R-projective resolution of G. So if H can be chosen to be R-projective (e.g., when $R = p^\alpha$), then R-purity has enough projectives. It can be shown that whenever this happens, then R is actually a radical.

In a slightly different direction, we say a sequence $0 \to A \to X \xrightarrow{\pi} G \to 0$ is p^α-*balanced* if $p^\alpha G[p] = \pi(p^\beta X[p])$ for every $\beta < \alpha$. If α is a limit, then this condition is equivalent to requiring that $p^\beta A = A \cap p^\beta X$ for all $\beta < \alpha$ (i.e., A is a p^α-*isotype* subgroup of X) and $p^\beta G = \pi(p^\beta X)$ for all $\beta < \alpha$ (i.e., A is a p^α-*nice* subgroup of X). Basically the same proof which shows that the balanced-projectives are the totally projective groups also shows that the p^α-balanced-projectives are the p^α-bounded totally projective groups, and that there are enough of these projectives.

Since any p^α-bounded totally projective group is p^α-projective, it follows that any p^α-pure sequence is p^α-balanced. Conversely, if $0 \to A \to X \to G \to 0$ is p^α-balanced and G is divisible, then $H_\alpha \circledT G \cong \oplus H_\alpha$ is p^α-balanced projective. This implies that $\partial_G : H_\alpha \circledT G \to G$ necessarily factors through X, so that the sequence is also p^α-pure. This implies, for example, that a p^α-high subgroup of X (i.e., one maximal with respect to intersecting $p^\alpha X$ trivially) will always be $p^{\alpha+1}$-pure in X.

LEMMA 11 *If J is a p^α-high subgroup of B, then J maps to a p^α-pure subgroup of $B/p^\alpha B$.*

Proof: We may clearly assume α is infinite, so that $K = B/J$ is divisible. If $\beta < \alpha$, $w \in B$ and $pw \in J + p^\alpha B$, then $pw = j + py$, where $j \in J$, $y \in p^\beta B$ and $py \in p^\alpha B$. Therefore, $p(w - y) \in J$, so there is an $x \in J$ and a $z \in p^\beta B[p]$ such that $w - y = x + z$. Therefore, $y + z + p^\alpha B \in p^\beta(B/p^\alpha B)[p]$ and $w + J + p^\alpha B = y + z + J + p^\alpha B \in B/(J + p^\beta B)$. It follows that $0 \to J \to B/p^\alpha B \to B/(J + p^\alpha B) \to 0$ is p^α-balanced, and hence p^α-pure, as required.

If λ is a limit ordinal, we say a group G is a $\mathcal{C}_\lambda$ if $G/p^\beta G$ is totally projective for every $\beta < \lambda$ (see [24]). Note that any divisible group is trivially a $\mathcal{C}_\lambda$.

LEMMA 12 *If λ is a limit ordinal and G is a $\mathcal{C}_\lambda$, then $E : 0 \to A \to X \to G \to 0$ is p^λ-pure iff it is p^λ-balanced.*

Proof: We have already observed that if E is p^λ-pure then it is p^λ-balanced. To prove the converse we induct on λ. Note that if λ is a limit of limits, then the result follows by induction. Otherwise $\lambda = \alpha + \omega$ for some ordinal α. Let L be a basic subgroup of $p^\alpha G$. Then $p^\alpha G/L$ is divisible, so that $G/L \cong (p^\alpha G/L) \oplus (K/L)$ for some subgroup K of G. It can be checked that K is a p^λ-balanced subgroup of G with divisible cokernel, and hence, that it is also p^λ-pure in G. One also verifies that $K/p^\alpha K \cong G/p^\alpha G$ is totally projective and $p^\alpha K$ is Σ-cyclic, so that K is totally projective.

Let $B^\lambda \text{Ext}(C, E)$ denote the p^λ-balanced sequences. There is an exact sequence

$$B^\lambda \text{Ext}(G/K, A) \to B^\lambda \text{Ext}(G, A) \to B^\lambda \text{Ext}(K, A).$$

Since K is p^λ-bounded totally projective, the last group vanishes. Therefore, there is an extension $F \in B^\lambda \text{Ext}(G/K, A)$ which maps to E. However, since G/K is divisible, $B^\lambda \text{Ext}(G/K, A) = p^\lambda \text{Ext}(G/K, A)$. Therefore F, and hence E, is p^λ-pure, as desired.

Suppose m, n are non-negative integers, λ is a limit ordinal and $\alpha = \lambda + n$. We say a group G is an *$\alpha * m$-GP* if there is a p^λ-balanced projective resolution:

$$0 \to G \to X_m \to X_{m-1} \to \cdots \to X_1 \to X_0 \to 0,$$

such that for $j \geq 0$, X_j is a totally projective group, $p^\lambda X_0 \cong \mathbf{Z}_{p^n}$ and for $j > 0$, $p^\lambda X_j = 0$. It follows from induction using Lemma 11 that an $\alpha * m$-GP group is a $\mathcal{C}_\lambda$, and that we could equally well have required that this long exact sequence be p^λ-pure. Note that an $\alpha * 0$-GP group is totally projective and an $\alpha * 1$-GP group is an A-group (see [9] for a definition of this term), so that there is a structure theory for groups which appear as direct sums of $\alpha * 1$-GP groups.

A short exact sequence $0 \to G \to X \xrightarrow{\pi} A \to 0$ determines a *c-valuation* on A by letting $v(a) = \sup\{ht_X(x) + 1 : \pi(x) = a\}$. The importance of this concept is contained in the following result of [10]: If $0 \to G_i \to X_i \to A_i \to 0$ $(i = 1, 2)$ are isotype sequences and the X_i are totally projective, then there is an isomorphism $X_1 \cong X_2$ taking G_1 to G_2 iff X_1 and X_2 have the same Ulm function and there is an isomorphism $A_1 \cong A_2$ which preserves c-valuations.

Though not actually utilized in the sequel, the following shows that there is a structure theory for $\alpha * m$-GP groups. It does not say that we can extend this to a structure theory for the direct sums of such groups.

PROPOSITION 13 *Two $\alpha * m$-GP groups are isomorphic iff their Ulm functions agree. A function f from the ordinals to the cardinals is the Ulm function of some $\alpha * m$-GP group iff it is admissible and for $\beta \geq \lambda$, $f(\beta) = 0$, unless $m = 0$ and $\beta = \alpha$, in which case $f(\beta) = 1$.*

Proof: We prove this by induction on m. The result is clear for $m = 0$ by the theory of totally projective groups. Now assume the result for $m - 1 \geq 0$, and let G and G' be $\alpha * m$-GP groups whose Ulm functions agree. There are p^λ-balanced sequences $0 \to G \to X_m \to L \to 0$ and $0 \to G' \to X'_m \to L' \to 0$, where X_m and X'_m are p^λ-bounded totally projective groups, and L and L' are $\alpha * (m-1)$-GP groups. By adding totally projective summands, we may assume X_m and X'_m, and L and L' have equal Ulm functions. By induction, then, there is an isomorphism $L \cong L'$. The fact that these sequences are p^λ-balanced can be used to show that this isomorphism preserves c-valulations. It follows from [10] that there is an isomorphism $X_m \cong X'_m$ which takes G to G'; in particular, they are isomorphic. The statement regarding Ulm function will be left to the reader.

An even simpler consequence of [10] is the following:

LEMMA 14 *If G is an IT group and Y is a totally projective group such that the Ulm functions of G and $G \oplus Y$ agree, then $G \cong G \oplus Y$.*
Proof: This follows from considering the isotype sequences $0 \to G \to X \to A \to 0$ and $0 \to G \oplus Y \to X \oplus Y \to A \to 0$, where X is totally projective, and noting that X and $X \oplus Y$ must have the same Ulm function.

We mention the following slightly vaguely stated result, which is essentially an example of *Shanuel's Lemma.*

LEMMA 15 *For one of the above variations of purity, if G is a group and $0 \to X \to Y \to G \to 0$ and $0 \to X' \to Y' \to G \to 0$ are pure projective resolutions of G, then $X \oplus Y' \cong X' \oplus Y$.*
Proof: Consider the following pull-back diagram:

$$
\begin{array}{ccccccccc}
 & & & & 0 & & 0 & & \\
 & & & & \downarrow & & \downarrow & & \\
 & & & & X' & = & X' & & \\
 & & & & \downarrow & & \downarrow & & \\
0 & \to & X & \to & Z & \to & Y' & \to & 0 \\
 & & \| & & \downarrow & & \downarrow & & \\
0 & \to & X & \to & Y & \to & G & \to & 0 \\
 & & & & \downarrow & & \downarrow & & \\
 & & & & 0 & & 0 & &
\end{array}
$$

It follows from purity and projectivity that $X \oplus Y' \cong Z \cong X' \oplus Y$.

We will use Shanuel's Lemma in a decisive way in the proof of our main Theorem. An immediate application of the result is contained in the following important observation.

LEMMA 16 *If G is an $\alpha * m$-GP group, then G is p^{α}-projective.*
Proof: We prove this by induction on m. If $m = 0$, it follows from the definition of total projectivity. Assume the result for $m - 1$. There is a p^{λ}-pure sequence $0 \to G \to X_m \to L \to 0$, where X_m is p^{λ}-bounded totally projective and L is $\alpha * (m-1)$-GP. In particular, L is also p^{α}-projective. Note that since X_m is p^{λ}-projective and L is a $\mathcal{C}_{\lambda}$, this is also a p^{λ}-pure projective resolution of L. There is a second p^{λ}-pure projective resolution $0 \to L \circledT M_{\lambda} \to L \circledT H_{\lambda} \to L \to 0$ and from Shanuel's lemma, $G \oplus (L \circledT H_{\lambda}) \cong X_m \oplus (L \circledT M_{\lambda})$. Since L is p^{α}-projective, so is $L \circledT M_{\lambda}$. Therefore, G is a summand of a p^{α}-projective, and hence G is p^{α}-projective.

Finally, our transfinite induction will utilize the following, which essentially allows us to replace groups by other groups of shorter length.

LEMMA 17 *If A is p^α-projective and B is fully starred, then $A \textcircled{T} B \cong A \textcircled{T} B/p^\alpha B$.*

Proof: Let J be a p^α-high subgroup of B, $K = B/J$ and $K' = B/(J + p^\alpha B)$. There are p^α-pure sequences $0 \to J \to B \xrightarrow{\pi} K \to 0$ and $0 \to J \to B/p^\alpha B \to K' \to 0$. Note that $\pi(p^\alpha B) \cong p^\alpha B$ is an essential subgroup of K, so they have the same rank. There is a subgroup $K'' \subseteq K$ which is a divisible hull of a basic subgroup C of $\pi(p^\alpha B)$, and it can be verified that $K = K'' + \pi(p^\alpha B)$ and $C = K'' \cap \pi(p^\alpha B)$. It follows that $K' \cong K/\pi(p^\alpha B) \cong K''/C$. Since B is fully starred, the rank of K, which agrees with the rank of $p^\alpha B$, equals the rank of C, which agrees with the rank of K'' and K'. Call this common rank ρ. There are p^α-pure sequences

$$0 \to A \textcircled{T} J \to A \textcircled{T} B \to A \textcircled{T} K \to 0$$

and

$$0 \to A \textcircled{T} J \to A \textcircled{T} B/p^\alpha B \to A \textcircled{T} K' \to 0.$$

Since $A \textcircled{T} K \cong \oplus_\rho A \cong A \textcircled{T} K'$ is p^α-projective, these sequences split, so that $A \textcircled{T} B \cong (A \textcircled{T} J) \oplus (\oplus_\rho A) \cong A \textcircled{T} B/p^\alpha B$.

THEOREM 18 *If A_1, A_2,... A_k are totally projective groups and $T = \textcircled{T}_{j=1}^k A_j$, then T is a direct sum of β, m-GP groups. In fact, if f is the Ulm function of T, then*

$$T \cong \oplus_\beta \oplus_{m<k} G_{\beta,m}$$

where $G_{\beta,m} \cong \oplus_{i \in I} L_i$ is a direct sum of $\beta * m$-GP groups, and where $|I| = 0$ when $f(\beta) = 0$, $f(\beta) \leq |I| < \omega$ when $f(\beta)$ is finite and $f(\beta) = |I|$ whenever $f(\beta)$ is infinite.

Proof: We first induct on k, noting that the result is trivial for $k = 1$. Assume next that the result holds for $k - 1$. We now induct on the length of T. Note that this length is the minimum of the lengths of the A_j. By Lemma 17, we may assume that A_1, A_2,..., A_k all have the same length. Furthermore, if this length is a limit ordinal, then A_1, and hence T, is isomorphic to a sum of groups of smaller length; by induction on the length of T, the result would be established in this case. Now suppose that A_1 has length $\alpha = \lambda + n$, where λ is a limit and n is a positive integer. Note that each A_j is isomorphic to a direct sum $\oplus U_i$, where each U_i has the property that $p^\lambda U_i$ is cyclic. Since these decompositions lead to a decomposition of T, we may assume that each $p^\alpha A_j$ is cyclic. In addition, by induction on lengths, we may assume each $p^\lambda A_j \cong \mathbf{Z}_{p^n}$.

Let J be a $p^{\alpha-1}$-high subgroup of A_1, so that $0 \to J \to A_1 \to \mathbf{Z}_{p^\infty} \to 0$ is p^α-pure. If $T' = \textcircled{T}_{j=2}^k A_j$, then there is a p^α-pure sequence

$$0 \to J \textcircled{T} T' \to T \to T' \to 0.$$

Since T' is p^α-projective, this sequence splits, so that $T \cong T' \oplus (J \textcircled{T} T')$. By induction on k,

$$T' \cong \oplus_\beta \oplus_{m<k-1} G'_{\beta,m}$$

where each $G'_{\beta,m}$ is a direct sum of the requisite number of $\beta * m$-GP groups. We therefore need to concentrate on the term $J \textcircled{T} T'$.

Since $p^\lambda A_1$ is cyclic, $J \cap p^\lambda A_1 = 0$, so that J is also p^λ-high in A_1. If $\overline{A} = A_1/p^\lambda A_1$, it follows that there is a p^λ-pure sequence $0 \to J \to \overline{A} \to \mathbf{Z}_{p^\infty} \to 0$, which determines a p^λ-pure exact sequence

$$0 \to J \textcircled{T} T' \to \overline{A} \textcircled{T} T' \to T' \to 0.$$

Since $\overline{A}$ is p^λ-projective, so is $\overline{A} \textcircled{T} T'$, so that this sequence is a p^λ-pure projective resolution of T'.

We now utilize our induction hypothesis to construct a second p^λ-pure projective resolution of T'. If f' denotes the Ulm function of T' and $f'(\beta) \neq 0$, then either $\beta < \lambda$ or $\beta = \alpha$. In the first case, each $G'_{\beta,m}$ is p^β-projective and hence p^λ-projective. If $\beta = \alpha$, then $f'(\alpha) = 1$, so we have the terms,

$$G'_{\alpha,0} \oplus G'_{\alpha,1} \oplus \cdots \oplus G'_{\alpha,k-2},$$

where $G'_{\alpha,m}$ is the direct sum of a finite number of $\alpha * m$-GP groups for each $m < k-1$. There are p^λ-pure sequences

$$0 \to G_{\alpha,m+1} \to X_m \to G'_{\alpha,m} \to 0,$$

where X_m is a p^λ-bounded totally projective group. In fact, it can be shown that we may choose X_m to have the same Ulm invariants as $G'_{\alpha,m}$ for all ordinals less than λ. We next claim that if

$$V = (\oplus_{\beta<\lambda} \oplus_{m<k-1} G'_{\beta,m}) \oplus (\oplus_{m<k-1} X_m) \stackrel{\text{def}}{=} V' \oplus X',$$

then $T \cong T \oplus V$. If $k = 2$, then T is an IT-group, V is totally projective and the Ulm invariants T and $T \oplus V$ agree, implying the result. If $k > 2$, then for each $\beta < \lambda$, $f'(\beta)$ is infinite, so it follows that $G'_{\beta,m} \cong G'_{\beta,m} \oplus G'_{\beta,m}$, which implies that $T' \cong T' \oplus V'$, which implies that $T \cong T \oplus V'$. Once again, X' is totally projective and the Ulm functions of T and $T \oplus X'$ agree, verifying the claim.

We now construct a p^λ-projective resolution of T', $0 \to U \to V \to T' \to 0$, where $U \stackrel{\text{def}}{=} \oplus_{m=1}^{k-1} G_{\alpha,m}$. By another application of Shanuel's Lemma, there is an isomorphism

$$(J \textcircled{T} T') \oplus V \cong U \oplus (\overline{A} \textcircled{T} T').$$

In addition, by induction on lengths

$$T'' \stackrel{\text{def}}{=} \overline{A} \textcircled{T} T' \cong \oplus_\beta \oplus_{m<k} G''_{\beta,m},$$

where the $G''_{\beta,m}$ are as required.

In conclusion, then, we have isomorphisms:

$$T \cong T \oplus V \cong T' \oplus (J \textcircled{T} T') \oplus V \cong T' \oplus U \oplus T''.$$

As we observed before, the result holds for both T' and T'', and U contains a finite number of $\alpha * m$-GP summand, for each $0 < m < k$, so the result follows.

COROLLARY 19 *Suppose $k > 1$ and $A_1, \ldots, A_k$ are totally projective groups of length exceeding ω_1. Then $\textcircled{T}_{j=1}^{k} A_k$ is not totally projective.*

Proof: This follows from the observation that $\textcircled{T}_{j=1}^{k} A_k$ has a summand which is a $(\omega_1 + m) * 1$-GP group, and such a group fails to be totally projective.

In [12] it was shown that when A and B are totally projective of length exceeding ω_1, then $A \textcircled{T} B$ is not totally projective. In fact, by a result of [32], $\textcircled{T}_{j=1}^{k} A_k$ is not totally projective whenever $A_1, \ldots, A_k$ are reduced groups of length exceeding ω_1.

4 OTHER WORK ON THE TORSION PRODUCT

The purpose of this section is to give an overview of other fairly recent work on the torsion product. Suppose R is a p-coprimary preradical represented by the sequence $0 \to \mathbf{Z} \to T \to H \to 0$. Recall that if R has infinite height (i.e., $1 \in \mathbf{Z}$ has infinite height in T), and H is R-projective, then for every group G there is an R-projective resolution

$$0 \to M \textcircled{T} G \to H \textcircled{T} G \to G \to 0,$$

where M is the torsion subgroup of T. Actually, more than this can be shown. If R is a representable p-coprimary preradical with enough projectives, then there must be such a representing sequence with H an R-projective. An outline of the verification of this fact proceeds as follows: if $0 \to M \to H \xrightarrow{\pi} \mathbf{Z}_{p^\infty} \to 0$ is an R-projective resolution of $\mathbf{Z}_{p^\infty}$, then R is represented by the image of $0 \to \mathbf{Z} \to Z[1/p] \to \mathbf{Z}_{p^\infty} \to 0$ under $\mathrm{Ext}(\pi, \mathbf{Z}) : \mathrm{Ext}(\mathbf{Z}_{p^\infty}, \mathbf{Z}) \to \mathrm{Ext}(H, \mathbf{Z})$. If, in addition, the height of R is $n < \omega$, then R must actually equal p^n.

When Nunke wrote his fundamental papers on R-purity, the only known examples of representable p-coprimary radicals with enough projectives were p^α, for α an ordinal or the symbol ∞. Oxford's question, therefore, is implicit in these papers, though it was only specifically stated in [30]. It was also shown in [30] that if R has height α and is represented by the sequence $0 \to \mathbf{Z} \to T \to H \to 0$ where H is an R-projective p-group, then the height of R equals the length of the group H.

THEOREM 20 ([23]) *Suppose R is a representable p-coprimary radical with enough projectives, whose length is α. If $\alpha \leq \omega$, then $R = p^\alpha$. This does not necessarily hold for ordinals $\alpha \geq \omega + 1$.*

Proof: This was noted above when α is finite, so suppose $\alpha \geq \omega$. If $0 \to M \to H \to \mathbf{Z}_{p^\infty} \to 0$ is a corepresenting sequence for R with H an R-projective, then

$$0 \to M \textcircled{T} H \to H \textcircled{T} H \to H \to 0$$

splits. If $\alpha = \omega$, then H is separable, and since H is a summand of $H \textcircled{T} H$, which is a summand of $H \textcircled{T} H \textcircled{T} H$, etc., we can conclude that H is in $\mathcal{T}_\infty$. By Theorem 2 we can conclude that H is Σ-cyclic, which, in turn, implies that $R = p^\omega$. On the other hand, in [23], a representable p-coprimary radical of height $\omega + 1$ with enough projectives was constructed such that H is not $p^{\omega+1}$-projective. It follows that $R \neq p^{\omega+1}$, completing the result.

In fact, the construction in [23] produced a class of examples whose diversity virtually guarantees that a complete description of all such representable radicals is impossible. Similar behavior was noted in [3] for the class of finitely cogenerated radicals.

Nunke's question was addressed in [28], and it was resolved in the case where A and B have different lengths. If A has length $\alpha \leq \omega_1$ and $p^\alpha B \neq 0$, it was shown that $A \circledT B$ is a dsc iff A is a dsc and B is a $\mathcal{C}_\alpha$ (when α is an isolated limit ordinal, then G is a $\mathcal{C}_\alpha$ if it has a $p^{\alpha-1}$-high subgroup which is a dsc). We will say a $\mathcal{C}_\alpha$-group G is *normal* if $p^\alpha G = 0$. In general, whenever $\alpha \leq \omega_1$ and $A \circledT B$ is a $\mathcal{C}_\alpha$, then we can conclude that both A and B are $\mathcal{C}_\alpha$ groups.

As Theorem 7 shows, even the question of when $A \circledT B$ is Σ-cyclic necessarily involves set-theoretic considerations. In general, Nunke's problem separates into two quite distinct subcases: When is $A \circledT B$ a dsc of countable length, and when is it a dsc of length ω_1? Considering the case of groups of countable length, a complete characterization is far from known. In [8], separate necessary and sufficient conditions are given for $A \circledT B$ to be Σ-cyclic, but these conditions have the defect of depending upon the selection of particular well-behaved filtrations of the groups involved. Another significant result is contained in [18], where it is shown that when $A_1, \ldots, A_k$ are separable groups, then a subgroup $G \subseteq \circledT_{j=1}^{k} A_j$ of cardinality $< \omega_k$ is always Σ-cyclic, and if $|G| = \omega_k$, then G is Σ-cyclic iff for every filtration $\{W_i\}_{i<\omega_k}$ of G, $S = \{i < \omega_k : p^\alpha(W_{i+1}/W_i) \neq 0\}$ is a stationary subset of ω_k. These results all generalize in natural ways to groups of arbitrary countable length.

Several surprises occur in the consideration of when $A \circledT B$ is a dsc of length ω_1. One begins with the observation that $A \circledT B$ is a dsc whenever both A and B are normal $\mathcal{C}_{\omega_1}$ groups of cardinality ω_1. (this is proven by representing A and B as ascending unions of p^α-high subgroups, for $\alpha < \omega_1$; see [8]). To begin a more general discussion of groups of larger cardinality it is necessary to employ some topological considerations. The ω_1-*topology* on a group G utilizes $\{p^\beta G : \beta < \omega_1\}$ as a neighborhood base of 0. This topology exhibits many pathological properties; for example, unlike the case of limit ordinals of countable cofinality, a cotorsion group need not be complete in the ω_1-topology (see [25]). Even more exotic is the behavior of this topology with respect to completions: given a collection of groups, each of which is complete in the ω_1-topology, then so is their direct sum. This implies, for example, that a dsc group is complete in the ω_1-topology. Another surprising result regarding this topology is contained in [15], where it is shown that if G is a normal $\mathcal{C}_{\omega_1}$ group and $\{A_i\}_{i\in I}$ is a smoothly ascending chain of p^{ω_1}-pure subgroups of G, each of which is closed in the ω_1-topology, then their union is also closed in this topology. This result is utilized in constructing filtrations of $\mathcal{C}_{\omega_1}$ groups consisting of closed subgroups. It means, essentially, that in such a construction one need not concern oneself with the case of limit ordinals, since closure is always preserved at such points. To appreciate this result, compare it with the case of the standard p-adic topology. If G is a separable group with basic subgroup B, then B is the ascending union of a collection of closed, bounded subgroups of G which are, in fact, summands. Their union, however, is actually dense in

G.

By a *Kurepa extension* we mean a p^{ω_1}-pure sequence $0 \to X \to G \xrightarrow{\pi} Y \to 0$ where G is a normal $\mathcal{C}_{\omega_1}$, $|X| = \omega_1$ and $|\overline{X}| = |p^{\omega_1}Y| > \omega_1$; in this case we say X is a Kurepa subgroup of G. In other words, a Kurepa subgroup of G is a p^{ω_1}-pure subgroup of cardinality ω_1 whose closure is strictly bigger.

A *Kurepa family* is a collection, $\mathcal{T} \subseteq \mathcal{P}(\omega_1)$, such that for every $\beta < \omega_1$,

$$\mathcal{T} \cap \beta = \{C \cap \beta : C \in \mathcal{T}\},$$

is countable. *Kurepa's Hypothesis* (KH) is the statement asserting the existence of a Kurepa family. Kurepa's Hypothesis is independent of ZFC, though it is valid in the constructible universe (see [4]).

The following striking result is contained in [14]:

THEOREM 21 *The following are equivalent:*
(a) KH,
(b) there is a Kurepa extension,
(c) there are normal $\mathcal{C}_{\omega_1}$ groups A and B such that $A \circledT B$ fails to be a dsc.

In fact, these is an even longer list of natural statements which are equivalent to the three mentioned. The proof that (a) is equivalent to (b) involves a detailed examination of the socles of the groups. In [2] this argument was translated into the language of valuated vector spaces. This not only provides a simplification of the argument appearing in [14], but also a method of generalizing the result to larger cardinalities.

Proving that (b) implies (c) is not difficult. Suppose $0 \to X \to G \to Y \to 0$ is a Kurepa extension. It is easy to see that there is no loss of generality in assuming that $|G| = \omega_2$. Given any filtration $\{X_i\}_{i\in I}$ of G by p^{ω_1}-pure subgroups, the presence of a Kurepa subgroup will imply that $p^{\omega_1}(G/X_i) \neq 0$ for a stationary set of $i < \omega_1$. If $G \circledT G$ were actually a dsc, then for some i we could conclude that $X_i \circledT X_i$ is a summand of $G \circledT G$, and hence of $X_{i+1} \circledT X_i$, which is also a dsc (since it is balanced subgroup of $G \circledT G$ of cardinality ω_1). This would imply that the quotient $(X_{i+1}/X_i) \circledT X_i$ is a dsc, contradicting Nunke's result concerning the case of groups of differing lengths. Conversely, if there are no Kurepa extensions, then given any $\mathcal{C}_{\omega_1}$ groups A and B, our lack of Kurepa extensions, together with the above result on the closures of unions of p^{ω_1}-pure subgroups of $\mathcal{C}_{\omega_1}$ groups implies that there are filtrations $\{A_i\}_{i\in I}$ and $\{B_i\}_{i\in I}$ consisting of closed p^{ω_1}-pure subgroups of A and B. It will then follow that there are exact sequences

$$0 \to A_i \circledT B_i \to A_{i+1} \circledT B_i \to (A_{i+1}/A_i) \circledT B_i \to 0$$

and

$$0 \to A_{i+1} \circledT B_i \to A_{i+1} \circledT B_{i+1} \to A_{i+1} \circledT (B_{i+1}/B_i) \to 0.$$

Since these filtrations consist of closed subgroups, these quotients are dsc groups because of their cardinality. Hence these sequences split, and we can conclude that $A \circledT B$ is always a dsc.

It might seem logical to expand Nunke's Problem to ask when such a torsion product is totally projective. However, it is a direct consequence of the results of [32] that for reduced groups A and B, $A \textcircled{T} B$ can only be totally projective when it is a dsc.

The torsion product continues to be a useful object of study. For example, the $\oplus_c$-*topology* on a group G is the linear topology generated by the collection of subgroups $X \subseteq G$ such that G/X is Σ-cyclic. A group is Hausdorff in this topology iff it is separable. The class of groups which are $\oplus_c$-complete is quite large, and in particular, it is closed under direct sums, summands and t-products (the t-product of a collection of groups is the torsion subgroup of their direct product). It follows that both the Σ-cyclic groups and the torsion complete groups are $\oplus_c$-complete. In a forthcoming paper the torsion product is utilized to give another method to construct groups which are $\oplus_c$-complete. It is shown that if G is any group which is $\oplus_c$-complete and A is any group such that $A/p^\omega A$ is Σ-cyclic, then $A \textcircled{T} G$ is $\oplus_c$-complete. This result, in turn, is applied to the computation of a certain double dual.

We conclude with some open problems:

(1) Suppose $\mathcal{C}$ is some interesting class of groups. For example, $\mathcal{C}$ could be the class of reduced groups of cardinality at most α or the class of p^α-projectives. Relate A and B if $A \textcircled{T} G \cong B \textcircled{T} G$ for every $G \in \mathcal{C}$.

(2) Is it true in ZFC that any groups A and B such that $A \textcircled{T} G \cong B \textcircled{T} G$ for all reduced G must differ by a Σ-cyclic summand?

(3) Find a classification schema for direct sums of $\alpha * m$-GP groups (where both α and m are allowed to vary). This class contains all iterated torsion products of totally projective groups.

(4) If G is an $\alpha * m$-GP group and G is not totally projective, does it follow that G is a proper p^α-projective, in other words, does G fail to be $p^{\alpha-1}$-projective. If this result holds, then the global homological dimension of p^α-purity is $n+1$ whenever $\alpha = \lambda + k$, where λ is a limit ordinal of cofinality ω_n. This result is already known when $k = 0$ (see [16])

(5) In V=L, when $A \textcircled{T} B$ is a dsc? It seems entirely probable that Nunke's Problem for the case of groups of uncountable length can be entirely settled in this model.

(6) When is $A \textcircled{T} A$ a dsc? When is it Σ-cyclic? In other words, by restricting to a single group, perhaps Nunke's problem becomes somewhat more tractable.

(7) One can define the torsion product of valuated vector spaces. To what extent can these

results be generalized or improved upon in this context?

REFERENCES

1. D. Cutler, Abelian p-groups A and B such that $\text{Tor}(A, G) \cong \text{Tor}(B, G)$, G reduced, *Proc. Amer. Math. Soc., 91*: 12–14 (1984).
2. D. Cutler and R. Dimitric, *Rocky Mount. J. Math., 23*: 1253–1266 (1993).
3. M. Dugas, T. Fay, and W. Shelah, Singly cogenerated annihilator classes, *Math. Ann., 261* : 359–385 (1982).
4. P. Eklof and A. Mekler, Almost Free Modules and Set-Theoretic Methods, North-Holland, Amsterdam (1990).
5. L. Fuchs, Infinite Abelian Groups, Vol. 1 and 2, Academic Press, New York (1970 and 1973).
6. P. Griffith, Infinite Abelian Groups, Chicago Lectures in Mathematics, Chicago and London (1970).
7. P. Hill, Two problems of Fuchs concerning tor and hom, *Jour. of Alg., 19*: 379-383 (1971).
8. P. Hill, The balance of Tor, *Math. Zeit., 122*: 179-188 (1983).
9. P. Hill, On the structure of abelian p-groups, *Trans. Amer. Math. Soc., 288*: 505–525 (1985).
10. P. Hill and C. Megibben, On the theory and classification of abelian p-groups, *Math. Zeit., 190*: 17–38 (1985).
11. J. Irwin, R. Cellars and T. Snabb, A functorial generalization of Ulm's theorem, *Comment. Math. Univ. St. Pauli, 27*: 155–177 (1978).
12. J. Irwin, T. Snabb & T. Cellars, The torsion product of totally projective p-groups, *Comment. Math. Univ. St. Pauli, 29*: 1–5 (1980).
13. P. Keef, On the Tor functor and some classes of abelian groups, *Pac. J. Math., 132*: 63–84 (1988).
14. P. Keef, "On set theory and the balanced-projective dimension of C_Ω-groups," *Contemporary Mathematics*, vol. 89: pp. 31–41, Amer. Math. Soc., Providence (1989).
15. P. Keef, A theorem on the closure of Ω-pure subgroups of C_Ω groups in the Ω-topology, *Jour. of Alg., 125*: 150–163 (1989).
16. P. Keef, S-groups, A-groups and the Tor functor, *Comm. Math. Univ. St. Paul., 39*: 1–14 (1990).
17. P. Keef, Some isomorphisms of abelian groups involving the Tor functor, *Proc. Amer. Math. Soc., 110*: 27–37 (1990).
18. P. Keef, On iterated torsion products of abelian p-groups, *Rocky Mount. J. Math., 21*: 1035–1055 (1991).
19. P. Keef, The torsion product in abelian groups, University of Padova Press, Padova, Italy (1991).
20. P. Keef, An application of set theory to the torsion product of abelian groups, *Proc. Amer. Math. Soc., 118*: 735–738 (1993).
21. P. Keef, On generalizations of purity in primary abelian groups, *Jour. of Alg., 167(2)*: 309–329 (1994).

22. P. Keef, On p^α-injective abelian groups, *Contemporary Mathematics*, vol. 171, pp. 205–216, Amer. Math. Soc., Providence, RI (1994).
23. P. Keef, Representable preradicals with enough projectives, to appear.
24. C. Megibben, A generalization of the classical theory of primary groups, *Tohoku Math. J., 22*: 347–356 (1970).
25. R. Mines, A family of functors defined on generalized primary groups, *Pacific J. Math. 26*: 349–360 (1967).
26. R. Nunke, "Purity and subfunctors of the identity, *Topics in Abelian Groups*, pp. 121–171, Chicago, Ill. (1963).
27. R. Nunke, "On the structure of Tor", *Proc. Colloquium on Abelian Groups*, pp. 115–124, Budapest, Hung. (1964).
28. R. Nunke, On the structure of Tor II, *Pac. J. Math., 22*: 453-467 (1967),
29. R. Nunke, Homology and direct sums of countable groups, *Math. Zeit., 101*: 182–212 (1967).
30. E. Oxford, p-Coprimary Functors With Enough Projectives, *Jour. of Alg., 53*: 129–135 (1978).
31. R. Warfield, A classification theorem for abelian p-groups, *Trans. Amer. Math. Soc., 210*: 149–168 (1975).
32. B. Wick, The Calculation of an invariant for Tor, *Pac. J. Math., 112*: 4445–450 (1984).

Intuitionistic Abelian Group Theory

FRED RICHMAN Florida Atlantic University, Boca Raton 33431

1 Intuitionistic logic

We are all familiar with the idea that different theorems about abelian groups hold in different set-theoretic worlds. Perhaps the simplest example, if not the most famous, is that $\mathrm{Tor}(\bar{B}, \bar{B})$ is a direct sum of cyclics if and only if the continuum hypothesis holds. Here $\bar{B}$ is the torsion-complete p-group with all Ulm invariants equal to 1. As a result, abelian group theory can progress along distinctly different lines depending on what additional set-theoretic axioms, if any, one chooses to assume.

Of course we don't want to assume the continuum hypothesis to prove a theorem that we can prove without it. If we prove a theorem without using any additional axioms, then it holds in all the set-theoretic worlds—it is a better theorem. What I want to suggest is that our theorems would be even better if we didn't use some of the *standard* axioms.

On the one hand, this is quite obvious—if I prove something from fewer assumptions, then I have proved something more general. On the other hand, the standard axioms are *true*, so what is the point of not using them? To put it less naively, are there any interesting set-theoretic worlds in which some standard axioms are false?

The only standard axiom that is well known to have received serious criticism, and which affects our daily lives, is the **axiom of choice**. But a set-theoretic world in which the axiom of choice fails seems foreign to our conception of what sets are and how we operate with them. The axiom of choice is inherent in our assumption of omniscience with respect to the mathematical universe.

Actually, I want to go much further and drop the axiom that is the formal reflection of the assumption of omniscience: the **law of excluded middle**—p or not p. When you forgo the law of excluded middle, you enter the domain of **intuitionistic logic**, the setting for constructive mathematics as described by Errett Bishop [1]. Although the law of excluded middle is normally considered a *logical* axiom as opposed to a set-theoretic one, the line is not so clear. In fact, the axiom of choice has also been considered a logical axiom. In any event, an axiom is an axiom; as algebraists, it doesn't make much difference to us exactly how it is classified.

I want to emphasize that the objects of study have not changed: we are still studying abelian groups in general, not some smaller "constructive" class of groups. The logic, not the subject matter, has been restricted. This is a key point: if you restrict the class of objects, you get less general theorems, but if you restrict the logic—the axioms—you get more general theorems. For further comments on this phenomenon see [12].

How can I be cool toward rejecting the axiom of choice, yet hot to drop the far more pervasive and acceptable law of excluded middle? The reason is that there are set-theoretic worlds that I find quite interesting in which the law of excluded middle fails, but none that I find interesting in which the law of excluded middle holds and the axiom of choice fails.

The main impetus for rejecting the law of excluded middle is the idea that mathematics is grounded on computation, a relatively uncontroversial idea that is contested only by mathematicians. We appeal to the law of excluded middle when confronted with a situation that we cannot resolve by computation. A classical computational setting in which we can sort of see the failure of the law of excluded middle is when dealing with recursive, or Turing computable, functions. If we get serious about restricting ourselves to such functions, then we interpret the statement "every Turing machine either halts or doesn't halt" to mean that some Turing machine can make that determination. That is false—the well-known unsolvability of the halting problem.

The recursive universe is constructed in the context of the law of excluded middle, in much the same way that we can construct a model of set theory where the continuum hypothesis fails within a model in which it holds. We can study this universe using intuitionistic logic by adopting an axiom—called *Church's thesis*, in this context—saying that every function from the natural numbers to the natural numbers is recursive. This axiom is incompatible with the law of excluded middle as we can prove the existence of nonrecursive functions using that law, such as the function that takes the value 1 on Turing machines that halt and 0 on those that don't halt.

The recursive universe, whether approached classically or intuitionistically, is interesting but too special. In order to appreciate life without the law of excluded middle in its full generality, we will simply observe through examples how its use deprives our mathematics of computational meaning, and how we operate without it.

2 Subgroups of cyclic groups

Consider the theorem that subgroups of cyclic groups are cyclic. It's pretty easy to see that it suffices to prove this for subgroups of the integers. As we go through a proof of this theorem, it is helpful to keep an example in mind. Let S be the subgroup generated by the perfect numbers, a countably generated subgroup of the integers.

How does the proof go? First we say that either S contains a positive integer or it doesn't. That's one use of the law of excluded middle. In our example, we know that 6 is in S, so we don't need the law of excluded middle, but we would have needed it if we didn't know any perfect numbers. If there is no positive integer in S, then S is cyclic with 0 as a generator. If there is a positive integer in S, we select the smallest one.

The principle that every nonempty set of positive integers contains a smallest one relies on the law of excluded middle. Either 1 is in the set or it is not. If 1 is in the set, then 1 is the smallest positive integer in the set. If not, then either 2 is in the set or it is not, and so on until we find the smallest element—the search is bounded because we already found an element of the set. In our example, we ask successively whether 1, 2, 3, 4, and 5 are in S. Now actually we know that $2 = 5 \times 6 - 28$ is in S, so the only issue is whether 1 is in S. To answer that we must either appeal to the law of excluded middle, or resolve the odd perfect number question.

Contrast this proof to the proof that a finitely generated subgroup of the integers is cyclic, a computational proof which uses the Euclidean algorithm. No excluded middle there, although we use the fact that every integer is either equal to 0 or it is not. But that is a property of the integers, they are **discrete**—either $m = n$ or $m \neq n$—not the consequence of a general principle. In contrast, the real numbers are *not* discrete: consider the sum $\sum_n a_n/10^n$ where $a_n = 1$ if n is an odd perfect number, and $a_n = 0$ otherwise.

3 The fundamental theorem of abelian group theory

Similar considerations hold for the fundamental theorem of abelian group theory, that a finitely generated group is a direct sum of cyclics. Here the problem is a little subtler. Suppose we know the group is generated by two elements x and y. How can we rule out the possibility, for example, that $x = ny$ for some large integer n? If that happens, then the group is cyclic and generated by y. If it doesn't happen (and if that were the only kind of relation that could hold between x and y), then the group is a direct sum of the infinite cyclic groups generated by x and y. But we don't know that either it happens or it doesn't happen, so we can't write the group as a direct sum of cyclics. Why a *large* integer n? We could just as well not know whether $x = 2y$, but in that case the group would not be discrete, and the possibility would remain that the theorem held for discrete groups.

The real fundamental theorem of abelian group theory is that a finitely *presented* group is a direct sum of finite and infinite cyclic groups. This can be proved by producing an algorithm to reduce a matrix of integers to Smith normal form [2, Thm. V.7.2], and other ways as well. In the proof, we don't appeal to the theorem that every ideal in the integers is principal, but rather to the fact that finitely generated ideals in the integers are principal—that is, the integers are a **Bézout domain**. That's not quite enough: last time I looked, it was still an open question whether you could diagonalize matrices over a Bézout domain. You need some sort of chain condition also.

I think the right idea here is to define a **Noetherian ring** to be one that satisfies the ascending chain condition on *finitely generated* ideals. In the presence of the law of excluded middle, this is equivalent to the usual definition. We must be a little careful to phrase the ascending chain condition so that it has some computational content—is not purely negative—and so that the integers satisfy it. Probably best is to say that if $I_1 \subset I_2 \subset I_3 \subset \cdots$ is a (weakly) ascending chain of finitely generated ideals, then there exists n such that $I_n = I_{n+1}$. The notion of a Noetherian Bézout domain is equivalent to that of a principal ideal domain in the presence of the law of excluded middle. Without the law of excluded middle, we cannot establish that the integers form a principal ideal domain, but we can show that they form a Noetherian Bézout domain.

These two ideas come together in diagonalizing a matrix. Using the Bézout property we make all but the first element of the first row zero by invertible column operations. Then we do the same for the first column with invertible row operations, but in doing so we might mess up the first row. We are okay if the first element of the first column divides the rest of the entries in that column; if not, then the ideal generated by that element will be strictly bigger next time around. That's where the chain condition comes in—to assure that the procedure will terminate with all entries in the first row and in the first column equal to zero, except the one in the corner.

Two remarks about this sketch of a proof. Although I didn't say it, a Bézout domain is discrete, and we certainly use that in the proof. Actually, it's not clear what the definition of a domain should be in the absence of discreteness. The real numbers ought to be a domain, but "$ab = 0$ implies $a = 0$ or $b = 0$" does not hold for them. The other point is that it looks like we are assuming, for two elements a and b of the domain, that either a divides b or a does not divide b. But that's only because I didn't indicate the proof accurately. What we do is proceed with our row and column operations until the ideal generated by the corner entry remains unchanged. When that happens, the previous corner entry had to divide the other row or column entries, so we back up and use elementary operations in the last step, and we are through.

4 More advanced topics

I don't want to give the impression that intuitionistic abelian group theory is preoccupied with such elementary topics. Although there is still a lot of work to do, substantial topics in abelian group theory have been developed without the law of excluded middle. Outside of the abelian groups community, Rick Smith showed that divisible subgroups of countable discrete groups are direct summands [15]. This was done in the context of recursive function theory, but the proof is fairly easily translated into an intuitionistic proof. Somewhat surprisingly, this kind of translation is quite rare—there seem to be two quite different ballgames going on.

The structure theory for countable p-groups was developed in [8]. One of the spin-offs of this endeavor was the first formulation of a valuated group as an independent entity—a group with a subheight function. This was dictated by the method for proving Zippin's theorem: a countable group with prescribed Ulm invariants was built piece by piece by successively enlarging a finite valuated group. A similar technique was used by Hunter, Richman and Walker [4] to prove Zippin's theorem for Warfield groups.

A start on the theory of torsion-free groups is contained in [11] including a proof of the equivalence of the two basic characterizations of Butler groups. The connection with Bext was established by Mines and Vinsonhaler in [6].

The structure of mixed groups has also been investigated. The theory of countable KT-modules was developed in [9], the theory of countable Warfield groups in [13]. The former is more elegant because KT-modules, unlike Warfield groups, are characterized among countably generated modules by the structure of their finitely generated valuated submodules.

All the theory I've discussed so far deals with discrete groups. Nondiscrete groups arise naturally when you take products. It is easy to see that direct sums of discrete groups, over a discrete index set, are discrete, but to show that the group of sequences of integers is discrete you would have to be able to determine whether or not a given sequence was all zeros—in the recursive universe this amounts to solving the halting problem. If your preferred subject is p-groups, you run into this phenomenon with $\bar{B}$.

Countably presented groups, which need not be discrete, arise as cohomology groups of compact metric spaces [5]. The only results I know regarding the structure of nondiscrete abelian groups are in Rogers' work [14] on basic subgroups of p-groups. I will use the main result of that paper to illustrate how one operates in abelian group theory without the law of excluded middle. In the process, I will even improve the result a little bit.

5 Basic subgroups

A pure subgroup B of a p-group G is a **basic subgroup** if B is a direct sum of finite cyclics and G/B is divisible. Notice the word "finite" in that definition. I could just as well have said "discrete" because G is a p-group. However, I would have defined something different had I simply omitted the word "finite". The group $\mathbf{Z}/S$, where S is generated by the perfect numbers, is certainly cyclic, and bounded by 2; but to show that it is finite we would have to construct a one-to-one correspondence with $\{1, 2, \ldots, n\}$ for some n. Note that $\mathbf{Z}/S$ is not discrete, or, more precisely, we haven't shown that it is discrete because we haven't established "$0 = 1$ or $0 \neq 1$".

I think that it is probably right to put discreteness into the definition of a basic subgroup. In some sense, the basic subgroup is our discrete hold on these continuous objects—particularly when there are no elements of infinite height.

Rogers showed in [14] that a p-group G has a countable basic subgroup if and only if

(1) G is discretely separable, and

(2) $p^n G$ is detachable for all n.

The first condition says that G contains a countable discrete dense subset. As for the second condition, which is trivial in the presence of the law of excluded middle, a subset S of a set X is **detachable** if for each x in X, either $x \in S$ or $x \notin S$. Note that the perfect numbers form a detachable subset of the integers. What about the subgroup generated by the perfect numbers?

Why have we restricted ourselves to *countable* basic subgroups? The usual reason given is that nobody knows any interesting uncountable discrete sets. What this means is that nobody has constructed a set that they could prove was both uncountable and discrete without adding axioms. On the other hand, any subset of the integers is discrete, but an intuitionistic proof that a particular subset of the integers is countable will translate into a proof that it is recursively enumerable. So there are discrete sets that we don't know how to count. The practical reason for restricting to countable basic subgroups is that something like Zorn's lemma seems to be needed to prove the existence of basic subgroups in the general case.

Condition (1) clearly holds for a group with a countable basic subgroup. To prove Condition (2) you approximate a given element x in G by an element b in B to within $p^n G$. Then you look to see whether $b \in p^n B$ or $b \notin p^n B$. (Exercise: $p^n B$ is detachable from B.) A neater proof is to observe that purity and density make the natural map $B/p^n B \to G/p^n G$ an isomorphism, and that $B/p^n B$ is discrete (because $p^n B$ is detachable), so $G/p^n G$ is discrete (whence $p^n G$ is detachable). Proving the *sufficiency* of Conditions (1) and (2) is, of course, where the real work lies.

It occurred to me, upon rereading this paper, that no mention was made as to whether summands of groups with countable basic subgroups have countable basic subgroups. The theorem itself seems designed to answer such a question: we need only verify that the two conditions are inherited by summands. Certainly the second one is, because if $G = H \oplus K$, then $p^n G = p^n H \oplus p^n K$, so $H/p^n H$ is a subgroup of the discrete group $G/p^n G$, hence is discrete.

Suppose now that $G = H \oplus K$ is discretely separable, with a discrete dense subset $(h_1, k_1), (h_2, k_2), \ldots$. Then $\{h_1, h_2, \ldots\}$ is certainly dense in H, but why should it be discrete?

After all, we might have $k_1 \neq k_2$, so $(h_1, k_1) \neq (h_2, k_2)$, without having $h_1 = h_2$ or $h_1 \neq h_2$. The trick is to extract a discrete dense subset out of $\{h_1, h_2, \ldots\}$, and this is actually a consequence of a general construction that shows that any separable p-group is discretely separable. I'll now outline that construction.

Enumerate the elements of a dense subset in such a way that each element appears an infinite number of times (the standard computer-science dovetailing technique)

$$h_1, h_1, h_2, h_1, h_2, h_3, h_1, h_2, h_3, h_4, \ldots = x_1, x_2, x_3, x_4, \ldots$$

At each stage we will have constructed a sequence $y_1, \ldots, y_{n-1}$ so that $\{y_1, \ldots, y_{n-1}\}$ is a discrete subset of $\{x_1, \ldots, x_{n-1}\}$. Now examine x_n. If x_n is equal to any of $y_1, \ldots, y_{n-1}$ modulo $p^n G$, set $y_n = y_{n-1}$; if not, set $y_n = x_n$. We can do this because $G/p^n G$ is discrete. It is easy to verify that $\{y_1, y_2, \ldots\}$ is discrete and dense in G.

This argument can be modified to apply to any pseudometric space, but it doesn't stay quite so clean because we have no detachability. Note that the detachability of $p^n G$ is needed to construct the p-adic pseudometric on G.

References

[1] Bishop, E., *Foundations of constructive analysis*, McGraw-Hill 1967.

[2] Eilenberg, S. and N. Steenrod, *Foundations of algebraic topology*, Princeton University Press, 1952.

[3] Feferman, S., Impredicativity of the existence of the largest divisible subgroup of an abelian p-group, *Model theory and algebra*, Springer Lecture Notes **498**(1975), 117–130.

[4] Hunter, R., F. Richman and E. A. Walker, Existence theorems for Warfield groups, *Trans. Amer. Math. Soc.*, **235**(1978), 345–362.

[5] Julian, W., R. Mines and F. Richman, Alexander duality, *Pacific J. Math.* **106**(1983), 115–127.

[6] Mines, R. and C. Vinsonhaler, Butler groups and Bext: a constructive view, *Contemporary Math.*, **130**(1992), 289–299.

[7] Mines, R., F. Richman and W. Ruitenburg, *A course in constructive algebra*, Springer Verlag 1988.

[8] Richman, F., The constructive theory of countable abelian p-groups, *Pacific J. Math.* **45**(1973), 621–637.

[9] ———, The constructive theory of KT-modules, *Pacific J. Math.* **61**(1975), 263–274.

[10] ———, Computing heights in Tor, *Houston J. Math.* **3**(1977) 267–270.

[11] ———, The constructive theory of torsion-free abelian groups, *Communications in Algebra*, **18**(1990), 3913–3922.

[12] ———, Intuitionism as generalization, *Philosophia Mathematica* **5**(1990), 124–128.

[13] ———, The constructive theory of countably generated Warfield modules, *Contemporary Math.*, **130**(1992), 371–384.

[14] Rogers, L., Basic subgroups from a constructive viewpoint, *Communications in algebra* **8**(1980), 1903–1925.

[15] Smith, R. L., Two theorems on autostability in p-groups, *Logic year* 1979–1980, Springer Lecture Notes **859**(1981), 302–311.

Socle Projective Representations of Partially Ordered Sets and Tits Quadratic Forms with Applications to Lattices Over Orders

Daniel Simson

Faculty of Mathematics and Informatics, Nicholas Copernicus University
87-100 Toruń, Poland, e-mail: simson@mat.uni.torun.pl

Summary. The main aim of this survey article is to present various concepts of linear representations of finite posets and their applications to the study of lattices over orders. Our main results assert that the finite representation type and tameness for categories of poset representations and for categories of lattices over a class of orders is equivalent to the weakly positivity and weakly non-negativity (respectively) of associated integral Tits quadratic forms (see Theorems 2.5, 2.11, 3.3 and 3.4). Following an idea of Drozd, Zavadskij and Kirichenko [10] we define and study in Section 5 filtered subprojective representations of finite posets over a commutative artinian principal ideal ring, and we present in Corollary 5.18 a Zavadskij-Kirichenko type criterion [48] for the finite representation type. This is an alternative approach to the problem solved by Plahotnik [28].

1 Introduction

A notion of a matrix representation of a partially ordered set S (abbr. poset) over a field K was introduced by Nazarova and Roiter [25] in a connection with the Second Brauer-Thrall Conjecture (see [26] and [40]). The main aim of their study was to determine indecomposable objects of the category Mat_S^{ad} (see [40, Chapter 2] and [12]) of all matrix representations of S over a fixed field K. The category Mat_S^{ad} is additive and each of its objects has a decomposition into a finite direct sum of indecomposable objects, which is unique up to isomorphism and permutation of indecomposable summands. The objects of Mat_S^{ad} can be viewed as partitioned matrices with coefficients in K and the classification of the indecomposable objects in Mat_S^{ad} is equivalent to the problem of putting any partitioned matrix in Mat_S^{ad} into a normal form by admissible elementary row and column transformations defined by means of

*Partially supported by Polish Scientific Grand KBN 1221/2/91

the partial order $\preceq$ of S. Indecomposable objects correspond to the indecomposable normal partitioned cells.

Aside from matrix representations of a poset S, the concept of an S-space (or a filtered K-linear representation of S [40, Section 3.1]) was introduced by Gabriel [11] in connection with the investigation of finite quivers (i.e. oriented graphs) having finitely many isomorphism classes of indecomposable K-linear representations.

We recall that an I-space is a system $\mathbf{M} = (M_i;\ M)_{i \in S}$, where M is a finite dimensional K-vector space together with subspaces $M_i \subseteq M$ such that $M_i \subseteq M_j$ if $i \preceq j$ in S. By a morphism $f : \mathbf{M} \to \mathbf{M}'$ between S-spaces $\mathbf{M}$ and $\mathbf{M}'$ we shall mean a K-linear map $f : M \to M'$ such that $f(M_j) \subseteq M'_j$ for all $j \in S$. The direct sum of $\mathbf{M}$ and $\mathbf{N}$ in S-sp is defined by formula $\mathbf{M} \oplus \mathbf{N} = (M_i \oplus N_i\,; M \oplus N)_{i \in S}$. An S-space $\mathbf{M}$ is said to be **indecomposable** if $\mathbf{M}$ is non-zero and is not a direct sum of two non-zero S-spaces (see [12],[40]).

We denote by S-sp the category of S-spaces (over a fixed field K). This category is sometimes denoted by $\mathrm{Rep}_K(S)$ or by $\mathrm{Rep}(K, S)$ (see [1], [4], [5]).

The category S-sp is said to be of finite representation type if the set of the isomorphism classes of indecomposable S-spaces is finite.

There is a close functorial connection between Mat_S^{ad} and S-sp given by a full additive dense functor (see [40, Section 3.1])

$$H : \mathrm{Mat}_S^{ad} \longrightarrow S\text{-sp}$$

preserving the representation type and vanishing only on a finite set of isomorphism classes of indecomposable objects. Therefore the study of the category Mat_S^{ad} is equivalent to the study S-sp from the representation theory point of view.

It was shown by Kleiner [22] (see [40, Section 10.1]) that the category S-sp is of finite representation type if and only if the poset S does not contain a subposet isomorphic to one of the following critical posets of Kleiner

$$\mathcal{K}_1 : (1,1,1,1) = \{\bullet\ \bullet\ \bullet\ \bullet\},\quad \mathcal{K}_2 : (2,2,2) = \left\{\begin{smallmatrix}\bullet & \bullet & \bullet\\ \uparrow & \uparrow & \uparrow\\ \bullet & \bullet & \bullet\end{smallmatrix}\right\},\quad \mathcal{K}_3 : (1,3,3) = \left\{\begin{matrix}\bullet & \bullet & \bullet\\ & \uparrow & \uparrow\\ & \bullet & \bullet\\ & \uparrow & \uparrow\\ & \bullet & \bullet\end{matrix}\right\}$$

$$\mathcal{K}_4 :\quad (N,4) = \left\{\begin{matrix}\bullet & & \bullet & \bullet\\ \uparrow & \nwarrow & \uparrow & \uparrow\\ \bullet & & \bullet & \bullet\\ & & & \uparrow\\ & & & \bullet\\ & & & \uparrow\\ & & & \bullet\end{matrix}\right\},\qquad \mathcal{K}_5 :\quad (1,2,5) = \left\{\begin{matrix}\bullet & \bullet & \bullet\\ & \uparrow & \uparrow\\ & \bullet & \bullet\\ & & \uparrow\\ & & \bullet\\ & & \uparrow\\ & & \bullet\\ & & \uparrow\\ & & \bullet\end{matrix}\right\}.$$

There is also a concept of tame representation type for the category S-sp of S-spaces (see [40, Section 14.4]) and according to the theorem of Nazarova [24] (see [40, Section 15.1]) the category S-sp is of tame representation type if and only if the poset S does not contain a subposet isomorphic to one of the following hypercritical posets of Nazarova

$$\mathcal{N}_1 : (1,1,1,1,1) = (\bullet\ \bullet\ \bullet\ \bullet\ \bullet),\qquad \mathcal{N}_2 : (1,1,1,2) = \left(\begin{matrix} & & & \bullet\\ & & & \uparrow\\ \bullet & \bullet & \bullet & \bullet\end{matrix}\right)$$

$\mathcal{N}_3 : (2,2,3) =$ 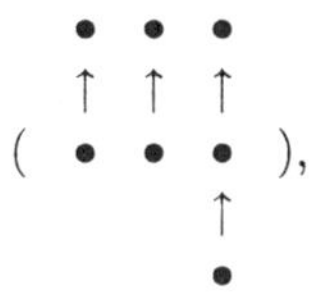, $\mathcal{N}_4 : (1,3,4) =$ 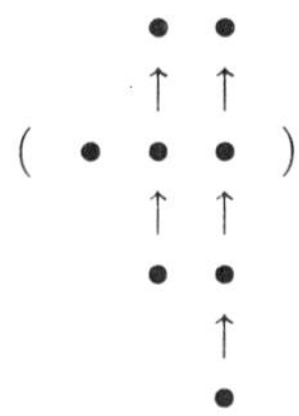

$\mathcal{N}_5 : (N,5) =$ 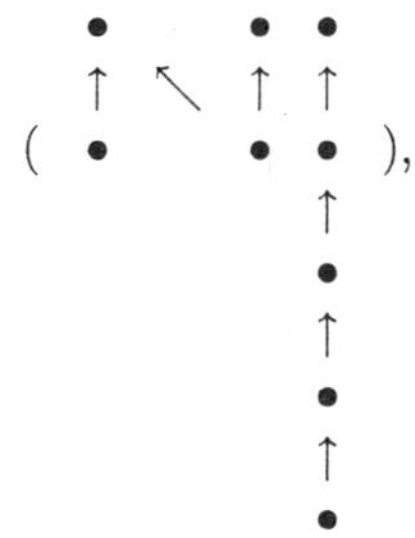, $\mathcal{N}_6 : (1,2,6) =$ 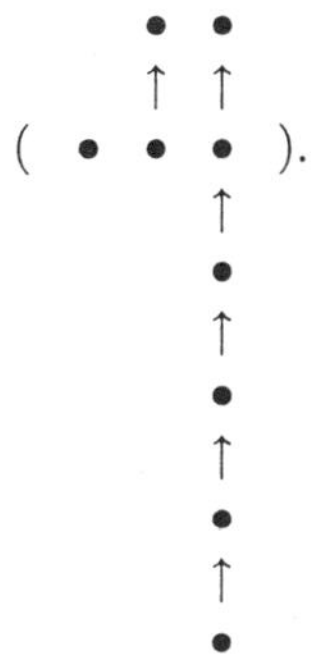.

Many of the representation theory problems can be reduced to the corresponding problems for S-spaces over a finite poset S. An elementary explanation of such reductions of some classification problems for classes of abelian groups to the category of S-spaces is given by Arnold [1] (see also [2],[3],[4]).

Another important reductions of this kind appear in the theory of lattices over orders (see [14], [32],[3],[40, Section 13.1], [44]). They are briefly discussed below.

Here we assume that D is a complete discrete valuation domain with the field of fractions D_0, the maximal ideal $\mathfrak{p}$ and the residue field $K = D/\mathfrak{p}$. We recall that a D-**order** in a semisimple D_0-algebra A is a D-subalgebra Λ of A which is a finitely generated free D-module and contains a D_0-basis of A [29].

By a **right Λ-lattice** we mean a finitely generated right Λ-module which is D-free. The category latt(Λ) of all right Λ-lattices is additive, has the finite unique decomposition property and is closed under taking kernels and extensions. Moreover latt(Λ) has Auslander-Reiten sequences as well as source maps and sink maps [31].

We say that Λ is of **finite lattice type** if latt(Λ) has a finite number of isomorphism classes of indecomposable modules. Tame and wild lattice types are defined in Section 3.

Let us recall from [40, Section 13.1] that there is a functorial connection between the category latt(Λ) of right Λ-lattices over a class of D-orders Λ in a simple algebra A and the categories S-sp of S-spaces.

If Λ is a D-order in a product of at least two simple algebras then the study of latt(Λ) is in general not reducible to the study of the category of S-spaces. However for a class of such orders Λ one can reduce the study of latt(Λ) to the study of the category I-spr of peak I-spaces (defined below) over the field K, where I is a poset. The idea is explained in Example 1.2 below.

One of the aim of this survey article is to discuss the importance of the category I-spr of peak I-spaces. We shall show in Section 2 that the representation type (defined below) of the category I-spr can be characterised by means of the Tits quadratic form $q_I : \mathbb{Z}^I \to \mathbb{Z}$ of

I defined by the formula

$$q_I(x) = \sum_{i \in I} x_i^2 + \sum_{\substack{i \prec j \\ j \in I \setminus \max I}} x_i x_j - \sum_{p \in \max I} (\sum_{i \prec p} x_i) x_p. \tag{1.1}$$

In particular, it is shown in Theorem 2.5 that I-spr is of finite representation type if and only if q_I is weakly positive, that is, $q_I(v) > 0$ for any non-zero vector $v \in \mathbb{N}^I$. Theorem 2.9 asserts that I-spr is of fully wild representation type if and only if there exists $v \in \mathbb{N}^I$ such that $q_I(v) < 0$.

We recall from [43] that the category I-spr (resp. latt(Λ)) is **fully wild** (resp. **wild**) if and only if there exists a full faithful exact functor (resp. a representation embedding functor) $\mathrm{mod}\begin{pmatrix} K & K^3 \\ 0 & K \end{pmatrix} \longrightarrow I\text{-spr}$ (resp. $\mathrm{mod}\begin{pmatrix} K & K^3 \\ 0 & K \end{pmatrix} \longrightarrow \mathrm{latt}(\Lambda)$).

We show in Theorem 2.11 that under some assumption on a two-peak poset I the category I-spr is of tame representation type (see [17], [42] and Section 2) if and only if the Tits form q_I is weakly non-negative.

In Section 3 we show that for a class of D-orders Λ the representation type of the category latt(Λ) can also be simply determined by means of a Tits quadratic form $q_\Lambda : \mathbb{Z}^{n(\Lambda)} \to \mathbb{Z}$ associated with Λ. A structure of the Auslander-Reiten quiver of latt(Λ) can be also described (see [44, Corollary 3.2]). We do it by associating with every such an order Λ a finite poset I_Λ^{*+} with exactly two maximal elements and by applying a reduction functor from the category latt(Λ) to the category I_Λ^{*+}-spr of peak I_Λ^{*+}-spaces.

In Section 4 we present an idea of Zavadskij-Kirichenko [47] of a study of the category latt(Λ) for any tiled D-order in terms of the category I_Λ-sp, where I_Λ is an infinite poset associated to Λ. We study the tameness of latt(Λ) in terms of I_Λ.

Following this idea and an idea of Ju. Drozd (see also [10]) we define in Section 5 the category $\mathbf{fspr}(I, D/\mathfrak{p}^m)$ of filtered subprojective representations of a finite poset I over the ring $D/\mathfrak{p}^m$, $m \geq 1$. Moreover we associate to $(I, D/\mathfrak{p}^m)$ an infinite poset $\widehat{I_m}$ with an action of the infinite cyclic group $\mathbb{Z}$, and we prove that the category $\mathbf{fspr}(I, D/\mathfrak{p}^m)$ is of finite representation type if and only if the poset $\widehat{I_m}$ does not contain the critical posets $\mathcal{P}_1, \dots, \mathcal{P}_5$ of Kleiner shown above. This is an alternative approach to the problem solved by Plahotnik [28].

One of the motivations for the study of the category the category $\mathbf{fspr}(I, D/\mathfrak{p}^m)$ are the results in [3] showing that block rigid accd groups are functorially related with filtered subprojective representations of finite posets over the factor rings of $\mathbb{Z}$ [3, Theorem 1.4].

Let us finish the introduction by explaining the idea of a reduction of the study of latt(Λ) to I-spr by the following example.

EXAMPLE 1.2 Consider the following D-suborders

$$\Lambda_1 = \begin{bmatrix} D & D & \mathfrak{p} \\ \mathfrak{p} & D & \mathfrak{p} \\ \mathfrak{p} & \mathfrak{p} & D \end{bmatrix}, \quad \Lambda_2 = \begin{bmatrix} D & D & D & D \\ \mathfrak{p} & D & \mathfrak{p} & \mathfrak{p} \\ \mathfrak{p} & \mathfrak{p} & D & D \\ \mathfrak{p} & \mathfrak{p} & \mathfrak{p} & D \end{bmatrix}, \quad \Lambda_3 = \begin{bmatrix} D & D & D \\ \mathfrak{p} & D & D \\ \mathfrak{p} & \mathfrak{p} & D \end{bmatrix}$$

of $\Gamma_1 = \mathbb{M}_3(D)$, $\Gamma_2 = \mathbb{M}_4(D)$ and $\Gamma_3 = \mathbb{M}_3(D)$, respectively. Denote by Λ the suborder of $\Lambda_1 \times \Lambda_2 \times \Lambda_3$ consisting of all triples $(a, b, c) \in \Lambda_1 \times \Lambda_2 \times \Lambda_3$ of matrices a, b, c satisfying the following conditions:

(i) $a_{ij} \equiv b_{ij}$ for all $1 \leq i \leq j \leq 2$,

(ii) $b_{11} \equiv c_{11}$, $b_{13} \equiv c_{12}$, $b_{14} \equiv c_{13}$, $b_{33} \equiv c_{22}$, $b_{44} \equiv c_{33}$, $b_{34} \equiv c_{23}$,
where $x \equiv y$ means $x - y \in \mathfrak{p}$.

If I is the following poset

$$I: \quad \begin{array}{ccccccc} 5 & \rightarrow & p_1 & & & & \\ & & \uparrow & & & & \\ & & 4 & \longrightarrow & p_2 & & \\ & & \uparrow & & \uparrow & & \\ & & 1 & \rightarrow 2 \rightarrow & 3 & \rightarrow & p_3 \end{array}$$

having exactly three maximal elements p_1, p_2, p_3, then there exists a full additive functor

$$T : \mathrm{latt}(\Lambda) \longrightarrow I\text{-spr} \tag{1.3}$$

establishing a one to one correspondence between the isoclasses of indecomposable Λ-lattices and the isoclasses of indecomposable objects in the category I-spr not isomorphic to the simple projective representations $\mathbf{S}(p_1), \mathbf{S}(p_2), \mathbf{S}(p_3)$ corresponding to the maximal elements p_1, p_2, p_3.

By constructing the preprojective component of I-spr or by applying the Tits quadratic form argument given in Theorem 2.7, one can show that the category I-spr has 100 isoclasses of indecomposable objects and therefore the order Λ is of finite lattice type and latt(Λ) has 97 isoclasses of indecomposable lattices. The structure of latt(Λ) and its Auslander-Reiten quiver are well understood because of Theorem 3.11.

The reduction functor T is defined as follows. First we note that Λ is contained in the hereditary D-order $\Gamma = \Gamma_1 \times \Gamma_2 \times \Gamma_3$ and the Jacobson radical $\mathfrak{A} = \mathrm{rad}(\Gamma) = \mathbb{M}_3(\mathfrak{p}) \times \mathbb{M}_4(\mathfrak{p}) \times \mathbb{M}_3(\mathfrak{p})$ of Γ coincides with the Jacobson radical of Λ. It is easy to see that the ring

$$\overline{\Lambda}_\Gamma = \begin{bmatrix} \Lambda/\mathfrak{A} & \Gamma/\mathfrak{A} \\ 0 & \Gamma/\mathfrak{A} \end{bmatrix}$$

is a finite dimensional K-algebra. Then according to [14] and [32] there exists a reduction functor

$$\mathbb{F} : \mathrm{latt}(\Lambda) \longrightarrow \mathrm{mod}_{\mathrm{sp}}(\overline{\Lambda}_\Gamma)$$

where $\mathrm{mod}_{\mathrm{sp}}(\overline{\Lambda}_\Gamma)$ is the full subcategory of $\mathrm{mod}(\overline{\Lambda}_\Gamma)$ consisting of modules having a projective socle, and $\mathbb{F}(X) = (X/X\mathfrak{A}, X\Gamma/X\mathfrak{A}, u)$, $X\Gamma$ is the Γ-submodule of $X \otimes_D D_0$ generated by X, D_0 is the field of fractions of D and $u : X/X\mathfrak{A} \to X\Gamma/X\mathfrak{A}$ is the $\Lambda/\mathfrak{A}$-monomorphism induced by the natural monomorphism $X \hookrightarrow X\Gamma$ (see [14] and [32]). We view $\mathbb{F}(X)$ as a right $\overline{\Lambda}_\Gamma$-module in a natural way.

Since $\Gamma/\mathfrak{A}$ is Morita equivalent to $K \times K \times K$, one can show that the algebra $\overline{\Lambda}_\Gamma$ is Morita equivalent to the incidence K-algebra

$$KI = \begin{bmatrix} K & K & K & 0 & 0 & K & K & K \\ & K & K & 0 & 0 & 0 & K & K \\ & & K & 0 & 0 & 0 & K & K \\ & & & K & 0 & K & K & 0 \\ & & & & K & K & 0 & 0 \\ & & & & & K & 0 & 0 \\ & 0 & & & & & K & 0 \\ & & & & & & & K \end{bmatrix} \subseteq \mathbb{M}_8(K)$$

of the poset I, and therefore there exists an equivalence of categories $\mathrm{mod}_{\mathrm{sp}}(\overline{\Lambda}_\Gamma) \cong \mathrm{mod}_{\mathrm{sp}}(KI) \cong I$-spr. The composition of $\mathbb{F}$ with the composed equivalence above defines the functor (1.3) with the required properties.

The reader is referred to Theorems 4.1 and 4.2 of [42] for details. We remark that the order Λ defined above is a stratification subposet order in the sense of [42, Section 4]. □

2 Finite representation type, tameness and fully wildness of categories of peak I-spaces

Throughout this paper K is a field and we denote by $(I, \preceq)$ a finite **poset** (i.e. partially ordered set) with respect to the partial order $\preceq$. We shall write $i \prec j$ if $i \preceq j$ and $i \neq j$. For the sake of simplicity we write I instead of $(I, \preceq)$. We denote by $\max I$ the set of all maximal elements of I and I will be called an t-**peak poset** if $|\max I| = t$.

We recall from [42] that the category I-spr of **peak I-spaces** (or **socle projective representations** of I) over the field K is defined as follows. The objects of I-spr are systems

$$X = (X_j)_{j \in I}$$

of finite dimensional K-vector spaces X_j such that $X_j \subseteq X^\bullet = \bigoplus_{p \in \max I} X_p$ for all $j \in I$, $\pi_p(X_j) = 0$ for $j \not\preceq p \in \max I$ and $\pi_j(X_i) \subseteq X_j$ for $i \prec j \in I$, where π_j is the composed map

$$X^\bullet \xrightarrow{\pi_j'} \bigoplus_{j \preceq p \in \max I} X_p \hookrightarrow X^\bullet \tag{2.1}$$

and π_j' is the direct summand projection.

By a morphism $f : X \to X'$ in I-spr we mean a system $f = (f_p)_{p \in \max I}$ of K-linear maps $f_p : X_p \to X_p'$, $p \in \max I$, such that $(\bigoplus_{p \in \max I} f_p)(X_j) \subseteq X_j'$ for all $j \in I$. It is clear that I-spr is an additive category with enough projective objects and with the finite unique decomposition property. Moreover I-spr is closed under taking kernels and is of finite global dimension [42, Lemma 2.1]. According to [27] and [42] the category I-spr has Auslander-Reiten sequences, source maps and sink maps (see [31]).

If I has a unique maximal element $*$, then $I\text{-spr} \cong (I \setminus \{*\})\text{-sp}$. Conversely, if S is a poset and $S^* = S \cup \{*\}$ is the poset-enlargement of S by a unique maximal element $*$, i.e. we set $s \prec *$ for any $s \in S$, then $S\text{-sp} = S^*\text{-spr}$. This means that the categories S-sp are precisely the categories I-spr, where I are posets with a unique maximal element.

In order to see the difference between peak I-spaces and I-spaces consider the following example.

EXAMPLE 2.2 Assume that I is the poset

$$I: \quad \begin{array}{ccccc} 1 & & & & 4 \\ & \searrow & & \nearrow & \\ & & 3 & & \\ & \nearrow & & \searrow & \\ 2 & & & & 5 \end{array}$$

with exactly two maximal elements 4 and 5. Then the category I-sp consists of systems $\mathbf{M} = (M_1, M_2, M_3, M_4, M_5\,; M)$ of the vector space embeddings

$$\mathbf{M}: \quad \begin{array}{ccccccccc} M_1 & & & & M_4 & & & & \\ & \searrow & & \nearrow & & \searrow & & & \\ & & M_3 & & & & M & & \\ & \nearrow & & \searrow & & \nearrow & & & \\ M_2 & & & & M_5 & & & & \end{array}$$

whereas I-spr consists of systems $X = (X_1, X_2, X_3, X_4, X_5)$ of vector spaces X_i, $i \in I$ forming the diagram of vector space embeddings

$$\begin{array}{ccccc} X_1 & & & & \\ & \searrow & & & \\ & & X_3 & \hookrightarrow & X_4 \oplus X_5 \,. \\ & \nearrow & & & \\ X_2 & & & & \end{array}$$

Note that every such a system X can be viewed as a commutative diagram of linear maps

$$\mathbf{X}: \quad \begin{array}{ccccc} X_1 & & & & X_4 \\ & \searrow & & \nearrow \varphi & \\ & & X_3 & & \\ & \nearrow & & \searrow \psi & \\ X_2 & & & & X_5 \end{array}$$

where the left hand arrows represent embeddings, and φ, ψ are K-linear maps such that $(\operatorname{Ker}\varphi) \cap (\operatorname{Ker}\psi) = 0$. This means that $\mathbf{X}$ is a K-linear socle projective representation of the poset I viewed as a quiver (see [40, Section 14.1]). The maps φ and ψ are the compositions of the embedding $X_3 \hookrightarrow X_4 \oplus X_5$ with the projection on X_4 and X_5, respectively.

Since the poset I contains no triple of incomparable elements, then the category I-sp is of finite representation type and the number of isoclasses of indecomposable I-spaces is equal to 8 (see [40, Lemma 6.1]).

The category I-spr is of infinite (tame) representation type. To see this we fix $m \geq 1$, $\lambda \in K$ and a K-linear endomorphism $\varphi_\lambda : K^m \to K^m$ such that the matrix of φ_λ in the standard basis of K^m is the simple Jordan cell $J(m, \lambda) \in \mathbb{M}_m(K)$ with the eigenvalue λ. Define the peak I-space

$$X^{(m,\lambda)} = (X_1^{(m,\lambda)}, X_2^{(m,\lambda)}, X_3^{(m,\lambda)}, X_4^{(m,\lambda)}, X_5^{(m,\lambda)})$$

where $X_4^{(m,\lambda)} = X_5^{(m,\lambda)} = K^m$, $X_3^{(m,\lambda)} = K^m \oplus K^m$, $X_1^{(m,\lambda)} = \{(v, v) \in X_3^{(m,\lambda)};\ v \in K^m\}$ and $X_2^{(m,\lambda)} = \{(v, \varphi_\lambda(v)) \in X_3^{(m,\lambda)};\ v \in K^m\}$.

A simple calculation shows that the endomorphism algebra $\operatorname{End}(X^{(m,\lambda)})$ of $X^{(m,\lambda)}$ is isomorphic to the local algebra $K[t]/(t - \lambda)^m$ and therefore for every $m \geq 1$ and $\lambda \in K$ the peak I-space $X^{(m,\lambda)}$ is indecomposable. This shows that the category I-spr is of infinite representation type. □

In order to formulate a criterion for the finite representation type of I-spr we assume that $\max I = \{p_1, \ldots, p_t\}$ and $I \setminus \max I = \{1, \ldots, n\}$. We associate with any vector $v \in \mathbb{N}^I$ the affine K-variety

$$\mathbf{Mat}_v^I$$

of all bipartitioned matrices of the form (compare with [40, Chapter 2])

$$A = \begin{array}{|c|c|c|c|l} \hline A_{1p_1} & A_{2p_1} & \cdots & A_{np_1} & \}v(p_1) \\ \hline \vdots & \vdots & \cdots & \vdots & \quad\vdots \\ \hline A_{1p_t} & A_{2p_t} & \cdots & A_{np_t} & \}v(p_t) \\ \hline \underbrace{}_{v(1)} & \underbrace{}_{v(2)} & \cdots & \underbrace{}_{v(n)} & \end{array} \tag{2.3}$$

with coefficients in K, where $A_{ip_r} = 0$ if $i \not\preceq p_r$, $r = 1, \ldots, t$. Moreover we define the algebraic group $\mathbf{G}_v^I$ to be the group generated by the following I-elementary transformations on matrices in $\mathbf{Mat}_v^I$.

(E_1) *All elementary transformations on rows inside each horizontal block of the matrix* A.

(E_2) *All elementary transformations on columns from the* i*-th block to the* j*-th block for all* $i \preceq j \in I \setminus \max I$.

It is easy to see that given a vector $v \in \mathbb{N}^I$

$$\mathbf{G}_v^I = H_v^I \times \mathrm{Gl}(v(p_1), K) \times \cdots \times \mathrm{Gl}(v(p_t), K)$$

where H_v^I is the group consisting of all matrices of the form

$$h = \begin{bmatrix} h_{11} & h_{12} & \ldots & h_{1n} \\ h_{21} & h_{22} & \ldots & h_{2n} \\ \vdots & \vdots & \ddots & \vdots \\ h_{n1} & h_{n2} & \ldots & h_{nn} \end{bmatrix} \in \mathrm{Gl}(v(1) + \cdots + v(n),\, K)$$

with $h_{ii} \in \mathrm{Gl}(v(i), K)$ and $h_{ij} = 0$ if $i \not\preceq j$ in $I \setminus \max I$.

We also define the algebraic action

$$* : \mathbf{G}_v^I \times \mathbf{Mat}_v^I \longrightarrow \mathbf{Mat}_v^I \tag{2.4}$$

by the formula $(h, C_{p_1}, \ldots C_{p_t}) * A = \mathrm{diag}(C_{p_1}, \ldots C_{p_r})(A \circ h^{-1})$, where $A \circ h' = A'$ is a bipartitioned matrix of the form (2.3) with $A'_{sp} = (Ah')_{sp}$ for $s \preceq p \in \max I$ and $A'_{sp} = 0$ for $s \not\preceq p \in \max I$.

The first main result of this section is the following.

THEOREM 2.5 *Suppose that I is a finite poset,* $\max I$ *is the set of all maximal elements of I and K be an infinite field. Then the following conditions are equivalent.*

(a) *The category I-*spr *of peak I-spaces is of finite representation type.*

(b) *The algebraic K-variety* $\mathbf{Mat}_v^I$ *has finitely many* $\mathbf{G}_v^I$*-orbits with respect to the action* (2.4) *for every vector* $v \in \mathbb{N}^I$

(c) *The Tits quadratic form* q_I (1.1) *is weakly positive, i.e.* $q(v) > 0$ *for any non-zero vector* $v \in \mathbb{N}^I$.

(d) *The poset I does not contain as a peak subposet any of the posets presented in Tables* 1 *below.*

(e) *There is no fully faithful exact functor* $\mathrm{mod}\begin{pmatrix} K & K^2 \\ 0 & K \end{pmatrix} \longrightarrow I\text{-spr}$.

Here a subposet I' of I is said to be **a peak subposet**, *if* $I' \cap \max I = \max I'$.

Proof: The equivalence of the statements (a) - (b) is proved in [42, Section 3].

In order to prove that (a) $\Leftrightarrow$ (e) holds it is sufficient to show that I-spr is of infinite representation type if and only if there exists fully faithful exact functor $\mathrm{mod}\begin{pmatrix} K & K^2 \\ 0 & K \end{pmatrix} \longrightarrow I$-spr (as announced in [45, Theorem 2.4]). In view of [30, Lemmata 1.5 and 8.6] and the results in [45, Section 3] it is sufficient to show that for any minimal poset J presented in Table 1 there exists a pair (U, V) of peak J-spaces U and V such that (i) $\mathrm{End}(U) \cong \mathrm{End}(V) \cong K$, (ii) $\mathrm{Hom}(U, V) = \mathrm{Hom}(V, U) = 0$ and (iii) $\dim_K(\mathrm{Ext}^1_{KJ}(U, V)) = 2$. One can prove this by applying the arguments from the proof of Theorem 1.3 in [18] (see also [13]). Details of the proof will be presented in a subsequent paper. □

Tables 1. Minimal posets $\mathcal{P}_1, \dots, \mathcal{P}_{110}$ of infinite prinjective type

In the following tables the numbers marked in the diagram $\mathcal{P}_j$ are the coordinates of a vector $\boldsymbol{\mu}_{\mathcal{P}_j} \in \mathbb{N}^{\mathcal{P}_j}$ which generates the infinite cyclic abelian group

$$\operatorname{Ker} q_{\mathcal{P}_j} := \{x \in \mathbb{Z}^{\mathcal{P}_j}; q_{\mathcal{P}_j}(x) = 0\} = \mathbb{Z}\boldsymbol{\mu}_{\mathcal{P}_j}.$$

called the **kernel** of the quadratic form $q_{\mathcal{P}_j} : \mathbb{Z}^{\mathcal{P}_j} \to \mathbb{Z}$ (see [40, 15.74]).

Part 1. The infinite series $\mathcal{P}_{1,n}$, $\mathcal{P}_{2,n+1}$, $\mathcal{P}'_{2,n+1}$, $\mathcal{P}''_{2,n}$, $\mathcal{P}_{3,n}$, $\mathcal{P}'_{3,n}$, $\mathcal{P}''_{3,n+1}$, $n \geq 0$

$\mathcal{P}_{1,n}$:  (n+2 star vertices),

$\mathcal{P}_{2,n+1}$: 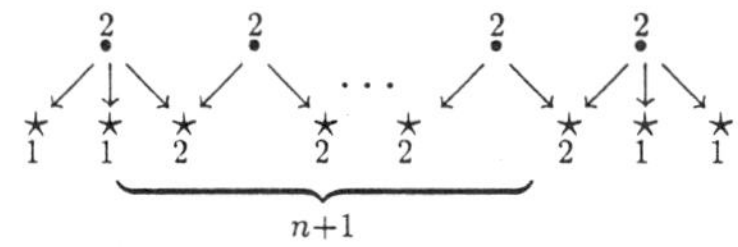 $\mathcal{P}'_{2,n+1}$: 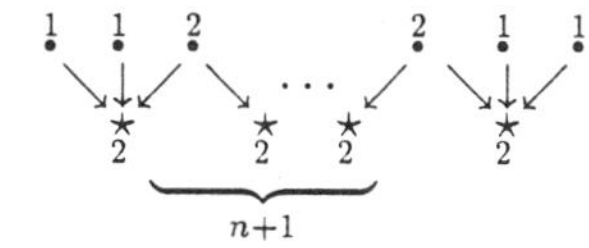

$\mathcal{P}''_{2,n}$: 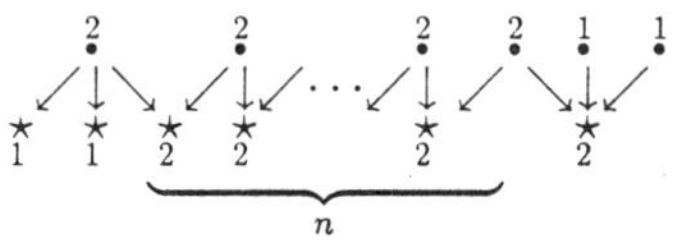 $\mathcal{P}_{3,n}$: 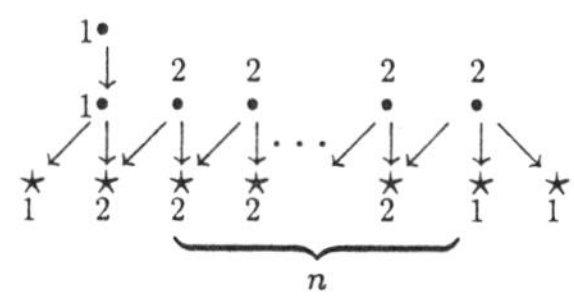

$\mathcal{P}'_{3,n}$: 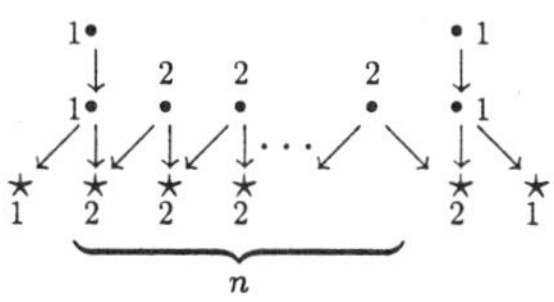 $\mathcal{P}''_{3,n}$: 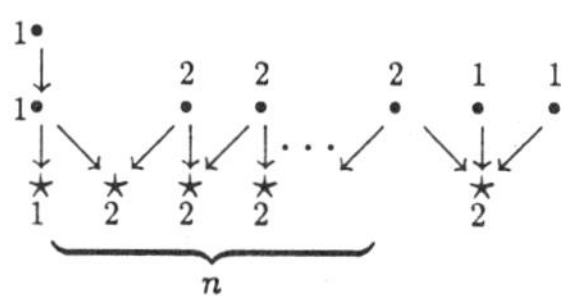

We complete the series $\mathcal{P}_{2,n+1}$ for $n = -1$ by the following four-peak poset

$\mathcal{P}_{2,0}$: 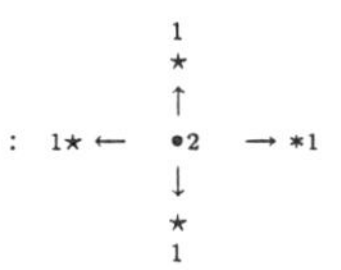 .

Part 2. One-peak enlargements $\mathcal{P}_4 - \mathcal{P}_8$ of Kleiner's posets

$\mathcal{P}_4 = \mathcal{K}_1^*$:  $\mathcal{P}_5 = \mathcal{K}_2^*$: 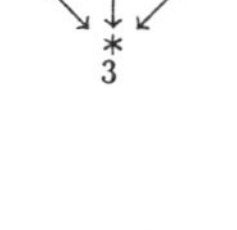 $\mathcal{P}_6 = \mathcal{K}_3^*$: 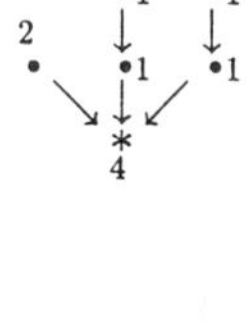

$\mathcal{P}_7 = \mathcal{K}_4^*$: 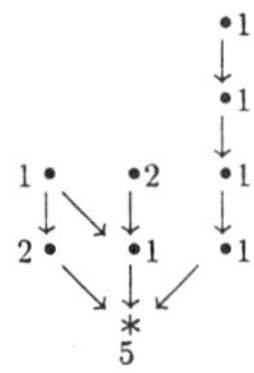 $\mathcal{P}_8 = \mathcal{K}_5^*$: 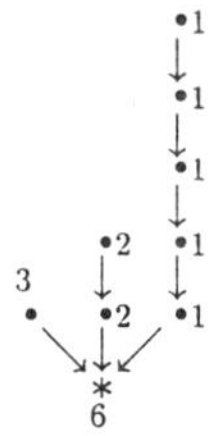

Part 3. Two-peak posets $\mathcal{P}_9 - \mathcal{P}_{31}$

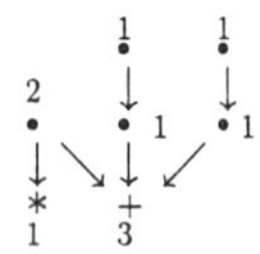 9 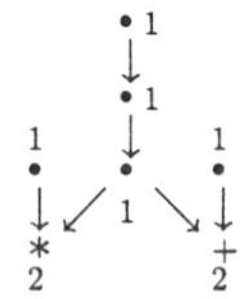 10 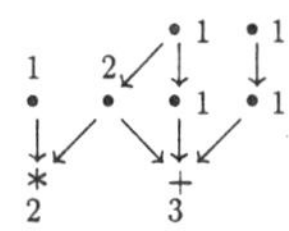 11 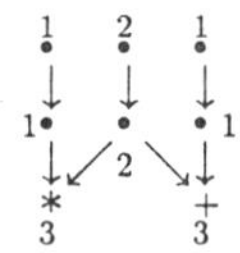 12

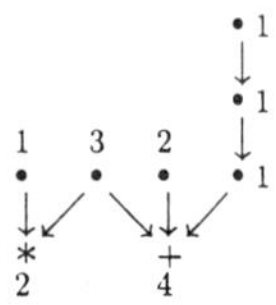 13 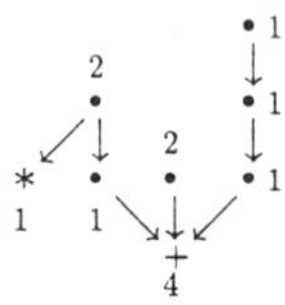 14 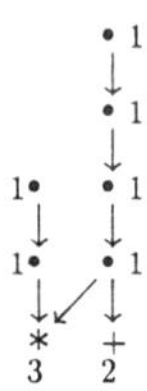 15 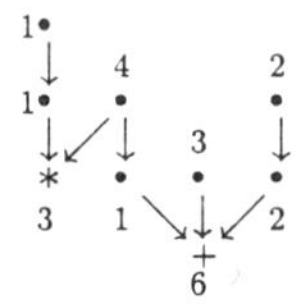 16

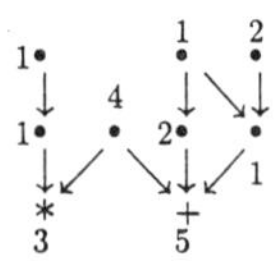 17 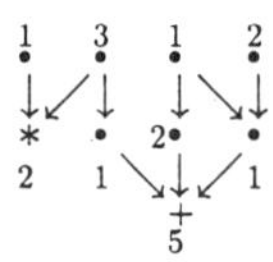 18 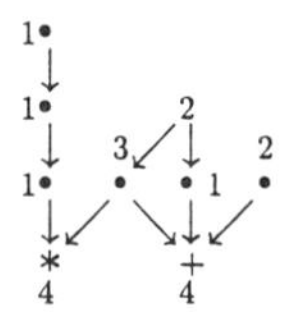 19 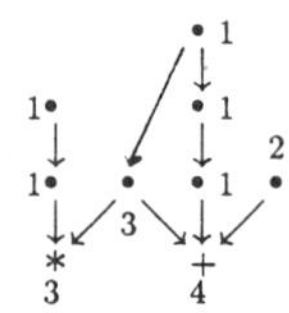 20

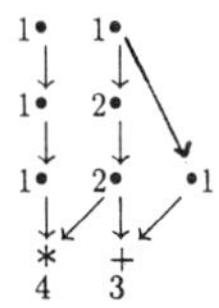 21 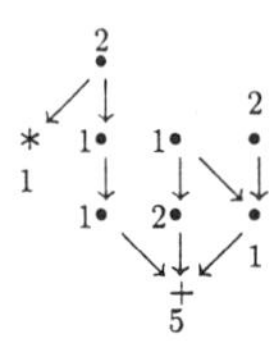 22 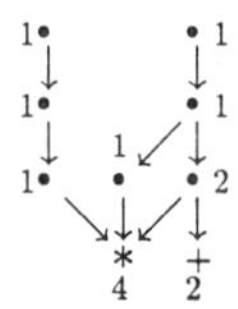 23 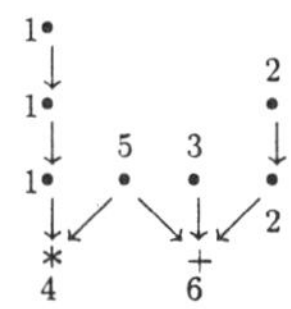 24

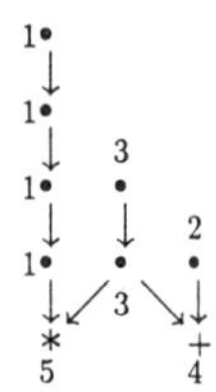 25 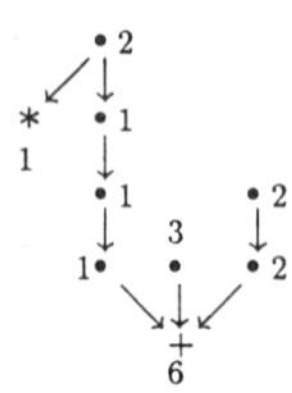 26 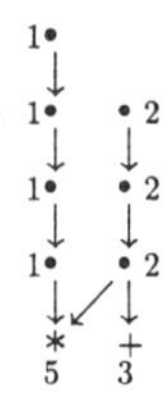 27 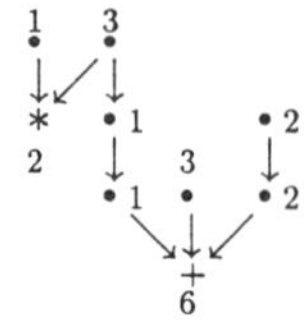 28

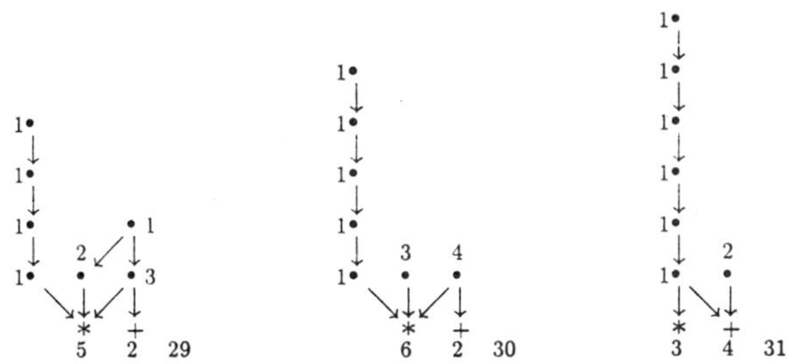

Part 4. Posets $\mathcal{P}_{32}, \dots, \mathcal{P}_{110}$ with $3 \le |\max \mathcal{P}_j| \le 5$

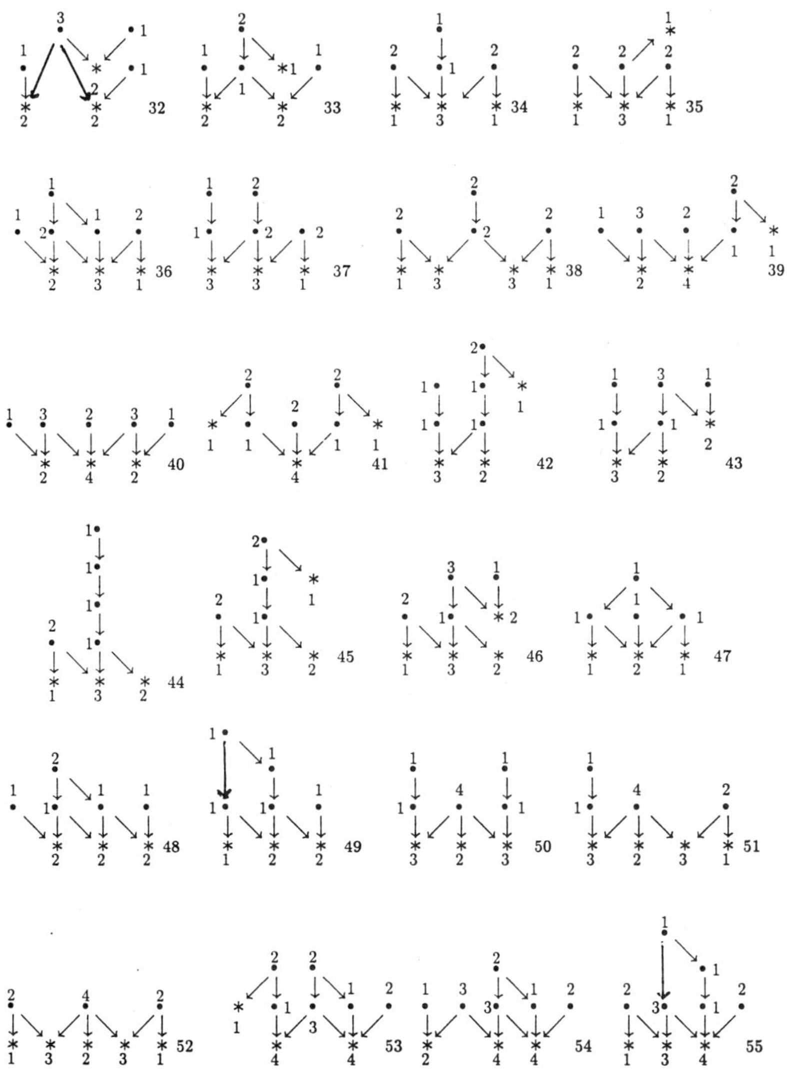

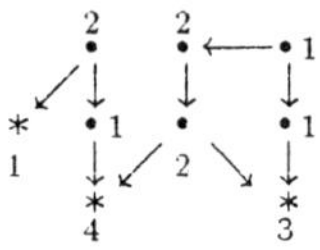
56

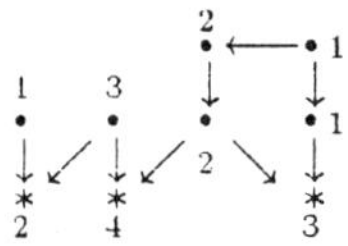
57

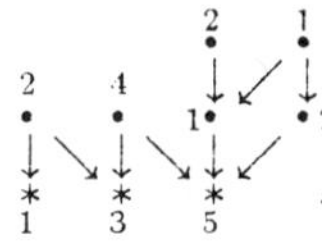
58

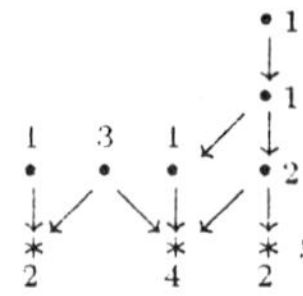
59

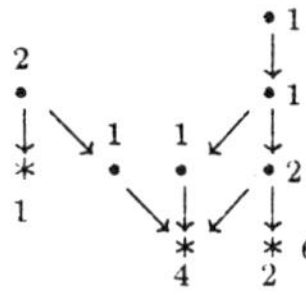
60

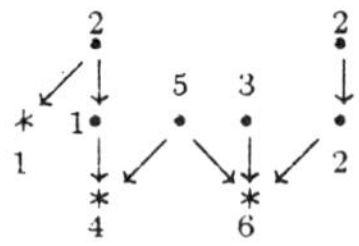
61

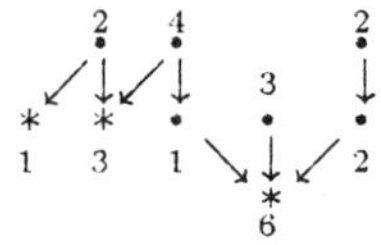
62

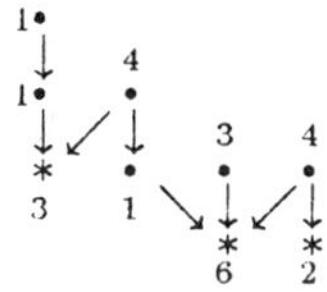
63

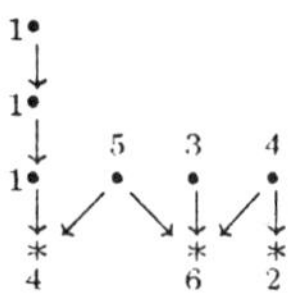
64

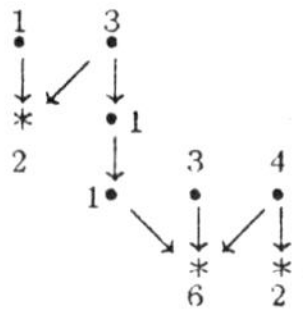
65

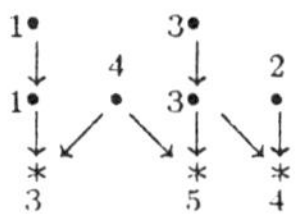
66

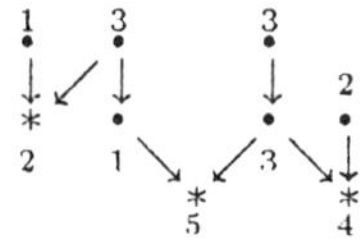
67

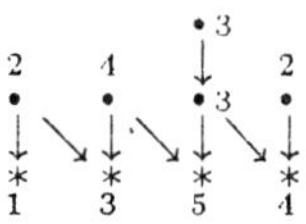
68

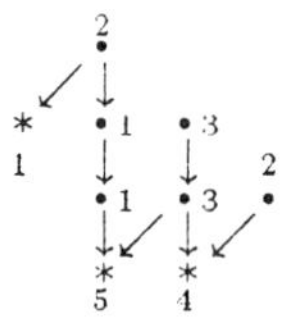
69

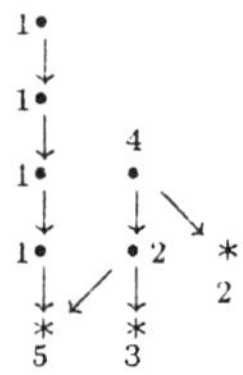
70

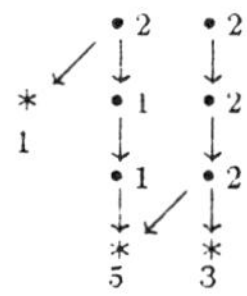
71

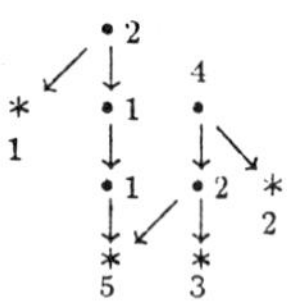
72

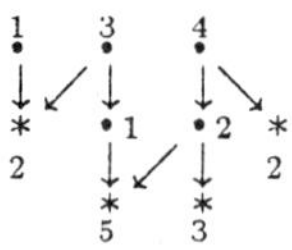
73

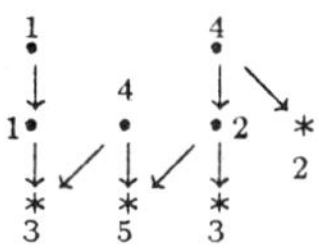
74

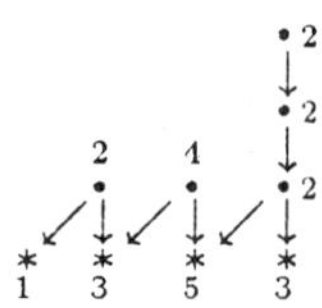
75

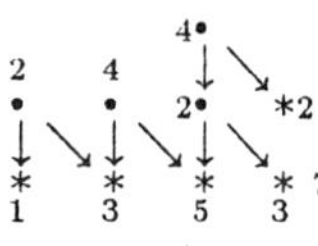
76

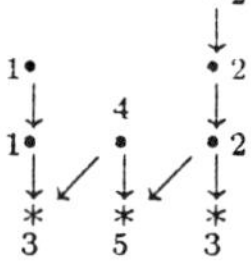
77

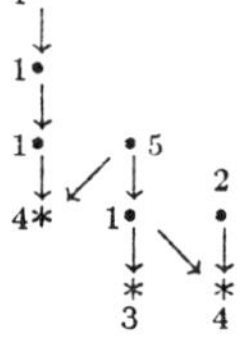
78

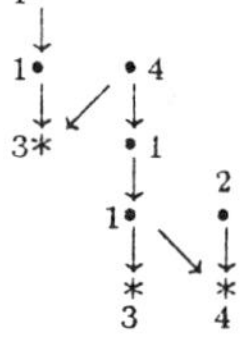
79

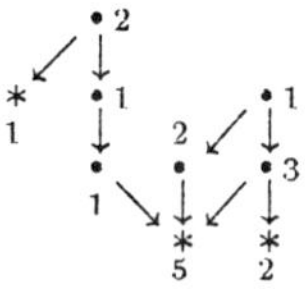
80

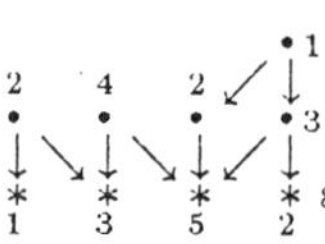
81

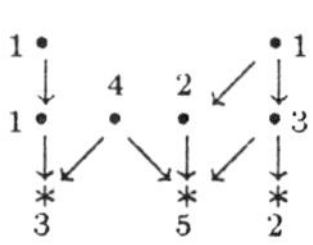
82

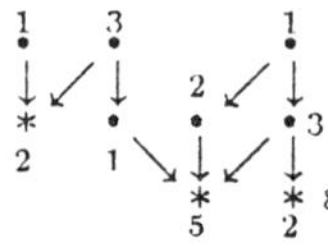
83

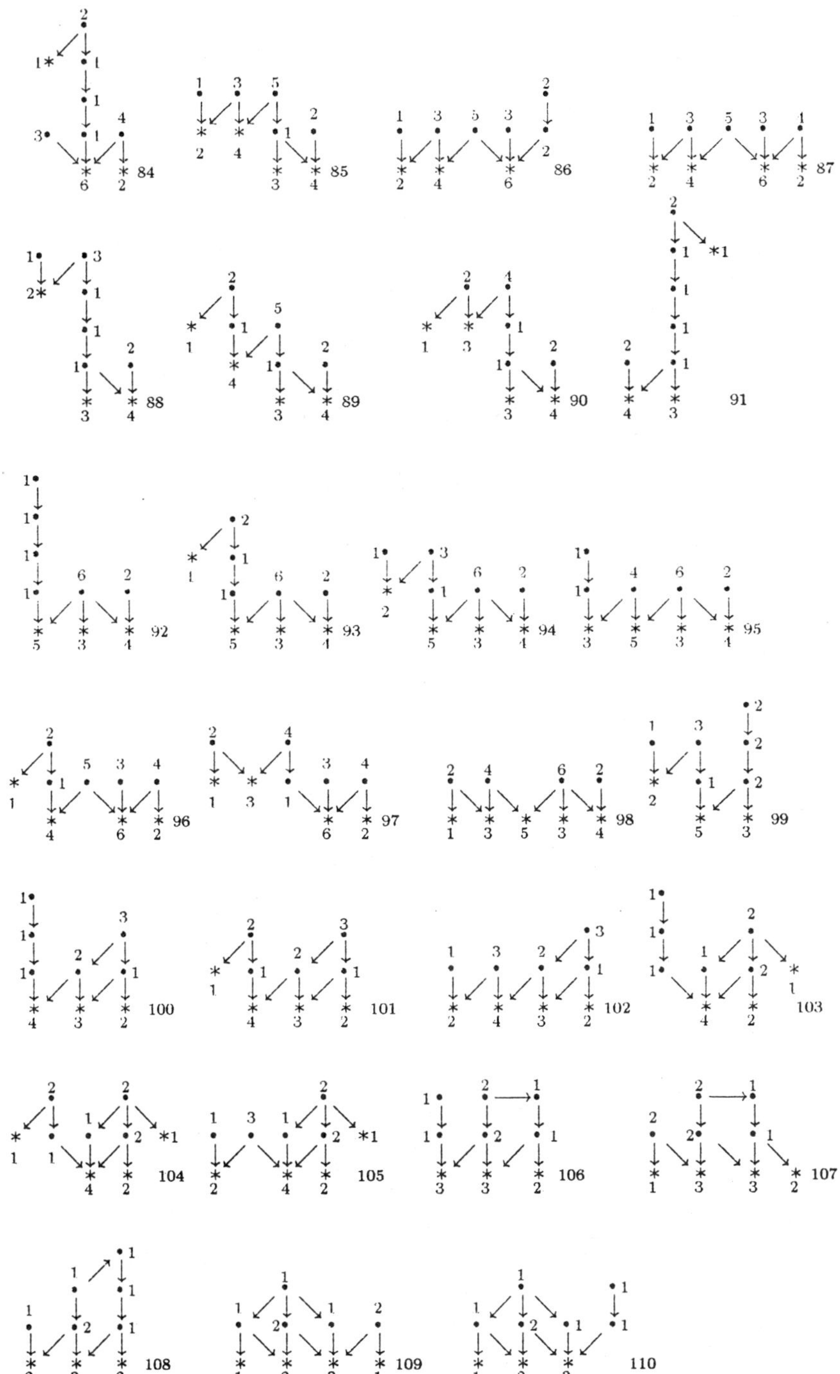
84
85
86
87
88
89
90
91
92
93
94
95
96
97
98
99
100
101
102
103
104
105
106
107
108
109
110

In case I-spr is of finite representation type there is a description of indecomposable objects in I-spr by means of the Tits quadratic form q_I. In order to present it we recall from [27] and [42] that to any peak I-space $X = (X_j)_{j\in I}$ one associates the **coordinate vector**

$$\mathbf{cdn}\, X \in \mathbb{N}^I \tag{2.6}$$

of X defined by the formula

$$\begin{aligned}(\mathbf{cdn}\, X)(j) &= \dim_K X_j & \text{for } j \in \max I,\\ &= \dim_K (X_j / \sum_{i \prec j} \pi_j(X_i)) & \text{for } j \in I \setminus \max I.\end{aligned}$$

where π_j is the composed map (2.1). We view $\mathbf{cdn}\, X$ as a map $\mathbf{cdn}\, X : I \to \mathbb{N}$. It is easy to see that every non-simple peak I-space X (viewed as a module in $\mathrm{mod}(KI)$) has a projective cover $P(X)$ of the form

$$P(X) = \bigoplus_{j \in I \setminus \max I} (e_j KI)^{(\mathbf{cdn} X)(j)}.$$

Note that if I is the poset in Example 2.2 and $X^{(m,\lambda)}$ is the peak I-space defined there, then $\mathbf{cdn}\ X^{m,\lambda} = (m, m, 0, m, m)$.

We recall that the set $\mathcal{R}^+_{q_I} = \{v \in \mathbb{N}^I;\ q_I(v) = 1\}$ is called the set of **positive roots** of the quadratic form q_I (see [40]).

Now we are able to formulate the following important addition to Theorem 2.5 (see [42, Corollary 3.2]).

THEOREM 2.7 *If I is a poset and the equivalent conditions* (a)-(e) *in Theorem* 2.5 *are satisfied then the following statements hold:*

(a) *The Auslander-Reiten translation quiver $\Gamma(I\text{-spr})$ of the category I-spr has a unique preprojective component $\mathcal{P}(I\text{-spr})$ (see* [40, Section 11.9]*). Moreover $\Gamma(I\text{-spr}) = \mathcal{P}(I\text{-spr})$.*

(b) *The map $X \mapsto \mathbf{cdn}\, X$ establishes a bijection between the isomorphism classes of indecomposable objects in I-spr and the positive roots of the quadratic form q_I which are distinct from the standard basis vectors $\varepsilon_j, j \in I \setminus \max I$ of $\mathbb{Z}^I$.*

(c) *If X is an indecomposable object in I-spr, then $\mathrm{End}(X) \cong K$, $\mathrm{Ext}^1_{KI}(X, X) = 0$ and $(\mathbf{cdn}\, X)(j) \le 6$ for all $j \in I$.*

The construction of a preprojective component $\mathcal{P}(I\text{-spr})$ of $\Gamma(I\text{-spr})$ is presented in (see [40, Section 11.9]) for the category S-sp of S-spaces, that is for the category I-spr, where I is a poset with a unique maximal element. The construction and the proof of Theorem 11.52 in [40] can be extended almost verbatim to the case $|\max I| \ge 2$, when I-spr is of finite representation type.

Since every indecomposable object X in I-spr is uniquely determined up to isomorphism by the coordinate vector $\mathbf{cdn}\, X$ of X we shall usually construct the preprojective component $\mathcal{P}(I\text{-spr})$ by means of coordinate vectors instead of indecomposable peak I-spaces. For a convenience in calculations we shall write the coordinate vector $v = \mathbf{cdn}\, V$ of the indecomposable peak I-space V in the exponential notation of coordinate vectors introduced in [40, 11.88], that is, the vector $v = \mathbf{cdn}\, V = (v_1, \ldots, v_m) \in \mathbb{N}^I$, $m = |I|$, is written in the form

$$v = \mathbf{cdn}\, V = 1^{v_1} 2^{v_2} \ldots m^{v_m}$$

where we omit j^{v_j} if $v_j = 0$, and we set $j^{v(j)} = j$ if $v(j) = 1$. We shall illustrate the construction by the following example.

EXAMPLE 2.8. Let L be the following poset

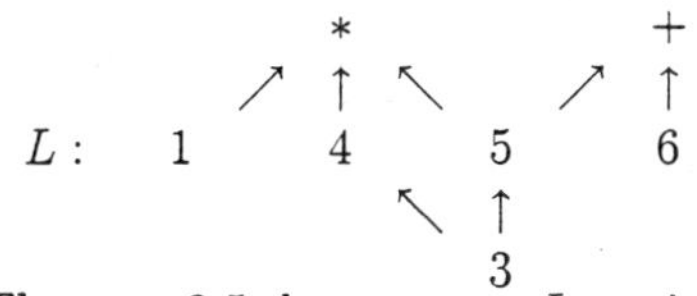

with $\max L = \{*, +\}$. By Theorem 2.5 the category L-spr is of finite representation type. By applying the preprojective component construction and Theorem 2.7 we conclude that the number of the isoclasses of indecomposable objects in L-spr is equal to 51. The Auslander-Reiten quiver $\Gamma(L\text{-spr})$ of the category L-spr is presented in Figure 1 below in the exponential notation of coordinate vectors $\mathbf{cdn}\,V$ of indecomposable peak L-spaces V. □

FIGURE 1. Auslander-Reiten quiver of L-spr

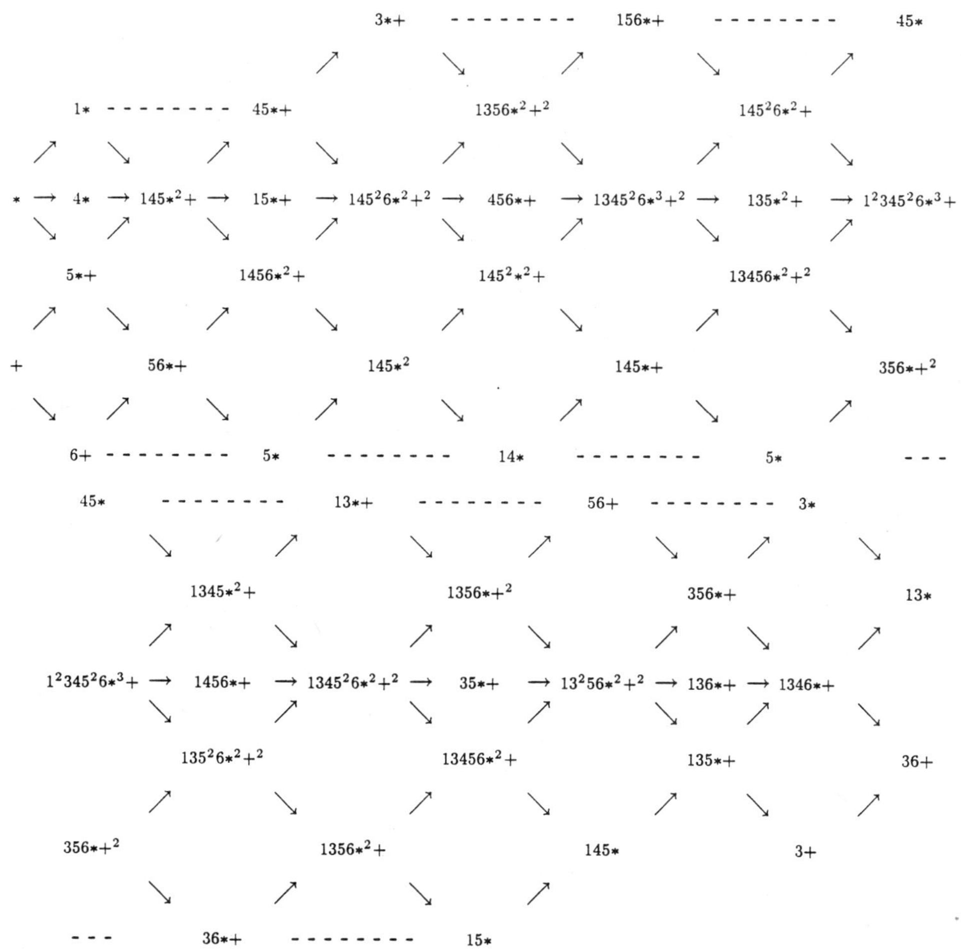

The second main result of this section is the following.

THEOREM 2.9 *Let K be an arbitrary commutative field and let I be a finite poset. The following statements are equivalent.*

(a) *The category I-spr is of fully wild representation type type.*

(b) *There exists a vector $v \in \mathbb{N}^I$ such $q_I(v) < 0$, where $\mathbb{N} = \{0, 1, 2, 3, \ldots\}$.*

(c) *There exists a pair (U, V) of peak I-spaces U and V satisfying the following three conditions*

(i) $\mathrm{End}(U) \cong \mathrm{End}(V) \cong K$,

(ii) $\mathrm{Hom}(U, V) = \mathrm{Hom}(V, U) = 0$,

(iii) $\dim_K(\mathrm{Ext}^1_{KI}(U, V)) \geq 3$.

Here we view U and V as right modules over the incidence K-algebra KI of I (see [42]).

(d) *The poset I contains as a peak subposet any of the hypercritical irreducible posets listed in* [18, Table 1] *(see also Tables 2 below), or a poset which is peak-reducible to any of the above ones (see* [17, Section 3]*).*

For the proof we refer to [18]. □

Let KI be the incidence K-algebra of I. It was shown in [42, Section 2] that I-spr is equivalent with the full subcategory $\mathrm{mod}_{\mathrm{sp}}(KI)$ of $\mathrm{mod}(KI)$ formed by all modules with a projective socle. We recall from [42] and [40, Section 14.4] that the category I-spr is of **tame representation type**) if for every $m \geq 1$ there exist functors

$$\widehat{N}^{(1)}, \ldots, \widehat{N}^{(r)} : \mathrm{mod}\, K[t] \longrightarrow \mathrm{mod}(KI)$$

such that $\widehat{N}^{(j)} = (-) \otimes_{K[t]} N^{(j)}$, $N^{(j)}$ is a $K[t]$-KI-bimodule which is finitely generated as a $K[t]$-module for $j = 1, \ldots, r$ and $\mathrm{Im}\, \widehat{N}^{(1)} \cup \ldots \cup \mathrm{Im}\, \widehat{N}^{(r)}$ contains up to isomorphism all but finitely many indecomposable objects $X = (X_j)_{j \in I}$ of I-spr $\cong \mathrm{mod}_{\mathrm{sp}}(KI)$ with $\sum\limits_{j \in I} \dim_K X_j = m$ (see also [17]).

It follows from [42, Lemma 2.1 (c)] and [17, Theorem 3.18] that if K is algebraically closed then I-spr is tame if and only if it is not wild.

In view of the results of Nazarova [24] and Kasjan-Simson [19],[21] one could expect that I-spr is tame iff it is not fully wild or equivalently, iff the Tits quadratic form q_I of I is weakly non-negative, that is, $q_I(v) \geq 0$ for any vector $v \in \mathbb{N}^I$. By Theorem 2.9, this would mean that wildness of I-spr is equivalent to the fully wildness. Unfortunately this is not the case, because of the following example due to S. Kasjan [16].

EXAMPLE 2.10 Let J be the poset

$J :$

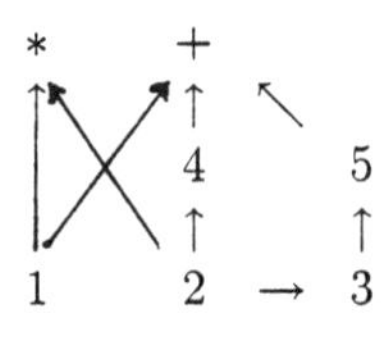

with $\max J = \{*, +\}$. It follows from Theorem 2.9 that q_J is weakly non-negative and therefore J-spr is not of fully wild representation type. On the other hand, by applying the peak reflection functor introduced in [20] or by looking at a Galois covering of the poset J

viewed as a quiver with commutativity relations (see [37], [40, Section 14.1] and [41]) one can prove that J-spr is of wild representation type. □

It is known from [27] and [17] that the tameness of I-spr implies that q_I is weakly non-negative. The above example shows that the inverse implication does not hold in general. However, Theorem 2.11 below shows that the inverse implication holds under some assumption on I.

THEOREM 2.11 *Suppose that K is an algebraically closed field, I is a finite poset having exactly two maximal elements $\star$ and $+$, and the following two conditions are satisfied:*

(i) *there is no pair a, b of incomparable elements in I such that $a \prec *$, $a \prec +$, $b \prec *$ and $b \prec +$,*

(ii) *each of the posets $I_* = \{i \in I;\ i \npreceq *\}$, $I_+ = \{i \in I;\ i \npreceq +\}$ does not contain a subposet isomorphic to one of the critical posets $\mathcal{K}_1, \dots, \mathcal{K}_5$ of Kleiner listed in Introduction.*

Then the following conditions are equivalent.

(a) *The category I-spr is of tame representation type.*

(b) *The Tits quadratic form $q_I : \mathbb{Z}^I \to \mathbb{Z}$ is weakly non-negative, that is, $q_I(v) \geq 0$ for any vector $v \in \mathbb{N}^I$.*

(c) *The poset I does not contain as a peak subposet a poset isomorphic to any of the one-peak enlargements*

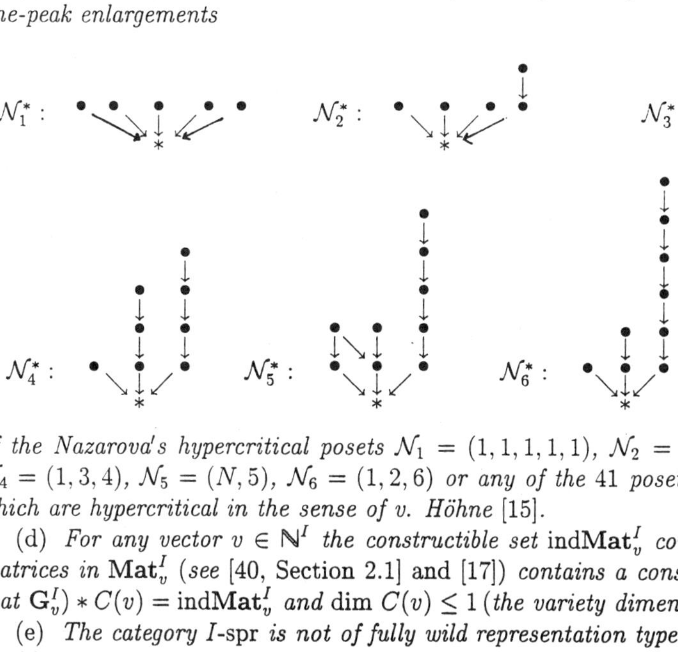

of the Nazarova's hypercritical posets $\mathcal{N}_1 = (1,1,1,1,1)$, $\mathcal{N}_2 = (1,1,1,2)$, $\mathcal{N}_3 = (2,2,3)$, $\mathcal{N}_4 = (1,3,4)$, $\mathcal{N}_5 = (N,5)$, $\mathcal{N}_6 = (1,2,6)$ or any of the 41 posets listed in Tables 2 below, which are hypercritical in the sense of v. Höhne [15].

(d) *For any vector $v \in \mathbb{N}^I$ the constructible set* $\mathrm{ind}\mathbf{Mat}_v^I$ *consisting of indecomposable matrices in* $\mathbf{Mat}_v^I$ *(see* [40, Section 2.1] and [17]) *contains a constructible subset $C(v)$ such that $\mathbf{G}_v^I) * C(v) = \mathrm{ind}\mathbf{Mat}_v^I$ and $\dim C(v) \leq 1$ (the variety dimension).*

(e) *The category I-spr is not of fully wild representation type.*

(f) *The category I-spr is not of wild representation type.*

Proof: The equivalence of the statements (a) - (c) is proved in [21]. The implication (a)⇒(f) follows from [17, Theorem 3.15], (f)⇒(e) is obvious, (e)⇒(b) is a consequence of Theorem 2.9. Since the equivalence (a)⇔(d) follows from Proposition 3.11 and Theorems 3.17, 3.18 in [17] then the conditions (a) - (f) are equivalent. □

REMARK 2.12 (a) The conditions (a), (b) and (c) in Theorem 2.11 are equivalent for every infinite field K.

(b) We hope that Theorem 2.11 remains valid without the assumption (ii).

In the following section we shall essentially use Theorem 2.11 for a characterisation of tameness of a class of D-orders Λ by means of weakly non-negativity of a reduced Tits quadratic form associated with Λ.

Tables 2. Minimal two-peak posets of wild prinjective type

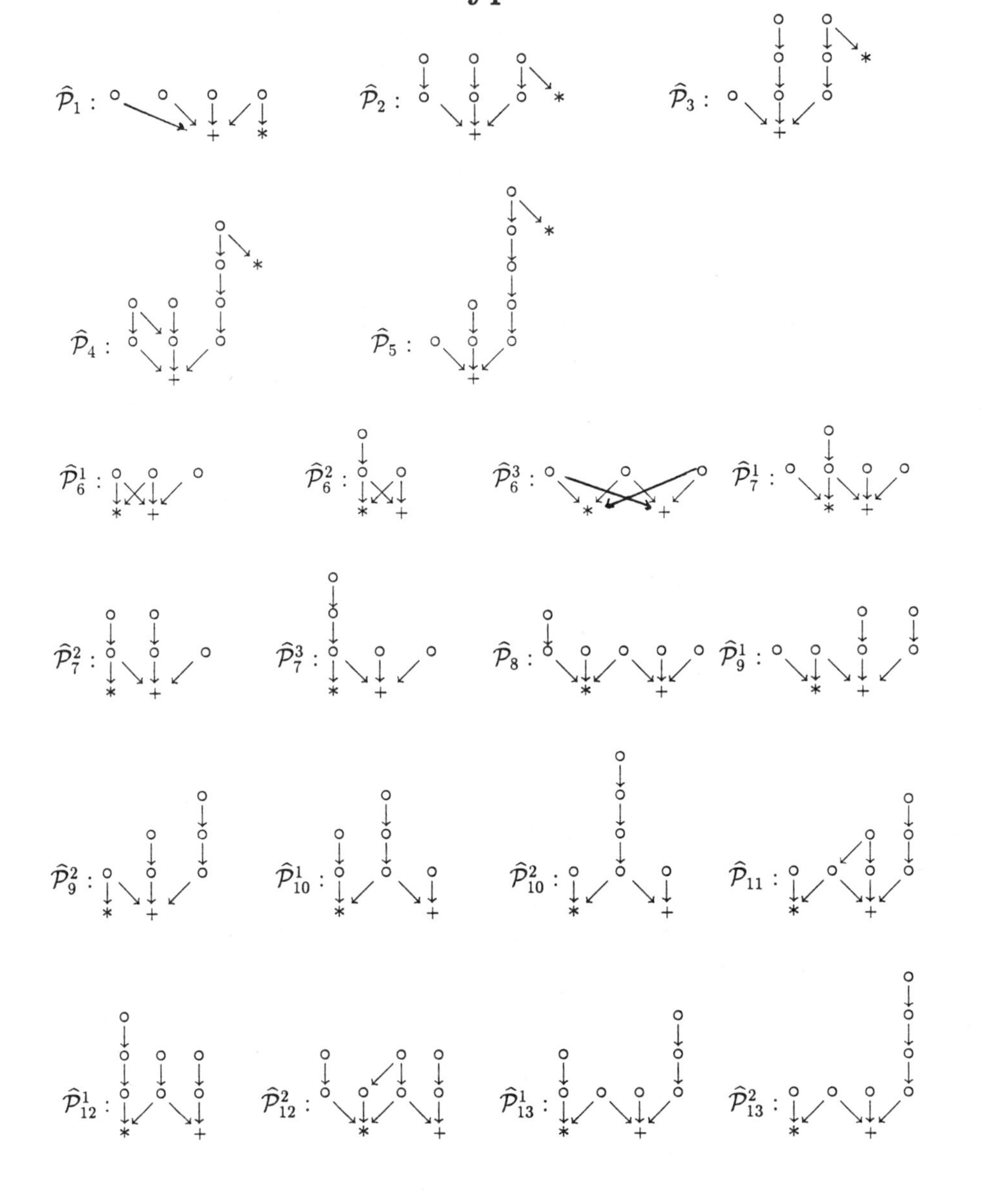

$\widehat{\mathcal{P}}^3_{13}$: 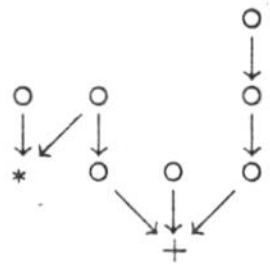

$\widehat{\mathcal{P}}_{14}$: 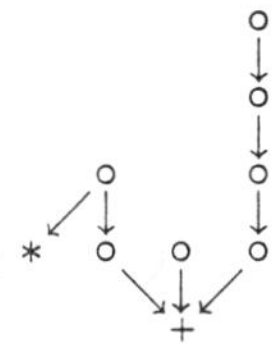

$\widehat{\mathcal{P}}^1_{15}$: 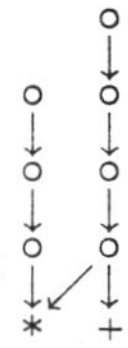

$\widehat{\mathcal{P}}^2_{15}$: 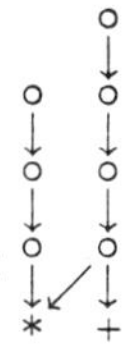

$\widehat{\mathcal{P}}^1_{16}$: 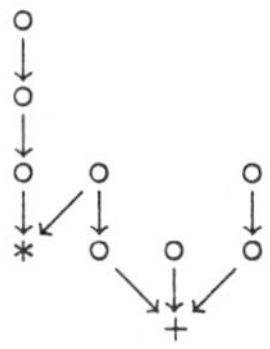

$\widehat{\mathcal{P}}^2_{16}$: 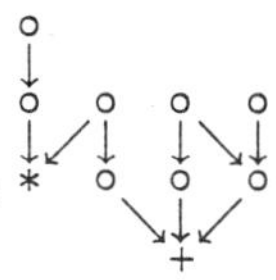

$\widehat{\mathcal{P}}^3_{16}$: 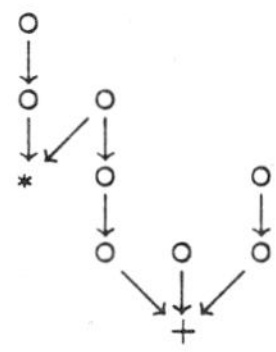

$\widehat{\mathcal{P}}_{17}$: 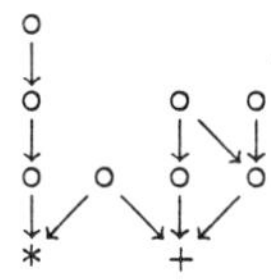

$\widehat{\mathcal{P}}_{18}$: 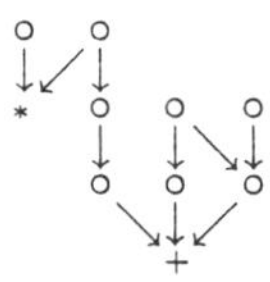

$\widehat{\mathcal{P}}_{19}$: 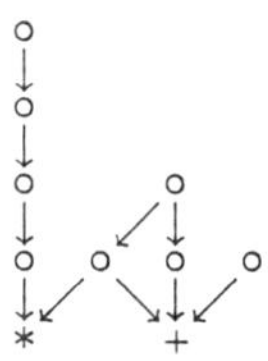

$\widehat{\mathcal{P}}_{23}$: 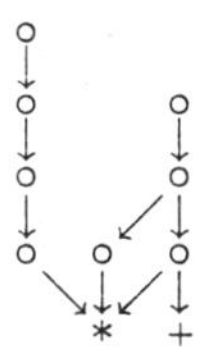

$\widehat{\mathcal{P}}^1_{24}$: 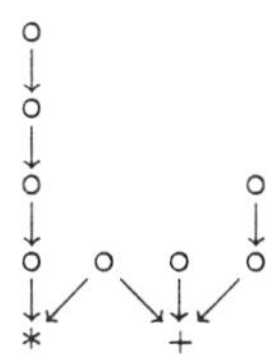

$\widehat{\mathcal{P}}^2_{24}$: 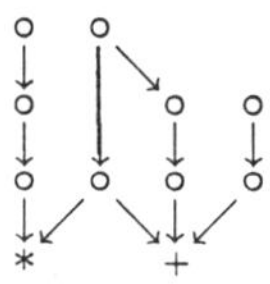

$\widehat{\mathcal{P}}^1_{25}$: 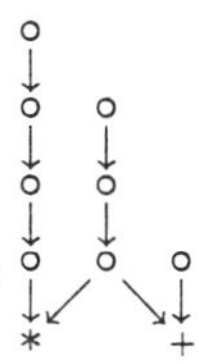

$\widehat{\mathcal{P}}^2_{25}$: 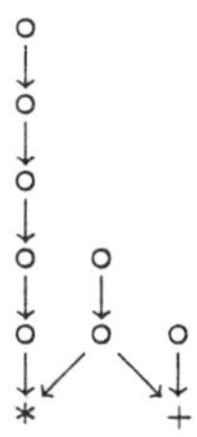

$\widehat{\mathcal{P}}_{26}$: 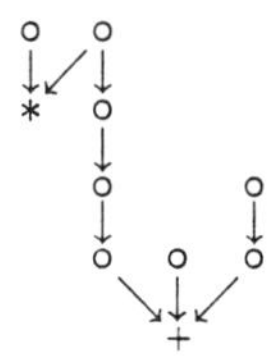

$\widehat{\mathcal{P}}_{27}$: 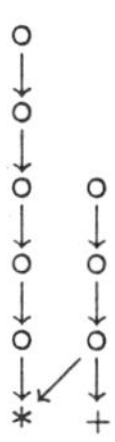

$\widehat{\mathcal{P}}_{29}$: 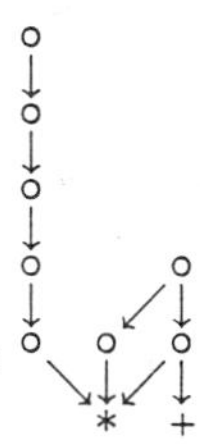

$\widehat{\mathcal{P}}_{30}$: 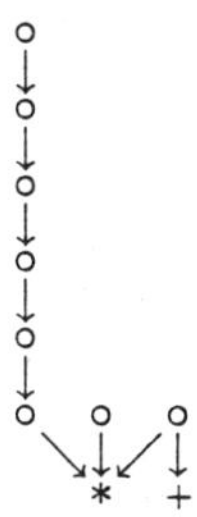

$\widehat{\mathcal{P}}_{31}$: 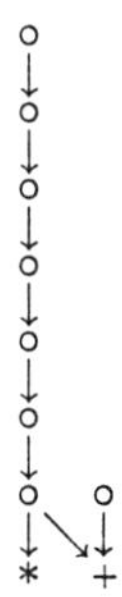

3 A class of orders of tame lattice type

Throughout this section we assume that D is a a complete discrete valuation domain with the field of fractions $F = D_0$. We denote by $\mathfrak{p}$ the unique maximal ideal of D and we set $K = D/\mathfrak{p}$. We also assume that D is a K-algebra.

A D-order Λ is said to be of **tame lattice type** if the indecomposable Λ-lattices of any fixed D-rank form a finite set of at most one-parameter families. A more precise definition can be found in [9], [40, Section 15.12], [43, Definition 3.1], [46, Section 7].

We fix a natural number $n \geq 2$ and we suppose that Λ_0 is a D-suborder of $\mathbb{M}_n(D)$ of the form

$$\Lambda_0 = \begin{pmatrix} D & D_{12} & \dots & D_{1n} \\ \mathfrak{p} & D & \dots & D_{2n} \\ \vdots & \vdots & & \vdots \\ \mathfrak{p} & \mathfrak{p} & \dots & D_{n-1\,n} \\ \mathfrak{p} & \mathfrak{p} & \dots & D \end{pmatrix} \quad \subseteq \quad \mathbb{M}_n(D)$$

where D_{ij} is either D or $\mathfrak{p}$. We denote by $J(\Lambda_0)$ the Jacobson radical of Λ_0.

We define Λ' to be the pull-back D-suborder

$$\Lambda' = \Lambda_0 \times_{\Lambda_0/J(\Lambda_0)} \Lambda_0 = \{(\lambda_1, \lambda_2) \in \Lambda_0 \times \Lambda_0;\ \lambda_1 - \lambda_2 \in J(\Lambda_0)\}$$

of $\Lambda_0 \times \Lambda_0 \subseteq \mathbb{M}_n(D_0) \times \mathbb{M}_n(D_0)$, and following [44] we define a D-suborder Λ of

$$\Gamma_0 = \begin{pmatrix} \Lambda_0 & \mathbb{M}_n(D) \\ \mathbb{M}_n(\mathfrak{p}) & \Lambda_0 \end{pmatrix}$$

to be the order

$$\Lambda = \begin{pmatrix} \Lambda_0 & \mathbb{M}_n(D) \\ \mathbb{M}_n(\mathfrak{p}) & \Lambda_0 \end{pmatrix} = \left\{ \left(\begin{smallmatrix} \lambda' & x \\ y & \lambda'' \end{smallmatrix}\right) \in \Gamma_0;\ \ \lambda' - \lambda'' \in \mathbb{M}_n(\mathfrak{p}) \right\} \quad \subseteq \quad \mathbb{M}_{2n}(D) \tag{3.1}$$

We associate to Λ the quadratic form

$$q_\Lambda : \mathbb{Z}^{n+2} \longrightarrow \mathbb{Z} \tag{3.2}$$

in the indeterminates $x_1, \dots, x_n, x_*, x_+$ defined by the formula

$$q_\Lambda(x_1, \dots, x_n, x_*, x_+) \ = \ \sum_{i=1}^{n} x_i^2 + x_*^2 + x_+^2 + \sum_{\substack{D_{ij}=D \\ i \neq j}} x_i x_j - (x_1 + \dots + x_n)(x_+ + x_+).$$

We call q_Λ the **reduced Tits quadratic form** of the order Λ.

Our main results of this section are the following two theorems.

THEOREM 3.3 *Assume that D is a a complete discrete valuation domain and $K = D/\mathfrak{p}$. Let $m \geq 2$, Λ_0, Λ' and Λ be as above. Then the following conditions are equivalent.*

(a) *The D-order Λ is of finite lattice type.*

(b) *The D-order Λ' is of finite lattice type.*

(c) *The reduced Tits quadratic form q_Λ (3.2) is weakly positive.*

(d) *The D-order Λ_0 is hereditary.*

THEOREM 3.4 *Assume that K is a field and D is a K-algebra which is a complete discrete valuation domain. Let $m \geq 2$, Λ_0, Λ' and Λ be as above. Then the following conditions are equivalent.*

(a) *The D-order Λ is of tame lattice type.*
(b) *The D-order Λ' is of tame lattice type.*
(c) *The reduced Tits quadratic form (3.2) is weakly non-negative.*
(d) *The D-order Λ_0 does not contain a 3×3-minor D-order isomorphic to one of the orders*

$$\Delta_0 = \begin{pmatrix} D & \mathfrak{p} & \mathfrak{p} \\ \mathfrak{p} & D & \mathfrak{p} \\ \mathfrak{p} & \mathfrak{p} & D \end{pmatrix}, \quad \Delta_1 = \begin{pmatrix} D & \mathfrak{p} & D \\ \mathfrak{p} & D & \mathfrak{p} \\ \mathfrak{p} & \mathfrak{p} & D \end{pmatrix}, \quad \Delta_2 = \begin{pmatrix} D & D & \mathfrak{p} \\ \mathfrak{p} & D & \mathfrak{p} \\ \mathfrak{p} & \mathfrak{p} & D \end{pmatrix}, \quad \Delta_3 = \begin{pmatrix} D & \mathfrak{p} & \mathfrak{p} \\ \mathfrak{p} & D & D \\ \mathfrak{p} & \mathfrak{p} & D \end{pmatrix}.$$

EXAMPLE 3.5 If Λ_0 is the D-order

$$\Gamma_0 = \begin{pmatrix} D & \mathfrak{p} & D & D \\ \mathfrak{p} & D & D & D \\ \mathfrak{p} & \mathfrak{p} & D & \mathfrak{p} \\ \mathfrak{p} & \mathfrak{p} & \mathfrak{p} & D \end{pmatrix}$$

then Λ and Λ' are of wild lattice type, because by omitting the second row and the third column in Λ_0 we get the 3×3-minor D-order Δ_1 and Theorem 3.4 applies. □

Outline of the proof: Following [44, Section 2] we shall associate to Λ a two-peak poset I_Λ^{*+} with two maximal elements $*, +$, and an additive reduction functor

$$\mathbb{H} : \operatorname{latt}(\Lambda) \longrightarrow I_\Lambda^{*+}\text{-spr} \cong \operatorname{mod}_{\mathrm{sp}}(KI_\Lambda^{*+}) \tag{3.6}$$

with the following properties:

(i) $\mathbb{H}$ is full, reflects isomorphisms and $\operatorname{Im}\mathbb{H}$ contains up to isomorphism all indecomposable objects of I_Λ^{*+}-spr except from the following two objects: the simple projective peak I_Λ^{*+}-space P_* corresponding to the maximal element $*$ of I_Λ^{*+}, and its injective envelope $E(P_*)$,

(ii) $\mathbb{H}$ preserves and reflects tameness, wildness and the finite representation type,

(iii) the Tits quadratic forms q_Λ and $q_{I_\Lambda^{*+}}$ coincide.

The poset I_Λ^{*+} is defined combinatorially in terms of Λ_0 in a simple way (see [44, (2.2)]). In order to construct the functor $\mathbb{H}$ we note that the order Λ is a D-suborder of the hereditary order $\Gamma = \mathbb{M}_{2n}(D)$ and $J = J(\Gamma) = J(\Lambda) = \mathbb{M}_{2n}(\mathfrak{p})$. It is not difficult to check that

$$\overline{\Lambda}_\Gamma = \begin{pmatrix} \Lambda/J & \Gamma/J \\ 0 & \Gamma/J \end{pmatrix}$$

is a finite dimensional right peak K-algebra (see [40]) with a unique simple right ideal P'_* up to isomorphism, and according to [14] and [32] the reduction functor

$$\mathbb{F} : \operatorname{latt}(\Lambda) \longrightarrow \operatorname{mod}_{\mathrm{sp}}(\overline{\Lambda}_\Gamma) \tag{3.7}$$

defined by the formula $\mathbb{F}(X) = (X/XJ, X\Gamma/XJ, u)$, is full, reflects isomorphism and $\operatorname{Im}\mathbb{F}$ contains up to isomorphism all indecomposable objects of $\operatorname{mod}_{\mathrm{sp}}(\overline{\Lambda}_\Gamma)$ except from P'_*. Here $X\Gamma$ is the Γ-submodule of $X \otimes_D D_0$ generated by X (see [29]), D_0 is the field of fractions of D and $u : X/XJ \to X\Gamma/XJ$ is the Λ/J-monomorphism induced by the natural monomorphism $X \hookrightarrow X\Gamma$. We view $\mathbb{F}(X)$ as a right $\overline{\Lambda}_\Gamma$-module in a natural way (see [14] and [32]). By the results of Section 7 in [46] the functor $\mathbb{F}$ preserves and reflects tameness, wildness and the finite representation type.

In [44, p.439] we associate to (Λ, Γ) a bipartite stratified poset

$$J_\rho = (J, \rho)$$

in the sense of [41, Definition 4.1], where J is a poset with a unique maximal element $*$ and ρ is an equivalence relation on the set $\blacktriangle J = \{(i,j) \in J \times J;\ i \preceq j\}$. We recall that the incidence K-algebra of J_ρ is the subalgebra KJ_ρ of KJ consisting of all matrices $\lambda = [\lambda_{pq}]_{p,q\in J}$ such that $\lambda_{ij} = \lambda_{st}$ if $(i,j),(s,t) \in \blacktriangle J$ and $(i,j)\rho(s,t)$. It was shown in [41] that KJ_ρ is a basic right peak algebra and the right socle of KJ_ρ is isomorphic to a direct sum of the simple projective right ideal $P'_* = e_* KJ_\rho$, called a right peak of KJ_ρ. A simple analysis shows that the algebra $\overline{\Lambda}_\Gamma$ defined above is Morita equivalent with the incidence algebra KJ_ρ and therefore there exists an equivalence of categories

$$G : \mathrm{mod_{sp}}(\overline{\Lambda}_\Gamma) \xrightarrow{\ \simeq\ } \mathrm{mod_{sp}}(KJ_\rho)$$

preserving tameness and wildness.

Let $(Q,\Omega) = (Q(J_\rho), \Omega(J_\rho))$ be the bound quiver associated canonically with J_ρ in [41, Definition 2.5]. It follows from [41, Proposition 2.8] that there exists a K-algebra isomorphism $K(Q,\Omega) \cong KJ_\rho$. Let

$$f : (\widetilde{Q}, \widetilde{\Omega}) \longrightarrow (Q, \Omega)$$

be the bound quiver Galois covering [41, (3.1)] of (Q,Ω) (see also [37, Section 4]. It follows from [41, Proposition 3.8] that f is a universal covering with the covering group $\mathbb{Z}$. Moreover, it follows from the construction that the two-peak bound subquiver J_ρ^{*+} of $(\widetilde{Q}, \widetilde{\Omega})$ associated with J_ρ in [41, (4.3)] is just the poset I_Λ^{*+} (3.2) associated with Λ. By [41, Theorem 4.19] the push down functor $f_\lambda : \mathrm{Mod}\, K(\widetilde{Q}, \widetilde{\Omega}) \to \mathrm{Mod}\, K(Q,\Omega)$ induces the push-down functor

$$\mathrm{mod_{sp}} K(\widetilde{Q}, \widetilde{\Omega}) \xrightarrow{\ f_{sp}\ } \mathrm{mod_{sp}} K(Q,\Omega) \cong \mathrm{mod_{sp}}(KJ_\rho)$$

and we get the following diagram

$$\begin{array}{ccc} \mathrm{mod_{sp}} K(Q,\Omega) & \xleftarrow{\ f_{sp}\ } & \mathrm{mod_{sp}} K(\widetilde{Q}, \widetilde{\Omega}) \\ \uparrow \cong & & \uparrow \Phi \\ \mathrm{mod_{sp}}(KJ_\rho) & \underset{f^-}{\overset{f^+}{\leftrightarrows}} & \mathrm{mod_{sp}}(KI_\Lambda^{*+}) \\ \uparrow G\circ\mathbb{F} & & \cong \uparrow T \\ \mathrm{latt}(\Lambda) & \xrightarrow{\ \mathbb{H}\ } & I_\Lambda^{*+}\text{-spr} \end{array} \tag{3.8}$$

where f^+ is the glueing functor [41, (4.14)], $\Phi = T_v \circ L_\xi$ is the embedding defined in [41, Proposition 4.23], f^- is the section functor [41, (5.1)], and in view of the identification $J_\rho^{*+} \equiv I_\Lambda^{*+}$ the functor T is the equivalence of categories $\rho : I_\Lambda^{*+}\text{-spr} \to \mathrm{mod_{sp}}(KI_\Lambda^{*+})$ defined in [42, Lemma 2.1 (b)]. We set

$$\mathbb{H} = T^{-1} \circ f^- \circ G \circ \mathbb{F}.$$

It follows from [41, Proposition 4.23] that every indecomposable module in $\mathrm{mod}_{\mathrm{sp}}K(\widetilde{Q},\widetilde{\Omega})$ is isomorphic up to a $\mathbb{Z}$-shift to a module of the form $\Phi(Y)$, where Y is an indecomposable module in $\mathrm{mod}_{\mathrm{sp}}(KJ_{\rho}^{*+})$. It was shown in [44, p. 440] that the functor $\mathbb{H}$ has the required properties.

It follows that Λ is of finite (resp. tame) lattice type if and only if the category I_{Λ}^{*+}-spr is of finite (resp. tame) representation type. Since the Tits quadratic forms q_{Λ} and $q_{I_{\Lambda}^{*+}}$ coincide Theorem 2.5 applies and proves Theorem 3.3 for the order Λ. By applying Theorem 2.9 we also get Theorem 3.4 for the order Λ (see [44, p.442] for details. The proof of Theorems 3.3 and 3.4 for the order Λ' follows by applying the arguments above to the reduction functor $T : \mathrm{latt}(\Lambda') \to I_{\Lambda}^{*+}$-spr defined in [39, 2.4] and [42, Section 4]. □

The following example illustrates the idea of the above proof.

EXAMPLE 3.9 Let D be a complete discrete valuation domain with the maximal ideal π and $D/\pi = K$. Consider the D-suborder

$$\Lambda = \left[\begin{array}{ccc|ccc} D & D & D & D & D & D \\ \pi & D & \pi & D & D & D \\ \pi & \pi & D & D & D & D \\ \hline \pi & \pi & \pi & D & D & D \\ \pi & \pi & \pi & \pi & D & \pi \\ \pi & \pi & \pi & \pi & \pi & D \end{array}\right] \subseteq \mathbb{M}_6(D)$$

of the order

$$\Gamma_0 = \begin{bmatrix} D & D & D & D & D & D \\ \pi & D & \pi & D & D & D \\ \pi & \pi & D & D & D & D \\ \pi & \pi & \pi & D & D & D \\ \pi & \pi & \pi & \pi & D & \pi \\ \pi & \pi & \pi & \pi & \pi & D \end{bmatrix} \subseteq \mathbb{M}_6(D)$$

consisting of all matrices $\gamma = (\gamma_{ij}) \in \Gamma_0$ satisfying the following relations

$$\gamma_{11} \equiv \gamma_{44}, \quad \gamma_{22} \equiv \gamma_{55}, \quad \gamma_{33} \equiv \gamma_{66}, \quad \gamma_{12} \equiv \gamma_{45}, \quad \gamma_{13} \equiv \gamma_{46}$$

modulo π. We shall show that the order Λ is of tame lattice type and there exists a reduction functor

$$\mathbb{H} : \mathrm{latt}(\Lambda) \longrightarrow I_{\Lambda}^{*+}\text{-spr}$$

preserving tameness, where I_{Λ}^{*+} is the poset

I_{Λ}^{*+} :

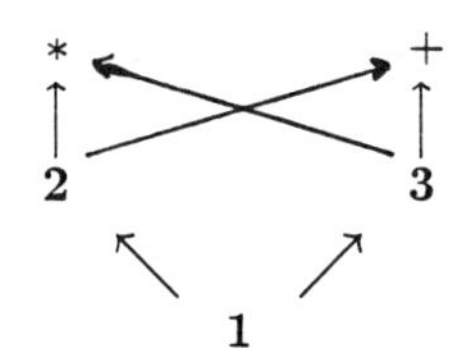

(3.10)

In order to prove tameness of $\mathrm{latt}(\Lambda)$ we note that the reduced Tits quadratic form $q_{\Lambda} : \mathbb{Z}^5 \to \mathbb{Z}$ (3.3) is the following

$$q_{\Lambda}(x_1, x_2, x_3, x_4, x_5) = \sum_{j=1}^{5} x_j^2 + x_1(x_2 + x_3) - (x_1 + x_2 + x_3)(x_4 + x_5).$$

Since q_Λ is the Tits form of the two-peak poset I_Λ^{*+}, then according to Theorem 2.9 the quadratic form q_Λ is weakly non-negative. It then follows from Theorem 3.4 that the order Λ is of tame lattice type.

The reduction functor $\mathbb{H}$ will be constructed as in the proof of Theorem 3.4 by constructing the diagram (3.8). For this purpose we note that $\Gamma = \mathbb{M}_6(D)$ is a hereditary order containing Λ and $J(\Gamma) = \mathbb{M}_6(\pi) = J(\Lambda)$. We shall associate to (Λ, Γ) the following:

(i) a bipartite stratified poset $J_{\boldsymbol{\rho}} = (J, \boldsymbol{\rho})$, where J is a one-peak poset and $\boldsymbol{\rho}$ is an equivalence relation on the set $\blacktriangle J = \{(i,j) \in J \times J;\ i \preceq j\}$ such that the algebra

$$\overline{\Lambda}_\Gamma = \begin{pmatrix} \Lambda/J(\Gamma) & \Gamma/J(\Gamma) \\ 0 & \Gamma/J(\Gamma) \end{pmatrix}$$

is Morita equivalent to the incidence K-algebra $KJ_{\boldsymbol{\rho}}$ of $J_{\boldsymbol{\rho}}$,

(ii) a bound quiver (Q, Ω) (i.e. a quiver Q together with an admissible set Ω of relations in Q [40, Chapter 14]) such that there is a K-algebra isomorphism $KJ_{\boldsymbol{\rho}} \cong K(Q,\Omega) := KQ/(\Omega)$,

(iii) a two-peak poset I_Λ^{*+} and a Galois covering $f : (\widetilde{Q}, \widetilde{\Omega}) \longrightarrow (Q, \Omega)$ (of quivers with relations) with a Galois group $\mathbb{Z}$ such that the poset I_Λ^{*+} is a subbound quiver of $(\widetilde{Q}, \widetilde{\Omega})$ and every indecomposable socle projective representation of $(\widetilde{Q}, \widetilde{\Omega})$ is isomorphic up to a $\mathbb{Z}$-shift to a representation induced from a peak I_Λ^{*+}-space.

It was shown in the proof of Theorem 3.4 that by applying (i)-(iii) we can construct the reduction functor $\mathbb{H}$ with the required properties.

For (i) we note that the ring $\overline{\Lambda}_\Gamma$ is a finite dimensional K-algebra isomorphic to the algebra

$$R' = \left[\begin{array}{ccc|ccc|cccccc}
K & K & K & K & K & K & K & K & K & K & K & K \\
0 & K & 0 & K & K & K & K & K & K & K & K & K \\
0 & 0 & K & K & K & K & K & K & K & K & K & K \\
\hline
0 & 0 & 0 & K & K & K & K & K & K & K & K & K \\
0 & 0 & 0 & 0 & K & 0 & K & K & K & K & K & K \\
0 & 0 & 0 & 0 & 0 & K & K & K & K & K & K & K \\
\hline
0 & 0 & 0 & 0 & 0 & 0 & K & K & K & K & K & K \\
0 & 0 & 0 & 0 & 0 & 0 & K & K & K & K & K & K \\
0 & 0 & 0 & 0 & 0 & 0 & K & K & K & K & K & K \\
0 & 0 & 0 & 0 & 0 & 0 & K & K & K & K & K & K \\
0 & 0 & 0 & 0 & 0 & 0 & K & K & K & K & K & K \\
0 & 0 & 0 & 0 & 0 & 0 & K & K & K & K & K & K
\end{array}\right] \subseteq \mathbb{M}_{12}(K)$$

which is Morita equivalent to the algebra

$$R'' = \left[\begin{array}{ccc|cccc}
K & K & K & K & K & K & K \\
0 & K & 0 & K & K & K & K \\
0 & 0 & K & K & K & K & K \\
\hline
0 & 0 & 0 & K & K & K & K \\
0 & 0 & 0 & 0 & K & 0 & K \\
0 & 0 & 0 & 0 & 0 & K & K \\
0 & 0 & 0 & 0 & 0 & 0 & K
\end{array}\right] \subseteq \mathbb{M}_7(K).$$

It is easy to see that R'' is isomorphic with the incidence K-algebra KJ_{ρ} of the bipartite stratified poset $J_{\rho} = (J, \rho)$, where J is the one peak poset

$J:$

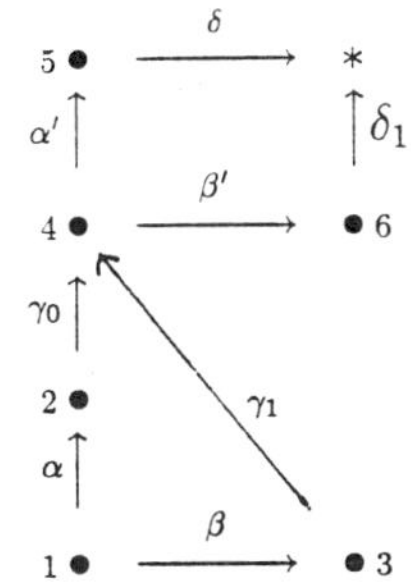

The relation ρ is defined by the formulas $2\rho 5$, $1\rho 4$, $3\rho 6$, $\alpha\rho\alpha'$ and $\beta\rho\beta'$, where we identify the pair (j, j) with the point j for every $j \in J$.

(ii) We define (Q, Ω) to be the bound quiver obtained from the poset J by making the identifications $1 \equiv 4$, $2 \equiv 5$, $3 \equiv 5$, $\alpha \equiv \alpha'$ and $\beta \equiv \beta'$ (see [37], [41]). This means that Q is the quiver

$Q:$

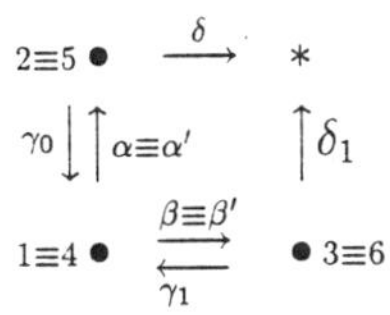

and Ω consists of the relations $\alpha\delta = \beta\delta_1$ and $\gamma_i\xi\gamma_j = 0$, where ξ is either α or β and $i, j \in \{0, 1\}$. It follows from [37] and [41, Proposition 2.8] that $KJ_{\rho} \cong KQ/(\Omega)$.

(iii) The poset I_{Λ}^{*+} was defined above. It is easy to see that the fundamental group of (Q, Ω) is isomorphic to $\mathbb{Z}$ and the universal covering $(\tilde{Q}, \tilde{\Omega})$ of (Q, Ω) is the following (see [41] and [37, Section 4])

$(\tilde{Q}, \tilde{\Omega}):$

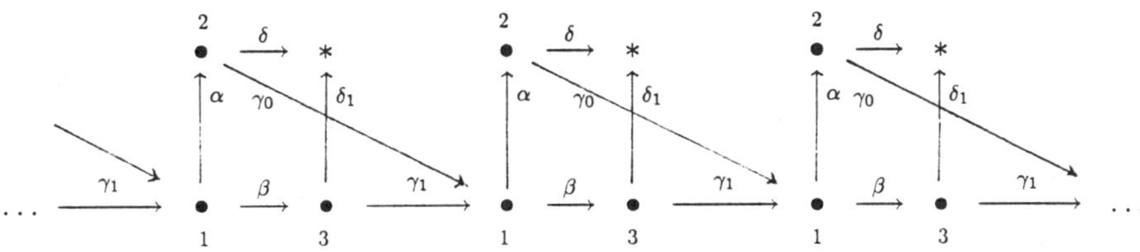

where $\tilde{\Omega} = \Omega$. The group $\mathbb{Z}$ acts on $(\tilde{Q}, \tilde{\Omega})$ by shift and the quotient bound quiver $(\tilde{Q}, \tilde{\Omega})/\mathbb{Z}$ is isomorphic to (Q, Ω).

Let J_{ρ}^{*+} be the subbound quiver of $(\tilde{Q}, \tilde{\Omega})$ formed by the points $1, 2, 3, *$ of the left hand section and by the point $*$ in the central section. It is clear that $J_{\rho}^{*+} \cong I_{\Lambda}^{*+}$ and according to [41, Proposition 4.23] the conditions in (iii) are satisfied.

Note that if I_{Λ}^{*+} is the poset (3.10) and $L = \left\{ \begin{smallmatrix} *' & \mathbf{2} & & \mathbf{3} & +' \\ & \nwarrow & & \nearrow & \\ & & \mathbf{1} & & \end{smallmatrix} \right\}$, then the study of the category I_{Λ}^{*+}-spr reduces to the study of L-sp via the functor $\mathbb{T} : I_{\Lambda}^{*+}\text{-spr} \longrightarrow L\text{-sp}$ defined by attaching to $X = (X_1, X_2, X_3, X_*, X_+)$ in I_{Λ}^{*+}-spr the L-space $X' = (X'_1, X'_2, X'_3, X'_{*'}, X'_{+'}; X')$, where $X' = X_2 + X_3 \subseteq X_* \oplus X_+$, $X'_1 = X_1$, $X'_2 = X_2$, $X'_3 = X_3$, $X'_{*'} = \mathrm{Ker}\,(X' \hookrightarrow$

$X_* \oplus X_+ \xrightarrow{\pi_*} X_*)$, $X'_{+'} = \mathrm{Ker}\,(X' \hookrightarrow X_* \oplus X_+ \xrightarrow{\pi_+} X_+)$ and π_*, π_+ are the natural projections.

Let us finish this section by a description of the Auslander-Reiten quiver of latt(Λ), where Λ is a D-order of the form (3.1) of tame lattice type.

THEOREM 3.11. *Assume that Λ is a D-order* (3.1) *of tame lattice type and let* $\mathbb{H}$: latt$(\Lambda) \to I_\Lambda^{*+}$-spr *be the reduction functor* (3.6).

(a) *There exist a preprojective component $\mathcal{P}$ in I_Λ^{*+}-spr and a preinjective component $\mathcal{I}$ in I_Λ^{*+}-spr (see Section* 11.9 in [40]). *The Auslander-Reiten quiver $\Gamma(I_\Lambda^{*+}$-spr) of I_Λ^{*+}-spr has the form shown in Figure 2.*

FIGURE 2

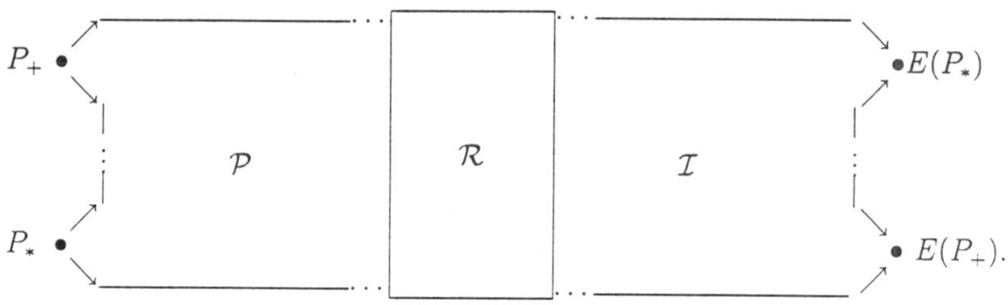

(b) *By applying the functor* $\mathbb{H}$, *the Auslander-Reiten quiver* $\Gamma(\mathrm{latt}(\Lambda))$ *of* latt(Λ) *can be constructed form* $\Gamma(I_\Lambda^{*+}$-spr) *by the identification of the simple projective module $P_* = e_* K I_\Lambda^{*+}$ with the injective envelope $E(P_+)$ of the simple projective module $P_+ = e_+ K I_\Lambda^{*+}$, and the identification of $E(P_*)$ with P_+ shown in Figure 2.*

(c) *If Λ is of finite lattice type then $\mathcal{R}$ is empty,* I^{*+}-spr $= \mathcal{P} = \mathcal{I}$ *and the map $X \mapsto$* $\mathbf{cdn}\,\mathbb{H}(X)$ *establishes a one to one correspondence between the isomorphism classes of indecomposable modules in* latt(Λ) *and the positive roots of the quadratic form*

$$q_\Lambda : \mathbb{Z}^{n+2} \to \mathbb{Z}$$

(3.2) *which are distinct from the roots $\varepsilon_1, \ldots, \varepsilon_n, \varepsilon_*, \widehat{\varepsilon}_*$, where $\varepsilon_i(j) = 1$ for $j = i$ and $\varepsilon_i(j) = 0$ for $j \neq i$, and $\widehat{\varepsilon}_*(j) = 1$ if $j = *$ or if j is a minimal element of the poset $*^\nabla$, and $\widehat{\varepsilon}_*(j) = 0$ otherwise.*

For the proof we refer to [44, Corollary 3.2]. Note also that the arguments given in [40, Corollary 11.96] prove that every indecomposable object X in the preprojective component of I_Λ^{*+}-spr is uniquely determined up to isomorphism by its coordinate vector $\mathbf{cdn}(X)$, $\mathrm{Ext}_{KI_\Lambda^{*+}}(X, X) = 0$ and $\mathrm{End}(X) \cong K$. Therefore $\mathcal{P}$ can be easily constructed by the method presented in [42, 11.57 and 11.62]. This together with Theorem 3.10 (b) gives a simple procedure for constructing the Auslander-Reiten quiver $\Gamma(\mathrm{latt}(\Lambda))$ of latt(Λ) in case Λ is of finite lattice type, because $\Gamma(I_\Lambda^{*+}$-spr$) = \mathcal{P}$.

The reader is referred to [44] for a generalisation of Theorems 3.3, 3.4, 3.11 and for details and examples (see also [41, Section 6]).

4 Tiled orders and associated covering posets

Let D be a complete discrete valuation domain with the field of fractions $F = D_0$. Let $\mathfrak{p}$ be the unique maximal ideal of D and $K = D/\mathfrak{p}$.

By a **tiled** D**-order** we mean the order of the form

$$\Lambda = \begin{pmatrix} D & \mathfrak{p}^{m_{12}} & \cdots & \mathfrak{p}^{m_{1n}} \\ \mathfrak{p}^{m_{21}} & D & \cdots & \mathfrak{p}^{m_{2n}} \\ \vdots & \vdots & \ddots & \vdots \\ \mathfrak{p}^{m_{n1}} & \mathfrak{p}^{m_{n2}} & \cdots & D \end{pmatrix} \subseteq \mathbb{M}_n(D_0) \tag{4.1}$$

in the simple matrix algebra $A = \mathbb{M}_n(D_0)$, where $m_{ij} \in \mathbb{Z}$ are integers $m_{ij} \in \mathbb{Z}$ such that $m_{ij} + m_{jt} \geq m_{it}$ for all $i, j, t = 1, \ldots, n$. We set $p^0 = D$.

Throughout we suppose that Λ is basic, or equivalently, the inequality $m_{ij} + m_{ji} > 0$ holds for all $i, j = 1, \ldots, m$.

Following Zavadskij and Kirichenko [48] we associate to any basic tiled order Λ of the form (4.1) an infinite poset $\mathbb{I}(\Lambda)$ consisting of all indecomposable projective right ideals of Λ contained in a fixed simple right ideal **S** of the simple algebra $A = \mathbb{M}_n(D_0)$. We take for the partial order in $\mathbb{I}(\Lambda)$ the inclusion.

It is not difficult to check (see [48], [40, Section 13.2]) that the poset $\mathbb{I}(\Lambda)$ is isomorphic to the $\mathbb{Z}$-graded poset

$$I(\Lambda) = \{1, ..., n\} \times \mathbb{Z} = \{1\} \times \mathbb{Z} \cup \cdots \cup \{n\} \times \mathbb{Z} \tag{4.2}$$

which is a disjoint union of n (= the number of rows in Λ) infinite chains

$$\begin{array}{ccccccccccc} \cdots & \longrightarrow & (1,-1) & \longrightarrow & (1,0) & \longrightarrow & (1,1) & \longrightarrow & (1,2) & \longrightarrow & \cdots \\ \cdots & \longrightarrow & (2,-1) & \longrightarrow & (2,0) & \longrightarrow & (2,1) & \longrightarrow & (2,2) & \longrightarrow & \cdots \\ & & \vdots & & \vdots & & \vdots & & \vdots & & \\ \cdots & \longrightarrow & (n,-1) & \longrightarrow & (n,0) & \longrightarrow & (n,1) & \longrightarrow & (n,2) & \longrightarrow & \cdots \end{array}$$

connected by the following additional relations

$$(i,t) \prec (j,s) \quad \Leftrightarrow \quad m_{ij} \leq s - t.$$

We set $\deg(j.s) = s$. The additive group $\mathbb{Z}$ of integers acts on the poset $I(\Lambda)$ by the shift $\sigma(i,t) = (i, t+1)$ as an automorphism group.

The following surprising result was proved by Zavadskij and Kirichenko [48].

THEOREM 4.3 Let D be a complete discrete valuation domain with the unique maximal ideal $\mathfrak{p}$ and $K = D/\mathfrak{p}$. A **tiled** D**-order** Λ of the form (4.1) is of finite lattice type if and only if the infinite poset $I(\Lambda)$ does not contain a poset isomorphic to any of the critical posets of Kleiner listed in the Introduction.

EXAMPLE 4.4 The D-order $\Lambda = \begin{pmatrix} D & \mathfrak{p} & \mathfrak{p}^2 \\ \mathfrak{p}^3 & D & \mathfrak{p} \\ \mathfrak{p}^2 & \mathfrak{p}^3 & D \end{pmatrix}$ is of finite lattice type, because the infinite poset $I(\Lambda)$ has the form

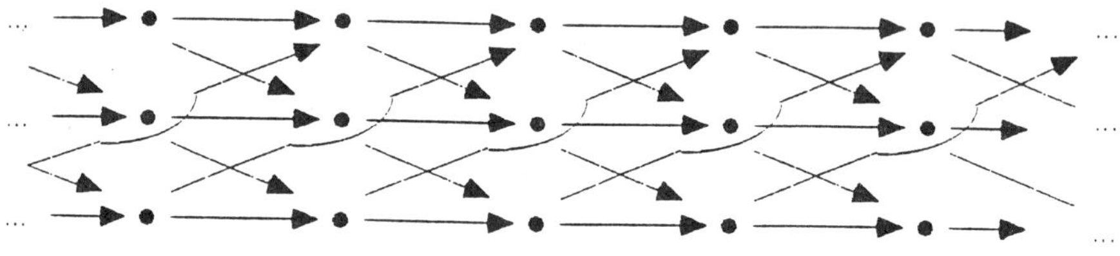

and does not contain any critical posets of Kleiner. □

In case D is arbitrary, a connection between $I(\Lambda)$-spaces and lattices over Λ is studied in the proof of the main theorem of [48] and by Rump [36, Theorems 4.2, 5.1 and 6.1]. In case $D = K[[t]]$ is a power series K-algebra there exists a covering type functor

$$\mathbb{F} : I(\Lambda)\text{-}\widetilde{\mathrm{sp}} \longrightarrow \mathrm{latt}(\Lambda) \tag{4.5}$$

defined by Roggenkamp and Wiedemann in [35] (see also [40, Section 13.3]). This is a covering type result providing us with a simple tool for the study of latt(Λ) by means of $I(\Lambda)$-spaces, because of the following result proved in [35].

THEOREM 4.5 *Let Λ be a tiled $k[[t]]$-order (4.1) and let $\widetilde{I} = I(\Lambda)$ be the infinite $\mathbb{Z}$-graded poset (4.2) associated to Λ. Denote by $\widetilde{I}\text{-}\widetilde{\mathrm{sp}}$ the full subcategory of $\widetilde{I}$-sp consisting of objects $\mathbf{M} = (M_\beta; M)$ such that $M_\beta = M$ for β sufficiently large. Then the following hold.*

(a) *If $\mathbf{M}$ is an indecomposable object in $\widetilde{I}\text{-}\widetilde{\mathrm{sp}}$, then the Λ-lattice $\mathbb{F}(\mathbf{M})$ is indecomposable and $\mathbb{F}(\mathbf{M}) \cong \mathbb{F}(\sigma\mathbf{M})$, where $\sigma\mathbf{M}$ is the shift of $\mathbf{M}$.*

(b) *The object $\mathbf{M}$ in $\widetilde{I} - \widetilde{\mathrm{sp}}$ is projective (resp. sp-injective) if and only if $\mathbb{F}(\mathbf{M})$ is projective (resp. relatively injective) in latt(Λ).*

(c) *If $\mathbf{f} : \mathbf{M} \to \mathbf{N}$ is an irreducible morphism, then $\mathbb{F}(\mathbf{f})$ is irreducible. If $\mathbb{X}$ is an Auslander-Reiten sequence in $\widetilde{I}\text{-}\widetilde{\mathrm{sp}}$, then $\mathbb{F}(\mathbb{X})$ is an Auslander-Reiten sequence in latt(Λ).*

(d) *Suppose that $\widetilde{I}$ does not contain the critical posets of Kleiner. Then the functor $\mathbb{F}$ is dense. Moreover, if $\mathbf{M}$ and $\mathbf{N}$ are indecomposable $\widetilde{I}$-spaces and $\mathbb{F}(\mathbf{M}) \cong \mathbb{F}(\mathbf{N})$, then $\mathbf{M} \cong \sigma^{\mathbf{t}}\mathbf{N}$ for some $t \in \mathbb{Z}$.*

(e) *The functor $\mathbb{F}$ induces an isomorphism of translation quivers*

$$\Gamma(\widetilde{I}\text{-}\widetilde{\mathrm{sp}})/\sigma^{\mathbb{Z}} \cong \Gamma(\mathrm{latt}\Lambda).$$

The following result was proved in [40, Section 15.12].

THEOREM 4.6 *Let K be an algebraically closed field, Λ be a tiled $K[[t]]$-order (4.1) and let $\widetilde{I} = I(\Lambda)$ be the infinite $\mathbb{Z}$-graded poset (4.2) associated to Λ. If the category latt(Λ) is of tame representation type then the poset $I(\Lambda)$ does not contain the hypercritical posets of Nazarova.*

Problem 4.7 **[40, Section 15.12]** *Does the converse of Theorem 4.6 hold?*

5 Filtered subprojective representations of a poset over a commutative artinian principal ideal ring

Assume that I is a finite poset with a unique maximal element p, i.e. $\max I = \{p\}$. We set

$$I^{-} = I \setminus \max I.$$

Assume that F is a commutative artinian principal ideal ring, $J(F)$ is the unique maximal ideal of F and $m \geq 1$ is such that $J(F)^m = 0$ and $J(F)^{m-1} \neq 0$.

It follows from the well-known Cohen structure theorem that there exist a complete discrete valuation domain D with the unique maximal ideal $\mathfrak{p}$ and ring isomorphisms

$$F \cong D/\mathfrak{p}^m, \quad \text{and} \quad F/J(F) \cong D/\mathfrak{p}. \tag{5.0}$$

DEFINITION 5.1 A **filtered subprojective F-representation** of a poset I is a system $X = (X_j)_{j\in I}$ of finitely generated F-modules satisfying the following conditions:

(a) X_p is a projective F-module,

(b) X_j is a submodule of X_p for every $j \in I$, and $X_i \subseteq X_j$, if $i \preceq j$ in I.

If in addition the following condition is satisfied

(c) X_j is a projective F-module and the embedding $X_j \hookrightarrow X_p$ splits for every $j \in I$,

we call X a **filtered projective splitting F-representation** of the poset I.

A filtered subprojective F-representation X is called **complete**, if $X_p = \sum_{j\in I^-} X_j = \sum_{j\prec p} X_j$.

By a morphism $f : X \to X'$ of filtered subprojective F-representations X and X' of the poset I we shall mean an F-homomorphism $f : X_p \to X'_p$ such that $f(X_j) \subseteq X'_j$ for every $j \in I$.

We denote by $\mathbf{fspr}(I, F)$ the category of filtered subprojective F-representations of the poset I over F, and by $\mathbf{fpr}(I, F)$ and $\widehat{\mathbf{fspr}}(I, F)$ the full subcategory of $\mathbf{fspr}(I, F)$ formed by all filtered projective splitting F-representations and all complete filtered subprojective F-representations.

LEMMA 5.2 (a) *The categories* $\mathbf{fspr}(I, F)$, $\mathbf{fpr}(I, F)$ *and* $\widehat{\mathbf{fspr}}(I, F)$ *are additive Krull-Schmidt categories.*

(b) *The category* $\mathbf{fspr}(I, F)$ *has Auslander-Reiten sequences.*

(c) *If the poset* $I^- = I \setminus \max I$ *has a unique maximal element, then the restriction functor*

$$\mathrm{res}_{I^-} : \widehat{\mathbf{fspr}}(I, F) \longrightarrow \mathbf{fspr}(I^-, F), \qquad X \mapsto (X_j)_{j\in I^-}, \tag{5.3}$$

is an equivalence of categories.

Proof: The statements (a) and (c) are obvious. The statement (b) follows from [6] and [27, Theorem 3.4], because we shall show below that the category $\mathbf{fspr}(I, F)$ is equivalent with the category of injectively cogenerated modules over the incidence F-algebra FI of I, that is, FI-modules Y such that the socle of Y is isomorphic to a direct sum of copies of the unique simple FI-module $S(p)$ corresponding to the unique maximal element p of I. □

It will be useful to have a module theory interpretation of the categories $\mathbf{fspr}(I, F)$ and $\widehat{\mathbf{fspr}}(I, F)$. For this purpose we recall the definition of the incidence F-algebra FI of I.

Convention 5.4 Without loss of generality we can assume that I consists of natural numbers $1, 2, \ldots, n, n+1$, the maximal element p of I is the number $n+1$, and the partial order $\preceq$ in I has the property that $i \prec j$ implies $i < j$ (in the natural order), for all $i, j \in I$.

The **incidence F-algebra** of the poset I is the upper triangular $I \times I$ matrix subalgebra

$$FI = \begin{pmatrix} F & {}_1F_2 & {}_1F_3 & \cdots & {}_1F_n & {}_1F_{n+1} \\ & F & {}_2F_3 & \cdots & {}_2F_n & {}_2F_{n+1} \\ & & F & \cdots & {}_3F_n & {}_3F_{n+1} \\ & & & \ddots & \vdots & \vdots \\ & 0 & & & F & {}_nF_{n+1} \\ & & & & & F \end{pmatrix} \tag{5.5}$$

of the full $(n+1) \times (n+1)$ matrix algebra $\mathbb{M}_{n+1}(F)$, where ${}_iF_j = F$ for $i \preceq j$ in I and ${}_iF_j = 0$ otherwise. Display the algebra FI in the triangular form

$$FI = \begin{pmatrix} T & {}_TN_F \\ 0 & F \end{pmatrix} \tag{5.6}$$

where $T = FI^-$ is the incidence algebra of the poset $I^- = I \setminus \max I = I \setminus \{p = n+1\}$ and

$$_TN_F = \left.\begin{bmatrix} F \\ \vdots \\ F \end{bmatrix}\right\} n$$

is viewed as an T-F-bimodule in a natural way. Right FI-modules can be identified with triples

$$X = (X'_T, X''_F, \varphi : X' \otimes_T N_F \to X''_F)$$

where X'_T is a T-module, X'_F is an F-module and φ is an F-homomorphism. Following [46, Section 2] we call X **propartite**, if X'_T is projective over T and X''_F is projective over F. The module X is said to be **complete** if φ is surjective. We denote by

$$\mathrm{mod}^{\mathrm{pr}}_{\mathrm{pr}}(FI) = \mathrm{mod}^{\mathrm{pr}}_{\mathrm{pr}}(FI)^{FI^-}_F$$

the category of finitely generated propartite right FI-modules, and by $\widehat{\mathrm{mod}}^{\mathrm{pr}}_{\mathrm{pr}}(FI)$ the full subcategory of $\mathrm{mod}^{\mathrm{pr}}_{\mathrm{pr}}(FI)$ consisting of complete modules.

Note that the ring $F = D/\mathfrak{p}^m$ is self-injective and therefore the category $\mathrm{mod}^{\mathrm{pr}}_{\mathrm{pr}}(FI)$ is the category $\mathrm{prin}(FI)^T_F$ of prinjective modules studied in [27]. The reader is referred to [46] for some properties and applications of propartite modules over arbitrary bipartite algebra.

Following [39] and [27] we call a module $X = (X'_T, X''_F, \varphi)$ in $\mathrm{mod}(FI)$ **projectively adjusted** if X''_F is F-projective and the T-homomorphism $\overline{\varphi} : X'_T \to \mathrm{Hom}_F({}_TN_F, X''_F$ adjoint to φ is injective. We denote by $\mathrm{mod}_{\mathrm{pr}}(FI) = \mathrm{mod}_{\mathrm{pr}}(FI)^T_F$ and $\widehat{\mathrm{mod}}_{\mathrm{pr}}(FI) = \widehat{\mathrm{mod}}_{\mathrm{pr}}(FI)^T_F$ the category of projectively adjusted FI-modules and complete projectively adjusted FI-modules, respectively. Consider the commutative diagram

$$\begin{array}{ccc} \widehat{\mathrm{mod}}^{\mathrm{pr}}_{\mathrm{pr}}(FI) & \hookrightarrow & \mathrm{mod}^{\mathrm{pr}}_{\mathrm{pr}}(FI) \\ \widehat{\Theta}^T\downarrow & & \downarrow\Theta^T \\ \widehat{\mathrm{mod}}_{\mathrm{pr}}(FI) & \hookrightarrow & \mathrm{mod}_{\mathrm{pr}}(FI) \\ \widehat{\rho}\uparrow\simeq & & \simeq\uparrow\rho \\ \widehat{\mathbf{fspr}}(I,F) & \hookrightarrow & \widehat{\mathbf{fspr}}(I,F) \end{array} \tag{5.7}$$

where Θ^A is the adjustment functor defined by the formula $\Theta^T(X) = (\operatorname{Im}\overline{\varphi}, X''_T, \widetilde{\varphi})$ and $\widetilde{\varphi} : \operatorname{Im}\overline{\varphi} \otimes_T N_F \to X''_F$ is the map adjoint to the inclusion $\operatorname{Im}\overline{\varphi} \hookrightarrow \operatorname{Hom}_T({}_TN_F, X''_F)$ (see [27], [38]).

The functors ρ and $\widehat{\rho}$ are defined as follows (see [40, Section 5.2]). Given a filtered subprojective F-representation $X = (X_j)_{j\in I}$ of the poset I we set

$$\rho(X) = \widehat{\rho}(X) = \bigoplus_{j\in I} X_j.$$

The F-module $\rho(X)$ is equipped with the right FI-algebra multiplication

$$\cdot : \rho(X) \times FI \to \rho(X)$$

defined by attaching to every $x = (x_j)_{j\in I} \in \oplus_{j\in I} X_j = \rho(X)$ and $\lambda = [\lambda_{ij}]_{i\preceq j\in I} \in KI \subseteq \mathbb{M}_{n+1}(F)$ the vector $x \cdot \lambda = x' = (x'_j)_{j\in I} \in \oplus_{j\in I} X_j = \rho(X)$, where $x'_i = \sum_{i\preceq j} x_i\lambda_{ij}$.

If $f : X \to X'$ is a morphism in $\mathbf{fspr}(I, F)$ defined by an F-homomorphism $f : X_p \to X'_p$, we define a homomorphism $\rho(f) = \widehat{\rho}(f) : \rho(X) \to \rho(X')$ of KI-modules by the formula $\rho(f)(x) = \widehat{\rho}(f)(x) = (f(x_i))_{i\in I}$ for every $x = (x_j)_{j\in I} \in \rho(X)$. It is easy to see that we have defined covariant additive functors ρ and $\widehat{\rho}$.

LEMMA 5.8 Let F be as above and let FI be the incidence F-algebra of a finite poset I with $\max I = \{p\}$.

(a) *The adjustment functors Θ^T and $\widehat{\Theta}^T$ in* (5.7) *are full dense, and* $\operatorname{Ker}\Theta^T = [\operatorname{pr}(T)]$, $\operatorname{Ker}\widehat{\Theta}^T = [\operatorname{pr}(T)]$, *that is,* $\operatorname{Ker}\Theta^T$ *and* $\operatorname{Ker}\widehat{\Theta}^T$ *are the two-sided ideals in* $\operatorname{mod}^{\mathrm{pr}}_{\mathrm{pr}}(FI)$ *and in* $\widehat{\operatorname{mod}}_{\mathrm{pr}}(FI)$, *respectively, consisting of all homomorphisms having a factorisation through a projective module over T viewed as a factor algebra of FI. The functors Θ^T and $\widehat{\Theta}^T$ preserve representation types.*

(b) *The functors ρ and $\widehat{\rho}$ are equivalences of categories.*

(c) *The category* $\mathbf{fspr}(I, F)$ *has enough projective objects. Every object X in* $\mathbf{fspr}(I, F)$ *is of finite length and admits a projective cover $P(X)$ in* $\mathbf{fspr}(I, F)$. *If $\{e_j\}_{j\in I}$ is the complete set of standard primitive matrix idempotents in FI, then the representations $P_j = \rho^{-1}(e_jFI)$, $j \in I$, form a complete set of indecomposable projective objects in* $\mathbf{fspr}(I, F)$ *up to isomorphism.*

Proof: The statement (a) follows from [46, Proposition 2.4].

(b) Following [40, Chapter 5] we consider the category $\operatorname{rep}(I, F)$ of all F-linear representations of the poset I, that is, systems

$$\mathbb{X} = (X_i, {}_j\varphi_i)_{i\preceq j\in I}$$

of modules X_i in $\operatorname{mod}(F)$ connected by F-homomorphisms ${}_j\varphi_i : X_i \to X_j$, $i \preceq j$, satisfying the following two conditions:

(i) ${}_i\varphi_i$ is the identity map on X_i for every $i \in I$,

(ii) $({}_s\varphi_j) \circ ({}_j\varphi_i) = {}_s\varphi_i$, if $i \preceq j \preceq s$.

A morphism from $\mathbb{X}$ to $\mathbb{X}'$ in the category $\operatorname{rep}(I, F)$ is a system $f = (f_j)_{j\in I}$ of F-homomorphisms $f_j : X_j \to X'_j$ such that $f_j \circ ({}_j\varphi_i) = ({}_j\varphi'_i) \circ f_i$ for all $i \preceq j$ in I.

By the arguments given in the proof of [40, Lemma 5.1] there exists an equivalence of categories

$$\operatorname{rep}(I, F) \cong \operatorname{mod}(FI).$$

It is easy to see that a representation $\mathbb{X}$ of $\mathrm{rep}(I, F)$ corresponds to a projectively adjusted module in $\mathrm{mod}_{\mathrm{pr}}(FI)$ if and only if $\mathbb{X}$ satisfies the conditions:

(a) the F-module X_p is projective, and

(b) the F-homomorphism ${}_p\varphi_i : X_i \to X_p$ is injective for every $i \in I$, where p is the unique maximal element of I.

A representation $\mathbb{X}$ of $\mathrm{rep}(I, F)$ corresponds to a projectively adjusted complete module in $\widehat{\mathrm{mod}}_{\mathrm{pr}}(FI)$ if and only if $\mathbb{X}$ satisfies the conditions (a), (b) and the following one

(c) $\sum_{j \prec p} {}_p\varphi_j(X_j) = X_p$.

It then easily follows as in the proof of [40, Lemma 5.1] that there exist the required equivalences of categories $\boldsymbol{\rho}$ and $\widehat{\boldsymbol{\rho}}$.

(c) Apply (b) and the arguments from the proof of [40, Lemma 5.1]. □

Let us associate to $(I, D/\mathfrak{p}^m)$ the infinite poset

$$\widehat{I_m} = \bigcup_{s \in \mathbb{Z}} I \times \{s\} \tag{5.9}$$

with the partial order relation $\preceq$ defined by the formulae:

(i) $(u, s) \preceq (v, s) \quad \Leftrightarrow \quad u \preceq v \in I$,

(ii) $(i, t) \prec (i, s)$ for all $s < t$ in $\mathbb{Z}$ and $i \in I$,

(iii) $(j, t) \prec (i, t + m)$, for all $j \not\preceq i$ in I and $t \in \mathbb{Z}$.

This means that the poset $I \times \{s\}$ is isomorphic to I and $\widehat{I_m}$ is a disjoint union of countably many copies of the poset $I \cong I \times \{s\}$, $s \in \mathbb{Z}$, with the relations (ii) and (iii). We view $\widehat{I_m}$ as follows:

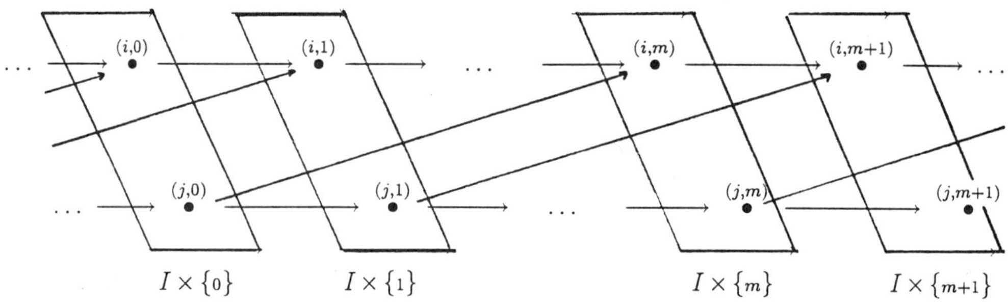

where we draw the skew arrow from (j, r) to $(i, r + m)$, if $j \not\preceq i$ in I. The infinite cyclic group $\mathbb{Z}$ acts on $\widehat{I_m}$ by shift in a natural way.

Following an idea of Ju. Drozd (communicated to me by A. G. Zavadskij) and by applying the idea presented in [10] we shall study the category $\mathbf{fspr}(I, D/\mathfrak{p}^m)$, $m \geq 1$, by means of the infinite poset $\widehat{I_m}$ as well as with a help of the category $\mathrm{latt}(\Lambda(I, D/\mathfrak{p}^m))$ lattices over the $I \times I$ tiled D-order (see Convention 5.4)

$$\Lambda(I, D/\mathfrak{p}^m) = \begin{pmatrix} D & \mathfrak{p}^{m_{12}} & \cdots & \mathfrak{p}^{m_{1n+1}} \\ \mathfrak{p}^m & D & \cdots & \mathfrak{p}^{m_{2n+1}} \\ \vdots & \vdots & \ddots & \vdots \\ \mathfrak{p}^m & \mathfrak{p}^m & \cdots & D \end{pmatrix} \subseteq \mathbf{M}_{n+1}(D) \tag{5.10}$$

where $n+1=|I|$, $\mathfrak{p}^{m_{ij}}=D$ for $i \preceq j$ in I, and $\mathfrak{p}^{m_{ij}}=\mathfrak{p}^m$ for $i \not\preceq j$ in I. We define a pair of functors

$$\begin{aligned} \mathbb{G}^- &: \operatorname{latt}(\Lambda(I^-, D/\mathfrak{p}^m)) \longrightarrow \widehat{\mathbf{fspr}}(I, D/\mathfrak{p}^m) \\ \mathbb{G} &: \operatorname{latt}(\Lambda(I, D/\mathfrak{p}^m)) \longrightarrow \mathbf{fspr}(I, D/\mathfrak{p}^m) \end{aligned} \tag{5.11}$$

as follows. We note that the hereditary order $\Gamma^- = \mathbb{M}_{n+1}(D)$ contains $\Lambda^- = \Lambda(I^-, D/\mathfrak{p}^m)$ and $\mathfrak{A} = \mathbb{M}_{n+1}(\mathfrak{p}^m)$ is a two-sided Γ^--ideal and a two-sided Λ^--ideal contained in $J(\Gamma^-) = J(\Lambda^-) = \mathbb{M}_{n+1}(\mathfrak{p})$. Let

$$\overline{\Lambda^-_{\Gamma^-}} = \begin{pmatrix} \Lambda^-/\mathfrak{A} & \Gamma^-/\mathfrak{A} \\ 0 & \Gamma^-/\mathfrak{A} \end{pmatrix}.$$

Following [10], [14] and [32] we define the reduction functor

$$\mathbb{G}_1^- : \operatorname{latt}(\Lambda(I^-, D/\mathfrak{p}^m)) \longrightarrow \widehat{\operatorname{mod}}_{\mathrm{pr}}(\overline{\Lambda^-_{\Gamma^-}}) \tag{5.12}$$

by the formula $\mathbb{G}_1^-(X) = (X/X\mathfrak{A}, X\Gamma^-/X\mathfrak{A}, u)$, where $X\Gamma^-$ is the Γ^--submodule of $X \otimes_D D_0$ generated by X (see [29]), D_0 is the field of fractions of D and $u : X/X\mathfrak{A} \to X\Gamma^-/X\mathfrak{A}$ is the $\Lambda^-/\mathfrak{A}$-monomorphism induced by the natural monomorphism $X \hookrightarrow X\Gamma^-$. We view $\mathbb{G}_1^-(X)$ as a right $\overline{\Lambda^-_{\Gamma^-}}$-module in a natural way. It follows from [10], [14] and [32] that $\mathbb{G}_1^-$ is a representation equivalence, that is, it is full dense and reflects isomorphisms.

Since $\Lambda^-/\mathfrak{A} \cong FI^-$, where $F = D/\mathfrak{p}^m$, and $\Gamma^-/\mathfrak{A} \cong \mathbb{M}_{n+1}(F)$ is Morita equivalent with F, then the algebra $\overline{\Lambda^-_{\Gamma^-}}$ is Morita equivalent with the F-algebra

$$\begin{pmatrix} FI^- & F^n \\ 0 & F \end{pmatrix} \cong FI.$$

It follows that there are equivalences of categories (see [40, Lemma 5.1 (c)])

$$\widehat{\operatorname{mod}}_{\mathrm{pr}}(\overline{\Lambda^-_{\Gamma^-}}) \xrightarrow{\simeq} \widehat{\operatorname{mod}}_{\mathrm{pr}}(FI) \xrightarrow{\simeq} \widehat{\mathbf{fspr}}(I, F) \tag{5.13}$$

We take for $\mathbb{G}^-$ the composition of $\mathbb{G}_1^-$ with the composed functor (5.13).

In order to define $\mathbb{G}$ we denote by $I^* = I \cup \{*\}$ an enlargement of the poset I by a new unique maximal element, that is, we set $i \prec *$ for every $i \in I$. We denote by

$$\mathbb{G}_1 : \operatorname{latt}(\Lambda(I, D/\mathfrak{p}^m)) \longrightarrow \widehat{\mathbf{fspr}}(I^*, D/\mathfrak{p}^m) \tag{5.14}$$

the functor $\mathbb{G}^-$ in (5.11) constructed for the poset I^*. We define $\mathbb{G}$ to be the composition of $\mathbb{G}_1$ with the equivalence $\operatorname{res}_I : \widehat{\mathbf{fspr}}(I^*, D/\mathfrak{p}^m) \xrightarrow{\simeq} \mathbf{fspr}(I, D/\mathfrak{p}^m)$ (see Lemma 5.2).

LEMMA 5.15 (a) The functors $\mathbb{G}$ and $\mathbb{G}^-$ are representation equivalences of categories preserving tameness, wildness and the finite representation type.

(b) The category $\operatorname{latt}(\Lambda(I, D/\mathfrak{p}^m))$ and $\operatorname{latt}(\Lambda(I^-, D/\mathfrak{p}^m))$ is of finite representation type (resp. tame, wild) if and only if the category $\mathbf{fspr}(I, D/\mathfrak{p}^m)$ and $\widehat{\mathbf{fspr}}(I, D/\mathfrak{p}^m)$ is of finite representation type (resp. tame, wild), respectively.

Proof: It follows from [10], [14] and [32] that $\mathbb{G}_1^-$ is a representation equivalence. Hence the first part of (a) easily follows. The remaining part follows from [9] and [46, Section 6].□

In case $F = K[[t]]/(t^m)$ we define also two functors

$$\begin{aligned} \widetilde{\mathbb{F}}^- &: \widehat{I_m^-}\text{-}\widetilde{\operatorname{sp}} \longrightarrow \widehat{\mathbf{fspr}}(I, F) \\ \widetilde{\mathbb{F}} &: \widehat{I_m}\text{-}\widetilde{\operatorname{sp}} \longrightarrow \mathbf{fspr}(I, F) \end{aligned} \tag{5.16}$$

to be the composition of the covering $\mathbb{F}$ (4.5) with the functors $\mathbb{G}^-$ and $\mathbb{G}$ in (5.11), respectively.

COROLLARY 5.17 *Assume that $F = K[[t]]/(t^m)$ and $\widehat{I_m}$ is the infinite poset (5.9). Then the functors $\tilde{\mathbb{F}}$ and $\tilde{\mathbb{F}}^-$ have the following properties.*

(a) *If* M *is an indecomposable object in* $\widehat{I_m}$-$\widetilde{\mathrm{sp}}$ *(resp. in* $\widehat{I_m^-}$-$\widetilde{\mathrm{sp}}$*), then the object* $\tilde{\mathbb{F}}(\mathrm{M})$ *(resp.* $\tilde{\mathbb{F}}^-(\mathrm{M})$*) is indecomposable and* $\tilde{\mathbb{F}}(\mathrm{M}) \cong \tilde{\mathbb{F}}(\sigma\mathrm{M})$ *(resp.* $\tilde{\mathbb{F}}^-(\mathrm{M}) \cong \tilde{\mathbb{F}}^-(\sigma\mathrm{M})$*), where* $\sigma\mathrm{M}$ *is the shift of* M.

(b) *If* M *and* N *are indecomposable objects in* $\widehat{I_m}$-$\widetilde{\mathrm{sp}}$ *(resp. in* $\widehat{I_m^-}$-$\widetilde{\mathrm{sp}}$*) and* $\tilde{\mathbb{F}}(\mathrm{M}) \cong \tilde{\mathbb{F}}(\mathrm{N})$ *(resp.* $\tilde{\mathbb{F}}^-(\mathrm{M}) \cong \tilde{\mathbb{F}}^-(\mathrm{N})$ *), then* $\mathrm{M} \cong \sigma^t\mathrm{N}$ *for some* $t \in \mathbb{Z}$.

(c) *If the category* $\mathbf{fspr}(I,F)$ *(resp.* $\widehat{\mathbf{fspr}}(I,F)$*) is of finite representation type, then the functor* $\tilde{\mathbb{F}}$ *(resp.* $\tilde{\mathbb{F}}^-$ *) is dense.*

Proof: Apply Theorem 4.5 and Lemma 5.15 (a). □

The following result in a slight different form was proved by Ju. Drozd (unpublished).

COROLLARY 5.18 *Assume that I is a poset, F is a commutative artinian principal ideal ring, $J(F)$ is the unique maximal ideal of F and $m \geq 2$ is such that $J(F)^m = 0$ and $J(F)^{m-1} \neq 0$.*

(a) *The category* $\mathbf{fspr}(I,F)$ *is of finite representation type if and only if the poset* $\widehat{I_m}$ *does not contain the critical posets* $\mathcal{K}_1, \ldots, \mathcal{K}_5$ *of Kleiner listed in the Introduction.*

(b) *The category* $\widehat{\mathbf{fspr}}(I,F)$ *is of finite representation type if and only if the poset* $\widehat{I_m^-}$ *does not contain the critical posets* $\mathcal{K}_1, \ldots, \mathcal{K}_5$ *of Kleiner.*

Proof: By the well-known Cohen structure theorem there exist a complete discrete valuation domain D with the unique maximal ideal $\mathfrak{p}$ and ring isomorphisms $F \cong D/\mathfrak{p}^m$ and $F/J(F) \cong D/\mathfrak{p}$.

By Lemma 5.15 the category $\mathrm{latt}(\Lambda(I, D/\mathfrak{p}^m))$ and $\mathrm{latt}(\Lambda(I^-, D/\mathfrak{p}^m))$ is of finite representation type if and only if the category $\mathbf{fspr}(I, D/\mathfrak{p}^m)$ and $\widehat{\mathbf{fspr}}(I, D/\mathfrak{p}^m)$ is of finite representation type, respectively. Then the theorem is a consequence of Theorem 4.3, because it is easy to see that the infinite poset (4.2) associated to $\Lambda(I, D/\mathfrak{p}^m)$ (resp. to $\Lambda(I^-, D/\mathfrak{p}^m)$) is the poset $\widehat{I_m}$ (resp. $\widehat{I_m^-}$). □

A characterisation of posets I for which the category $\mathbf{fspr}(I,F)$ is of finite representation type was given by Plahotnik [28]. By our alternative approach in Corollary 5.18 we get the following.

COROLLARY 5.19 *Assume that I is a poset, $m \geq 2$ and D is a complete discrete valuation domain with the unique maximal ideal $\mathfrak{p}$.*

(a) *If I is a poset containing a triple of pairwise incomparable elements, then the category* $\widehat{\mathbf{fspr}}(I, D/\mathfrak{p}^m)$ *is of infinite representation type.*

(b) *If I is linearly ordered, then the category* $\mathbf{fspr}(I, D/\mathfrak{p}^m)$ *is of finite representation type if and only if one of the following conditions is satisfied:*

(i) $|I| \leq 2$,

(ii) $|I| = 3$ *and* $m \leq 5$,

(iii) $4 \leq |I| \leq 5$ *and* $m \leq 3$,

(iv) $m = 2$.

Proof: (a) It is sufficient to prove (a) in case the poset $I^- = I \setminus \max I$ consists of three pairwise incomparable elements. In this case the infinite poset $\widehat{I_m^-}$ contains a subposet

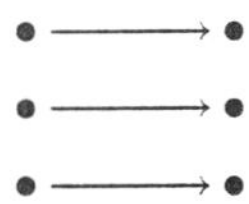

of type $(2,2,2)$, because we assume $m \geq 2$. It follows from Corollary 5.18 that the category $\widehat{\mathbf{fspr}}(I, D/\mathfrak{p}^m)$ is of infinite representation type.

(b) Assume that I is the chain $1 \longrightarrow 2 \longrightarrow \cdots \longrightarrow n \longrightarrow n+1$. Then the infinite poset $\widehat{I_m}$ is obtained from the poset

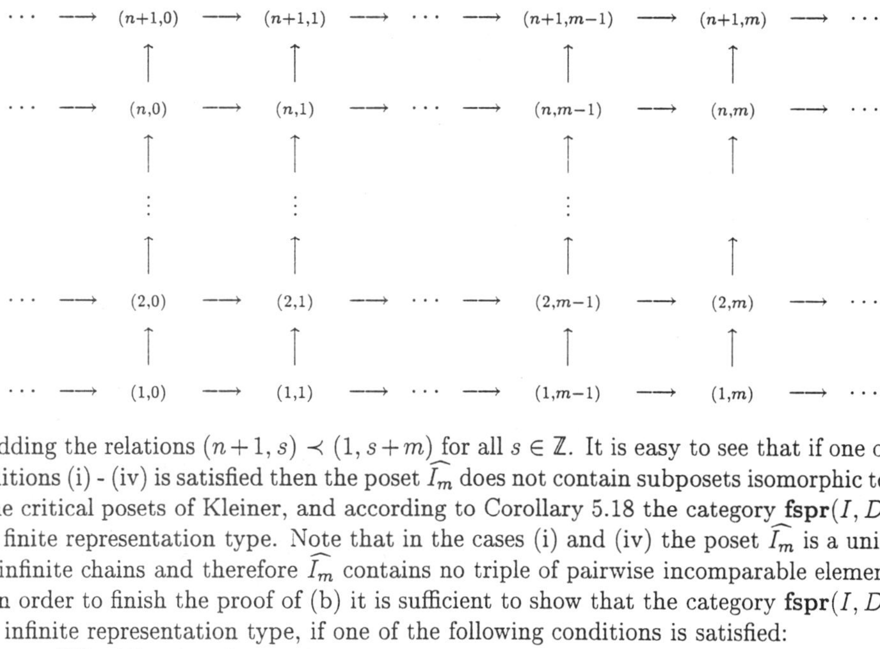

by adding the relations $(n+1, s) \prec (1, s+m)$ for all $s \in \mathbb{Z}$. It is easy to see that if one of the conditions (i) - (iv) is satisfied then the poset $\widehat{I_m}$ does not contain subposets isomorphic to any of the critical posets of Kleiner, and according to Corollary 5.18 the category $\widehat{\mathbf{fspr}}(I, D/\mathfrak{p}^m)$ is of finite representation type. Note that in the cases (i) and (iv) the poset $\widehat{I_m}$ is a union of two infinite chains and therefore $\widehat{I_m}$ contains no triple of pairwise incomparable elements.

In order to finish the proof of (b) it is sufficient to show that the category $\widehat{\mathbf{fspr}}(I, D/\mathfrak{p}^m)$ is of infinite representation type, if one of the following conditions is satisfied:

(ii′) $|I| = 3$ and $m \geq 6$,
(iii′) $|I| = 4$ and $m \geq 4$,
(iii″) $|I| = 6$ and $m \geq 3$.

It follows from the above description of the infinite poset $\widehat{I_m}$ that $\widehat{I_m}$ contains

- a critical poset of type $(2,2,2)$ in case (ii′),
- a critical poset of type $(1,1,1,1)$ in case (iii′),
- a critical poset of type $(2,2,2)$ in case (iii″),

More precisely, in case (ii′) the subposet $\{(1,4) \prec (1,5), (2,2) \prec (2,3), (3,0) \prec (3,1)\}$ of $\widehat{I_m}$ is of type $(2,2,2)$, in case (iii′) the subposet $\{(1,4), (2,3), (3,2), (4,1)\}$ of $\widehat{I_m}$ is of type $(1,1,1,1)$, and in case (iii″) the subposet $\{(1,2) \prec (2,2), (3,1) \prec (4,1), (5,0) \prec (6,0)\}$ of $\widehat{I_m}$ is of type $(2,2,2)$. In view of Corollary 5.18, this implies that the category $\widehat{\mathbf{fspr}}(I, D/\mathfrak{p}^m)$ is of infinite representation type in each of the cases (ii′) - (iv′). The proof is complete. □

COROLLARY 5.20 *Assume that I is the chain $1 \to 2 \to 3$, $m \geq 6$ and D is a complete discrete valuation domain with the unique maximal ideal $\mathfrak{p}$. Then category $\mathbf{fspr}(I, D/\mathfrak{p}^m)$ is of infinite representation type, whereas its full subcategory $\widehat{\mathbf{fspr}}(I, D/\mathfrak{p}^m)$ is of finite representation type.*

Proof: Since I is a chain then according to Lemma 5.2 there is an equivalence of categories $\widehat{\mathbf{fspr}}(I, D/\mathfrak{p}^m) \cong \mathbf{fspr}(I^-, D/\mathfrak{p}^m)$. Then the corollary follows from Corollary 5.19 (b). □

Problems 5.21 *Let I be a finite poset, D a complete discrete valuation domain and let $m \geq 2$ be an integer.*

(a) *Give a simple criterion in terms of the poset I and $m \geq 2$ for the finite representation type of the category $\mathbf{fpr}(I, D/\mathfrak{p}^m)$ of filtered projective splitting $D/\mathfrak{p}^m$-representations of I.*

(b) *Prove that the categories $\mathbf{fpr}(I, D/\mathfrak{p}^m)$ and $\widehat{\mathbf{fspr}}(I, D/\mathfrak{p}^m)$ have Auslander-Reiten sequences (see* [40, Definition 11.10]), *that is, for every indecomposable relatively non-projective object X in $\mathbf{fpr}(I, D/\mathfrak{p}^m)$ (resp. in $\widehat{\mathbf{fspr}}(I, D/\mathfrak{p}^m)$) and for every relatively non-injective object Y in $\mathbf{fpr}(I, D/\mathfrak{p}^m)$ (resp. in $\widehat{\mathbf{fspr}}(I, D/\mathfrak{p}^m)$) there exist Auslander-Reiten sequences*

$$0 \longrightarrow X'' \longrightarrow X' \longrightarrow X \longrightarrow 0$$
$$0 \longrightarrow Y \longrightarrow Y' \longrightarrow Y'' \longrightarrow 0$$

in $\mathbf{fpr}(I, D/\mathfrak{p}^m)$ (resp. in $\widehat{\mathbf{fspr}}(I, D/\mathfrak{p}^m)$).

(c) *Let $F = D/\mathfrak{p}^m$ and $K = D/\mathfrak{p}$. Then*

$$F' = D/\mathfrak{p} \oplus \mathfrak{p}/\mathfrak{p}^2 \oplus \cdots \oplus \mathfrak{p}^{m-1}/\mathfrak{p}^m \cong K[[t]]/(t^m)$$

is the graded K-algebra associated to F, and according to Corollary 5.18 the category $\mathbf{fspr}(I, F)$ is of finite representation type if and only if the category $\mathbf{fspr}(I, F')$ is of finite representation type.

Is there a nice functorial connection between the categories $\mathbf{fspr}(I, F)$ and $\mathbf{fspr}(I, F')$, provided they are of finite representation type? Are the Auslander-Reiten quivers of the categories $\mathbf{fspr}(I, F)$ and $\mathbf{fspr}(I, F')$ isomorphic? □

ACKNOWLEDGMENTS I would like to thank to J. Haefner, K.M. Rangaswamy and M. Siddoway for their invitation to Colorado Springs and for financial support.

References

[1] D. M. Arnold, Representations of partially ordered sets and abelian groups. *Contemporary Math.* 87 (1989), 91-109.

[2] D. M. Arnold and M. Dugas, Representations of finite posets and near-isomorphism of finite rang Butler groups, *Rocky Mountain J.* 25(1995), 591-609.

[3] D. M. Arnold and M. Dugas, Block rigid almost completely decomposable groups and lattices over multiple pullback rings, *J. Pure Appl. Alg.* 87(1993), 105-121.

[4] D. M. Arnold, F. Richman and C. Vinsonhaler, Representations of finite posets and valuated groups, *J. Algebra*, 155(1993), 110-126.

[5] D. M. Arnold and C. Vinsonhaler, Invariants for classes of indecomposable representations of finite posets, *J. Algebra*, 147(1992),245-264.

[6] M. Auslander and S. O. Smalø, Almost split sequences in subcategories, *J. Algebra* 69 (1981), 426-454.

[7] E. Dieterich, Tame orders, *in "Topics in Algebra, Part I: Rings and Representations of Algebras"*, Banach Center Publications, Vol. 26, PWN Warszawa, 1990, pp. 233 - 261.

[8] Ju. A. Drozd, Tame and wild matrix problems, *in "Representations and Quadratic Forms"*, Kiev 1979, 39-74 (in Russian).

[9] J. A. Drozd and M. G. Greuel, *Tame-wild dichotomy for Cohen-Macaulay modules*, *Math. Ann.* 294(1992), 387-394.

[10] J. Drozd, A. G. Zavadskij and V. V. Kirichenko, Matrix problems and integral representations, *Izv. Akad. Nauk SSSR* 38 (1974) 291 - 293 (in Russian).

[11] P. Gabriel, Indecomposable representations I, *Manuscripta Math.* 6 (1972),71-103.

[12] P. Gabriel and A. V. Rojter, *"Representations of Finite Dimensional Algebras"* Algebra VIII, Encyclopedia of Math. Sc., Vol 73, Springer-Verlag, 1992.

[13] R. Göbel and D. Simson, Endomorphism ring problem via embeddings of Kronecker modules into the category of prinjective modules, preprint, 1994.

[14] E. L. Green and I. Reiner, Integral representations and diagrams, *Michigan Math. J.* 25 (1978), 53-84.

[15] H. J. v. Höhne, On weakly non-negative unit forms and tame algebras, *Proc. London Math. Soc.* 1995, in print.

[16] S. Kasjan, *Ph. D. Thesis*, Nicholas Copernicus University, Toruń, 1996, (in Polish).

[17] S. Kasjan and D. Simson, Varieties of poset representations and minimal posets of wild prinjective type, *in* Proceedings of the Sixth International Conference on Representations of Algebras, *Canadian Mathematical Society Conference Proceedings*, Vol. 14, 1993, 245-284.

[18] S. Kasjan and D. Simson, Fully wild prinjective type of posets and their Tits form, *J. Algebra*, 172(1995), 506-529.

[19] S. Kasjan and D. Simson, Tame prinjective type and Tits form of two-peak posets I, *J. Pure Appl. Algebra* 106(1996), in press.

[20] S. Kasjan and D. Simson, Peak reduction functor for socle projective representations, J. Algebra, to appear.

[21] S. Kasjan and D. Simson, Tame prinjective type and Tits form of two-peak posets II, J. Algebra, to appear.

[22] M. M. Kleiner, Partially ordered sets of finite type, *in* Zap. Nauchn. Sem. Leningrad. Otdel. Mat. Inst. Steklov. (LOMI), 28 (1972), 32 - 41 (in Russian)..

[23] H. Lenzing and D. Simson, Cohen-Macaulay modules, the completion functor, representations of posets and rational tiled orders, in preparation.

[24] L. A. Nazarova, Partially ordered sets of infinite type, *Izv. Akad. Nauk SSSR* 39 (1975), 963 - 991 (in Russian).

[25] L. A. Nazarova and A.V. Roiter, Representations of partially ordered sets, *in* Zap. Nauchn. Sem. Leningrad. Otdel. Mat. Inst. Steklov. (LOMI), 28 (1972), 5 - 31 (in Russian).

[26] L. A. Nazarova and A.V. Roiter, Kategorielle Matrizen-Probleme und die Brauer-Thrall-Vermutung, *Mitt. Math. Sem. Giessen* 115 (1975), 1 - 153.

[27] J. A. de la Peña and D. Simson, Prinjective modules, reflection functors, quadratic forms and Auslander-Reiten sequences, *Trans. Amer. Math. Soc.* 329(1992), 733-753.

[28] V. V. Plahotnik, Representations of partially ordered sets over commutative rings, *Izv. Akad. Nauk SSSR* 40 (1976), 527 - 543 (in Russian).

[29] I. Reiner, *"Maximal Orders"*, Academic Press, London - New York - San Francisco, 1975.

[30] C. M. Ringel, Representations of K-species and bimodules, *J. Algebra*, 41 (1976), 269 - 302.

[31] C. M. Ringel, *"Tame Algebras and Integral Quadratic Forms"*, Lecture Notes in Math. Vol. 1099, Springer-Verlag, Berlin-Heidelberg-New York-Tokyo, 1984.

[32] C. M. Ringel and K. W. Roggenkamp, Diagrammatic methods in the representation theory of orders, *J. Algebra* 60(1979), 11 - 42.

[33] K. W. Roggenkamp, Indecomposable representations of orders, *in "Topics in Algebra, Part I: Rings and Representations of Algebras"*, Banach Center Publications, Vol. 26, PWN Warszawa, 1990, pp. 449 - 491.

[34] K. W. Roggenkamp and J. Schmidt, Almost split sequences for integral group rings and orders, *Comm. Algebra* 4 (1976), 893 - 917.

[35] K. W. Roggenkamp and A. Wiedemann, Auslander-Reiten quivers of Schurian orders, *Comm. Algebra* 12 (1984), 2525 - 2578.

[36] W. Rump, Enlacements and representation theory of completely reducible orders, *in* Lecture Notes in Math. No. 1178, Springer-Verlag, Berlin-Heidelberg-New York-Tokyo, (1986), pp. 272 - 308.

[37] D. Simson, Representations of bounded stratified posets, coverings and socle projective modules, *in "Topics in Algebra, Part I: Rings and Representations of Algebras"*, Banach Center Publications, Vol. 26, PWN Warszawa, 1990, pp. 499 - 533.

[38] D. Simson, Moduled categories and adjusted modules over traced rings, *Dissertationes Math.* 269 (1990), 1 - 67.

[39] D. Simson, Two-peak posets of finite prinjective type, *in* Proceedings of the Tsukuba International Conference on Representations of Finite Dimensional Algebras, *Canadian Mathematical Society Conference Proceedings,* 11(1991), pp. 287 - 298.

[40] D. Simson, *"Linear Representations of Partially Ordered Sets and Vector Space Categories"* in: Algebra, Logic and Applications, Vol. 4, Gordon & Breach Science Publishers, 1992.

[41] D. Simson, Right peak algebras of two-separate stratified posets, their Galois coverings and socle projective modules, *Comm. Algebra*, 20(1992), 3541-3591.

[42] D. Simson, Posets of finite prinjective type and a class of orders, *J. Pure Appl. Algebra* 90(1993), 77-103.

[43] D. Simson, On representation types of module subcategories and orders, *Bull. Pol. Acad. Sci. Math.* 41(1993), 77-93.

[44] D. Simson, A reduction functor, tameness and Tits form for a class of orders, *J. Algebra*, 174(1995), 430-452.

[45] D. Simson, Triangles of modules and non-polynomial growth, *C. R. Acad. Sci. Paris*, Série I, 321 (1995), 33-38.

[46] D. Simson, Prinjective modules, propartite modules, representations of bocses and lattices over orders, *J. Math. Soc. Japan* 48(1996), to appear.

[47] A. G. Zavadskij and V. V. Kirichenko, Torsion-free modules over prime rings,*in* Zap. Nauchn. Sem. Leningrad. Otdel. Mat. Inst. Steklov. (LOMI), 57(1976), 100-116 (in Russian)..

[48] A. G. Zavadskij and V. V. Kirichenko, Semimaximal rings of finite type, *Mat. Sbornik* 103 (1977), 323 - 345 (in Russian).

A Survey of Balanced Butler Groups and Representations

C. VINSONHALER Department of Mathematics, University of Connecticut, Storrs, CT 06269, E–mail: vinson@uconnvm.uconn.edu

It was the great swami, Ranga, who first made me aware of how assiduously we seek virtue in our research. Our papers are filled with terms like **pure**, **neat**, **decent** and **nice**. In this survey, I want to highlight another word in the virtuous category. I want to convince you that there should be more **balance** in your life, especially in your exact sequences. Throughout history, sages have sung the praises of this property.

"A false balance is an abomination in the eyes of the Lord." (contrapositive, Proverbs 11:1)

"Imagination reveals itself in the balance..." (Samuel Coleridge, 1817)

"Just as there must be balance in what a community produces, so there must also be balance in what the community consumes." (John Kenneth Galbraith, **The Affluent Society**, 1958)

"Balanced subgroups play an important role in the theory of torsion–free abelian groups." (Laszlo Fuchs, 1993)

Let's review some definitions. Unless otherwise obvious, "group" means torsion–free abelian group. A (finite rank) **Butler group** is a pure subgroup of a (finite rank) completely decomposable group. A **type** is an isomorphism class of subgroups of the additive rationals Q. A Butler group has a finite **typeset** (set of types of rank one pure subgroups). If the Butler group B has typeset contained in a lattice L of types, we call B an L–Butler group. An element τ in the lattice L is called join irreducible if $\tau = \tau_1 \vee \tau_2$ implies $\tau_1 = \tau$ or $\tau_2 = \tau$. If

$$E: 0 \to A \to B \to C \to 0$$

is an exact sequence of L–Butler groups, then it is easy to show that

$$E(\tau): \ 0 \to A(\tau) \to B(\tau) \to C(\tau) \to 0$$

is exact for each join–irreducible type $\tau \in L$. The sequence E is called **balanced** if $E(\tau)$ is exact for all $\tau \in L$. Since $\tau = \tau_1 \vee \tau_2$ implies $B(\tau) = B(\tau_1) \cap B(\tau_2)$, E is balanced if and only if

$$0 \to \cap_{\tau \in T} A(\tau) \to \cap_{\tau \in T} B(\tau) \to \cap_{\tau \in T} C(\tau) \to 0$$

is exact for all finite subsets T of the join irreducible elements of L. Of course there are many equivalent definitions of balanced – those sequences with respect to which the rank–one groups are projective, for example.

Now let's look at representations. A representation $V = (V; V(i): i \in S)$ of a poset S over a field k is a finite dimensional k–vector space V together with distinguished subspaces V(i) that respect the ordering in S: $i \leq j$ in S implies $V(i) \subseteq V(j)$. It will be convenient to use V for both the underlying vector space and the representation. The collection of representations becomes a category Rep(k,S) if we define the morphisms from V to W to be the linear transformations that map each V(i) into W(i). Recall that Butler defined a category equivalence from the category of L–Butler groups and quasi–homomorphisms to the category Rep(Q,S) where S is the set of join irreducible elements of L (with the opposite ordering). The correspondence is

$$B \to (QB; QB(\tau):\tau\in S)$$

Keeping this correspondence in mind we define a **type** to be an element of Rep(k,S) whose underlying vector space is k. An exact sequence of representations $E: 0 \to U \to V \to W \to 0$ is an exact sequence of vector spaces such that

$$E(i): \quad 0 \to U(i) \to V(i) \to W(i) \to 0$$

is exact for each $i \in S$. Under Butler's correspondence, E(i) corresponds to $E(\tau)$ above, when τ is join irreducible. Moreover, we are motivated to call the exact sequence E of representations **balanced** if

$$0 \to \cap_{i\in T}U(i) \to \cap_{i\in T}V(i) \to \cap_{i\in T}W(i) \to 0$$

is exact for each subset T of S. Since a subset T of S naturally determines a type (how?), we are just saying that $0 \to U(T) \to V(T) \to W(T) \to 0$ is exact for each type T, once $V(T) = \cap_{i\in T}V(i)$ is defined. It is easy to check that the balanced sequences are precisely those with respect to which the types are projective.

Our next definitions were originally given by Kravchenko [Kr], but the treatment in [NV–1] is more accessible. Denote by $\mathcal{K}(0)$ the class of Butler groups and inductively let $\mathcal{K}(n)$, $n\geq 1$, denote the class of Butler groups A that appear in a balanced exact sequence $0 \to A \to B \to C \to 0$ where B is completely decomposable and $C \in \mathcal{K}(n-1)$. We call such a sequence a $\mathcal{K}(n)$–sequence. An equivalent definition is that a group A is in $\mathcal{K}(n)$ if and only if it appears as the first term in a balanced exact sequence:

$$0 \to A \to C_n \to C_{n-1} \to \cdots \to C_1 \to G \to 0,$$

with each C_i finite rank completely decomposable. Kravchenko showed that the chain of classes $\mathcal{K}(n)$ is strictly descending with $\cap_{n\in\omega}\mathcal{K}(n)$ the class of completely decomposable groups. The same results hold for representations, with a **completely decomposable representation** defined as a direct sum of types. Thus, the classes $\mathcal{K}(n)$ can be thought of as a ladder from the mysterious Butler groups or representations to the easily understood completely decomposable groups.

Completely Decomposables

$\mathcal{K}(3)$

$\mathcal{K}(2)$

$\mathcal{K}(1)$

Butler Groups, Representations

F. Richman has pointed out that the construction here is quite general. In any suitable category, one can form projective resolutions of the objects in a class $\mathcal{K}(0)$ and let $\mathcal{K}(n)$ be the objects that appear as kernels after n terms. As long as $\mathcal{K}(0) \subseteq \mathcal{K}(1)$ the sequence will be nested and one can ask, for example, if $\cap_{n\in\omega}\mathcal{K}(n)$ is the class of projectives.

In the abelian group setting, an important aspect of Kravchenko's work was to give internal characterizations for the classes $\mathcal{K}(n)$. For convenience, I'll just work with Butler groups. But all the results hold for representations as well, once a few more definitions are made.

Theorem 1 [Kr], [NV–1,2]. Let B be a Butler group (or representation). The following are equivalent for $n \geq 1$.

(a) $B \in \mathcal{K}(n)$

(b) $B(\tau_1)+ \cdots +B(\tau_n)$ is balanced in B for all types $\tau_1,\ldots,\tau_n$

(c) $B(\tau_1)+ \cdots +B(\tau_n)+B(\tau_{n-1})$ is pure in B for all $\tau_1,\ldots,\tau_{n+1}$

(d) $B(\tau)\cap[B(\tau_1)+\cdots+B(\tau_{n+1})] = B(\tau)\cap B(\tau_1) + \cdots + B(\tau)\cap B(\tau_{n+1})$ for all $\tau,\tau_1,\ldots\tau_{n+1}$

The last condition, a distributive law for socles is equivalent to the others only in the representation setting. Conditions (b)–(d) make sense in any torsion–free abelian group (as well as in valuated groups and certain mixed groups). At this point an example seems appropriate.

Example 2. Let $A_1,\dots,A_{n+3}$ be subgroups of Q with incomparable types and common intersection $A_i \cap A_j = A_0$. Form the "bracket group" $G = G[A_1,\dots,A_{n+3}] = \oplus A_i/D$ where $D = \{(a,a,\dots,a):a \in A_0\}$. These groups have been studied extensively (see [FM], [GUV], [HM], [AV–2]). One can check from Theorem 1(c) that $G \in \mathcal{K}(n)$. The canonical $\mathcal{K}(n)$–sequence containing G is obtained from the pure embedding $G \to \oplus_{i<j}(A_i+A_j)$, where $G \to A_i+A_j$ is given by $(a_1,\dots,a_{n+3}) + D \to a_i-a_j$ (see [AV–1]). Note that the completely decomposable group $\oplus_{i<j}(A_i+A_j)$ has $\binom{n+3}{2}$ terms.

Next let me give you some arguments on why the conditions of Theorem 1 should be studied.

NEAR ISOMORPHISM. It is not hard to show ([NV–1]) that the classes $\mathcal{K}(n)$ (in Butler groups) are closed under near–isomorphism as defined by Lady [L] or numerous other references. Near isomorphism has been an increasing force in the study of Butler groups, particularly almost completely decomposables [AD], [MV], [R]. I'll mention a few more recent results.

Faticoni and Schultz [FS] have shown that an almost completely decomposable group with a p–primary regulating quotient has unique decomposition up to near–isomorphism. They use Schultz's notion of a near–endomorphism ring [S] that originates in the work of Richman [R].

Wickless and I [VW] have developed a near–isomorphism duality. It can be used to explain the many results in the theory of Butler groups that are "dual" but cannot be obtained by Warfield duality. (Remember that Warfield duality only applies in a locally free setting). To provide an example of near–isomorphism duality, the $\mathcal{K}(n)$ definitions above can all be dualized. The notion of a cobalanced exact sequence is well–known (see [AV–1]) and won't be repeated here. Now let's look at a theorem from [NV–1].

Theorem 3A. Let $0 \to A \to B \to C \to 0$ be a $\mathcal{K}(n)$–sequence. If there are fewer than $\binom{n+3}{2}$

non–isomorphic rank one summands of B, then A is completely decomposable.

Example 2 shows that the number $\binom{n+3}{2}$ is minimal. The next theorem is an "obvious dual."

Theorem 3B. Let $0 \to A \to B \to C \to 0$ be a co–$\mathcal{K}(n)$–sequence. If there are fewer than $\binom{n+3}{2}$ non–isomorphic rank one summands of B, then C is completely decomposable.

The equivalence of the two theorems is easy if Warfield duality [Wa] applies. Now even in the general case, equivalence follows from the duality in [VW].

THE LADDER PHENOMENA. Theorem 3A provides an example of another property that makes the study of balanced Butler groups and representations intriguing. Each rung in the $\mathcal{K}(n)$ ladder seems to allow us to view a new result by changing n in a result on the previous rung to n+1. In particular, we can grasp structure results on the n+1 rung that were out of reach on rung n. Here are some more examples.

Butler Decomposition [Kro]. Let $G \in \mathcal{K}(n)$. If $\tau_1,\dots,\tau_{n+2}$ are incomparable types in typeset(G) and $k \leq n+2$, then $G(\tau_1) + G(\tau_2) + \dots + G(\tau_k) = G_1 \oplus G_2 \oplus \cdots \oplus G_k \oplus (\Sigma_{i=1}^{k} G^{\#}(\tau_i))$ where $G_i \simeq G(\tau_i)/G^{\#}(\tau_i)$ is homogeneous completely decomposable of type τ_i.

Here $G^{\#}(\tau)$ is the pure subgroup generated by elements of type greater than τ. P. Krog has a dual version of this theorem also.

Small Typesets [NV–1]. Let $G \in \mathcal{K}(n)$, $n \geq 0$, have at most n+2 types in the critical typeset. Then G is a direct sum of rank–ones and indecomposable, almost completely decomposables of rank n+2.

This theorem was proved in the case $n = 0$ by Arnold, Dugas, Lewis, Mader

Benabdallah–Ouldbeddi and by Files–Goebel for countable rank.

Representations of Antichains. It has been known since the 1970's that if S is an antichain (poset with no relations) then Rep(k,S) has finite representation type for $|S| = 1,2,3$, tame representation type for $|S| = 4$ and wild representation type for $|S| \geq 5$ (see [A] or [Si]). The ladder phenomenon in this setting is illustrated in the following diagram [NV–3]:

|S|	Vector Space Dimensions of Indecomposable Representations				
	$\mathcal{K}(0)$	$\mathcal{K}(1)$	$\mathcal{K}(2)$	...	$\mathcal{K}(n)$
1	1	1	1	...	1
2	1	1	1	...	1
3	1,2	1	1	...	1
4	tame	1,3	1	...	1
5	wild	**wild**	1,4	...	1
6	wild	wild	**wild**	...	1
7	wild	wild	wild	...	1
⋮	⋮	⋮	⋮	...	⋮
n+2					1
n+3					1,n+2
n+4					**wild**
n+5					wild

For example, the table says that the only indecomposable $\mathcal{K}(1)$–representations of the 4–point antichain have dimension either 1 or 3, while the $\mathcal{K}(1)$–representations of the 5–point antichain have wild representation type (see [KNV]). The tame entry in the $\mathcal{K}(0)$ column is now a slight anomaly. The diagram was completed after the Colorado Springs conference, by the boldface "wild" entries for the $\mathcal{K}(n)$ representations of antichains with n+4 points [KNV]. Coxeter functors provide quick examples of such representations. The Coxeter functor C^- is defined as follows. If U is a representation of an antichain S define $C^-U = (V;V(i))$ where V is the cokernel of $0 \to U \to \oplus_{i \in S} U/U(i)$ and V(i) is the image in V of V/V(i). For more details, see [A] or [NV–2]. If $|S| = n+4$ and U is any indecomposable representation with $\dim U > 1$, then $C^{-m}U$ is an indecomposable element of $\mathcal{K}(n)$ for all $m \geq 2$. This immediately gives infinite $\mathcal{K}(n)$ representation type for $|S| = n+4$.

I hope these few results have given you an indication of the potential for using the $\mathcal{K}(n)$ ladder to expand our understanding of Butler groups and representations. In browsing through the proceedings of the last Oberwolfach conference, I observed that over 1/3 of the papers could produce interesting problems in the balanced setting. I'll conclude with a few suggested areas in which balanced properties could be especially fruitful.

Almost Completely Decomposables. Obvious situations in which the ladder phenomenon could apply include ACD's with cyclic regulating quotient [MV] and ACD's with p–primary regulating quotient [FS]. The classification work by Arnold and Dugas [AD] also holds promise for extension.

Finite Rank Butler Groups. The work on bracket groups should provide a host of techniques in the $\mathcal{K}(n)$ setting. A bracket group is a completely decomposable group factored by a rank–one ($B^{(1)}$–group in the notation of [FM]). It seems reasonable to investigate groups in $B^{(n)}$ (completely decomposable factored by rank–n) that also belong to $\mathcal{K}(n-1)$ for $n \geq 1$.

Infinite Rank Butler groups. The definition of $\mathcal{K}(n)$–groups is easy to extend (via Theorem 1) to the variously defined classes of infinite rank Butler groups. As yet we have no definition corresponding to $\mathrm{Bext}(G,T) = 0$. Even so, one might be able to resolve the famous B_1 implies B_2 question if the conditions (b)–(d) of Theorem 1 are added. Any definition that uses finite rank Butler groups, such as the notion of prebalanced, could be changed to a $\mathcal{K}(n)$ statement. In any case, there are different views of the work of Bican, Dugas, Fuchs, Goebel, Hill, Rangaswamy, Salce and others from the $\mathcal{K}(n)$ ladder (see [F]).

Mixed Groups. The conditions of Theorem 1 have meaning in the class of mixed groups that has received attention from Albrecht, Files, Fomin, Glaz, Wickless and others. A $\mathcal{K}(n)$ look at their results, in particular those on endomorphism rings, should prove stimulating (see [Wi]).

Representations. As indicated already, the $\mathcal{K}(n)$ conditions produce interesting theorems in the representation setting, theorems that immediately specialize to quasi–homomorphism categories of Butler groups. The question of finite representation type can be revisted on the n–th rung of the ladder. The work of Arnold, Dugas, Goebel, May, Simson and others is full of good ideas for $\mathcal{K}(n)$ applications. Indeed the representations in [DGM] turn out to be balanced.

References

[A] D.M. Arnold, Representations of partially ordered sets and abelian groups, Abelian Group Theory, Contemporary Mathematics **87**, Amer. Math. Soc.(1989), 91–110.

[AD] D.M. Arnold and M. Dugas, Representations of finite posets and near–isomorphism of finite rank Butler groups, (to appear)

[AV–1] D.M. Arnold and C.Vinsonhaler, Pure subgroups of finite rank completely decomposable groups II, Lecture Notes in Mathematics **1006**, Springer–Verlag, New York, 1983, 97–143.

[AV–2] ________________, Isomorphism invariants for abelian groups, Trans. Amer.Math.Soc. 330(1992), 711–724

[B] M.C.R. Butler, A class of torsion–free abelian groups of finite rank, Proc. London Math. Soc. **15**(1965), 680–698.

[DGM] M. Dugas, R. Goebel and W. May, Free modules with two distinguished submodules, preprint.

[FS] T. Faticoni and P. Schultz, Direct decompositions of ACD groups with primary regulating quotient, preprint.

[F] L. Fuchs, A survey of Butler groups of infinite rank, Contemporary Math. **171**, Abelian Group Theory and Related Topics, Providence, 1994, 121–140.

[FM] L. Fuchs and C. Metelli, On a class of Butler groups, Manuscripta Math. **71**(1991), 1–28.

[GUV] H.P. Goeters, W. Ullery and C. Vinsonhaler, Numerical invariants for a class of Butler groups, Contemporary Math. **171**, Abelian Group Theory and Related Topics, Providence, 1994, 159–172.

[HM] P. Hill and C. Megibben, Classification of certain Butler groups, J. Algebra **160**(1993), 524–551.

[KNV] M. Kleiner, L.G. Nongxa and C. Vinsonhaler, Balanced representations of antichains, preprint.

[Kr] A.A. Kravchenko, Balanced and cobalanced Butler groups, Math. Notes Acad.Sci. USSR **45**(1989), 369–373.

[Kro] K.P. Krog, The structure of balanced and cobalanced Butler groups, preprint.

[L] E.L. Lady, Nearly isomorphic torsion free abelian groups, J. Algebra **35**(1974), 235–238.

[MV] A. Mader and C. Vinsonhaler, Almost completely decomposable groups with cyclic regulating quotient, to appear in J. Algebra.

[N] Loyiso G. Nongxa, Balanced subgroups of finite rank completely decomposable Abelian groups, Trans. Am. Soc. **301**(1987), 637–648.

[NV–1] Loyiso G. Nongxa and C. Vinsonhaler, Balanced Butler groups, to appear in J. Algebra.

[NV–2] Loyiso G. Nongxa and C. Vinsonhaler, Balanced representations of partially ordered sets, preprint.

[R] F. Richman, Isomorphism of Butler groups at a prime, Contemporary Math. 171, Abelian Group Theory and Related Topics, Providence, 1994, 333–338.

[Sc] P. Schultz, Near–endomorphism rings of ACD groups, to appear in the Proceedings of the 1994 Padova Conference.

[Si] D. Simson, Linear Representations of Partially Ordered Sets and Vector Space Categories, Gordon & Breach, London, 1992.

[VW] C. Vinsonhaler and W. Wickless, A near–isomorphism duality for torsion–free abelian groups, preprint.

[Wa] R.B. Warfield, Jr., Homomorphisms and duality for torsion–free groups, Math Z. **107**(1968), 189–200.

[Wi] W. Wickless, Functors from locally completely decomposable groups to mixed groups, preprint.

Mixed Abelian Groups with the Summand Intersection Property

ULRICH ALBRECHT Department of Mathematics, Auburn University, Auburn, AL 36849, albreuf@mail.auburn.edu

JUTTA HAUSEN Department of Mathematics, University of Houston, Houston, TX 77204-3476, hausen@uh.edu

INTRODUCTION

An abelian group is said to have the *the summand intersection property* (SIP for short) if the intersection of any two direct summands is again a direct summand. A problem in L. Fuchs' monograph [10] asks for a characterization of the abelian groups with SIP. Wilson considered the summand intersection property for modules and described the torsion modules with SIP over Dedekind domains [16] (cf 2.3 and 2.4 below). Hausen showed that a module M has SIP if and only if, given any decomposition $M = K \oplus L$ and any module homomorphism $\varepsilon : K \to L$, the kernel of ε is a direct summand of K. This result, the "only if"-part of which was established by Wilson [16], proved a powerful tool to describe classes of modules with SIP, especially indecomposable modules M with the property that $M \oplus M$ (or, more generally, the direct sum of copies of M) has SIP (cf [12], [9]). For example, an indecomposable torsion-free abelian group A of finite rank has the property that A^n has SIP for all natural numbers n if and only if its endomorphism ring $E(A)$ is right semi-hereditary [12, 4.1].

For torsion-free abelian groups, the *quasi-summand-intersection property* (q-SIP) was introduced by Albrecht and Hausen in [5] as a natural analogue to the summand intersection property. Similar to the characterization of groups with SIP, a torsion-free abelian group G was shown to have q-SIP if and only if, given any quasi-decomposition $G \doteq K \oplus L$ and any homomorphism $\varepsilon : K \to L$, the kernel of ε is a quasi-direct summand of K. It turned out that the (quasi-)endomorphism ring of G alone does not suffice to reveal whether or not G has the (quasi-)intersection property. However, in the case that G is flat as a module over its endomorphism ring, the absence or presence of (q-)SIP is, in fact, divulged by $E(G)$ (cf [9, Corollary 11], [5, Corollary 2.8]).

It is the purpose of this article to consider SIP and q-SIP for mixed abelian groups of finite rank. In this setting, a class $\mathcal{G}$ of groups introduced by Glaz and Wickless in [11] arises naturally (cf 3.1). We shall recall the definition of $\mathcal{G}$ in §2. Goeters, Wickless, and Albrecht showed that the elements of $\mathcal{G}$ are precisely the mixed self-small abelian groups which have finite torsion-free rank [4]. Furthermore, a group $A \in \mathcal{G}$ is flat as an $E(A)$-module if and only if A/tA is flat as an $E(A)/tE(A)$-module, and this is the case whenever $E(A)/tE(A)$ is semi-simple Artinian [4].

After a preliminary section collecting basic results on quasi-summands of mixed groups, groups with SIP and q-SIP, and groups in the class $\mathcal{G}$, we prove that for groups in $\mathcal{G}$, SIP implies q-SIP (cf 3.2). This is in contrast to the torsion-free case where both properties are independent [5].

We then characterize the groups $A \in \mathcal{G}$ for which $E(A)/tE(A)$ is semi-simple and show that all A-free groups of finite rank have q-SIP (cf 4.3); similarly, we characterize those $A \in \mathcal{G}$ for which $E(A)$ is a regular ring and show that the A-free groups of finite rank have SIP (cf 4.5). It follows that, for an essentially indecomposable group $A \in \mathcal{G}$, A^n has q-SIP for all positive integers n if and only if $E(A)/tE(A)$ is semi-simple Artinian, and A^n has SIP for all $n < \omega$ iff $E(A)/tE(A)$ is a regular ring (cf 4.6). Methods for constructing mixed abelian groups $A \in \mathcal{G}$ with prescribed $E(A)/tE(A)$ can be found in [4] and [15].

PRELIMINARIES

For torsion-free groups, the notions of quasi-equality and quasi-isomorphism are familiar. Given a *mixed* group A, we call a subgroup U of A quasi-equal to A, in symbol $U \doteq A$, if $mA \subseteq U \subseteq A$ for some positive integer m. Two mixed groups A and A' are quasi-isomorphic if they contain quasi-equal subgroups U and U' which are isomorphic. A quasi-direct summand (or quasi-summand) of A is a subgroup U of A such that $A \doteq U \oplus W$ for some $W \leq A$. One easily verifies the following facts which are well known in the torsion-free setting.

LEMMA 2.1 *Let G be an abelian group and let $B \leq C \leq G$.*

a) *If B is a quasi-summand of C and C is a quasi-summand of G, then B is a quasi-summand of G.*
b) *If B is a quasi-summand of G and $B \subseteq C$, then B is a quasi-summand of C.*

A characterization of the groups with SIP was given in [12]. In [5], an analogue was proved for torsion-free groups which is valid in the mixed setting as well. We have

LEMMA 2.2 *Let G be an abelian group.*

a) *G has SIP if and only if, whenever $G = B \oplus C$ and $\varepsilon : B \to C$ is a homomorphism, then $\ker\varepsilon$ is a direct summand of B; and*
b) *G has q-SIP if and only if, whenever $G \doteq B \oplus C$ and $\varepsilon : B \to C$ is a homomorphism, then $\ker\varepsilon$ is a quasi-summand of B.*

Proof: In view of [12], we need to concern ourselves with *b*) only, and only in the case that G is mixed. Suppose G has q-SIP, $mG \leq B \oplus C \leq G$, and $\varepsilon : B \to C$. Define $T = \{b + \varepsilon(b) \mid b \in B\}$. Then $B \oplus C = T \oplus C$, and $\ker\varepsilon = B \cap T$. Hence $\ker\varepsilon$ is a quasi-summand of G. For the converse, suppose $mG \leq B_i \oplus C_i \leq G$ and let $\pi_i : B_i \oplus C_i \to B_i$ be the projection along C_i, $i = 1, 2$. Define $\varepsilon = m^2(1 - \pi_1)\pi_2|B_1$. Then $\varepsilon : B_1 \to C_1$ is a homomorphism. One verifies that $m^2\ker\varepsilon \leq (B_1 \cap B_2) \oplus (B_1 \cap C_2) \leq \ker\varepsilon$. Hence, $B_1 \cap B_2$ is a quasi-summand of $\ker\varepsilon$ and, by hypothesis, $\ker\varepsilon$ is a quasi-summand of B_1. Apply 2.1 *a*).

COROLLARY 2.3 [16, Proposition 9] *Let G be reduced abelian group with SIP. Then each of its primary components is a direct summand of G which is either cyclic or elementary.*

COROLLARY 2.4 A reduced abelian torsion group has SIP if and only if each of its primary components is either cyclic or elementary.

We shall be concerned with groups belonging to the class $\mathcal{G}$. In order to define $\mathcal{G}$, we first consider the class Γ of mixed groups G with the property that G is isomorphic to a pure subgroup of the product $\Pi_p\ G_p$ containing the direct sum $\oplus_p\ G_p$. (G_p denotes the p-primary component of G.) The symbol Γ_∞ denotes the groups in Γ which have finite torsion-free rank. Every $G \in \Gamma_\infty$ contains a finite independent subset X such that $F = \langle X \rangle$ is a free subgroup of G with G/F torsion. We view G as a pure subgroup of $\Pi_p\ G_p$, and write $X = \{x_i = (x_{ip}) \mid i = 1 \ldots n\}$. Glaz and Wickless investigated the class $\mathcal{G}$ of groups in Γ_∞ for which G_p is finite for all p and satisfies $G_p = \langle x_{1p} \ldots x_{np} \rangle$ for all but finitely many p. Observe that every element of $\mathcal{G}$ is an honest, i.e. neither torsion nor torsion-free, mixed group of finite torsion-free rank. Glaz and Wickless showed in [11] that a group $G \in \Gamma_\infty$ such that G_p is finite for all primes p is in $\mathcal{G}$ if and only if $\mathrm{Hom}(G, tG)$ is a torsion group. In particular, for every $G \in \mathcal{G}$, $E(G)/tE(G)$ is a finite dimensional Q-algebra. For ease of reference, we list some of the properties of the groups in $\mathcal{G}$.

LEMMA 2.5 *Let $A \in \mathcal{G}$.*

a) *Every direct summand of A is either finite or belongs to $\mathcal{G}$.*
b) *If $A \in \mathcal{G}$ and $n \in \mathbf{N}$, then $A^n \in \mathcal{G}$.*

Proof: For *a*), see [15, p. 413]. It follows from results of Glaz and Wickless that, for a group A in Γ_∞ with all p-components finite, A belongs to $\mathcal{G}$ if and only if $Hom(A, tA)$ is torsion [11, 3.1 and 3.4]. This fact proves *b*).

The following notation will be convenient and is used throughout this paper. Let $\{p_1, p_2 \dots\}$ be an enumeration of the positive rational primes in their natural order. Given a group $A \in \mathcal{G}$ and a natural number k, let $A_{(k)} = \oplus_{i=1}^{k} A_{p_i}$. Then $A_{(k)}$ is finite, and there exists a subgroup $A^{(k)}$ of A such that $A = A_{(k)} \oplus A^{(k)}$. Since $\mathrm{Hom}(A^{(k)}, A_{(k)}) = 0$, both $A_{(k)}$ and $A^{(k)}$ are fully invariant subgroups of A. We let $m_k = m_k(A)$ denote the order of $A_{(k)}$. Hence $m_k A = m_k A^{(k)} = A^{(k)}$; in fact, $A^{(k)} = p_i A^{(k)}$ for $i = 1, \dots, k$. Also, $A^{(k)} = (A^{(k)} \cap A_{(\ell)}) \oplus A^{(\ell)}$ for all $\ell \geq k$. Observe that, for every subgroup B of A and for every k, $B = (B \cap A_{(k)}) \oplus B'$ for some $B' \leq B$.

LEMMA 2.6 *Let $A \in \mathcal{G}$, let B_i be subgroups of A, let $m \in \mathbf{N}$, and suppose $mA \leq B_1 \oplus B_2 \leq A$. Let $k \in \mathbf{N}$ such that all prime divisors of m are among the first k primes. Then, for $i = 1, 2$, $B_i = (B_i \cap A_{(k)}) \oplus (B_i \cap A^{(k)})$, and $A^{(k)} = (B_1 \cap A^{(k)}) \oplus (B_2 \cap A^{(k)})$.*

Proof: Since $A^{(k)} = mA^{(k)} \subseteq B_1 \oplus B_2$ and $B_i = (B_i \oplus A_{(k)}) \oplus B_i'$ for some $B_i' \leq B_i$, $i = 1, 2$, we have $A = A_{(k)} \oplus B_1' \oplus B_2'$. The fact that $A^{(k)}$ is fully invariant implies $A^{(k)} = B_1' \oplus B_2'$. By modularity, $B_i' = A^{(k)} \cap B_i$.

LEMMA 2.7 *Let $A \in \mathcal{G}$. The following conditions are equivalent for a subgroup U of A.*

a) *U is a quasi-summand of A.*
b) *$U = T \oplus V$ for some finite group T and some direct summand V of A.*
c) *There exists a positive integer n such that, for all integers $k \geq n$, we have $U = (U \cap A_{(k)}) \oplus (U \cap A^{(k)})$ and $U \cap A^{(k)}$ is a direct summand of $A^{(k)}$.*

Proof: In view of 2.6, only the implication from b) to a) requires a proof. Suppose $U = T \oplus V$ with T finite and V a direct summand of A. Then $T \subseteq A_{(k)}$ for some integer k. By 2.6, $V = T' \oplus V'$ such that $T' \leq A_{(k)}$ and $V' \oplus W = A^{(k)}$. Hence $m_k A = A^{(k)} \subseteq U \oplus W$ completing the proof.

COROLLARY 2.8 *Let $A \in \mathcal{G}$. Then every quasi-summand of A is a direct summand of A if and only if all primary components of A are elementary.*

MIXED GROUPS WITH SIP

Let A be an abelian group such that A^2 has SIP. By Lemma 2.2, the kernel of every endomorphism of A must be a direct summand. Clearly, this has a strong bearing on the endomorphism ring of A. Our first result shows that the class $\mathcal{G}$ arises naturally in the discussion of non-split mixed abelian groups with this property.

PROPOSITION 3.1 *Let A be a reduced mixed group such that A/tA is divisible of finite rank. If A^m has SIP for some $m \geq 2$, then*

a) *A_p is elementary for all primes p.*
b) *$A = T \oplus G$ for some torsion group T and some group $G \in \mathcal{G}$ such that G^m has SIP.*

Proof: For every prime p, the p-component A_p^m of A^m has SIP which implies A_p^m and hence A_p is elementary [16] (cf 2.3 above). Thus, each A_p is a direct summand, and $A \in \Gamma_\infty$ by [4, Theorem 2.2]. We consider A as being embedded in $\Pi_p A_p$ as a pure subgroup containing $\oplus_p A_p$. Let $X = \{x_1 \dots x_k\} \subseteq A$ be an independent subset of A such that $A/tA = \oplus_{i=1}^k Q(x_i + tA)$. We write $x_i = (x_{ip})$, and set $X_p = \langle x_{1p} \dots x_{kp} \rangle \subseteq A_p$. By what has been shown, $A_p = X_p \oplus T_p$. Let G be the subgroup of $\Pi_p X_p \leq \Pi_p A_p$ with the property that $\oplus_p X_p \leq G$ and $G/(\oplus_p X_p) = \oplus_{i=1}^k Q(x_i + \oplus_p X_p)$. Then $tG = \oplus_p X_p = \oplus_p G_p$, and $G \in \mathcal{G}$ [4, Corollary 2.4]. Let $T = \oplus_p T_p$. Since $X \subseteq T \oplus G$ and $(T \oplus G)/tA = \oplus_{i=1}^k Q(x_i + tA)$, we have $A = T \oplus G$. The final assertion follows from the fact that SIP is inherited by direct summands.

In [5], the authors showed that for torsion-free groups the summand intersection property and the quasi-summand intersection property are independent. This is not the case for mixed groups. As a direct consequence of our characterization of the quasi-summands of groups in $\mathcal{G}$ we obtain the fact that, for these groups, SIP implies q-SIP. We have

THEOREM 3.2 *An abelian group $A \in \mathcal{G}$ has SIP if and only if it has q-SIP and each primary component A_p is cyclic or elementary.*

Proof: Let $A \in \mathcal{G}$. Suppose A has SIP. Let U_i, $i = 1, 2$, be two quasi-summands of A. By Lemma 2.7, we can find a positive integer k such that $U_i = (U_i \cap A_{(k)}) \oplus (U_i \cap A^{(k)})$ with $U_i \cap A^{(k)}$ a direct summand of A, for $i = 1, 2$. Hence, $U_1 \cap U_2 = (U_1 \cap U_2 \cap A_{(k)}) \oplus (U_1 \cap U_2 \cap A^{(k)})$. Since A has SIP, $U_1 \cap U_2 \cap A^{(k)}$ is a direct summand, and $U_1 \cap U_2$ is a quasi-summand, by 2.7. The statement about the p-components follows from 2.3. Conversely, suppose A has q-SIP with all primary components cyclic or elementary. Let $A = B \oplus C$ and $\varepsilon : B \to C$ be a homomorphism. By 2.2, the kernel of ε is a quasi-summand of B, and B is finite or $B \in \mathcal{G}$, by 2.5. Applying 2.7 to B, there exists $n < \omega$ such that $\ker\varepsilon = T \oplus V$ with $T \leq B_{(n)}$ and $V \leq B^{(n)}$ a direct summand of $B^{(n)}$. If $pB_p \neq 0$ then $C_p = 0$ so that $T_p = B_p$. Hence, T is a direct summand of $B_{(n)}$ which implies $\ker\varepsilon$ is a direct summand of B. By 2.2, A has SIP.

A group A is quasi-indecomposable if every quasi-summand of A is either bounded or is quasi-equal to A. We call A essentially indecomposable if $A = B \oplus C$ implies B or C is bounded. If A is an essentially indecomposable group in $\mathcal{G}$, then every infinite direct summand must contain $A^{(k)}$ for some k. This observation provides a source of mixed groups with q-SIP:

COROLLARY 3.3 *Let $A \in \mathcal{G}$ such that A is essentially indecomposable. Then*

every quasi-summand of A is either finite or contains $A^{(k)}$ for some positive integer k. In particular, A has q-SIP.

MIXED GROUPS A SUCH THAT A^2 HAS SIP

Clearly, if A is a group such that A^m has SIP for some integer $m > 1$, then A^2 has SIP. The following result relates SIP and q-SIP for groups with this property. It is an immediate consequence of the preceding section.

COROLLARY 4.1 *Let $A \in \mathcal{G}$ and let $m \geq 2$ be an integer. Then A^m has SIP if and only if A^m has q-SIP and A_p is elementary for all primes p.*

As in [12] and [9], we shall use the homological machinery developed by Arnold, Lady and Murley (cf [8]) in order to investigate the endomorphism rings of groups A for which A^2 has SIP. The functors $H_A(\cdot) = \mathrm{Hom}(A, \cdot)$ and $T_A(\cdot) = \cdot \otimes_{E(A)} A$ form an adjoint pair of functors between the category of abelian groups and right $E(A)$-modules. We say that a group G is A-solvable if the induced map $\theta_G : T_A H_A(G) \to G$ is an isomorphism. In general, the image of θ_G is denoted by $S_A(G)$. We refer the reader to [2] for more information on A-solvable groups in $\mathcal{G}$. We will also use the fact that, for a mixed group A and an $E(A)$-module M, the torsion subgroup tM of M is a submodule so that we can view $\overline{M} = M/tM$ as an $E(A)/tE(A)$-module. Since $E(A)/tE(A)$ is a Q–algebra [11], $\overline{M}$ is torsion-free divisible whenever $A \in \mathcal{G}$.

We shall start with an auxiliary result. Let $A \in \mathcal{G}$ and $k < \omega$. The decomposition $A = A_{(k)} \oplus A^{(k)}$ into fully invariant subgroups of A gives rise to a pair ρ_k and $\pi_k = 1 - \rho_k$ of central idempotents in $E(A)$ such that $\rho_k(A) = A_{(k)}$ and $\pi_k(A) = A^{(k)}$. We obtain a ring-direct decomposition $E(A) = E(A)\rho_k \dotplus E(A)\pi_k$ of the endomorphism ring of A in which the ring $E(A)\rho_k = \rho_k E(A)\rho_k$ is finite. Every right $E(A)$-module M can be decomposed as $M = M\rho_k \oplus M\pi_k$. Given any exact sequence (any commutative diagram) of right $E(A)$-modules, multiplying each module with ρ_k will result in another exact sequence (commutative diagram) the terms of which may be regarded as $E(A)\rho_k$-modules. The analogous statement, of course, holds true when each module is multiplied by π_k.

PROPOSITION 4.2 *Let $A \in \mathcal{G}$ such that $E(A)/tE(A)$ is semisimple Artinian. Let M be a finitely generated right $E(A)$-module which is contained in a free $E(A)$-module F. Then there is a natural number k such that $M = M\rho_k \oplus M\pi_k$ with $M\rho_k$ finite and $M\pi_k$ a direct summand of $F\pi_k$. In particular, $M\pi_k$ is a projective $E(A)$-module.*

Proof: Since A is flat as an $E(A)$-module by [4], A-generated subgroups of A^n and kernels of morphisms between A-projective groups are A-solvable. Since M is finitely generated, we may assume that $F \cong E(A)^m$ has finite rank. We denote the inclusion $M \subseteq F$ by ι and obtain the commutative diagram

$$\begin{array}{ccccc} 0 & \longrightarrow & M & \xrightarrow{\ \iota\ } & F \\ & & {\scriptstyle \pi_1}\downarrow & & \downarrow{\scriptstyle \pi_2} \\ 0 & \longrightarrow & \overline{M} & \xrightarrow{\ \bar{\iota}\ } & \overline{F} \end{array}$$

where π_1 and π_2 are the canonical epimorphisms. Since $E(A)/tE(A)$ is semi-simple Artinian, the last row splits, and we can find a map $\delta : \overline{F} \to \overline{M}$ such that $\delta\bar{\iota} = 1_{\overline{M}}$. Because F is free, there is $\varepsilon : F \to M$ with $\pi_1\varepsilon = \delta\pi_2$. Observe that $\pi_1\varepsilon\iota = \delta\pi_2\iota = \delta\bar{\iota}\pi_1 = \pi_1$. Let $\sigma = \varepsilon\iota - 1_M$. Since $\pi_1\sigma = 0$, we may regard σ as a map from M into tM. By hypothesis, there exists an epimorphism $\eta : E(A)^n \to M$. Since T_A is right exact, $T_A(\eta)$ is an epimorphism from $T_A(E(A)^n) \cong A^n$ to $T_A(M)$; since A is flat, $T_A(\iota)$ is a monomorphism from $T_A(M)$ to $T_A(F) \cong A^m$. It follows from [2, Theorem 2.2] that $T_A(M) \in \mathcal{G}$. By [4], $\mathrm{Hom}(T_A(M), t(T_A(M))$ is torsion. Clearly, $T_A(tM)$ is torsion, and the flatness of A implies that it is embedded in $T_A(M)$ and hence in $t(T_A(M))$. Thus, $T_A(\sigma) : T_A(M) \to T_A(tM)$ induces a homomorphism from $T_A(M)$ to $t(T_A(M))$ which must have finite order; let q be a nonzero integer with $qT_A(\sigma) = 0$. Given a right $E(A)$-module N, we let $\phi_N : N \to H_AT_A(N)$ be the natural map. Since $tM \subseteq F$ and ϕ_F is an isomorphism, ϕ_{tM} is a monomorphism. This together with the fact that $\phi_{tM}q\sigma = H_AT_A(q\sigma)\phi_M = 0$ implies $q\sigma = 0$. Choose k such that all prime divisors of q are among the first k primes. Multiplication by q will induce an automorphism of $A^{(k)}$. Consider the central orthogonal idempotents $\rho = \rho_k$ and $\pi = \pi_k$. Since $E(A) = E(A)\rho \dotplus E(A)\pi$ with $E(A)\rho$ finite, every $E(A)$-module N decomposes into $N = N\rho \oplus N\pi$, and $N\rho$ is finite if N is finitely generated. We have $q \cdot 1_A = q\rho + q\pi$ with $q\pi$ a unit in $E(A)\pi$. Thus, there exists $\tau \in E(A)$ such that $q\pi\tau\pi = \pi$. This implies $\sigma(M\pi) = \sigma(Mq\tau\pi) \subseteq q\sigma(M) = 0$. Hence, for all $x \in M$, $(\varepsilon\iota - 1_M)\pi(x) = 0$. It follows that the restriction of ι to $M\pi$ splits.

It was shown in [2] that every group $A \in \mathcal{G}$ has the form $A \cong A_1^{n_1} \oplus \ldots \oplus A_m^{n_m} \oplus T$ for some finite group T, natural numbers n_i, and essentially indecomposable groups $A_1 \ldots A_m$ with the property that, for $i \neq j$, A_i and A_j are not quasi-isomorphic.

THEOREM 4.3 *The following conditions are equivalent for a group $A \in \mathcal{G}$:*

a) *$E(A)/tE(A)$ is semi-simple Artinian.*

b) *i) If U is a finitely A-generated subgroup of A^n for some $n < \omega$, then U a quasi-summand of A^n.*

ii) If $\varepsilon : A^m \to A^n$ for some $m, n < \omega$, then ker ε is a finitely A-generated group.

c) *i) A^n has q-SIP for all $n < \omega$ (for $n = 2$).*

ii) Whenever $A \cong A_1^{n_1} \oplus \ldots \oplus A_m^{n_m} \oplus T$ with T finite and $A_1 \ldots A_m$ pairwise non-quasi-isomorphic and essentially indecomposable, then, for each $i \neq j$, Hom(A_i, A_j) is torsion.

Proof: In order to simplify notation, we set $E = E(A)$.

a) $\Rightarrow$ b): Suppose E/tE is semisimple Artinian. Then A is E-flat and A^n is A-solvable for all $n < \omega$ [4].

We begin by considering an exact sequence of the form $0 \to V \to A^m \to G \to 0$ with $G \subseteq A^n$ for some $m, n < \omega$. Being the kernel of a homomorphism between A-projective groups, V is A-solvable [4], hence $V \cong T_A H_A(V)$. We obtain an exact sequence $0 \to H_A(V) \to H_A(A^m) \to M \to 0$ for some submodule M of $H_A(G)$. Note that M is finitely generated and embedded in the free module $H_A(A^n)$. By Proposition 4.2, there exists an integer k and orthogonal central idempotents $\rho = \rho_k$ and $\pi = \pi_k$ such that $M\rho$ is finite and $M\pi$ is a direct summand of the module $H_A(A^n)\pi$. Since $M\pi$ is projective, the induced sequence $0 \to H_A(V)\pi \to H_A(A^m)\pi \to M\pi \to 0$ splits. This shows that $V \cong T_A H_A(V) \cong T_A(H_A(V)\rho) \oplus T_A(H_A(V)\pi)$ where $T_A(H_A(V)\pi)$ is a direct summand of $T_A H_A(A^m) \cong A^m$ and, consequently, finitely A-generated. Since A is flat as an $E(A)$-module [4], $T_A(H_A(V)\rho)$ is embedded in $T_A(H_A(A^m)\rho) \cong T_A(E^m\rho) \cong (A \otimes E\rho)^m \cong A\rho = A_{(k)}$ which is finite. The proof of ii) is complete. Also note that $M\rho \subseteq H_A(A^n)\rho \cong E^n\pi$ is finite. In order to show i), we let $G = U$. One verifies that the evaluation map $\theta : T_A(M) \to U$ is surjective and, since A^n is A-solvable, an isomorphism. Because of the flatness of A we obtain a commutative diagram

$$\begin{array}{ccccc} 0 & \longrightarrow & T_A(M) & \longrightarrow & T_A H_A(A^n) \\ & & \theta \wr \downarrow & & \downarrow \wr \theta_G \\ 0 & \longrightarrow & U & \longrightarrow & A^n \end{array}$$

in which the vertical maps are isomorphisms. Thus, $U \cong \theta(T_A(M\rho)) \oplus \theta(T_A(M\pi))$ with $\theta(T_A(M\rho))$ finite. Since $M\pi$ is a direct summand of $H_A(A^n)\pi$ and $H_A(A^n) = H_A(A^n)\rho \oplus H_A(A^n)\pi$, $\theta(T_A(M\pi))$ is a direct summand of A^n.

$b) \Rightarrow c)$: In order to show i) is satisfied, let $A \in \mathcal{G}$, let $n \in \mathbf{N}$, and set $G = A^n$. By Lemma 2.5, $G \in \mathcal{G}$. Suppose $B_i \leq G$ and m is a positive integer such that $mG \subseteq B_1 \oplus B_2$, and let $\varepsilon : B_1 \to B_2$. By Lemma 2.6, there exists a natural number k such that, for $i = 1, 2$, $B_i = (B_i \cap G_{(k)}) \oplus (B_i \cap G^{(k)})$ and $G^{(k)} = (B_1 \cap G^{(k)}) \oplus (B_2 \cap G^{(k)})$. Hence $\varepsilon(B_1 \cap G_{(k)}) \subseteq (B_2 \cap G_{(k)})$ and $\varepsilon(B_1 \cap G^{(k)}) \subseteq (B_2 \cap G^{(k)})$. This implies $\ker\varepsilon = K \oplus K'$ with $K = \ker\varepsilon \cap G_{(k)}$ and $K' = \ker\varepsilon \cap G^{(k)}$. Define $\varphi : G \to G$ by $\varphi|(B_1 \cap G^{(k)}) = \varepsilon|(B_1 \cap G^{(k)})$ and $\varphi(G_{(k)} \oplus (B_2 \cap G^{(k)})) = 0$. Then $\ker\varphi = G_{(k)} \oplus K' \oplus (B_2 \cap G^{(k)})$. By hypothesis, $\ker\varphi$ is a quasi-summand of G which, using Lemma 2.1, implies K' is a quasi-summand of G and hence of $B_1 \cap G^{(k)}$. Thus, for some $q \in \mathbf{N}$ and $L \leq B_1 \cap G^{(k)}$, we have $q(B_1 \cap G^{(k)}) \subseteq K' \oplus L \subseteq B_1 \cap G^{(k)}$. Choose $q' \in \mathbf{N}$ such that $q'G_{(k)} = 0$. Then $qq'B_1 = qq'(B_1 \cap G^{(k)}) \subseteq K \oplus K' \oplus L \subseteq B_1$ proving $\ker\varepsilon$ is a quasi-summand of B. By 2.2, G has q-SIP. For the proof of ii), suppose $A = B \oplus C \oplus X$ with $B \cong A_i$, $C \cong A_j$ for some i, j. Let $\phi : B \to C$ be a homomorphism. Note that B and C are essentially indecomposable and belong to $\mathcal{G}$ (2.5), so that they are quasi-equal to their infinite quasi-summands (3.3). Having shown already that A has q-SIP, 2.2 implies that $\ker\phi$ is a quasi-summand of B. Suppose ϕ has infinite order. Then $B = T \oplus V$ with T finite and $V \cong \phi(V)$. Since B is finitely A-generated, so is $V \cong \phi(V)$. By bi), $\phi(V)$ is an infinite quasi-summand of A and hence of C. Thus, $B \doteq V \cong \phi(V) \doteq C$, proving $i = j$.

$c) \Rightarrow a)$: By [3, Theorem 2.4], it is enough to show that, for each i, $\overline{E(A_i)}$ is a division algebra. Set $A_i = B$, and suppose $\phi : B \to B$ has infinite order.

Since B^2 has q-SIP, ker ϕ is a quasi-summand of B which must be finite. No generality is lost if we assume that ϕ is one-to-one: if $B = T \oplus V$ with $T = B_{(k)}$ finite and ker $\phi \subseteq T$, replace ϕ by the map that acts as the identity on T and coincides with ϕ on V. A cardinality argument shows that $\phi(tB) = tB$. Moreover, $r_0(B) = r_0(\phi(B)) < \infty$. Since B/tB and its subgroup $\phi(B)/\phi(tB)$ are torsion-free divisible, we have $\phi(B) = B$, and ϕ is an isomorphism. This completes the proof.

A group is said to be A-projective of finite A-rank if it is a direct summand of a direct sum of finitely many copies of A.

COROLLARY 4.4 *Let A be a group in $\mathcal{G}$ such that $E(A)/tE(A)$ is semi-simple Artinian and let $n < \omega$. If U is a subgroup of A^n, then the following conditions are equivalent.*

a) *U is finitely A-generated.*
b) *$A^n/U \cong P \oplus T$ for some finite group T and some A-projective group P of finite A-rank.*
c) *A^n/U can be embedded into A^m for some $m < \omega$.*

Proof: $a) \Rightarrow b)$: By Theorem 4.3, U is a quasi-summand of A^n, and $A^n \in \mathcal{G}$, by 2.5. By 2.7, $U = T \oplus V$ such that $T \subseteq A^n_{(k)}$ and $A^n = A^n_{(k)} \oplus V \oplus P$. Since $A^n_{(k)}$ is finite, $b)$ holds.

$b) \Rightarrow c)$: Since T is a finite A-generated group, there exists an integer t such that T can be embedded into A^t; clearly, A-projective groups of finite A rank are embedded in A^m for some $m < \omega$.

$c) \Rightarrow a)$: Use $bii)$ of Theorem 4.3.

Another consequence of Theorem 4.3 is the following result concerning groups $A \in \mathcal{G}$ such that A^2 has SIP. Note that $E(A)$ regular implies that $E(A)/tE(A)$ is semisimple Artinian [14]. Rangaswamy [13] has shown that the endomorphism ring of an abelian group G is regular if and only if the kernel and the image of every endomorphism of G are direct summands of G (cf [10, 112.1, p. 239]).

THEOREM 4.5 *The following conditions are equivalent for a group $A \in \mathcal{G}$:*

a) *$E(A)$ is a regular ring.*
b) *i) If U is a finitely A-generated subgroup of A^n for some $n < \omega$, then U a direct summand of A^n.*
   ii) If $\varepsilon : A^m \to A^n$ for some $m, n < \omega$, then kerε is a finitely A-generated group.
c) *i) A^n has SIP for all $n < \omega$ (for $n = 2$).*
   ii) Whenever $A \cong A_1^{n_1} \oplus \ldots \oplus A_m^{n_m} \oplus T$ with T finite and $A_1 \ldots A_m$ pairwise non-quasi-isomorphic and essentially indecomposable groups, then, for each $i \neq j$, Hom(A_i, A_j) is torsion.

Proof: Observe that each of $a)$, $bi)$, and $ci)$ imply that $pA_p = 0$ for all primes p. By 2.8, the notions of direct summand and quasi-summand coincide. Apply Theorem 4.3.

For an essentially indecomposable group $A \in \mathcal{G}$, condition *cii)* is automatically satisfied. We obtain the following consequence of the theorems above.

COROLLARY 4.6 *Let $A \in \mathcal{G}$ be essentially indecomposable.*

a) *A^n has q-SIP for all $n < \omega$ iff $E(A)/tE(A)$ is a division algebra.*
b) *A^n has SIP for all $n < \omega$ iff $E(A)$ is regular.*

By [4, Theorem 4.2], we can find a group $A \in \mathcal{G}$ such that $pA_p \neq 0$ for all relevant primes p, and $E(A)/tE(A) \cong \mathbf{Q} \times \mathbf{Q}$. Then, A^2 has q-SIP, but not SIP. This shows that, while SIP always implies q-SIP for groups in $\mathcal{G}$, the converse, in general, fails.

REFERENCES

[1] U. Albrecht, *Endomorphism rings of faithfully flat abelian groups of infinite rank*, Results in Mathematics **17** (1990), 179–201.

[2] U. Albrecht, *An Azumaya Theorem for a class of mixed abelian groups*, preprint.

[3] U. Albrecht, *Mixed groups projective as modules over their endomorphism ring*, preprint.

[4] U. Albrecht, H. P. Goeters and W. Wickless, *The flat dimension of mixed abelian groups as E-modules*, Rocky Mountain Journal of Mathematics **25** (1995), 569–590.

[5] U. Albrecht and J. Hausen, *Modules with the quasi-summand-intersection property*, Bull. Austr. Math. Soc. **44** (1991), 189–201.

[6] U. Albrecht and J. Jeong, *Homomorphisms between A-projective groups and left Kasch rings*, preprint.

[7] F. W. Anderson and K. R. Fuller, *Rings and categories of modules*, Springer-Verlag, 1973.

[8] D. M. Arnold, *Finite Rank Torsion Free Abelian Groups and Rings, Lecture Notes in Mathematics* **931**, Springer–Verlag, 1982.

[9] D. M. Arnold and J. Hausen, *Modules with the Summand Intersection Property*, Comm. in Algebra **18** (1990), 519–528.

[10] L. Fuchs, *Infinite Abelian Groups*, Academic Press, New York, 1970/73.

[11] S. Glaz and W. Wickless, *Regular and principal projective endomorphism rings of mixed abelian groups*, Comm. in Algebra **22** (1994 pages 1161–1176).

[12] J. Hausen, *Modules with the summand intersection property*, Comm. in Algebra **17** (1989), 135–148.

[13] K. M. Rangaswamy, *Abelian groups with endomorphic images of special types*, J. Algebra **6** (1967), 271–280.

[14] L. Rowen, *Ring Theory, Vol I*, Academic Press New York (1988).

[15] W. J. Wickless, *A functor from mixed groups to torsion-free groups*, Abelian Group Theory and Related Topics, vol. 171, Amer. Math. Soc., Contempory Mathematics, 1994, pp. 407–417.

[16] G. V. Wilson, *Modules with the summand intersection property*, Comm. in Algebra **14** (1986), 21–38.

Remarks on B_2-Groups

ROBERT EL BASHIR KA MFF UK, Sokolovská 83, 186 00 Praha 8 – Karlín, Czech Republik

LADISLAV BICAN KA MFF UK, Sokolovská 83, 186 00 Praha 8 – Karlín, Czech Republik

ABSTRACT. Fuchs and Rangaswamy [FR2] investigated several types of smooth ascending unions of B_2-groups which are B_2-groups. In this remark we shall show that a B_1-group which is a smooth ascending union of B_2-groups is a B_2-group. From this we derive all the main results of [FR2] and we characterize B_2-groups as B_1-groups having, in certain sense, enough B_1-subgroups.

In the past few years the smooth ascending unions of Butler groups have been investigated several times. A B_1-group which is a union of a smooth ascending chain

This research has been partially supported by the grant GAUK 45/94 of the Charles University Grant Agency.

of pure subgroups with countable typesets is a B_2-group ([FR1; Th. 3.1]). Further, in [FR2; Th. 1.1] it is shown that the union $G = \bigcup_{n<\omega} G_n$ of pure subgroups G_n that are B_2-groups is a B_2-group. In the same paper (Th. 2.2 and 2.3) it is proved that a smooth ascending union $G = \bigcup_{\alpha<\lambda} G_\alpha$ of B_2-groups is a B_2-group provided that either $\lambda = \omega_1$ and each G_α is a decent subgroup of G of the cardinality at most $\aleph_1$, or each G_α is TEP in G.

In the first part of this note we are going to combine these two approaches by showing that a B_1-group that is a smooth ascending union of B_2-groups is a B_2-group. The second section is devoted to some basic results concerning the torsion extension property (TEP). As a corollary we obtain a simple proof of the fact that the class of B_2-groups is contained in that of B_1-groups. In the third section we shall build some $G(\aleph_0)$-families of subgroups over a countable collection of (pure) subgroups and we shall present two results on balanced-projective resolutions of torsion-free groups which are smooth ascending unions of B_2-groups. This leads us to the unification of all three theorems from [FR2] mentioned above by showing that they are consequences of theorem 1.5. As a further application of this theorem we shall characterize in the last section B_2-groups as B_1-groups with "enough" B_1-subgroups. The special case in theorem 4.5 involves axiom-3-λ-families and $G(\lambda)$-families owing to lemma 4.4. We do not know whether either of these two families is sufficient in the more general case in theorem 4.2.

0 INTRODUCTION

All groups in this paper are additively written abelian. If x is an element of a torsion-free group G then $\mathbf{t}_G(x) = \mathbf{t}(x)$ will denote the type of x in G. By a *smooth (ascending) union* $G = \bigcup_{\alpha<\tau} G_\alpha$ we mean a collection of pure subgroups G_α indexed by an initial segment of ordinals with the property that $G_\beta \leq G_\alpha$ whenever $\beta \leq \alpha$ and $G_\alpha = \bigcup_{\beta<\alpha} G_\beta$ for limit ordinals α. For unexplained terminology and notations see [F1].

Another relevant concept in the study of infinite rank Butler groups is the *torsion extension property* (TEP). A pure subgroup H of a torsion-free group G is said to have TEP in G or, briefly, H is TEP(-subgroup) in G, if every homomorphism $H \to T$ with T torsion extends to a homomorphism $G \to$T.

Let G be a torsion-free group and H a pure subgroup of corank 1 in G. The types $\mathbf{t}(J)$ of those pure rank 1 subgroups J of G which are not contained in H generate a lattice ideal $\mathcal{P}_{G|H}$ in the lattice $\mathcal{T}$ of all types. We say that H is *preseparative* ($\aleph_0$*-prebalanced* in the terminology of [BF]) in G, if the ideal $\mathcal{P}_{G|H}$ is countably generated. Equivalently, H is preseparative in G, if for each $g \in G \setminus H$ there is a countable subset $\{h_0, h_1, \dots\} \subseteq H$ such that for each $h \in H$ there are $m, n < \omega, m \neq 0$, with $\mathbf{t}(g + h) \leq \mathbf{t}(mg + h_0) \cup \mathbf{t}(mg + h_1) \cup ... \cup \mathbf{t}(mg + h_n)$. In this case we shall also say that $\{h_0, h_1, \dots\}$ is *a preseparative set for g over H*. If the corank of H in G is > 1, then H is defined to be *preseparative* in G if the ideal $\mathcal{P}_{K|H}$ is countably generated for every pure subgroup K of G that contains H as a corank 1 subgroup.

A smooth ascending union $G = \bigcup_{\alpha<\tau} G_\alpha$ is said to be *preseparative chain* of G if, for each $\alpha < \tau$, G_α is preseparative in G and the factor-groups $G_{\alpha+1}/G_\alpha$ are countable. Recall, that a pure subgroup H of a torsion-free group G is called *decent* if for each finite subset X of G there is a finite rank Butler group B such that $H+B$ contains X and it is a pure subgroup of G. Equivalently, for each finite rank pure subgroup L/H of G/H, L is of the form $H + B$ with B finite rank Butler. An exact sequence $E : 0 \to H \to G \xrightarrow{\beta} K \to 0$ with K torsion-free is *balanced* if the induced map β_*: $\mathrm{Hom}(J, G) \to \mathrm{Hom}(J, K)$ is surjective for each rank one torsion-free group J. Equivalently, E is balanced if all rank one (completely decomposable) torsion-free groups are projective with respect to E. Any balanced-exact sequence E with G completely decomposable is called *balanced-projective resolution* of K. A torsion-free group B is said to be *B_1-group* (*Butler group*) if $\mathrm{Bext}(B, T) = 0$ for all torsion groups T, where Bext is the subfunctor of Ext consisting of all balanced-exact extensions.

A torsion-free group B is called *B_2-group* if it has a *B-filtration*, i.e. B is a smooth union of pure subgroups $B_\alpha, \alpha < \lambda$, λ an ordinal, and $B_{\alpha+1} = B_\alpha + H_\alpha$ for all $\alpha < \lambda$ where H_α is a Butler group of finite rank, i.e. a pure subgroup of a completely decomposable group of finite rank, or, equivalently ([B]), a torsion-free homomorphic image of a completely decomposable group of finite rank. Recall, that a collection $\mathfrak{C}$ of subgroups of a group K is called $G(\aleph_0)$-family over a subgroup H, if it consists of subgroups containing H, contains H and K, it is closed under smooth ascending unions and for each $V \in \mathfrak{C}$ and each countable subset $X \subseteq K$ there is $U \in \mathfrak{C}$ containing both V and X and such that $|U/V| \le \aleph_0$.

1 A B_1-UNION OF B_2-GROUPS

1.1 LEMMA ([F2], cf. [F3; Th. 4.1]) A B_1-group is a B_2-group if and only if it has a preseparative chain. □

1.2 LEMMA ([F2], cf. [F3; Th. 11.1]) A pure subgroup H of a B_2-group G is a B_2-group if and only if there is a preseparative chain from H to G. □

Recall, that a torsion-free group G is said to be *absolutely preseparative* provided it is preseparative in each torsion-free group in which it is contained as a pure subgroup.

1.3 LEMMA ([R; Th. 2]) Every B_2-group is absolutely preseparative. □

1.4 LEMMA Let a torsion-free group G be a smooth ascending union $G = \bigcup_{\alpha<\tau} G_\alpha$ of its subgroups such that G_α is preseparative in $G_{\alpha+1}$ for each $\alpha < \tau$. Then G_α is preseparative in G for each $\alpha < \tau$.
Proof: The purity of G_α in G is obvious. Proving indirectly, let us assume that $g \in G \setminus G_\alpha$ is an element without a G_α-preseparative set with $g \in G_\delta$, $\alpha < \delta < \tau$ as small as possible. It follows that $\delta = \beta + 1$ for some $\alpha < \beta < \tau$. If $\{g_n \mid n < \omega\}$ is a G_β-preseparative set for g and $\{g_{nk} \mid k < \omega\}$ is a G_α-preseparative set for g_n, $n < \omega$, then it is a routine to show (cf. [B2]) that $\{-kg_{ij} \mid k, i, j < \omega, k \neq 0\}$ is a G_α-preseparative set for g, a contradiction finishing the proof. □

1.5 THEOREM A B_1-group B which is a smooth ascending union $B = \bigcup_{\alpha<\tau} B_\alpha$ of B_2-groups is a B_2-group.
Proof: Without loss of generality we may assume that $B_0 = 0$. By lemma 1.2 there is a preseparative chain from B_α to $B_{\alpha+1}$ for each $\alpha < \tau$. It follows now from lemma 1.3 and lemma 1.4 that B has a preseparative chain and an application of lemma 1.1 finishes the proof. □

2 TEP SUBGROUPS

2.1 LEMMA Consider the commutative diagram

$$\begin{array}{ccccccccc}
E' : 0 & \longrightarrow & H & \xrightarrow{\alpha'} & G' & \xrightarrow{\beta'} & K' & \longrightarrow & 0 \\
 & & \| & & \downarrow{\scriptstyle \delta} & & \downarrow{\scriptstyle \gamma} & & \\
E : 0 & \longrightarrow & H & \xrightarrow{\alpha} & G & \xrightarrow{\beta} & K & \longrightarrow & 0
\end{array}$$

with splitting exact rows. If $\gamma^* : \mathrm{Hom}(K, H) \to \mathrm{Hom}(K', H)$ is an epimorphism, then to each splitting map $\lambda' : G' \to H$ for E' there is a splitting map $\lambda : G \to H$ for E such that $\lambda' = \lambda\delta$.
Proof: There is $\mu' : K' \to G'$ with $\beta'\mu' = 1_{K'}$ and $\alpha'\lambda' + \mu'\beta' = 1_{G'}$. If $\rho : G \to H$ is any splitting map for E then there is $\sigma : K \to H$ with $\sigma\gamma = \rho\delta\mu', \gamma^*$ being an epimorphism. Now for $\lambda = \rho - \sigma\beta$ we have $\lambda\alpha = \rho\alpha = 1_H$ and $\lambda\delta = \rho\delta - \sigma\beta\delta = \rho\delta - \sigma\gamma\beta' = \rho\delta(1_{G'} - \mu'\beta') = \rho\delta\alpha'\lambda' = \rho\alpha\lambda' = \lambda'$, as desired. □

2.2 PROPOSITION Let $B = \bigcup_{\alpha<\lambda} B_\alpha$ be a smooth ascending union of pure and

B_1-subgroups of a torsion-free group B. If, for each $\alpha < \lambda$, B_α is TEP in $B_{\alpha+1}$, then B is a B_1-group.
Proof: Any balanced-exact sequence $E : 0 \to T \to G \to B \to 0$ with T torsion induces balanced-exact sequences $E_\alpha : 0 \to T \to G_\alpha \to B_\alpha \to 0$ in the natural way. By the preceding lemma we can inductively construct splitting maps $\sigma_\alpha : G_\alpha \to T$ for E_α such that $\sigma_\alpha \mid G_\beta = \sigma_\beta$ for each $\beta < \alpha < \lambda$. This collection obviously defines a splitting map $\sigma : G \to T$ for E. □

2.3 PROPOSITION Let a torsion-free group G be expressible as a sum $H + B$ of B_1-subgroups. If $H \cap B$ is TEP in B then G is a B_1-group.
Proof: Taking any balanced-exact sequence $0 \to T \to V \xrightarrow{\pi} G \to 0$ with T torsion we shall consider the commutative diagram

$$\begin{array}{lcccccccc}
E_H : 0 & \longrightarrow & T & \longrightarrow & U & \longrightarrow & H & \longrightarrow & 0 \\
 & & \| & & \downarrow & & \downarrow & & \\
E_G : 0 & \longrightarrow & T & \longrightarrow & V & \xrightarrow{\pi} & G & \longrightarrow & 0 \\
 & & \| & & \uparrow & & \uparrow & & \\
E_B : 0 & \longrightarrow & T & \longrightarrow & W & \longrightarrow & B & \longrightarrow & 0
\end{array}$$

with exact rows. The sequences E_H and E_B are obviously balanced and consequently they are splitting, B and H being B_1-groups. If $\omega : H \to U$ and $\alpha : B \to W$ are the respective splitting maps, then the restriction of $\omega - \alpha$ to $B \cap H$ has its image in T owing to $\pi(\omega - \alpha) = 0$ and consequently it extends to $\bar{\alpha} : B \to T$ by the hypothesis. The equality $\omega - \alpha = \bar{\alpha}$ on $B \cap H$ allows us to define a homomorphism $\varrho : G \to V$ by the formula $\varrho(h + b) = \omega(h) + (\bar{\alpha} + \alpha)(b)$, $h \in H, b \in B$, which is a splitting map for E_G. □

2.4 COROLLARY Let G be a torsion-free group of the form $G = H + B$, where H is a B_1-group pure in G and B is a finite rank Butler group. Then G is a B_1-group.
Proof: The intersection $H \cap B$ is TEP in B by [FMe; Cor. 3.2] and proposition 2.3 applies. □

2.5 LEMMA Let H be a pure subgroup of a (mixed) group G such that $G = H + B$ with B finite rank Butler. If $T = t(G) \leq H$, then H is TEP in G.
Proof: The purity of H in G yields the purity of H/T in G/T and consequently $B/B \cap H \cong G/H$ is torsion-free. By [FMe; Cor. 3.2] $B \cap H$ is TEP in B and so the restriction of $\varphi : H \to \tilde{T}$ for an arbitrary torsion group $\tilde{T}$ extends to $\eta : B \to \tilde{T}$. The desired extension ψ of φ is now given by the formula $\psi(h+b) = \varphi(h) + \eta(b)$. □

2.6 COROLLARY B_2-groups are B_1-groups.

Proof: Let $B = \bigcup_{\alpha<\lambda} B_\alpha$ be a B-filtration of a B_2-group B, $B_{\alpha+1} = B_\alpha + G_\alpha$ with G_α finite rank Butler for each $\alpha < \lambda$. Using the transfinite induction, $B_{\alpha+1}$ is a B_1-group by corollary 2.4, provided B_α is so. For α limit and $\beta < \alpha$ the subgroup B_β is TEP in $B_{\beta+1}$ by lemma 2.5 and it suffices to use proposition 2.2 □

3 UNIONS OF B_2-GROUPS

3.1 THEOREM ([FR2; Th.2.3]) The group B is a B_2-group whenever it is the union of smooth ascending chain of TEP subgroups $B = \bigcup_{\alpha<\tau} B_\alpha$ where each B_α is a B_2-group.
Proof: Since every B_2-group is a B_1-group by corollary 2.6, B is a B_1-group by proposition 2.2 and theorem 1.5 applies. □

3.2 LEMMA Let K be a subgroup of a torsion-free group G and $\mathfrak{C}, \tilde{\mathfrak{C}}$ be $G(\aleph_0)$-families of G, K, respectively. If $D \in \mathfrak{C}$ is such that $D \cap K \in \tilde{\mathfrak{C}}$ then for each $X \subseteq G$ countable there is $\bar{D} \in \mathfrak{C}$ such that $D \cup X \subseteq \bar{D}$, $\bar{D} \cap K \in \tilde{\mathfrak{C}}$ and $\bar{D}/D$ is countable.
Proof: There is $D_0 \in \mathfrak{C}$ containing $D \cup X$ and such that D_0/D is countable. Assuming $D_n \in \mathfrak{C}$ with D_n/D countable has been already constructed for some $n < \omega$, we have $(D_n \cap K)/(D \cap K) \simeq ((D_n \cap K) + D)/D \leq D_n/D$ and so there is $\tilde{D}_n \in \tilde{\mathfrak{C}}$ such that $D_n \cap K \leq \tilde{D}_n$ and $\tilde{D}_n/D \cap K$ is countable. Writing $\tilde{D}_n$ in the form $\tilde{D}_n = D \cap K + \tilde{L}_n$ with $\tilde{L}_n$ countable we can find $D_{n+1} \in \mathfrak{C}$ containing $D_n \cup \tilde{L}_n$ and such that D_{n+1}/D_n is countable. Now the union $\bar{D} = \bigcup_{n<\omega} D_n$ is in $\mathfrak{C}$, the factor $\bar{D}/D$ is countable and $\bar{D} \cap K = \bigcup_{n<\omega}(D_n \cap K) = \bigcup_{n<\omega} \tilde{D}_n \in \tilde{\mathfrak{C}}$, as desired. □

3.3 LEMMA Let $K_n, n < \omega$, be pure subgroups of a torsion-free group G, $\mathfrak{C}$ be a $G(\aleph_0)$-family of G and $\mathfrak{C}_n$ be a $G(\aleph_0)$-family of K_n, $n < \omega$. Then $\mathfrak{D} = \{H \in \mathfrak{C} \mid H \cap K_n \in \mathfrak{C}_n, n < \omega\}$ and $\mathfrak{D}_* = \{H \in \mathfrak{D} \mid H + K_n \leq_* G, n < \omega\}$ are $G(\aleph_0)$-families of G.
Proof: Obviously, 0 and G belong to $\mathfrak{D}$ and $\mathfrak{D}_*$ and both families are closed with respect to smooth ascending unions. Let $\{L_n | n < \omega\}$ be a sequence of subgroups of G such that each L_n is some K_m and every K_n occurs on infinitely many different places. Taking $H \in \mathfrak{D}$ and $X \subseteq G$ countable, lemma 3.2 yields the existence of $H_0 \in \mathfrak{C}$ with $H \cup X \subseteq H_0$, H_0/H countable and $H_0 \cap L_0 \in \tilde{\mathfrak{C}}_0$, $\tilde{\mathfrak{C}}_0$ being the given $G(\aleph_0)$-family $\mathfrak{C}_m$ corresponding to L_0. Assuming $H_n \in \mathfrak{C}$ with H_n/H countable and $H_n \cap L_n \in \mathfrak{C}_n$ has been already defined, lemma 3.2 yields the existence of $H_{n+1} \in \mathfrak{C}$ with H_{n+1}/H_n countable and $H_{n+1} \cap L_{n+1} \in \tilde{\mathfrak{C}}_{n+1}$. Now, $\bar{H} = \bigcup_{n<\omega} H_n$ is a countable extension of H, contains X and lies in $\mathfrak{C}$. Moreover, due to the choice of L_n, we have $\bar{H} \cap K_m \in \mathfrak{C}_m$, $m < \omega$, and $\mathfrak{D}$ is a $G(\aleph_0)$-family.

Proceeding to $\mathfrak{D}_*$ select $H \in \mathfrak{D}_*$ and $X \subseteq G$ countable arbitrarily. Clearly,

$\langle H + X\rangle_*$ is a countable extension of H and so it is contained in an $H_0 \in \mathfrak{D}$ with $|H_0/H| \leq \aleph_0$. Now, assuming $H_n \in \mathfrak{D}$ with $|H_n/H| \leq \aleph_0$ has been already constructed, the factor-group $(H_n + L_n)/(H + L_n)$ is countable and $\langle H_n + L_n\rangle_*$ is a countable extension of $H_n + L_n$. Thus there is $H_{n+1} \in \mathfrak{D}$ such that $\langle H_n + L_n\rangle_* \leq H_{n+1} + L_n$ and $|H_{n+1}/H| \leq \aleph_0$. Finally, setting $\bar{H} = \bigcup_{n<\omega} H_n$ we have $\bar{H} \in \mathfrak{D}_*$ the desired countable extension of $H + X$ and the proof is complete. □

3.4 LEMMA Let H be a decent subgroup of a torsion-free group G. If G/H is countable, then H is TEP in G.

Proof: If H is of finite corank in G, then $G = H + B$ for a finite rank Butler group B and lemma 2.5 applies. In the opposite case write $G/H = \bigcup_{n<\omega} L_n/H$ with L_n/H pure in G/H and of rank n. Then for each $n < \omega$, $L_n = H + B_n$ for a finite rank Butler group B_n, hence $L_{n+1} = L_n + B_{n+1}$. Thus L_n is TEP in L_{n+1} by lemma 2.5 and the assertion follows. □

3.5 THEOREM If a torsion-free group $G = \bigcup_{\alpha<\tau} G_\alpha$ is a smooth ascending union of B_2-groups, then the group K in any balanced-projective resolution $E : 0 \to K \to C \to G \to 0$ is a B_2-group.

Proof: Since the summands of B_2-groups are B_2-groups ([BF; L. 1.2]), a version of Schanuel's lemma restricts our situation to the following special one. Consider the commutative diagram

$$\begin{array}{lccccccccc}
E_\alpha & :0 & \longrightarrow & K_\alpha & \longrightarrow & C_\alpha & \longrightarrow & G_\alpha & \longrightarrow & 0 \\
 & & & \downarrow & & \downarrow & & \downarrow & & \\
E_{\alpha+1} & :0 & \longrightarrow & K_{\alpha+1} & \longrightarrow & C_{\alpha+1} & \longrightarrow & G_{\alpha+1} & \longrightarrow & 0 \\
 & & & \downarrow & & \downarrow & & \downarrow & & \\
E & :0 & \longrightarrow & K & \longrightarrow & C & \longrightarrow & G & \longrightarrow & 0
\end{array}$$

constructed inductively. Each row E_α is a balanced-projective resolution of G_α and $E_{\alpha+1}$ extends E_α naturally in the sense that C_α is a direct summand of $C_{\alpha+1}$. Note, that at limit ordinals the sequence is a direct limit of its predecessors. [FMe; Prop. 2.1] yields that E_α is TEP. Thus K_α is TEP in $K_{\alpha+1}$ and consequently K is a B_1-group by proposition 2.2, each K_α being B_2-group by [B1; Th. 13]. Hence K is a B_2-group by theorem 1.5. □

3.6 THEOREM If a torsion-free group $G = \bigcup_{\alpha<\tau} G_\alpha$ is a smooth ascending union of B_2-groups, then any balanced-projective resolution $E : 0 \to K \to C \to G \to 0$ of G is TEP provided either τ is countable, or $|G| = \aleph_1$ and the groups G_α are decent in G.

Proof: Looking at the diagram

$$\begin{array}{lccccccccc}
E_\alpha & :0 & \longrightarrow & K & \longrightarrow & U_\alpha & \longrightarrow & G_\alpha & \longrightarrow & 0 \\
 & & & \| & & \downarrow & & \downarrow & & \\
E_{\alpha+1} & :0 & \longrightarrow & K & \longrightarrow & U_{\alpha+1} & \longrightarrow & G_{\alpha+1} & \longrightarrow & 0 \\
 & & & \| & & \downarrow & & \downarrow & & \\
E & :0 & \longrightarrow & K & \longrightarrow & C & \longrightarrow & G & \longrightarrow & 0
\end{array}$$

constructed naturally via pullbacks, let $\mathfrak{C}_\alpha$ be a $G(\aleph_0)$-family of decent subgroups of U_α over K.

First, let $\tau = \omega$. The family $\mathfrak{D}_*$ from lemma 3.3 (here the choice of $\mathfrak{C}$ is not important) consists of decent subgroups of C since the decency of $H \cap U_n$ in U_n yields that of H in $H+U_n$ by [FR2; L. 2.1] and consequently in $\bigcup_{n<\omega}(H+U_n) = C$, according to the properties of $\mathfrak{D}_*$. Thus we can easily construct a smooth ascending union $C = \bigcup_{\beta<\nu} K_\beta$, $K_0 = K$, of members of $\mathfrak{D}_*$ with countable factors. By lemma 3.4 K_β is TEP in $K_{\beta+1}$ for each $\beta < \nu$, hence E is TEP.

For $|G| = \aleph_1$ we can suppose $\tau = \omega_1$. An application of lemma 3.3 gives us the $G(\aleph_0)$-families $\mathfrak{D}_\alpha = \{B \in \mathfrak{C}_\alpha | B \cap U_\beta \in \mathfrak{C}_\beta, B + U_\beta \leq_* U_\alpha, \beta \leq \alpha\}$ for α isolated. Now, obviously, we can assume that the completely decomposable group C in the row E is of the size $\aleph_1$ and consequently we can write $C = \bigoplus_{\alpha<\omega_1} X_\alpha$. Set $K_0 = K$ and assume that for some $\alpha < \omega_1$ the smooth ascending chain $\{K_\beta | \beta < \alpha\}$ of subgroups of C has been constructed in such a way that $|K_\beta / K| \leq \aleph_0$ and $K_\beta \in \mathfrak{D}_\beta$ for $\beta < \alpha$ isolated. For $\alpha = \beta + 1$ we select $K_\alpha \in \mathfrak{D}_\alpha$ that contains $K_\beta \cup (\bigoplus_{\gamma<\alpha} X_\gamma \cap U_\alpha)$, while for α limit we simply set $K_\alpha = \bigcup_{\beta<\alpha} K_\beta$. This way we obviously get $C = \bigcup_{\alpha<\omega_1} K_\alpha$ and with respect to lemma 3.4 it remains to verify the decency of K_α in C for every $\alpha < \omega_1$. But this is easy since K_α is decent in U_α, the limit case is analogous to the above one, and U_α is decent in C, G_α being so in G. □

3.7 THEOREM ([FR2; Th. 1.1 and 2.2]) If a torsion-free group $G = \bigcup_{\alpha<\tau} G_\alpha$ is a smooth ascending union of B_2-groups, then G is a B_2-group provided either τ is countable, or $|G| = \aleph_1$ and the groups G_α are decent in G.
Proof: A balanced-projective resolution $E : 0 \to K \to C \to G \to 0$ of G is TEP by theorem 3.6, hence G is a B_1-group by [FMe; Prop.2.1] and theorem 1.5 applies. □

4 B_1-GROUPS WITH "ENOUGH" B_1-SUBGROUPS ARE B_2-GROUPS

4.1 DEFINITION Let λ be an infinite cardinal and H be a subgroup of a torsion-free group G. A collection $\mathfrak{C} = \mathfrak{C}_\lambda(H, G)$ of pure subgroups of G containing H is said to be SL_λ*-family of* H *in* G if $H, G \in \mathfrak{C}$ and (i) $\sum_{\alpha \in I} H_\alpha \in \mathfrak{C}$ whenever H_α belongs to $\mathfrak{C}$ for each $\alpha \in I$; (ii) if $V \subseteq \tilde{V}$ are elements of $\mathfrak{C}$ and $X \subseteq \tilde{V}$ is any subset with $|X| \leq \lambda$ then there is $U \in \mathfrak{C}$ such that $V \cup X \subseteq U \leq \tilde{V}$ and $|U/V| \leq \lambda$.

If, instead of (i), a weaker condition (i') stating that $\mathfrak{C}$ is closed under arbitrary smooth ascending unions is satisfied, then we say that $\mathfrak{C}$ is a WL_λ*-family of* H *in* G. In both cases we shall also speak about families of G over H. Especially, for $H = 0$ we shall speak simply about families of G.

Similarly, such a collection $\mathfrak{C}$ is called *axiom-3-*λ*-family* if, instead of (ii), it satisfies a weaker condition (ii') stating that for each $V \in \mathfrak{C}$ and any subset $X \subseteq G$ of the cardinality at most λ there is $U \in \mathfrak{C}$ containing $V \cup X$ and such that $|U/V| \leq \lambda$. A collection $\mathfrak{C}$ satisfying (i') and (ii') is called $G(\lambda)$*-family.*

Furthermore, we say that a smooth ascending union $G = \bigcup_{\alpha<\mu} L_\alpha$ of G-pure subgroups is a λ*-chain from* L_0 *to* G if $|L_{\alpha+1}/L_\alpha| \leq \lambda$ for each $\alpha+1 < \mu$. Especially, for $L_0 = 0$ we will speak simply about a λ*-chain.*

4.2 THEOREM Assume, that for some infinite cardinal κ every B_1-group of cardinality at most κ is a B_2-group. Then the following conditions for a B_1-group B are equivalent:

(i) B is a B_2-group;

(ii) B has an SL_κ-family consisting of B_1-groups;

(iii) B has an WL_κ-family consisting of B_1-groups.

Proof: (i) implies (ii) by [B1; Th. 6] and (ii) implies (iii) trivially. Let $\mathfrak{C}$ be a WL_κ-family of B_1-subgroups of a B_1-group B which is not a B_2-group and has the smallest possible cardinality in the class of all such groups. Using transfinite induction we can construct a smooth ascending κ-chain $B = \bigcup_{\alpha<\mu} B_\alpha$ consisting of members of $\mathfrak{C}$ such that $|B_\alpha| < |B|$. Obviously, the set $\{U \in \mathfrak{C} \mid U \leq B_\alpha\}$ is a WL_κ-family of B_1-subgroups of B_α and consequently B_α is a B_2-group for each $\alpha < \mu$ by the choice of B. Now it suffices to use theorem 1.5. □

4.3 COROLLARY (CH) The following conditions for a B_1-group B are equivalent:

(i) B is a B_2-group;

(ii) B has an $SL_{\aleph_\omega}$-family consisting of B_1-groups;

(iii) B has an $WL_{\aleph_\omega}$-family consisting of B_1-groups;

Proof: By [DHR; Th. 7.7] under (CH) every B_1-group of the cardinality at most $\aleph_\omega$ is a B_2-group. □

4.4 LEMMA If H is a B_2-subgroup of a B_1-group G with $|G/H| \leq \aleph_1$, then G is a B_2-group.

Proof: Write G/H as a smooth ascending union $G/H = \bigcup_{\alpha<\omega_1} L_\alpha/H$ with L_α/H countable. All the subgroups L_α are G-preseparative, H being so by lemma 1.3. Applying lemma 1.1 to H, we see that G has a preseparative chain and lemma 1.1 applies once more. □

4.5 THEOREM The following conditions are equivalent for a B_1-group B and $\lambda = \aleph_0$ or $\lambda = \aleph_1$:

(i) B is a B_2-group;

(ii) B has an SL_λ-family consisting of B_1-groups;

(iii) B has an WL_λ-family consisting of B_1-groups;

(iv) B has an axiom-3-λ-family consisting of B_1-groups;

(v) B has an $G(\lambda)$-family consisting of B_1-groups;

(vi) B has a λ-chain consisting of B_1-groups;

Proof: Every B_2-group has an $SL_{\aleph_0}$-family of decent, TEP and B_2-subgroups by [B1; Th. 6]. Furthermore, if $B = \bigcup_{\alpha<\mu} B_\alpha$ is an $\aleph_1$-chain of B consisting of B_1-groups,then the simple transfinite induction shows that all the groups B_α are B_2-groups, applying lemma 4.4 and theorem 1.5. The rest is obvious. □

REMARK It has been proved in [BF; Th. 5.1] that under (V = L) every B_1-group is absolutely preseparative. By [B2; Th. 4.2] and lemma 1.1 we see that a B_1-group is a B_2-group if and only if it has an SL_λ-family of preseparative subgroups for $\lambda = \aleph_0$ or $\lambda = \aleph_1$. Thus theorem 4.5 can be viewed as a slight generalization of this fact.

REFERENCES

B1 L. Bican, Butler groups and Shelah's singular compactness, to appear.

B2 L. Bican, Families of preseparative subgroups, to appear.

B M.C.R. Butler, A class of torsion-free abelian groups of finite rank, *Proc. London Math. Soc., 15*: 686-698 (1965).

BF L. Bican and L. Fuchs, Subgroups of Butler groups, *Communications in Algebra, 22*: 1037-1047 (1994).

DHR M. Dugas, P. Hill and K.M. Rangaswamy, Infinite rank Butler groups II, *Trans. Amer. Math. Soc., 320*: 643-664 (1990).

F1 L. Fuchs, Infinite abelian groups, vol. I and II, *Academic Press, New York*, (1973 and 1977).

F2 L. Fuchs, Infinite rank Butler groups, preprint.

F3 L. Fuchs, A survey of Butler groups of infinite rank, *Contemporary Math., 171*: 121-139 (1994).

FMe L. Fuchs and C. Metelli, Countable Butler groups, *Contemporary Math. 130*: 133-143 (1992).

FR1 L. Fuchs and K.M. Rangaswamy, Butler groups that are unions of subgroups with countable typesets, *Arch. Math. 61*: 105-110 (1993).

FR2 L. Fuchs and K.M. Rangaswamy, Unions of chains of Butler groups, *Contemporary Math. 171*: 141-146 (1994).

R K.M. Rangaswamy, A property of B_2-groups, *Comment.Math.Univ.Carolinae, 35*: 627-631 (1994).

A Strong Stacked Bases Theorem for Finite Rank HCD Groups

K. BENABDALLAH AND M.A. OULDBEDDI
University de Montreal, Montreal, Quebec, Canada

ABSTRACT. Let A be a homogeneous completely decomposable(hcd) torsion free abelian group of finite rank k and let G be a torsion-free abelian group containing A such that $G/A = \bigoplus_{i=1}^{k} \langle x_i + A \rangle$ where $1 < o(x_i + A) | o(x_{i+1} + A), i = 1, \ldots, k-1$. We show that there exist $y_1, \ldots, y_k$ in G such that

(a) $G = \bigoplus_{i=1}^{k} \langle y_i \rangle_*$
(b) $A = \bigoplus_{i=1}^{k} d_i \langle y_i \rangle_*$ where $d_i = o(x_i + A)$
(c) $x_i + A = y_i + A, i = 1, \ldots, k$.

We call this result a strong stacked bases theorem for hcd groups of finite rank because of the additional property (c). We acknowledge that from a theorem by O. Mutzbauer[3, Theorem 3.3] one can derive a "weak" stacked bases theorem assuming that both G and A are hcd's of same type. Our proof is different , it is independant of the stacked bases theorem for finite rank free modules over PID's and yelds the stronger version which in essence says that every canonical decomposition of G/A can be lifted to a simultaneous decomposition of A and G.

1. INTRODUCTION

In a previous article[1] we introduced the notion of relative primeness between a $k \times k$ matrix with integral coefficients and a k-tuple of elements of a given torsion free abelian group G. This notion was then used to prove a controled substitution theorem for finite extensions of completely decomposable groups with two critical types and obtain a new proof of the well-known theorem that such groups decompose into direct sums of groups of rank 1 or 2. In this note we give yet another application of this substitution theorem.

1991 *Mathematics Subject Classification.* 20K15.

Acknowledgement: The authors wish to thank Professor A. Mader for suggesting that our subtitution theorem may imply a stacked bases theorem and for pointing out that O. Mutzbauer's theorem 3.3 in [3] implies also a stacked bases theorem for hcd's.

All groups considered here are abelian groups.

For the reader's convenience we recall first some notations and some definitions and the exact statement of the substitution theorem. Let G be a torsion free group and k a positif integer. We put

$$G^k = \{(g_1, \ldots, g_k) : \ g_i \in G, i = 1, \ldots, k\}$$

the element $(g_1, \ldots, g_k)$ of G^k, is denoted simply by $\boldsymbol{g}$. Thus if we write $\boldsymbol{h} \in G^k$ we mean $\boldsymbol{h} = (h_1, \ldots, h_k), h_i \in G$, $i = 1, \ldots, k$. G^k is a $\mathbb{Z}$–module with the usual addition and scalar multiplication. G^k is also naturally a left and right module over the ring $M_k(\mathbb{Z})$ of $k \times k$ matrices over the ring of integers $\mathbb{Z}$. If $\boldsymbol{g} \in G^k$ and $N \in M_k(\mathbb{Z}), N\boldsymbol{g}$ is the k-tuple obtained by formal line–column product of N by $\boldsymbol{g}$ considered as a column and similarly for $\boldsymbol{g}N$. For our purposes in the special case where $G = \mathbb{Z}$ it is convenient to write $\mathbb{Z}_k$ instead of $\mathbb{Z}^k$. The elements of $\mathbb{Z}_k$ will generally be denoted by greek letters such as $\boldsymbol{\alpha}, \boldsymbol{\beta}, \ldots$. Finally, for $\boldsymbol{\alpha} \in \mathbb{Z}_k$ and $\boldsymbol{g} \in G^k$ we define $\boldsymbol{\alpha g} = \sum_{i=1}^{k} \alpha_i g_i \in G$. This is a convenient way to write linear combinations of elements of G, in particular $\mathbb{Z}_k\boldsymbol{g} = \{\boldsymbol{\alpha g} : \ \boldsymbol{\alpha} \in \mathbb{Z}_k\}$ is simply the subgroup generated by $g_1, \ldots, g_k$.

Definition 1.1. Let $\boldsymbol{g} \in G^k, N \in M_k(\mathbb{Z})$, and p a prime number. We define the ***p–annihilator*** of $\boldsymbol{g}$ by

$$Ann_p(\boldsymbol{g}) = \{\boldsymbol{\alpha} \in \mathbb{Z}_k : \ \boldsymbol{\alpha g} \in pG\}$$

and the ***p–kernel*** of N by

$$K_p(N) = \{\boldsymbol{\alpha} \in \mathbb{Z}_k : \ \boldsymbol{\alpha} N \in p\mathbb{Z}_k\}.$$

These are subgroups of $\mathbb{Z}_k$ containing $p\mathbb{Z}_k$.

Definition 1.2. Let $N \in M_k(\mathbb{Z})$ and $\boldsymbol{g} \in G^k$ where G is a torsion free group. We say that N and $\boldsymbol{g}$ are ***relatively prime*** in G and we write $(N, \boldsymbol{g}) = 1$, if for every prime number p dividing $\det N$ we have

$$Ann_p(\boldsymbol{g}) \cap K_p(N) = p\mathbb{Z}_k.$$

This is a generalization of the usual relative primeness of integers wich correspond to the case $G = \mathbb{Z}$ and $k = 1$.

We are ready to state the substitution theorem proved in [1].

Theorem 1.3. *Let A be an hcd group of finite rank. Let $D = diag(d_1, \ldots, d_k) \in M_k(\mathbb{Z})$ and $\boldsymbol{a} \in A^k$ such that $1 < d_1 | \ldots | d_k$ and $(D, \boldsymbol{a}) = 1$. Then there exists $\boldsymbol{b} \in A^k$ such that:*

(1) $\boldsymbol{b} \equiv \boldsymbol{a} \mod D$, *i.e* $\quad b_i - a_i \in d_i A$, *and*

(2) $(\mathbb{Z}_k\boldsymbol{a})_* = (\mathbb{Z}_k\boldsymbol{b})_* = \bigoplus_{i=1}^{k} \langle b_i \rangle_*$

2. Strong stacked bases

Before giving our main theorem we need some lemmas.

Lemma 2.1. *Let G be an hcd and let A be a subgroup of G such that $G/A = \bigoplus_{i=1}^{k} \langle g_i + A\rangle$ and $d_i = o(g_i + A)$ where $1 < d_i | d_{i+1}, i = 1, \cdots k-1$. Then the k-tuple $\boldsymbol{g} = (d_1 g_1, \cdots, d_k g_k) \in A^k$ is relatively prime with $D = diag(d_1, \cdots, d_k)$ in A.*

Proof. Let $\boldsymbol{\alpha} \in \mathbb{Z}_k$ such that $\boldsymbol{\alpha} \in Ann_p(\boldsymbol{g}) \cap K_p(D)$ then, $\sum_{i=1}^{k} \alpha_i d_i g_i = pa$ for some $a \in A$ and $\alpha_i d_i = pn_i, i = 1, \cdots, k$. We want to show that α_i is divisible by p for all primes p that divide $\det D$. It suffices to do this for p dividing d_k. Suppose first that p divides all d_i's. Write

$$d_i = p^{m_i} c_i \quad \text{where} \quad (c_i, p) = 1,$$

$1 \leqslant m_i \leqslant m_{i+1}$, then $\sum_{i=1}^{k} \alpha_i p^{m_i} c_i g_i = pa$ i.e. $\sum_{i=1}^{k} \alpha_i p^{m_i - 1} c_i g_i = a \in A$. Therefore, $\sum_{i=1}^{k} \alpha_i p^{m_i-1} c_i (g_i + A) = 0$, hence $p^{m_i-1} c_i (g_i + A) = 0$ and $d_i | p^{m_i-1} c_i$. It follows that $p | \alpha_i c_i$ and since $(p, c_i) = 1$ we have $p | \alpha_i, i = 1, \cdots, k$.
Suppose now that p does not divide d_1, let j be the smallest index such that $p | d_j$ then, $p | \alpha_i, i = 1, \cdots, j-1$. Therefore:

$$\sum_{i=j}^{k} \alpha_i d_i a_i \in pA.$$

Applying the same argument as in the first case we have $p | \alpha_i, i = j, \cdots, k$. Hence $\boldsymbol{\alpha} \in p\mathbb{Z}^k$ i.e. $(D, \boldsymbol{a}) = 1$. $\square$

Lemma 2.2. *Let X be a rank one torsion-free group and Y a subgroup of X such that X/Y is finite of order n then $Y = nX$.*

Proof. It is well known that X/nX is a finite group and since X is rank one X/nX is cyclic whose order divides n. But since $nX \subset Y$ we have $|X/nX| = |X/Y|.|Y/nX|$. Therefore, $|Y/nX| = 1$ i.e. $Y = nX$. $\square$

Theorem 2.3. *(Strong stacked bases) Let A be a homogeneous completely decomposable torsion free group of rank k and let G be torsion free extension of A such that $G/A = \bigoplus_{i=1}^{k} \langle g_i + A\rangle$ and let $d_i = o(g_i + A)$ where $1 < d_i | d_{i+1}, i = 1, \cdots k-1$. Then, there exist $y_1, \cdots, y_k \in G$ such that*

(1) $G = \oplus \langle y_i \rangle_*$
(2) $A = \oplus d_i \langle y_i \rangle_*$
(3) $g_i + A = y_i + A, i = 1, \cdots, k$.

Proof. Let $\boldsymbol{a} = (d_1g_1, \cdots, d_kg_k), D = \operatorname{diag}(d_1, \cdots, d_k)$ then by Lemma 2.1 $(D, \boldsymbol{a}) = 1$ and by the substitution theorem there exists $\boldsymbol{b} \in A^k$ such that $(\mathbb{Z}_k\boldsymbol{a})_*^A = (\mathbb{Z}_k\boldsymbol{b})_*^A = \bigoplus_{i=1}^k \langle b_i\rangle_*^A$, where $\langle b_i\rangle_*^A$ is the purification of $\langle b_i\rangle$ in A. Furthermore, there exists $c_i \in A$ such that $b_i - a_i = d_ic_i$. But $a_i = d_ig_i$ therefore $b_i = d_i(c_i + g_i), i = 1, \ldots, k$. Put $y_i = c_i + g_i, i = 1, \ldots, k$. We claim that these are the y_i's we are looking for. Indeed, note first that $(\mathbb{Z}_k\boldsymbol{a})_*^A = A$ since the components of $\boldsymbol{a}$ are linearly independent in A and $\operatorname{rk}(A) = k$. Also since $b_i = d_iy_i$ we have that the y_i's are independent in G. Now,

$$\langle y_i\rangle_*^G = \langle b_i\rangle_*^G \supset \langle b_i\rangle_*^A, i = 1, \ldots k.$$

Thus $\bigoplus_{i=1}^k \langle y_i\rangle_*^G \supset \bigoplus_{i=1}^k \langle b_i\rangle_*^A = A$ and since $y_i = c_i + g_i$ where $c_i \in A$ we conclude that

$$\bigoplus_{i=1}^k \langle y_i\rangle_*^G \supset A + \langle g_i, \ldots, g_k\rangle = G$$

and therefore we have $G = \bigoplus_{i=1}^k \langle y_i\rangle_*^G$. Clearly $y_i + A = g_i + A, i = 1, \ldots, k$. It remains to show that $\langle b_i\rangle_*^A = d_i\langle y_i\rangle_*^G$. Observe that $\langle b_i\rangle_*^A = \langle b_i\rangle_*^G \cap A = \langle y_i\rangle_*^G \cap A$. therefore

$$\langle y_i\rangle_*^G / \langle b_i\rangle_*^A = \langle y_i\rangle_*^G / \langle y_i\rangle_*^G \cap A \cong (\langle y_i\rangle_*^G + A)/A \supset (\langle g_i\rangle + A)/A = \langle g_i + A\rangle.$$

Thus d_i divides the number of elements m_i in $\langle y_i\rangle_*^G / \langle b_i\rangle_*^A$ say $m_i = d_ie_i, i =, \ldots, k$ Now,

$$G/A = \frac{\bigoplus_{i=1}^k \langle y_i\rangle_*^G}{\bigoplus_{i=1}^k \langle b_i\rangle_*^A} \cong \bigoplus_{i=1}^k \frac{\langle y_i\rangle_*^G}{\langle b_i\rangle_*^A}.$$

So that

$$|G/A| = \prod_{i=1}^k m_i = \prod_{i=1}^k d_ie_i = \prod_{i=1}^k d_i$$

therefore $m_i = d_i$ and from Lemma 2.2 $\langle b_i\rangle_*^A = d_i\langle y_i\rangle_*^G$. This completes the proof. □

Note that a by product of this theorem is the well-known fact that finite torsion free extensions of hcd's of finite rank are also hcd's with the same type. For another corollary we need the following definition:

Definition 2.4. Let C be a finite abelian group. If $C = \bigoplus_{i=1}^k \langle g_i\rangle, o(g_i) = d_i > 1$, and $d_i | d_{i+1}, i = 1, \ldots, k-1$. The set $\{d_1, \cdots, d_k\}$ is uniquely determined by C. We denote this set by $\operatorname{Inv}(C)$ and call such a decomposition a ***canonical*** decomposition of C, the number k is called the ***width*** of C.
In general a decomposition $C = \bigoplus_{i=1}^m C_i$ is said to be ***normal*** if $\operatorname{Inv}(C_i), i =$

$1, \cdots, m$ form a partition of Inv(C). In particular canonical decompositions are normal decompositions corresponding to the partition of Inv(C) in singletons.

The following corollary to the strong stacked bases theorem is a generalization of a result in[2] ([2, 4.3.8]).

Corollary 2.5. *Let G be homogeneous completely decomposable torsion-free group of finite rank k and A a subgroup of G such that G/A is of width k. Then every normal decomposition of the finite group G/A lifts to a simultaneous decomposition of G and A.*

Finally, in the general situation we have the following comprehensive stacked bases theorem.

Theorem 2.6. *Let G be a homogeneous completely decomposable group of finite rank n and A a subgroup of G then there exist $y_1, \cdots, y_n \in G$ and nonnegative integers $d_1, \ldots, d_n$ such that $d_1|d_2|\cdots|d_n$ and $G = \bigoplus_{i=1}^{n}\langle y_i\rangle_*$, and $A_* = \bigoplus_{i\in S}\langle y_i\rangle_*$ and $A = \bigoplus_{i\in S} d_i\langle y_i\rangle_*$ where $S = \{i : d_i \geq 1\}$.*

Proof. We let H be a maximal pure subgroup of G contained in A then $G = H \oplus K$ for some subgroup K of G and

$$A = H \oplus K \cap A, \quad A_* = H \oplus (K \cap A_*) \quad \text{and} \quad K \cap A_* = (K \cap A)_*.$$

It is easy to verify that $K \cap A$ contains no proper pure subgroup of G and that $(K \cap A)_*/K \cap A$ is a finite group whose canonical decompositions contain the same number of cyclic factors as the rank of $K \cap A$. Finally, $G = H \oplus (K \cap A)_* \oplus B$ for some subgroup B of K. Now, let $H = \bigoplus_{i=1}^{s}\langle y_i\rangle_*$, $(K \cap A)_* = \bigoplus_{i=1}^{s+t}\langle y_i\rangle_*$ and $B = \bigoplus_{i=s+t+1}^{n}\langle y_i\rangle_*$ and let
$d_1 = \cdots = d_s = 1, \{d_{s+1}, \ldots, d_{s+t}\} = Inv((K \cap A)_*/K \cap A) and \quad d_{s+t+1} = \ldots d_n =$ 0, the $y_i's$ for $i = s+1, \ldots, s+t$ to be chosen as in $Theorem 2.1$. □

References

[1] K. Benabdallah and M. A. Ouldbeddi, *On finite essential extensions of torsion free abelian groups* (to appear).

[2] A. Mader, Almost completely decomposable abelian groups, Montreal Notes, 1992.

[3] O. Mutzbauer, *Regulating subgroups of Butler groups*, Abelian Groups (Proceedings, 1991 Curaçao Conference), Lecture Notes in Pure and Applied Mathematics, Vol. 146, Dekker, 1993, pp. 209–217.

[4] M. A. Ouldbeddi, Sur les extensions essentielles finies des groupes abéliens sans torsion, Thèse de Doctorat, Université de Montréal, 1995.

Département de Mathématiques et de Statistique,
Université de Montréal C.P 6128-Centre-ville, Montréal,
P.Q, Canada, H3C 3J7

E-mail address: benabdal@ere.umontreal.ca ouldbedd@ere.umontreal.ca

Families of Preseparative Subgroups

LADISLAV BICAN KA MFF UK, Sokolovská 83, 186 00 Praha 8 – Karlín, Czech Republik

ABSTRACT. Recently, axiom-3 and $G(\lambda)$-families of subgroups of torsion-free groups, especially in the case of Butler groups, have been investigated. In this paper we shall introduce two natural generalizations of these families and we shall prove the equivalence of the existence of either of these families. A "uniformity" of one of them is used to extend some characterizations of B_2-subgroups of B_2-groups, as well as B_1-groups that are B_2-groups.

1 INTRODUCTION

All groups in this paper are additively written abelian. If x is an element of a torsion-free group G then $|x|_G = |x|$ is the characteristic and $\mathbf{t}_G(x) = \mathbf{t}(x)$ is the

1991 *Mathematics Subject Classification.* 20K20.

Key words and phrases. B_1-group, B_2-group, prebalanced subgroup, torsion extension property, prespecial set, axiom-3 family.

This research has been partially supported by the grant GAUK 45/94 of the Charles University Grant Agency.

type of x in G. By a *smooth (ascending) union of a group G* we mean a collection of pure subgroups G_α indexed by an initial segment of ordinals with the property that $G_\beta \leq G_\alpha$ when $\beta < \alpha$ and $G_\alpha = \cup_{\beta<\alpha} G_\beta$ whenever α is a limit ordinal. For unexplained terminology and notations see [F1].

An exact sequence $E \ : \ 0 \to H \to G \xrightarrow{\beta} K \ \to 0$ with K torsion-free is *balanced* if the induced map $\beta_* : \operatorname{Hom}(J, G) \to \operatorname{Hom}(J, K)$ is surjective for each rank one torsion-free group J. Equivalently, E is balanced if all rank one (completely decomposable) torsion-free groups are projective with respect to E. If, in addition, G is completely decomposable, then E is said to be a balanced-projective resolution of K. A torsion-free group B is said to be a *B_1-group* (*Butler group*) if $\operatorname{Bext}(B, T) = 0$ for all torsion groups T, where Bext is the subfunctor of Ext consisting of all balanced-exact extensions. It is known [BS] that this definition coincides with the familiar one if B has finite rank, i.e. a pure subgroup of a completely decomposable group, or, equivalently [Bu], a torsion-free homomorphic image of a completely decomposable group of finite rank.

Recall that a subgroup H of a group G with a torsion-free quotient G/H is called *prebalanced* if for each rank one (pure) subgroup J of G/H there is a pair (X, ϕ) consisting of a finite rank completely decomposable group X and a homomorphism $\phi : X \to G$ such that $\beta\phi X = J$, β being the canonical projection $G \to G/H$. Note that a pure subgroup H of a torsion-free group G is prebalanced if and only if for each $g \in G$ there are a non-zero integer m and a finite subset $\{h_0, \dots, h_n\} \subseteq H$ such that $\mathbf{t}(g + H) = \cup_{i=0}^n \mathbf{t}(mg + h_i)$. Equivalently, H is prebalanced in G if for a suitable subset $\{h_0, \dots, h_n\} \subseteq H$ and a non-zero integer m the equality $|mg + H| = \cup_{i=0}^n |mg + h_i|$ holds. In either of these cases we shall also say that that $\{h_0, h_1, \dots\}$ *is a prebalanced set for g over H* or an *H-prebalanced set for g*. An exact sequence $0 \to H \xrightarrow{\alpha} G \xrightarrow{\beta} K \ \to 0$ is prebalanced if αH is a prebalanced subgroup of G. For the sake of brevity we shall say that a subgroup H is G-balanced, G-prebalanced, etc., instead it is balanced, prebalanced, etc. in G.

Another relevant concept in the study of infinite rank Butler groups is the *torsion extension property* (TEP). A pure subgroup H of a torsion-free group G is said to have TEP in G, or briefly, H is TEP(-subgroup) in G, if every homomorphism $H \to T$ with T torsion extends to a homomorphism $G \to$T.

Let G be a torsion-free group and H a pure subgroup of corank 1 in G. The types $\mathbf{t}(J)$ of those pure rank 1 subgroups J of G which are not contained in H generate an ideal $\mathcal{P}_{G|H}$ in the lattice $\mathcal{T}$ of all types. We say that H is *preseparative* (*$\aleph_0$-prebalanced* in the terminology of [BF]) in G, if the ideal $\mathcal{P}_{G|H}$ is countably generated. Equivalently, H is preseparative in G, if for each $g \in G \setminus H$ there is a countable subset $\{h_0, h_1, ...,\} \subseteq H$ such that for each $h \in H$ there are $m, n < \omega, m \neq 0$, with $\mathbf{t}(g + h) \leq \mathbf{t}(mg + h_0) \cup \mathbf{t}(mg + h_1) \cup ... \cup \mathbf{t}(mg + h_n)$. In this case we shall also say that that $\{h_0, h_1, \dots\}$ *is a preseparative set for g over H* or an *H-preseparative set for g*. If the corank of H in G is > 1, then H is defined to be *preseparative* in G if the ideal $\mathcal{P}_{K|H}$ is countably generated for every pure subgroup K of G that contains H as a corank 1 subgroup. A smooth ascending union $G = \bigcup_{\alpha<\mu} G_\alpha$ of preseparative subgroups with $G_0 = H$ and $|G_{\alpha+1}/G_\alpha| \leq \aleph_0$

for each $\alpha < \mu$ is said to be a *preseparative chain from* H *to* G. For $H = 0$ we speak about a *preseparative chain of* G.

A torsion-free group B is called a B_2*-group* if it has a B*-filtration*, i.e. B is a smooth union of pure subgroups $B_\alpha, \alpha < \lambda$, λ an ordinal, and $B_{\alpha+1} = B_\alpha + H_\alpha$ for all $\alpha < \lambda$, where H_α is a Butler group of finite rank. It is well-known that the class of B_2-groups coincides with that of B_3-groups, i.e. torsion-free groups possessing an axiom-3 family of decent subgroups (see [AH] or [FMa]).

Recall, that an *axiom-3 family* of a torsion-free group G over its subgroup H is a collection $\mathfrak{C}$ of pure subgroups of G containing H such that $H, G \in \mathfrak{C}$ and (i) $\sum_{\alpha \in I} H_\alpha \in \mathfrak{C}$ whenever H_α belongs to $\mathfrak{C}$ for each $\alpha \in I$; (ii) for every $U \in \mathfrak{C}$ and every countable subset X of G there is $V \in \mathfrak{C}$ such that $U \cup X \subseteq V$ and $|V/U| \leq \aleph_0$. Moreover, (see [DHR]) such a collection $\mathfrak{C}$ is said to be a $G(\lambda)$*-family* if, instead of (i), $\mathfrak{C}$ is closed under arbitrary smooth ascending unions of members of $\mathfrak{C}$ and in (ii) the countability is replaced by the infinite cardinal λ.

In this paper we are going to introduce generalizations of these two notions having some additional "ubiquity" property, which is useful in some constructions. We shall show that the appearance of any of these families consisting of preseparative subgroups in a torsion-free group is equivalent to the appearance of all others. It is well-known [BF] that in the constructible universe (V = L) the prebalanced subgroups of B_2-groups are B_2-groups and that (see [DHR] and [FMe]) the same is true under (CH) for torsion-free groups up to the cardinality $\aleph_\omega$. On the other hand, it has been shown by Fuchs [F3] that a pure subgroup H of a B_2-groups G is a B_2-group if and only if there exists a preseparative chain from H to G and that a B_1-group G is a B_2-group if and only if it has a preseparative chain. As an application, the characterizations of B_1-groups which are B_2-groups in terms of prebalanced subgroups are obtained. Since prebalanced subgroups of completely decomposable groups of cardinalities at most $\aleph_\omega$ are B_2-groups under the Continuum Hypothesis, we obtain in this case that a B_1-group is a B_2-group whenever it has a WL_λ-family of prebalanced subgroups for some $\aleph_0 \leq \lambda \leq \aleph_\omega$. For the sake of completeness we include the reduction theorems on preseparative subgroups of B_2-groups, although we do not know whether preseparative subgroups of B_2-groups of cardinalities greater than $\aleph_1$ are B_2-groups or not.

2 PRESEPARATIVE SUBGROUPS

2.1 LEMMA Let H be a pure subgroup of a torsion-free group G, $g \in G \setminus H$, $h \in H$.

(i) If, for some $0 < m < \omega$, the element mg has a preseparative set over H, then the same set is a preseparative set for g over H;

(ii) If $\{h_n \mid n < \omega\}$ is a preseparative set for g over H, then $\{-mh + h_n \mid n < \omega, 0 < m < \omega\}$ is that for $g + h$ over H.

Proof: (i) For a preseparative set $\{h_n \mid n < \omega\}$ for mg over H we have $\mathbf{t}(g+h) = \mathbf{t}(mg+mh) \leq \cup_{i=0}^{r} \mathbf{t}(kmg+h_i)$ for suitable $r, k < \omega$, $k \neq 0$.

(ii) For an arbitrary $x \in H$ there are $m, n < \omega$, $m \neq 0$ such that $\mathbf{t}(g+h+x) \leq \cup_{i=0}^{r} \mathbf{t}(mg+h_i) = \cup_{i=0}^{r} \mathbf{t}(m(g+h) - mh + h_i)$ □

The following construction is based on ideas of [F2] and [FMa].

2.2 CONSTRUCTION Let $G = \bigcup_{\alpha<\mu} L_\alpha$ be a smooth union of G-preseparative subgroups. For each $\alpha + 1 < \mu$ we fix a set R_α of representatives of cosets of $L_{\alpha+1}/L_\alpha$ and for each $g \in R_\alpha$ we fix a preseparative set $\{h_n^g \mid n < \omega\} \subseteq L_\alpha$ for g over L_α. Now if we set

$$B_\alpha = \langle (mg + h_n^g)_* \mid g \in R_\alpha, m, n < \omega, m \neq 0 \rangle$$

then it is easy to verify that

$$L_{\alpha+1} = L_\alpha + B_\alpha, \quad |B_\alpha| = |L_{\alpha+1}/L_\alpha|$$

and

$$L_\alpha = \sum_{\rho<\alpha} B_\rho + L_0$$

for all relevant α's. Finally we say that a subset $S \subseteq \mu$ is *closed*, if for each $\beta \in S$, $L_\beta \cap B_\beta \leq L_0 + \langle B_\gamma \mid \gamma \in S, \gamma < \beta \rangle$.

2.3 LEMMA If $S \subseteq \mu$ is a closed subset, then $G(S) = L_0 + \sum_{\beta \in S} B_\beta$ is a pure subgroup of G.

Proof: Proving indirectly, let us suppose the set $U = \{x \in G(S) \mid ny = x, y \notin G(S), n \text{ suitable}\}$ is non-empty. With each $x \in U$ we associate a finite chain $\beta_1 < \cdots < \beta_m$ of elements of S in such a way, that $x = l_0 + b_1 + \cdots + b_m$, $l_0 \in L_0, b_i \in B_{\beta_i}$, $i = 1, \ldots, m$, and we shall consider an element $x \in U$ with β_m the smallest possible.

So, $y = z + b'_m \in L_{\beta_m+1} = L_{\beta_m} + B_{\beta_m}$, $z \in L_{\beta_m}, b'_m \in B_{\beta_m}$, L_{β_m+1} being pure in G, and consequently $nz = l_0 + b_1 + \cdots + b_m - nb'_m \in L_0 + \langle B_\beta \mid \beta \in S, \beta < \beta_m \rangle$ owing to $b_m - nb'_m \in B_{\beta_m} \cap L_{\beta_m} \leq L_0 + \langle B_\beta \mid \beta \in S, \beta < \beta_m \rangle$. Hence $z \in G(S)$ by the choice of x and $y = z + b'_m \in G(S)$ yields the desired contradiction. □

2.4 LEMMA If $S \subseteq \mu$ is a closed subset, then $G(S) = L_0 + \sum_{\beta \in S} B_\beta$ is a preseparative subgroup of G.

Proof: By lemma 2.3, $G(S)$ is a pure subgroup of G. For $g \in G$ we denote $\nu(g) = \nu$ if $g \in L_{\nu+1} \setminus L_\nu$ and, proving indirectly, we shall take an element $g \in G \setminus G(S)$ having no preseparative set with respect to $G(S)$ and with $\nu(g) = \nu$ as small as possible.

Assuming $\nu \in S$ we can write $g = \tilde{g} + \tilde{b}, \tilde{g} \in L_\nu, \tilde{b} \in B_\nu$. By the choice of ν, $\tilde{g}$ has a $G(S)$-preseparative set and lemma 2.1 yields a contradiction. So, $\nu \notin S$ and we can choose an L_ν-preseparative set $\{a_n \mid n < \omega\} \cup \{b_n \mid n < \omega\}$ for g, where $a_n \notin G(S)$ and $b_n \in G(S)$ for each $n < \omega$. The choice of ν gives the existence of a $G(S)$-preseparative set $\{a_{nk} \mid k < \omega\}$ of a_n for each $n < \omega$ and it remains to show that $P_g = \{-ra_{nk} \mid r, n, k < \omega, r \neq 0\} \cup \{sb_n \mid s, n < \omega, s \neq 0\}$ is a $G(S)$-preseparative set for g.

Proving indirectly we infer from lemma 2.1 that P_g is not a $G(S)$-preseparative set for mg for any $0 \neq m \in Z$ and consequently for each such integer m we can fix an element $x_m \in G(S)$ with $\nu(x_m) = \mu_m$ as small as possible and such that the type $\mathbf{t}(mg + x_m)$ cannot be estimated by a finite union of types of elements from P_g. Now we select μ to be the smallest ordinal among $\{\mu_m \mid 0 \neq m \in Z\}$. In accordance with lemma 2.1 we can without loss of generality assume that $\mu = \mu_1, x_1 = x$ and $\mu_m \geq \mu$ for each $0 \neq m \in Z$. Now the proof splits into three parts.

Assuming $\mu = \nu$ we can write $x = y + z$, where $y \in L_0 + \langle B_\rho \mid \rho \in S, \rho < \nu \rangle$ and $z \in \langle B_\rho \mid \rho \in S, \rho > \nu \rangle$. Hence $z = z_1 + \cdots + z_k$, where $z_i \in B_{\rho_i}, \rho_1 < \cdots < \rho_k$, $\rho_i \in S$ and ρ_k is the smallest possible. Now $z_k = x - y - z_1 - \cdots - z_{k-1} \in B_{\rho_k} \cap L_{\rho_k} \leq L_0 + \langle B_\rho \mid \rho \in S, \rho < \rho_k \rangle$ yields $z = 0$ and consequently $x = y \in L_0 + \langle B_\rho \mid \rho \in S, \rho < \nu \rangle$. This means $\nu(x) < \nu = \mu$ contradicting the choice of μ.

Proceeding to the case $\mu < \nu$ we have $x \in L_{\mu+1} \leq L_\nu$ and so $\mathbf{t}(g + x) \leq \cup_{i=0}^{r} \mathbf{t}(mg + a_i) \cup \cup_{j=0}^{s} \mathbf{t}(mg + b_j)$ for suitable integers $m, r, s < \omega$, $m \neq 0$, from which $\mathbf{t}(g + x) \leq \cup_{i=0}^{r} \mathbf{t}(a_i - mx) \cup \cup_{j=0}^{s} \mathbf{t}(mg + b_j)$. Further, for each $i \in \{0, \ldots, r\}$ and suitable $0 < n_i < \omega$ we have $\mathbf{t}(a_i - mx) \leq \cup_{k=0}^{q_i} \mathbf{t}(n_i a_i + a_{ik}) = \cup_{k=0}^{q_i} \mathbf{t}(na_i + \tilde{n}_i a_{ik})$, where $n = n_i \tilde{n}_i$ is the least common multiple of $n_0, \ldots, n_r$. Thus $\mathbf{t}(a_i - mx) \leq \cup_{k=0}^{q_i} \mathbf{t}(nmx + \tilde{n}_i a_{ik})$, hence $\mathbf{t}(g + x) \leq \cup_{i=0}^{r} \cup_{k=0}^{q_i} \mathbf{t}(nmx + \tilde{n}_i a_{ik}) \cup \cup_{j=0}^{s} \mathbf{t}(nmg + nb_j)$, giving $\mathbf{t}(g + x) \leq \cup_{i=0}^{r} \cup_{k=0}^{q_i} \mathbf{t}(nmg - \tilde{n}_i a_{ik}) \cup \cup_{j=0}^{s} \mathbf{t}(nmg + nb_j)$, which contradicts the choice of x.

We continue to the last possibility, $\mu > \nu$. In this case $x \in L_{\mu+1} = L_\mu + B_\mu$ and $x \notin L_\mu$ shows that x has a non-zero component in B_μ and consequently $\mu \in S$, for otherwise (as in the case $\mu = \nu$) we obtain $x \in L_0 + \sum_{\rho < \mu} B_\rho \subseteq L_\mu$.

Assume that $\{c_n \mid n < \omega\}$ is a L_μ-preseparative set for x given by the following way. If $\tilde{x} \in R_\mu$ is the selected representative of the coset $x + L_\mu$ and $\{h_k^{\tilde{x}} \mid k < \omega\}$ is the corresponding L_μ-preseparative set for $\tilde{x}$, then with respect to lemma 2.1 every c_n is of the form $h_k^{\tilde{x}} - m(x - \tilde{x})$ for a suitable non-zero integer m. By the construction 2.2 $m\tilde{x} + h_k^{\tilde{x}} \in B_\mu$, hence $c_n \in G(S)$ and $h_k^{\tilde{x}}, x - \tilde{x} \in L_\mu$ yields $\nu(c_n) < \mu$ for each $n < \omega$. Now we have $\mathbf{t}(x + g) \leq \cup_{i=0}^{r} \mathbf{t}(mx + c_i)$ and so $\mathbf{t}(g + x) \leq \cup_{i=0}^{r} \mathbf{t}(mg - c_i)$ for some $r, m < \omega, m \neq 0$. Thus the choice of μ yields for each $i \in \{0, \ldots, r\}$ the following estimation $\mathbf{t}(mg - c_i) \leq \cup_{j=0}^{s_i} \mathbf{t}(n_i mg + \alpha_j b_j) \cup \cup_{k=0}^{q_i} \cup_{l=0}^{u_i} \mathbf{t}(n_i mg - \beta_{kl} a_{kl})$ for suitable $s_i, q_i, u_i < \omega$ and $n_i, \alpha_j, \beta_{kl} < \omega$ non-zero. Setting $n = n_i \tilde{n}_i$ the least common multiple of $n_0, \ldots, n_r$, we obtain $\mathbf{t}(g + x) \leq \cup_{i=0}^{r} (\cup_{j=0}^{s_i} \mathbf{t}(nmg + \tilde{n}_i \alpha_j b_j) \cup \cup_{k=0}^{q_i} \cup_{l=0}^{u_i} \mathbf{t}(nmg - \tilde{n}_i \beta_{kl} a_{kl})$, the final contradiction completing the proof of the lemma. □

2.5 LEMMA Any union of closed subsets of μ is closed in μ.
Proof: Let $S = \bigcup_{i\in I} S_i$ with S_i closed in μ. For $\beta \in S$ we have $\beta \in S_i$ for some $i \in I$ and so $L_\beta \cap B_\beta \le L_0 + \langle B_\gamma \mid \gamma \in S_i, \gamma < \beta\rangle \le L_0 + \langle B_\gamma \mid \gamma \in S, \gamma < \beta\rangle$. □

2.6 DEFINITION Let λ be an infinite cardinal. We say that a smooth ascending union $G = \bigcup_{\alpha<\mu} L_\alpha$ of G-pure subgroups is a *λ-chain from L_0 to G* if $|L_{\alpha+1}/L_\alpha| \le \lambda$ for each $\alpha + 1 < \mu$. If all L_α's are G-preseparative, then we speak about a *λ-preseparative chain from L_0 to G*. Especially, for $L_0 = 0$, we shall speak simply about a *λ-chain* or a *λ-preseparative chain of G*.

2.7 LEMMA Let $G = \bigcup_{\alpha<\mu} L_\alpha$ be a λ-preseparative chain from L_0 to G. If $\overline{S} \subseteq \mu$ is any closed subset and $F \subseteq \overline{S}$ is any subset with $|F| \le \lambda$, then there is a closed subset $S \subseteq \mu$ such that $F \subseteq S \subseteq \overline{S}$ and $|S| \le \lambda$.
Proof: If $|\overline{S}| \le \lambda$ then there is nothing to prove. Otherwise we set $S_0 = F$ and assume that $S_0 \subseteq S_1 \subseteq \cdots \subseteq S_n \subseteq \overline{S}$ of cardinalities $\le \lambda$ have been already defined. For each $\beta \in S_n$ the intersection $L_\beta \cap B_\beta$ has cardinality at most λ and since $|S_n| \le \lambda$, we can find $S_{n+1} \subseteq \overline{S}$ containing S_n and of cardinality $\le \lambda$ such that $L_\beta \cap B_\beta \le L_0 + \langle B_\gamma \mid \gamma \in S_{n+1}, \gamma < \beta\rangle$ for each $\beta \in S_n$. Setting $S = \bigcup_{n<\omega} S_n \subseteq \overline{S}$ we see that S is the desired closed subset of μ. □

3 CLOSED SUBSETS

3.1 DEFINITION Let λ be an infinite cardinal and H be a subgroup of a torsion-free group G. A collection $\mathfrak{C}_\lambda = \mathfrak{C}_\lambda(H, G)$ of pure subgroups of G containing H is said to be an *SL_λ-family of H in G* if $H, G \in \mathfrak{C}_\lambda$ and (i) $\sum_{\alpha\in I} H_\alpha \in \mathfrak{C}_\lambda$ whenever H_α belongs to $\mathfrak{C}_\lambda$ for each $\alpha \in I$; (ii) if $V \subseteq \tilde{V}$ are elements of $\mathfrak{C}_\lambda$ and $X \subseteq \tilde{V}$ is any subset with $|X| \le \lambda$ then there is $U \in \mathfrak{C}_\lambda$ such that $V \cup X \subseteq U \le \tilde{V}$ and $|U/V| \le \lambda$.

If, instead of (i), a weaker condition stating that $\mathfrak{C}_\lambda$ is closed under arbitrary smooth ascending unions is satisfied, then we say that $\mathfrak{C}_\lambda$ is a *WL_λ-family of H in G*. In both cases we shall also speak about families of G over H. Especially, for $H = 0$ we shall speak simply about families of G.

Before continuing we recall some further definitions. If A, B are subgroups of a torsion-free group G then $A\|B$ (Hill's compatibility) used in [AH] and [DHR] means that $A + B$ is G-pure (not necessarily in [DHR]) and for all $a \in A, b \in B$ there is $c \in A \cap B$ with $|a+b| \le |a+c|$. Since this notion is related to balancedness, we adapted it in [B] slightly for prebalancedness. Recall that $A\|B$ will mean that $A + B$ is pure in G and for each $a \in A$ and $b \in B$ there is $0 < m < \omega$ and $\{c_0, \ldots, c_n\} \subseteq A \cap B$ such that

$$\mathbf{t}(a+b) \leq \cup_{i=0}^{n} \mathbf{t}(ma + c_i).$$

Let K be a pure subgroup of a completely decomposable group $C = \oplus_{i\in M} X_i$ with X_i rank one and torsion-free. For each $J \subseteq M$ we shall use the following notations.

$$C(J) = \oplus_{i\in J} X_i, \quad K(J) = C(J) \cap K$$

and $\varphi_i : C \to X_i$ will denote the canonical projection. Moreover, J is said to be K-prespecial if $C(J) \| K$.

Now we are ready to start our investigations. Unifying the theories of the families of prebalanced and preseparative subgroups we shall speak about the family $\mathfrak{C}$ of pre-$*$ subgroups to express that the family $\mathfrak{C}$ consists of the prebalanced, or the preseparative subgroups, respectively. In the same sense we shall use the term pre-$*$ set of an element over a subgroup.

3.2 DEFINITION Let K be a subgroup of a completely decomposable group $C = \bigoplus_{i\in M} X_i$ having a WL_λ-family of pre-$*$ subgroups $\mathfrak{C}_\lambda$ in C. For $V \in \mathfrak{C}_\lambda$ we say that a subset $J \subseteq M$ is $\mathfrak{C}_\lambda(V)$*-closed* if $C(J) + V \in \mathfrak{C}_\lambda$ and each element from $C(J) \setminus V$ has a V-pre-$*$ set contained in $C(J) \cap V$.

In the sequel we shall need the following fact proved in [B; L.2.9] (cf. [DHR; L.7.2] in the "balanced" case): If A, B, C are subgroups of a torsion-free group G such that $A \| B$ and $(A + B) \| C$, then $B \| (A + C)$. The following simple lemma will be also useful.

3.3 LEMMA Let K be a subgroup of a torsion-free group C. Then

(i) if $\mathfrak{C}_\lambda$ is a WL_λ-family of pre-$*$ subgroups of K in C and if $U, V \in \mathfrak{C}_\lambda$, $U \leq V$, are arbitrary, then $\mathfrak{C}_\lambda(U,V) = \{\tilde{V} \in \mathfrak{C}_\lambda \mid U \leq \tilde{V} \leq V\}$ is an WL_λ-family of pre-$*$ subgroups of U in V;

(ii) if K is C-prebalanced and $K \leq H \leq C$ is arbitrary, then H has a WL_λ-family of pre-$*$ subgroups in C if and only if H/K has an WL_λ-family of pre-$*$ subgroups in C/K;

(iii) if $\mathfrak{C}_\lambda$ is a WL_λ-family of pre-$*$ subgroups of K in C and if L is C-prebalanced and $U \in \mathfrak{C}_\lambda$ is any member containing L, then $\mathfrak{C}_\lambda(U/L, C/L) = \{V/L \mid V \in \mathfrak{C}_\lambda, U \leq V\}$ is a WL_λ-family of pre-$*$ subgroups of U/L in C/L;

(iv) if $C = \bigoplus_{i\in M} X_i$ is completely decomposable, $\mathfrak{C}_\lambda$ is a WL_λ-family of pre-$*$ subgroups of K in C and $V \in \mathfrak{C}_\lambda$ is arbitrary then every $\mathfrak{C}_\lambda(V)$-closed subset $J \subseteq M$ is V-prespecial;

(v) if $C = \bigoplus_{i\in M} X_i$ is completely decomposable, $\mathfrak{C}_\lambda$ is a WL_λ-family of pre-$*$ subgroups of K in C, $V \in \mathfrak{C}_\lambda$ is arbitrary and $J \subseteq \tilde{J} \subseteq M$ are such that J is $\mathfrak{C}_\lambda(V)$-closed and $\tilde{J}$ is $\mathfrak{C}_\lambda(C(J)+V)$-closed, then $\tilde{J}$ is $\mathfrak{C}_\lambda(V)$-closed.

Proof: (i), (ii) and (iii) are obvious.

(iv) For $g \in C(J) \setminus V$ it suffices to consider the V-pre-$*$ set for g contained in $C(J) \cap V$.

(v) Clearly, $C(\tilde{J}) + V \in \mathfrak{C}_\lambda$ and so let $g \in C(\tilde{J}) \setminus V$ be arbitrary. Obviously, only the case $g \notin C(J)$ requires a verification. Assume first, that $\mathfrak{C}_\lambda$ consists of preseparative subgroups. By hypothesis, there is a $(C(J) + V)$-preseparative set $\{x_n + v_n \mid x_n \in C(J), v_n \in V, n < \omega\}$ for g contained in $C(\tilde{J}) \cap (C(J) + V)$. Further, each x_n has a V-preseparative set $\{w_{nm} \mid m < \omega\}$ contained in $C(J) \cap V$. Now $v_n = x_n + v_n - x_n \in C(\tilde{J})$, hence $\alpha v_n - \beta_n w_{nm} \in C(\tilde{J}) \cap V$ for all $n, m, \alpha, \beta_n < \omega, \alpha \neq 0, \beta_n \neq 0$ and for each $v \in V$ we have $\mathbf{t}(g + v) \leq \cup_{i=0}^{r} \mathbf{t}(lg + x_i + v_i)$ for suitable $0 < l < \omega, r < \omega$ and so $\mathbf{t}(g+v) \leq \cup_{i=0}^{r} \mathbf{t}(x_i + v_i - lv)$. Moreover, $\mathbf{t}(x_i + v_i - lv) \leq \cup_{k=0}^{q_i} \mathbf{t}(s_i x_i + w_{ik})$ for suitable $0 < s_i < \omega, q_i < \omega$ yields $\mathbf{t}(x_i + v_i - lv) \leq \cup_{k=0}^{q_i} \mathbf{t}(sv_i - slv - \tilde{s}_i w_{ik})$, where $s = s_i \tilde{s}_i$ is the least common multiple of $s_0, \ldots, s_r$. Finally, $\mathbf{t}(g + v) \leq \cup_{i=0}^{r} \cup_{k=0}^{q_i} \mathbf{t}(slg + sv_i - \tilde{s}_i w_{ik})$ shows that $\{\alpha v_i - \beta_i w_{ik} \mid i, k, \alpha, \beta_i < \omega, \alpha \neq 0, \beta_i \neq 0\}$ is the desired V-preseparative set for g.

We proceed now to the prebalanced subgroups case. There is $\{x_i + v_i \mid x_i \in C(J), v_i \in V, i \leq r\} \subseteq C(\tilde{J})$ such that $|lg + C(J) + V| = \cup_{i=0}^{r} |lg + x_i + v_i|$ for some $l =\neq 0$ and for each $0 \leq i \leq r$ there is $\{w_{ik} \mid \ \leq q_i\} \subseteq C(J) \cap V$ such that $|s_i x_i + V| = \cup_{k=0}^{q_i} |s_i x_i + w_{ik}|$. Setting $s = s_i \tilde{s}_i$ the least common multiple of $s_0, s_1, \ldots, s_r$ and taking $v \in V$ arbitrarily, we have $|lsg + v| \leq |lsg + C(J) + V| = \cup_{i=0}^{r} |lsg + sx_i + sv_i|$ and so $|lsg + v| \leq \cup_{i=0}^{r} |sx_i + sv_i - v|$. Furthermore, $|sx_i + sv_i - v| \leq |sx_i + V| = \cup_{k=0}^{q_i} |sx_i + \tilde{s}_i w_{ik}|$ for each $i = 0, \ldots, r$. Hence $|sx_i + sv_i - v| \leq \cup_{k=0}^{q_i} |sv_i - v - \tilde{s}_i w_{ik}|$ and consequently $|lsg + v| \leq \cup_{i=0}^{r} \cup_{k=0}^{q_i} |sv_i - v - \tilde{s}_i w_{ik}|$. Finally, $|lsg + v| \leq \cup_{i=0}^{r} \cup_{k=0}^{q_i} |lsg + sv_i - \tilde{s}_i w_{ik}|$ for each $v \in V$ easily yields that $|lsg + V| = \cup_{i=0}^{r} \cup_{k=0}^{q_i} |lsg + sv_i - \tilde{s}_i w_{ik}|$ and the proof is complete. □

3.4 LEMMA Let a subgroup K of a completely decomposable group $C = \bigoplus_{i \in M} X_i$ have an WL_λ-family of pre-$*$ subgroups $\mathfrak{C}_\lambda$ in C. If V is any member of $\mathfrak{C}_\lambda$ and $X \subseteq C$ is any subset with $|X| \leq \lambda$, then there is a subset $J \subseteq M$ such that $|J| \leq \lambda$, $V \cup X \subseteq C(J) + V$ and $C(J) + V \in \mathfrak{C}_\lambda$.

Proof: Set $J_0 = \{i \in M \mid \varphi_i(X) \neq 0\}$ and assume that $J_n \subseteq M$ has been already defined for some $n < \omega$ in such a way that $|J_n| \leq \lambda$. By the hypothesis there is $V_{n+1} \in \mathfrak{C}_\lambda$ such that $C(J_n) + V \leq V_{n+1}$ and $|V_{n+1}/V| \leq \lambda$. Now we can write $V_{n+1} = \tilde{V}_{n+1} + V$, where $|\tilde{V}_{n+1}| = |V_{n+1}/V| \leq \lambda$ and we set $J_{n+1} = J_n \cup \{i \in M \mid \varphi_i(\tilde{V}_{n+1}) \neq 0\}$. Setting $J = \bigcup_{n<\omega} J_n$ we clearly have $|J| \leq \lambda$ and $C(J) + V \in \mathfrak{C}_\lambda$ because $C(J) + V = \bigcup_{n<\omega}(C(J_n) + V) \leq \bigcup_{n<\omega} V_{n+1} \leq \bigcup_{n<\omega}(C(J_{n+1}) + V) = C(J) + V$. □

3.5 LEMMA Let a subgroup K of a completely decomposable group $C = \bigoplus_{i \in M} X_i$ have a WL_λ-family of pre-$*$ subgroups $\mathfrak{C}_\lambda$ in C. If $V \in \mathfrak{C}_\lambda$ and $J \subseteq \tilde{J} \subseteq M$ are such that $C(J) + V, C(\tilde{J}) + V$ are in $\mathfrak{C}_\lambda$ then for each subset $X \subseteq C(\tilde{J}) + V$ with $|X| \leq \lambda$ there is $I \subseteq \tilde{J}$ with $|I| \leq \lambda$, $X \subseteq C(I) + V$ and $C(J \cup I) + V \in \mathfrak{C}_\lambda$.

Proof: Write each element $x \in X$ in the form $x = \tilde{c}_x + v_x, \tilde{c}_x \in C(\tilde{J}), v_x \in V$ and set $I_0 = \{i \in \tilde{J} \mid \varphi_i(\tilde{c}_x) \neq 0, x \in X\}$. Assuming $I_n \subseteq \tilde{J}$ with $|I_n| \leq \lambda$ has been already constructed, there is $V_{n+1} \in \mathfrak{C}_\lambda$ with $C(J \cup I_n) + V \leq V_{n+1} \leq C(\tilde{J}) + V$ and $|V_{n+1}/(C(J) + V)| \leq \lambda$. Now we have $V_{n+1} = \tilde{V}_{n+1} + C(J) + V$ with $|\tilde{V}_{n+1}| \leq \lambda$. Writing each $x \in \tilde{V}_{n+1}$ in the form $x = \tilde{c}_x + v_x, \tilde{c}_x \in C(\tilde{J}), v_x \in V$, we can set $I_{n+1} = I_n \cup \{i \in \tilde{J} \mid \varphi_i(\tilde{c}_x) \neq 0, x \in \tilde{V}_{n+1}\}$. Obviously, $|I| \leq \lambda$, where $I = \bigcup_{n<\omega} I_n$ and $C(J \cup I) + V \in \mathfrak{C}_\lambda$ in view of $C(J \cup I) + V = \bigcup_{n<\omega}(C(J \cup I_n) + V) \leq \bigcup_{n<\omega} V_{n+1} \leq \bigcup_{n<\omega}(C(J) + C(I_{n+1}) + V) = C(J \cup I) + V$. □

3.6 LEMMA Let P be a pre-$*$ and K a pure subgroup of a torsion-free group G. If $P \parallel K$ then $P \cap K$ is a pre-$*$ subgroup of K.

Proof: For the K-prebalancedness of $P \cap K$ we refer to [B; L.2.9] (cf. [DHR; L.7.2] for the "balanced" case).

Proceeding to the K-preseparativity, take $x \in K \setminus P \cap K$ arbitrarily. There is P-preseparative set $\{k_n \mid n < \omega\}$ for x and the compatibility relation yields the existence of $\{c_{ij}^m \mid j \leq s_i, i, m < \omega, m \neq 0\} \subseteq P \cap K$ such that $\mathbf{t}(mx + k_i) \leq \bigcup_{j=0}^{s_i} \mathbf{t}(m_i mx + c_{ij}^m)$ for suitable $0 < m_i < \omega$. Now for $k \in P \cap K$ arbitrary we have $\mathbf{t}(x + k) \leq \cup_{i=0}^n \mathbf{t}(mx + k_i) \leq \cup_{i=0}^n \cup_{j=0}^{s_i} \mathbf{t}(\tilde{m}mx + \tilde{m}_i c_{ij}^m)$, where $\tilde{m} = m_i \tilde{m}_i$ is the least common multiple of $m_0, \ldots, m_n$ showing that $\{lc_{ij}^m \mid j \leq s_i, i, m, l < \omega, l \neq 0, m \neq 0\}$ is a $P \cap K$-preseparative set for x. □

3.7 LEMMA Let a subgroup K of a completely decomposable group $C = \bigoplus_{i \in M} X_i$ have a WL_λ-family of pre-$*$ subgroups $\mathfrak{C}_\lambda$ in C. If $J \subseteq M$ is a $\mathfrak{C}_\lambda(K)$-closed subset, then the set $\mathfrak{C}_\lambda^J = \{C(L) + K(J) \mid L \subseteq J, C(L) + K \in \mathfrak{C}_\lambda\}$ is a WL_λ-family of pre-$*$ subgroups of $K(J) = K \cap C(J)$ in $C(J)$.

Proof: The set $\mathfrak{C}_\lambda^J$ obviously contains $K(J), C(J)$ and since $C(\tilde{L}) + K = C(\tilde{L} \cup L) + K$ whenever $C(L) + K \leq C(\tilde{L}) + K$, $\mathfrak{C}_\lambda^J$ is closed under smooth ascending unions. Further, if $L \subseteq \tilde{L} \subseteq J$ are such that $C(L) + K, C(\tilde{L}) + K \in \mathfrak{C}_\lambda$ and if $X \subseteq C(\tilde{L}) + K(J) \leq C(J)$ is any subset with $|X| \leq \lambda$, then lemma 3.5 (and the beginning of its proof) yields the existence of $I \subseteq \tilde{L}$ with $|I| \leq \lambda$, $X \subseteq C(I) + K$ and $C(L \cup I) + K \in \mathfrak{C}_\lambda$. Thus, $X \subseteq ((C(L \cup I) + K) \cap C(J) = C(L \cup I) + K(J)$ and it remains to verify that the members of $\mathfrak{C}_\lambda^J$ are $C(J)$-pre-$*$. However, the $\mathfrak{C}_\lambda(K)$-closedness of J yields $C(J) \| K$, while $L \subseteq J$ gives $C(L) \| C(J)$. Hence $C(J) \| (C(L) + K)$ by [B; L.2.9] and lemma 3.6 applies. □

3.8 LEMMA Let a subgroup K of a completely decomposable group $C = \bigoplus_{i \in M} X_i$ have a WL_λ-family of pre-$*$ subgroups $\mathfrak{C}_\lambda$ in C. If $V \in \mathfrak{C}_\lambda$ is arbitrary and $X \subseteq C$ is any subset with $|X| \leq \lambda$ then there is a $\mathfrak{C}_\lambda(V)$-closed subset $J \subseteq M$ such that $|J| \leq \lambda$ and $X \subseteq C(J) + V$.

Proof: Set $J_0 = \{i \in M \mid \varphi_i(X) \neq 0\}$ and assume that J_n with $|J_n| \leq \lambda$ has been already defined. For each $g \in C(J_n)$ we select a V-pre-$*$ set $\{h_k^g \mid k < \omega\}$. Taking J_{n+1} by lemma 3.4 with respect to $X = C(J_n) \cup \{h_k^g \mid k < \omega, g \in C(J_n) \setminus V\}$, we see easily that $J = \bigcup_{n<\omega} J_n$ is the desired subset. □

3.9 PROPOSITION Let K be a subgroup of a completely decomposable group $C = \bigoplus_{i \in M} X_i$ such that $|K| = |C/K| = \kappa \geq \aleph_1$ and K has a WL_λ-family of pre-$*$ subgroups $\mathfrak{C}_\lambda$ in C. Then M is a smooth ascending union $M = \bigcup_{\alpha<\kappa} J_\alpha$, such that for each $\alpha < \kappa$

(i) $|J_\alpha| < \kappa$;

(ii) J_α is $\mathfrak{C}_\lambda(K)$-closed;

(iii) $J_{\alpha+1}$ is $\mathfrak{C}_\lambda(C(J_\alpha) + K)$-closed.

Proof: Let $M = \{m_\alpha \mid \alpha < \kappa\}$ be an arbitrary well-ordering of M, set $J_0 = \emptyset$ and assume that $J_\beta, \beta < \alpha$, has been already defined for some ordinal $\alpha < \kappa$. For α limit we simply set $J_\alpha = \bigcup_{\beta<\alpha} J_\beta$, while for $\alpha = \gamma + 1$ we select J_α in accordance to $V = C(J_\gamma) + K$ and $X = \{m_\gamma\}$. Obviously, $M = \bigcup_{\alpha<\kappa} J_\alpha$ is a smooth ascending union satisfying (i), (iii) follows from lemma 3.8 while (ii) is a consequence of lemma 3.3. □

3.10 THEOREM Let λ be an infinite cardinal number such that each pre-$*$ subgroup of a completely decomposable group of cardinality at most λ is a B_2-group. If a subgroup K has a WL_λ-family of pre-$*$ subgroups in a completely decomposable group C then K is a B_2-group.

Proof: Without loss of generality we can assume that $|K| = |C/K| = \kappa > \lambda$ is the smallest cardinality for which the assertion is false. Write M as a smooth ascending union $M = \bigcup_{\alpha<\kappa} J_\alpha$ by the preceding proposition. By lemma 3.7 every $K_\alpha = C(J_\alpha) \cap K$ has an WL_λ-family of pre-$*$ subgroups in $C(J_\alpha)$ and consequently it is a B_2-group by the choice of κ. Further, K_α is $K_{\alpha+1}$-prebalanced by lemma 3.6 and so with respect to [DHR; Cor.3.10] it remains to verify that $K_{\alpha+1}/K_\alpha$ is a B_2-group. However, $K_{\alpha+1}/K_\alpha \cong (C(J_\alpha) + K_{\alpha+1})/C(J_\alpha)$, where $C(J_\alpha) + K_{\alpha+1}$ is a member of an WL_λ-family of pre-$*$ subgroups of $K_{\alpha+1}$ in $C(J_{\alpha+1})$ by lemma 3.7. Thus $(C(J_\alpha) + K_{\alpha+1})/C(J_\alpha)$ has an WL_λ-family pre-$*$ subgroups in $C_{\alpha+1}/C_\alpha$ by lemma 3.3 and so it is a B_2-group by the choice of κ. □

3.11 COROLLARY Let λ be an infinite cardinal such that each pre-$*$ subgroup of a completely decomposable group of cardinality at most λ is a B_2-group. If a subgroup H has a WL_λ-family pre-$*$ subgroups in a B_2-group B, then H is a B_2-group.

Proof: Consider the commutative diagram

$$\begin{array}{ccccccccc} F:0 & \longrightarrow & K & \longrightarrow & D & \longrightarrow & H & \longrightarrow & 0 \\ & & \| & & \downarrow & & \downarrow & & \\ E:0 & \longrightarrow & K & \longrightarrow & C & \longrightarrow & B & \longrightarrow & 0 \end{array}$$

where E is a balanced-projective resolution of B. The sequence F is TEP, E being so, and hence by [R1; Th.3] H is a B_2-group, provided D and K are so. However, K is a B_2-group by [F2; Cor.2.5], B being a B_2-group, while D is a B_2-group by theorem 3.10 and lemma 3.3. □

3.12 COROLLARY If a subgroup H has a WL_λ-family pre-$*$ subgroups in a B_2-group B, where $\lambda = \aleph_0$ or $\lambda = \aleph_1$, then H is a B_2-group.
Proof: By [BR; Cor.5.7] every pre-$*$ subgroup of a B_1-group of cardinality at most $\aleph_1$ is a B_1-group and consequently a B_2-group by [DHR]. □

4 MAIN RESULTS

4.1 DEFINITION Let λ be an infinite cardinal and H be a subgroup of a torsion-free group G. A collection $\mathfrak{C}_\lambda = \mathfrak{C}_\lambda(H, G)$ of pure subgroups of G containing H is said to be an *axiom-3-λ-family of H in G* if $H, G \in \mathfrak{C}_\lambda$ and (i) $\sum_{\alpha \in I} H_\alpha \in \mathfrak{C}_\lambda$ whenever H_α belongs to $\mathfrak{C}_\lambda$ for each $\alpha \in I$; (ii) if V is any elements of $\mathfrak{C}_\lambda$ and $X \subseteq G$ is any subset with $|X| \leq \lambda$ then there is $U \in \mathfrak{C}_\lambda$ such that $V \cup X \subseteq U$ and $|U/V| \leq \lambda$.

If, instead of (i), a weaker condition stating that $\mathfrak{C}_\lambda$ is closed under arbitrary smooth ascending unions is satisfied, then we say that $\mathfrak{C}_\lambda$ is a *$G(\lambda)$-family of H in G*. In both cases we shall also speak about families of G over H. Especially, for $H = 0$ we shall speak simply about families of G.

Manifestly, every SL_λ-family is a WL_λ-family and an axiom-3-λ-family and each of the last two families is a $G(\lambda)$-family. Moreover, if a torsion-free group G has a $G(\lambda)$-family over H, then there is a λ-chain from H to G.

4.2 THEOREM The following conditions are equivalent for a subgroup H of a torsion-free group G and an infinite cardinal λ:

(i) there is an SL_λ-family of preseparative subgroups of G over H;
(ii) there is a WL_λ-family of preseparative subgroups of G over H;
(iii) there is an axiom-3-λ-family of preseparative subgroups of G over H;
(iv) there is a $G(\lambda)$-family of preseparative subgroups of G over H;
(v) there is a λ-preseparative chain from H to G.

Proof: Obviously, only the implication (v) $\Rightarrow$ (i) requires a verification. For a λ-preseparative chain $G = \bigcup_{\alpha<\mu} L_\alpha$ from H to G we take $\mathfrak{C}_\lambda = \{G(S) \mid S \subseteq \mu, S \text{ closed}\}$. Then $H, G \in \mathfrak{C}_\lambda$ and the condition (i) from the definition 3.1 follows from lemma 2.5. Further, if $\tilde{S} \subseteq \bar{S}$ are closed subsets of μ and $X \subseteq G(\bar{S})$ is of the cardinality at most λ, then $F = \{i \in \bar{S} \mid \varphi_i(X) \neq 0\} \subseteq \bar{S}$ has the same property and consequently lemma 2.7 yields the existence of a closed subset $S \subseteq \mu$ with $F \subseteq S \subseteq \bar{S}$ and $|S| \leq \lambda$. By lemma 2.5 the union $\tilde{S} \cup S \subseteq \bar{S}$ is closed in μ and the rest is easy to verify. □

4.3 THEOREM (CH) Let $\lambda \leq \aleph_\omega$ be an infinite cardinal. If a subgroup H has a WL_λ-family of prebalanced subgroups in a B_2-group G, then H is a B_2-group.

Proof: By [B; Cor.4.5] every prebalanced subgroup of a completely decomposable group of cardinality at most $\aleph_\omega$ is a B_2-group and it suffices to use corollary 3.11. □

4.4 THEOREM Let κ be an infinite cardinal such that every preseparative subgroup of a completely decomposable group of cardinality at most κ is a B_2-group. The following conditions are equivalent for a subgroup H of a B_2-group G and any cardinal λ with $\aleph_0 \leq \lambda \leq \kappa$:

(i) H is a B_2-group;

(ii) there is an SL_λ-family of preseparative subgroups of G over H;

(iii) there is a WL_λ-family of preseparative subgroups of G over H;

(iv) there is an axiom-3-λ-family of preseparative subgroups of G over H;

(v) there is a $G(\lambda)$-family of preseparative subgroups of G over H;

(vi) there is a λ-preseparative chain from H to G.

Proof: The conditions (ii)–(vi) are equivalent by theorem 4.2 the condition (i) follows from (ii) by corollary 3.11. So, let H be a B_2-group and let $0 \to K \to H \oplus C \to G \to 0$ be any relative balanced-projective resolution. Now there is an $\aleph_0$-preseparative chain from K to $H \oplus C$, G being a B_2-group, and consequently K is a B_2-group. An application of [F2; Th.2.4] (cf. [F3; Th.3.3]) gives the existence of an $\aleph_0$-preseparative chain and hence a λ-preseparative chain from H to G. □

4.5 COROLLARY The conditions (i)–(vi) from theorem 4.4 are equivalent for a subgroup H of a B_2-group G and $\lambda = \aleph_0$ or $\lambda = \aleph_1$.

Proof: By [BR; Cor.5.7] every preseparative subgroup of a B_1-group of cardinality at most $\aleph_1$ is a B_1-group and consequently a B_2-group by [DHR]. □

4.6 THEOREM Assume that for some infinite cardinal κ every prebalanced subgroup of a completely decomposable group of cardinality at most κ is a B_2-group. The following conditions for a B_1-group G and any cardinal λ with $\aleph_0 \leq \lambda \leq \kappa$ are equivalent:

(i) G is a B_2-group;
(ii) there is an SL_λ-family of prebalanced subgroups of G;
(iii) there is a WL_λ-family of prebalanced subgroups of G.

Proof: Let G be a B_1-group having a WL_λ-family of prebalanced subgroups and let $E : 0 \to K \to C \to G \to 0$ be its balanced-projective resolution. By lemma 3.3 C has a WL_λ-family of prebalanced subgroups over K and consequently K is a B_2-group by theorem 3.10. Moreover, E is TEP, G being a B_1-group and [R2; Th.3] applies. The rest is obvious. □

4.7 THEOREM (CH) The following conditions for a B_1-group G and any cardinal λ with $\aleph_0 \leq \lambda \leq \aleph_\omega$ are equivalent:
(i) G is a B_2-group;
(ii) there is an SL_λ-family of prebalanced subgroups of G;
(iii) there is a WL_λ-family of prebalanced subgroups of G.

Proof: It follows from [BF] that under (CH) every prebalanced subgroups of a completely decomposable group of cardinality at most $\aleph_\omega$ is a B_2-group. □

4.8 COROLLARY The conditions (i)–(iii) from theorem 4.7 are equivalent for any B_1-group B and $\lambda = \aleph_0$ or $\lambda = \aleph_1$.

Proof: By [DHR] every prebalanced subgroups of a completely decomposable group of cardinality $\aleph_1$ is a B_2-group. □

4.9 THEOREM Assume that for some infinite cardinal κ every preseparative subgroup of a completely decomposable group of cardinality at most κ is a B_2-group. The following conditions for a B_1-group G and any cardinal λ with $\aleph_0 \leq \lambda \leq \kappa$ are equivalent:
(i) G is a B_2-group;
(ii) there is an SL_λ-family of preseparative subgroups of G;
(iii) there is a WL_λ-family of preseparative subgroups of G;
(iv) there is an axiom-3-λ-family of preseparative subgroups of G;
(v) there is a $G(\lambda)$-family of preseparative subgroups of G;
(vi) there is a λ-preseparative chain of G.

Proof: If G is a B_2-group then by theorem 4.4 G has an SL_λ-family of preseparative subgroups. Let G be a B_1-group having a λ-preseparative chain and let $E : 0 \to K \to C \to G \to 0$ be its balanced-projective resolution. Then K has a λ-preseparative chain in C and consequently it is a B_2-group again by theorem 4.4. Moreover, E is TEP , G being a B_1-group and an application of [R2; Th.3] shows that G is a B_2-group. The rest is obvious. □

4.10 COROLLARY The conditions (i)–(vi) from theorem 4.9 are equivalent for any B_1-group B and $\lambda = \aleph_0$ or $\lambda = \aleph_1$.
Proof: By [BR; Cor.5.7] every preseparative subgroup of a B_1-group of cardinality at most $\aleph_1$ is a B_1-group and consequently a B_2-group by [DHR]. □

REFERENCES

AH U. Albrecht and P. Hill, Butler groups of infinite rank and axiom 3, *Czech. Math. J., 37*: 293-309 (1987).

B L. Bican, Ascending unions of prebalanced subgroups of Butler groups, to appear.

BF L. Bican and L. Fuchs, Subgroups of Butler groups, *Communications in Algebra, 22*: 1037-1047 (1994).

BR L. Bican and K. M. Rangaswamy, Subgroups of Butler groups, to appear.

BS L. Bican and L. Salce, Infinite rank Butler groups *Proc. Abelian Group Theory Conference, Honolulu Lecture Notes in Math., Springer-Verlag, 1006* 171-189 (1983).

Bu M.C.R. Butler, A class of torsion-free abelian groups of finite rank, *Proc. London Math. Soc., 15*: 686-698 (1965).

DHR M. Dugas, P. Hill and K.M. Rangaswamy, Infinite rank Butler groups II, *Trans. Amer. Math. Soc., 320*: 643-664 (1990).

F1 L. Fuchs, Infinite abelian groups, vol. I and II, *Academic Press, New York*, (1973 and 1977).

F2 L. Fuchs, Infinite rank Butler groups, preprint.

F3 L. Fuchs, A survey of Butler groups of infinite rank, *Contemporary Math., 171*: 121-139 (1994).

FMa L. Fuchs and M. Magidor, Butler groups of arbitary cardinality, *Israel J. Math., 130*: 239-263 (1993).

FMe L. Fuchs and C. Metelli, Countable Butler groups, *Contemporary Math.,130*: 133-143 (1992).

R1 K. M. Rangaswamy, A property of B_2-groups, *Comment. Math. Univ. Carolinae, 35*: 627-631 (1994).

R2 K. M. Rangaswamy, A homological characterization of abelian B_2-groups, *Proc. Amer. Math. Soc., 121* 409-415 (1994).

e-mail address: bican@karlin.mff.cuni.cz

Direct Decompositions of Almost Completely Decomposable Abelian Groups

E.BLAGOVESHCHENSKAYA Department of Mathematics.
St.Petersburg State Technical University, St.Petersburg, Russia

Abstract. Block-rigid almost completely decomposable groups which can be described as extensions of completely decomposable groups by elementary p-groups C are represented by matrices with elements from finite field of p elements (p is a prime). Necessary and sufficient conditions on the rank of C are obtained for such groups to be decomposable. These conditions completely describe the decomposability properties that are determined by the rank of C.

1 INTRODUCTION

Direct decompositions of torsion-free abelian groups of finite rank were investigated in many papers: [1-7]. This one is devoted to the same subject in the following case. We consider so-called block-rigid completely decomposable groups X of arbitrary rank n ($n \geq 3$), that contain a completely decomposable group A so that X/A is an elementary p-group of rank t. According to the definition, the critical types of X [4,6,7] are pairwise incomparable. In [7] these groups are described as the extensions of completely decomposable groups by finite groups C. By combining this approach with methods of linear algebra necessary conditions are obtained on $\operatorname{rk} C = t$ for X to be indecomposable. Moreover, these conditions on the values t are best possible. The author is much endebted to

A.Mader for the opportunity to read his unpublished notes [7] that influenced very much the style of this paper. Due to it the results are placed into the context of the current state of the theory of almost completely decomposable groups. From now on all groups are so. Our notation is standard and as in [1,4,6,7].

If a group X is isomorphic to Y, we write $X \cong Y$. For the group generated by some system of elements we use the symbol $\langle \ldots \rangle$. As usual Z is the group of all integers and sometimes we use $\overline{Z}$ to denote the group Z/pZ (we will use field properties of $\overline{Z} = Z/pZ$ too). Note that this symbol is convenient in various situations when we have the natural epimorphisms $Z \longmapsto \overline{Z}$ and $M_n \longmapsto \overline{M_n}$, where $D_n \in M_n$ is a matrix with elements $d_{ij} \in Z$ and $\overline{D_n} \in \overline{M_n}$ is a matrix with elements $\overline{d_{ij}} \in \overline{Z}$. The lower index indicates the number of rows (columns) of a square matrix. Let $F(K)$ denote a matrix F with elements from the field K.

We write $f \in \mathrm{Mon}(G, F)$ if $f : G \longmapsto H$ is an injective homomorphism. At last working with vector spaces over Z/pZ we use the fact that

$$\dim(A + B) = \dim A + \dim B - \dim(A \cap B),$$

which is fundamental for the main results.

2 PRELIMINARIES

We consider a block-rigid group X of rank n $(n \geq 3)$ which contains a completely decomposable group A, so that X/A is an elementary p-group of rank t. More precisely

$$A = A_{\tau_1} \oplus A_{\tau_2} \oplus \ldots \oplus A_{\tau_k}$$

is a decomposition of A into τ_i-homogeneous components, where $T_{cr}(A) = T_{cr}(X) = \{\tau_1, \ldots, \tau_k\}$ is a set of pairwise incomparable types that are not p-divisible.

Let

$$A_{\tau_i} = A^1_{\tau_i} \oplus A^2_{\tau_i} \oplus \ldots \oplus A^{m_i}_{\tau_i}$$

be an arbitrary decomposition of the group A_{τ_i} into rank-one summands, where $m_i = \mathrm{rk} A_{\tau_i}$, $m_1 + m_2 + \ldots + m_k = n$. The group X is supposed to be not rigid [4,7], that is $k < n$. There are rational groups A_i,

$i = 1, \ldots, k$, containing Z, so that $A^s_{\tau_i} \cong A_i$, $s = 1, \ldots, m_i$. Let

$$A = (A_1 a_1 \oplus \ldots \oplus A_1 a_{m_1}) \oplus (A_2 a_{m_1+1} \oplus \ldots \oplus A_2 a_{m_1+m_2}) \oplus \ldots \\ \oplus \left(A_k a_{m_1+\ldots+m_{k-1}+1} \oplus \ldots \oplus A_k a_n\right)$$

be a decomposition of A into rank-one summands, $\rho : A \longmapsto A/pA = \overline{A} = \overline{A_{\tau_1}} \oplus \ldots \oplus \overline{A_{\tau_k}}$ denote the natural epimorphism. Then $a\rho = \overline{a}$ for all $a \in A$ and

$$A/pA = \langle \overline{a_1} \rangle \oplus \ldots \oplus \langle \overline{a_n} \rangle, \qquad |\overline{a_i}| = p \tag{1}$$

Futhermore ρ will be used to denote as well the induced homomorphisms $\rho : \mathrm{End}A \longmapsto \mathrm{End}\overline{A}$ and $\rho : \mathrm{Aut}A \longmapsto \mathrm{Aut}\overline{A}$. We consider such almost completely decomposable groups as the extensions of completely decomposable group A by finite group

$$C = \langle s_1 \rangle \oplus \ldots \oplus \langle s_t \rangle$$

which is an elementary p-group of rank t $(C \cong X/A)$. It is well-known [7, 3.2.2] that

$$\mathrm{Ext}(C, A) \cong \mathrm{Hom}(C, \overline{A}).$$

More precisely $X \cong X_f$ where X_f, $f \in \mathrm{Mon}(C, \overline{A})$, is a torsion-free group determined under natural assumption that $C = X/A$ by the following [7, 3.2.3, 3.2.5]:

$$\begin{array}{ll} 1) & f : \quad X/A \longmapsto A/pA = \overline{A} : \quad (x + A)f = \overline{px}; \\ 2) & X_f = \{(a, c) : cf = \overline{a}\} \subset A \oplus C. \end{array}$$

It is clear [7, 3.2.4, 3.2.5] that X_f contains the group

$$A\epsilon = \{(pa, 0) : a \in A\} \cong A$$

for all $f \in \mathrm{Mon}(C, \overline{A})$, where $\epsilon : A \longmapsto X_f : a\epsilon = (pa, 0)$ is a natural map. From now on, instead of the group X we consider X_f. Since $s_i f$ is a linear combination of elements $\overline{a_1}, \ldots, \overline{a_n}$,

$$s_i f = \beta_{i1}\overline{a_1} + \ldots + \beta_{in}\overline{a_n}, \tag{2}$$

any map $f \in \mathrm{Mon}(C, \overline{A})$ and therefore X_f is represented by matrix of coefficients, which we introduce here:

$$B^f = \{\beta_{ij} : \beta_{ij} \in Z/pZ \quad (1 \leq i \leq t, \quad 1 \leq j \leq n)\}. \tag{3}$$

It contains n columns and t linear independent rows over Z/pZ. Evidently that $t \leq n$.

The first essential tool is the following Proposition [7. 3.2.6].

PROPOSITION 2.1 (Mader) Let $f, g \in \mathrm{Mon}(C, \overline{A})$. If $\eta \in \mathrm{End}A$ and $\xi \in \mathrm{End}C$ satisfy the condition $f\overline{\eta} = \xi g$, then $(a, c)\Psi = (a\eta, c\xi)$ defines a homomorphism from X_f to X_g with $A\epsilon\Psi \subset A\epsilon$. Conversely, if $\Psi \in \mathrm{Hom}(X_f, X_g)$ and $A\epsilon\Psi \subset A\epsilon$ then there exist unique maps $\eta \in \mathrm{End}A$ and $\xi \in \mathrm{End}C$ such that $f\overline{\eta} = \xi g$ and $(a, c)\Psi = (a\eta, c\xi)$ for each $(a, c) \in X_f$. $\triangle$

We need to use this result in the special case when $\xi \in \mathrm{Aut}C$. Then there is $\gamma \in \mathrm{Aut}C$, $\gamma = \xi^{-1}$ and the condition of 2.1 is equivalent to $\gamma f\overline{\eta} = g$. If in addition $\eta \in \mathrm{Aut}A$ then $\Psi : X_f \longmapsto X_g$ is an isomorphism with $A\epsilon\Psi = A\epsilon$.

The second essential ingredient of our approach is some facts, which are connected with the case, when Ψ is injective. In particular,

$$[X_g : X_f\Psi] = [A : A\eta]. \tag{4}$$

We could easily check this fact, but it was established in the proof of Lemma 4.2.2 in [7]. Our interest in injective homomorphisms from X_f to X_g is due to the following characterization of near-isomorphism [1, p.80, 143], established in [7, 4.1.3], which we will serve as definition here.

THEOREM 2.2 Let G and H be torsion-free groups of finite rank. Then G and H are near-isomorphic (in symbols $G \cong_n H$) if and only if for all primes q there is a monomorphism $\varphi_q : G \longmapsto H$ such that q does not divide $[H : G\varphi_q]$. $\triangle$

Applying this theorem to X_f and X_g, where $f, g \in \mathrm{Mon}(C, \overline{A})$, we obtain the following result, which was proved in [7, 4.2.3].

THEOREM 2.3 Let C be an elementary p-group of rank t and $f, g \in \mathrm{Mon}(C, \overline{A})$. Then $X_f \cong_n X_g$ if and only if there is injective map $\Psi \in \mathrm{Hom}(X_f, X_g)$ so that the condition

$$(p, [X_g : X_f\Psi]) = 1 \tag{5}$$

is satisfied. $\triangle$

Proof. $\Longrightarrow$ By definition of near-isomorphism there is a monomorphism $\Psi : X_f \longmapsto X_g$ such that $\gcd(p, [X_g : X_f\Psi]) = 1$.

$\Longleftarrow$ If $q = p$ then the monomorphism Ψ has the property

$$\gcd(q, [X_g, X_f\Psi]) = 1.$$

If $q \neq p$ then the composite map

$$\varphi : X_f \overset{\pi_A}{\longmapsto} A \overset{\epsilon}{\longmapsto} X_g$$

(where $\pi_A : X_f \longmapsto A$ is the canonical projection) is a monomorphism and

$$\gcd(q, [X_g : X_f\varphi]) = 1.$$

By Theorem 2.2 we obtain $X_f \cong_n X_g$. □

It is very important for us that two near-isomorphic torsion-free groups of finite rank are either both indecomposable or both decomposable into summands of the same ranks [1, p.144]. Our main goal are some theorems which constitute sufficient conditions for the group X_f to be decomposable (in terms of rank C).

3 DECOMPOSITIONS OF GROUPS

Return to the matrix B^f, which represents $f \in \mathrm{Mon}(C, \overline{A})$, where C and A are fixed above.

Let M_t denote the group of matrices

$$\Gamma = \{\gamma_{ij} : \gamma_{ij} \in Z/pZ \quad (1 \leq i, j \leq t), \quad \det \Gamma \neq \overline{0}\}$$

and M_n denote the group of matrices

$$D = \{d_{ij} : d_{ij} \in Z \quad (1 \leq i, j \leq n), \quad \det D \neq 0\}$$

of block-diagonal structure:

$$D = \begin{pmatrix} D_{m_1} & 0 & \dots & 0 \\ 0 & D_{m_2} & \dots & 0 \\ \dots & \dots & \dots & \dots \\ 0 & 0 & \dots & D_{m_k} \end{pmatrix} \tag{6}$$

where each submatrix D_{m_s} consists of m_s columns and m_s rows. It is well-known that $\mathrm{Aut}C \cong M_t$ and M_n is isomorphic to the subgroup of $\mathrm{End}A$.

Let $\overline{M_n}$ denote the group of block-diagonal matrices

$$\overline{D} = \{\overline{d_{ij}} : \overline{d_{ij}} \in Z/pZ \quad (1 \leq i, j \leq n), \quad \det \overline{D} \neq \overline{0}\}$$

having the structure (6), which is isomorphic to the subgroup of Aut$\overline{A}$ called TypAut$\overline{A}$ in [7].

From now on we write:

$$D = \{d_{ij}\}; \qquad \overline{D} = \{\overline{d_{ij}}\}; \qquad \Gamma = \{\gamma_{ij}\}; \qquad B^f = \{\beta_{ij}\}.$$

LEMMA 3.1 For each $\overline{D} = \{\overline{d_{ij}}\} \in \overline{M_n}$ there exists $D = \{d_{ij}\} \in M_n$ so that $\rho(d_{ij}) = \overline{d_{ij}}$ and $\det D$ is not divisible by p. $\triangle$
Proof. Let $\det \overline{D} = \overline{k} \neq \overline{0}$, $k \in Z$, and let d_{ij} be the least positive integer so that $\rho(d_{ij}) = \overline{d_{ij}}$. Construct matrix $D = \{d_{ij}\}$. Since $\det D \equiv k \bmod p$ and $k \neq 0$ we have $\det D \neq 0$ and moreover $\det D$ is not divisible by p. $\square$

It is clear that matrix D, constructed in 3.1, represents injective map $\eta \in$ End$A$. Furthermore $\gcd(p, [A : A\eta]) = 1$.

By usual methods of linear algebra we obtain that if $f, g \in$ Mon$(C, \overline{A})$, $\gamma \in$ Aut$C$, $\overline{\eta} \in$ TypAut$\overline{A}$ satisfy the condition $g = \gamma f \overline{\eta}$ of Proposition 2.1 then there are such matrices $\Gamma \in M_t, \overline{D} \in \overline{M_n}$ that

$$B^g = \Gamma B^f \overline{D}. \tag{7}$$

THEOREM 3.2 Let $f, g \in$ Mon$(C, \overline{A})$ are represented by matrices B^f and B^g, C is an elementary p-group. If there exist matrices $\Gamma \in M_t$ and $\overline{D} \in \overline{M_n}$ ($\det \Gamma \neq \overline{0}$, $\det \overline{D} \neq \overline{0}$) so that $B^g = \Gamma B^f \overline{D}$, then $X_f \cong_n X_g$. $\triangle$
Proof. Obvious or easy. By Proposition 2.1, Lemma 3.1 and equality (4) we show that the injective map $\Psi \in$ Hom(X_f, X_g) with $A\epsilon\Psi \subset A\epsilon$, satisfying (5), exists and then use Theorem 2.3. $\square$

REMARK Note that if X_f, X_g are connected by the condition of this theorem, they are indeed so-called type-isomorphic groups (in symbols $X_f \cong_t X_g$) from [7]. But the definition itself of type-isomorphic groups and majority of results (in particular that $X_f \cong_n X_g$ if and only if $X_f \cong_t X_g$) were made under additional assumption that each $A_{\tau_i}\epsilon$ is pure in X_f and X_g. It was assumed that $f, g \in$ ReMon$(C, \overline{A})$. From [7,4.2.3] and (7) we obtain immediately that $X_f \cong_n X_g$, $f, g \in$ ReMon$(C, \overline{A})$, if and only if there are matrices $\Gamma \in M_t$ and $\overline{D} \in \overline{M_n}$, so that $B^g = \Gamma B^f \overline{D}$.

Suppose for the moment that $X_g = X^1 \oplus X^2$, $g \in \mathrm{ReMon}(C, \overline{A})$. Then $A\epsilon$ is a fully invariant subgroup of X_g and $A\epsilon = A_1\epsilon \oplus A_2\epsilon$ where $A_1\epsilon \subset X^1$, $A_2\epsilon \subset X^2$ ([6], 9.3). It implies the decomposition $A = A_1 \oplus A_2$. Since g is injective, $(pa, c) \in X_g$ only if $c = 0$. For each $(a, c) \in X_g$, $c \neq 0$, we can write

$$(a, c) = (a_1, c_1) + (a_2, c_2),$$

where $(a_1, c_1) \in X^1$, $(a_2, c_2) \in X^2$. Therefore

$$p(a, c) = (pa, pc) = (pa_1, pc_1) + (pa_2, pc_2),$$

that is

$$(pa, 0) = (pa_1, 0) + (pa_2, 0)$$

is a sum of elements from X^1 and X^2 respectively. Since $c_1 g = \overline{a_1}$, $c_2 g = \overline{a_2}$ there is the decomposition $C = C_1 \oplus C_2$ so that $C_1 g \subset \overline{A_1}$ and $C_2 g \subset \overline{A_2}$. Therefore the special choice of generating systems of A and C assures that B^g has the following structure after suitable interchanging of its columns:

$$B^g = (B' \mid 0) \tag{8}$$

or

$$B^g = \begin{pmatrix} B' & | & 0 \\ - & - & - \\ 0 & | & B'' \end{pmatrix}. \tag{9}$$

It is clear from the meaning (2) of B^g that the form (8) corresponds to the case when $C_2 = 0$, that is $Cg \subset \overline{A_1}$ and X^2 is completely decomposable.

Thus if X_f, $f \in \mathrm{ReMon}(C, \overline{A})$, is decomposable then by 2.1 and Remark there are matrices $\Gamma \in M_t$, $\overline{D} \in \overline{M_n}$ so that $B^g = \Gamma B^f \overline{D}$ has the form (8) or (9), since $X_f \cong X_g$ implies $X_f \cong_n X_g$, where g is represented by B^g.

Note that $f \in \mathrm{Mon}(C, \overline{A})$ does not belong to $\mathrm{ReMon}(C, \overline{A})$ if and only if there exist matrices $\Gamma \in M_t$ and $\overline{D} \in \overline{M_n}$, so that $B^g = \Gamma B^f \overline{D}$ has the special block-diagonal form

$$B^g = \begin{pmatrix} \alpha & | & 0 \\ - & - & - \\ 0 & | & B'' \end{pmatrix}, \qquad \alpha \in Z/pZ, \qquad \alpha \neq \overline{0}$$

after suitable interchanging of its columns. This is straightforward consequence of the ReMon Criterion [7,3.2.11,3.2.13]. In fact, f does not belong to $\mathrm{ReMon}(C, \overline{A})$ if and only if there are $c \in C$ $(c \neq 0)$ and i

$(1 \leq i \leq k)$, so that $cf \in \overline{A_{\tau_i}}$, where $c = \alpha_1 s_1 + \ldots + \alpha_t s_t$, $\alpha_j \in Z/pZ$. By linear algebra, using (2) and (3), we obtain immediately this result. Thus if $f \notin \mathrm{ReMon}(C, \overline{A})$ then by 3.2 X_f is decomposable.

DEFINITION 3.3 Let $g \in \mathrm{Mon}(C, \overline{A})$. We call B^g a decomposed matrix if it has the form (8) or (9) (after suitable changing of its columns order).

We are now ready for the following theorem which establishes a very convenient matrix instrument for investigating direct decompositions for such groups.

THEOREM 3.4 Let A be a completely decomposable group of rank n, C be an elementary p-group of rank t and let $f \in \mathrm{Mon}(C, \overline{A})$ is represented by matrix B^f (Z/pZ). Then block-rigid group X_f is decomposable if and only if there exist matrix $\Gamma_t\,(Z/pZ)$, $\det \Gamma_t \neq \overline{0}$, and block-diagonal matrix $\overline{D_n}\,(Z/pZ)$, $\det \overline{D_n} \neq \overline{0}$, so that $B^g = \Gamma B^f \overline{D}$ is a decomposed matrix. $\triangle$

Proof. $\Longrightarrow$ Suppose to the contrary that B^f can not be reduced to the form (8) or (9). Then $f \in \mathrm{ReMon}(C, \overline{A})$ automatically and X_f is indecomposable. Since it contradicts the decomposability of X_f, the claimed condition is proved.

$\Longleftarrow$ By 3.2 the group X_g, represented by B^g, is near-isomorphic to X_f and at the same time X_g is decomposable. It is true for X_f. $\square$

DEFINITION 3.5 Let $f \in \mathrm{Mon}(C, \overline{A})$. We call B^f a decomposable matrix if and only if there are Γ and $\overline{D}$ as in 3.4 so that $B^g = \Gamma B^f \overline{D}$ is a decomposed matrix. In the opposite case B^f is an indecomposable matrix.

4 DECOMPOSITIONS OF MATRICES

The last theorem reduces the problem of group decomposability to a problem of matrix decomposability. With these tools we can find sufficient conditions for decomposability using only the rank of B^f.

Draw vertical lines inside B^f to separate from one another the stripes

$F^1, \ldots, F^k$ containing $m_1, m_2, \ldots, m_k$ columns, separated by so-called type-dividing lines marked by the symbol τ.

$$B^f = \Big(\underbrace{F^1}_{m_1} \overset{\tau}{|} \underbrace{F^2}_{m_2} \overset{\tau}{|} \ldots \overset{\tau}{|} \underbrace{F^k}_{m_k}\Big). \tag{10}$$

As usual, to differentiate the unit matrix and the zero matrix from arbitrary submatrices of B^f, we use the symbols 0 and E. Furthermore, $e_1, \ldots, e_t$ will be used to denote the columns of E_t. Without loss of generality we assume

$$m_1 \geq m_2 \geq \ldots \geq m_k \qquad (m_1 \geq 2).$$

Let $m = m_1$ denote rank of the homogeneous component of A having the greatest rank. The matrix ΓB^f, $\Gamma \in M_t$, is obtained from B^f by the finite set of the following operations:

1) adding any row to another;
2) multiplying any row by $\overline{\alpha} \in Z/pZ \quad (\overline{\alpha} \neq \overline{0})$;
3) changing of the order of rows.

Similarly, $B^f\overline{D}$, $\overline{D} \in \overline{M_n}$, is obtained from B^f by these operations with columns only inside of the same stripes. Such operations will be called admissible operations. They do not change ranks of matrices.

We need to use the following well-known result. Let $F(Z/pZ)$ be an $t \times m$ matrix. If $\operatorname{Rk} F = \varrho$ then by rows transformations 1) — 3) and interchanging columns the matrix F can be reduced to the form

$$F' = \begin{pmatrix} E_\varrho & | & H \\ - & - & - \\ 0 & | & 0 \end{pmatrix}. \tag{11}$$

In particular if $\varrho = m$ then

$$F' = \begin{pmatrix} E_m \\ 0 \end{pmatrix} \tag{12}$$

DEFINITION 4.1 B^{f_1} and B^{f_2}, $f_1, f_2 \in \operatorname{Mon}(C, \overline{A})$, are almost equal matrices if B^{f_1} may be obtained from B^{f_2} by changing columns order inside stripes and by changing rows order.

From now on B^f is the class of all matrices B^{f_1}, $f_1 \in \operatorname{Mon}(C, \overline{A})$, which are almost equal to B^f. Evidently that two almost equal matrices are either both indecomposable or both decomposable.

Let $\mathrm{Rk}\, B^f$ denote the rank of B^f and let $(b^i,\dots,b^j)$ denote the matrix, consists of columns number $i,\dots,j$ of matrix B^f. Then $\mathrm{Rk}\,(b^i,\dots,b^j)$ is equal to the dimension of the vector space $\langle b^i,\dots,b^j\rangle$, generated by $b^i,\dots,b^j$ over Z/pZ. Let $\dim\langle F\rangle$ denote the dimension of the vector space, generated by all columns of matrix F.

Let symbol r over arrow denote rows transformations by operations 1), 2), that is multiplying B^f by Γ, and let symbol c over arrow denote analogous columns transformations (inside of the stripes), that is multiplying B^f by $\overline{D}$.

DEFINITION 4.2 If B^f can be decomposed by transformations of its rows, it will be called r-decomposable matrix. Similarly, if B^f can be decomposed by transformations of its columns, it will be called c-decomposable matrix.

LEMMA 4.3 B^f is r-decomposable if and only if after suitable changing of order of its columns b^i (without any restriction) the following is true:

$$\mathrm{Rk}\,(b^1,\dots,b^s) + \mathrm{Rk}\,(b^{s+1},\dots,b^n) = \mathrm{Rk}\, B^f = t \qquad (s < n). \qquad \triangle$$

Proof. Let $r_1 = \mathrm{Rk}\,(b^1,\dots,b^s) > 0, \quad r_2 = \mathrm{Rk}\,(b^{s+1},\dots,b^n) > 0$ be given (in opposite case B^f is decomposed already).

$\Longrightarrow$ Suppose that $\widetilde{B^f} = (b^1,\dots,b^s,b^{s+1},\dots,b^n)$ is obtained from B^f by suitable changing of order of its columns. By the condition

$$\widetilde{B^f} \xrightarrow{r} \begin{pmatrix} F & | & 0 \\ - & - & - \\ \underbrace{0}_{s} & | & \underbrace{G}_{n-s} \end{pmatrix}$$

then $\mathrm{Rk}\, F + \mathrm{Rk}\, G = \mathrm{Rk}\,\widetilde{B^f} = \mathrm{Rk}\, B^f$. Evidently that

$$(b^1,\dots,b^s) \xrightarrow{r} \begin{pmatrix} F \\ 0 \end{pmatrix} \qquad \text{and} \qquad (b^{s+1},\dots,b^n) \xrightarrow{r} \begin{pmatrix} 0 \\ G \end{pmatrix},$$

that is $\mathrm{Rk}\, F = r_1$, $\mathrm{Rk}\, G = r_2$ and the condition $r_1 + r_2 = t$ is satisfied.

$\Longleftarrow$ Suppose that $\widetilde{B^f} = (F \mid G)$ is obtained from B^f by suitably interchanging columns and $\operatorname{Rk} F + \operatorname{Rk} G = \operatorname{Rk} \widetilde{B^f}$. Then

$$\widetilde{B^f} = \left(\underbrace{F}_{s} \mid \underbrace{G}_{n-s} \right) \xrightarrow{r} \begin{pmatrix} F' & \mid & G' \\ - & - & - \\ \underbrace{0}_{s} & \mid & \underbrace{G''}_{n-s} \end{pmatrix} \xrightarrow{r} \begin{pmatrix} F' & \mid & 0 \\ - & - & - \\ 0 & \mid & G'' \end{pmatrix}$$

because

$$\operatorname{Rk} F' = \operatorname{Rk} F = r_1 \qquad \text{and} \qquad \operatorname{Rk} G'' = t - r_1 = \operatorname{Rk} \begin{pmatrix} G' \\ G'' \end{pmatrix},$$

that is all rows of G' are combinations of rows of G''. This means that B^f is r-decomposable. □

Return to the matrix (10):

$$B^f = \left(\underbrace{F^1}_{m_1} \overset{\tau}{|} \underbrace{F^2}_{m_2} \overset{\tau}{|} \dots \overset{\tau}{|} \underbrace{F^k}_{m_k} \right)$$

$(m = m_1 \geq m_2 \geq \dots \geq m_k, \quad m \geq 2)$. Let ϱ denote $\operatorname{Rk} F^1 = \dim \langle F^1 \rangle$.

THEOREM 4.4 If

$$\dim \left(\langle F^1 \rangle \cap \langle F^2 \mid \dots \mid F^k \rangle \right) < m$$

then B^f is a decomposable matrix. △

Proof. Let

$$D = \langle F^1 \rangle \cap \langle F^2 \mid \dots \mid F^k \rangle.$$

Before using Lemma 4.3, by admissing operations we reduce B^f to the form $(F \mid G)$ so that $\operatorname{Rk} F + \operatorname{Rk} G = \operatorname{Rk} B^f$ (this vertical line is not a type-dividing line in general). Let $d = \dim D$.

1). $\varrho = m$. By (12) we have

$$B^f \xrightarrow{r} \begin{pmatrix} E_m & \overset{\tau}{|} & F \\ - & - & - \\ 0 & \mid & G \end{pmatrix} = B^{f_1}$$

for some $f_1 \in \operatorname{Mon}(C, \overline{A})$. Evidently that

$$\operatorname{Rk} E_m = m = \dim \langle F^1 \rangle; \quad \dim \langle \frac{F}{G} \rangle = \dim \langle F^2 \overset{\tau}{|} \dots \overset{\tau}{|} F^k \rangle.$$

Therefore
$$\dim\left(\langle\begin{matrix}E_m\\0\end{matrix}\rangle\cap\langle\begin{matrix}F\\G\end{matrix}\rangle\right)=d.$$
otherwise $\operatorname{Rk}B^f\neq t$. Let P be this space of dimension d. Construct arbitrary basis of P: $b^{m-d+1},b^{m-d+2},\ldots,b^m$ and add to these columns any another columns $b^1,\ldots,b^{m-d}$ so, that $\det(b^1,\ldots,b^m)\neq\overline{0}$ (here we naturally consider that all these columns consist of m elements). Then by columns transformations of B^{f_1} we obtain the following:
$$B^{f_1}\xrightarrow{c}\begin{pmatrix}b^1,\ldots,b^{m-d} & | & b^{m-d+1},\ldots,b^m & \overset{\tau}{|} & F\\ ------ & - & ------ & - & -\\ 0 & | & 0 & | & G\end{pmatrix}=B^{f_2}.$$
It is clear that the left vertical line (which is not a type-dividing line) divides B^{f_2} into two submatrices, so that sum of their ranks is equal to the rank B^f. By Lemma 4.3 B^f is decomposable. Moreover we conclude that $X_f=X^1\oplus X^2$, where X^1 is a completely decomposable homogeneous group of rank $(m-d)$, because $b^1,\ldots,b^{m-d}$ are linear independent.

2). $\varrho<m$. By (11) we obtain
$$B^f\xrightarrow{r}\begin{pmatrix}E_\varrho & | & H & \overset{\tau}{|} & F\\ - & - & - & - & -\\ 0 & | & 0 & | & G\end{pmatrix}=B^{f_1}$$
(this arrow marked r admit changing of columns order inside of the first stripe of B^f, it is foreseed in the definition of almost equal matrices).

Then
$$B^{f_1}\xrightarrow{c}\begin{pmatrix}E_\varrho & | & 0 & \overset{\tau}{|} & F\\ - & - & - & - & -\\ 0 & | & 0 & | & G\end{pmatrix}.$$
So, $(m-\varrho)$ zero-columns are separated. By 4.3 B^f is a decomposable matrix. □

Based on the preceding results, we can obtain the following theorems.

THEOREM 4.5 If $m>n/2$, that is $m>n-m$, then B^f is decomposable. △

Proof. The condition of 4.4 automatically satisfied. □

THEOREM 4.6 If $\operatorname{Rk}B^f\geq n-m+1$ then B^f is decomposable. △

Proof.

$$B^f = \left(\underbrace{F^1}_{m} \overset{\tau}{\mid} \underbrace{G}_{n-m} \right). \tag{13}$$

where

$$G = \left(F^2 \overset{\tau}{\mid} \ldots \overset{\tau}{\mid} F^k \right).$$

As before, $\varrho = \operatorname{Rk} F^1$ and $d = \dim\left(\langle F^1 \rangle \cap \langle G \rangle\right)$. Let $r = \operatorname{Rk} G$. Then $t = \varrho + r - d$. There are two variants.

1). $\varrho + r = t, \quad d = 0$. The condition of 4.3 is satisfied and B^f is decomposable.

2). $\varrho + r > t, \quad d > 0$. Since $\varrho \leq m, \quad r \leq n - m$ we have $d \leq n - t \leq n - (n - m + 1) = m - 1$, that is $d < m$. By 4.4 B^f is decomposable. □

COROLLARY 1 If $\operatorname{Rk} B^f = n - 1$ then B^f is decomposable, because $m > 1$.

THEOREM 4.7 If $\operatorname{Rk} B^f = t < m$ then B^f is decomposable. △
Proof. Since $\varrho < m$ this fact was proved in 4.4, 2). □

THEOREM 4.8 If $t = n$ then B^f is decomposable and X_f is completely decomposable. △
Proof. $B^f \xrightarrow{r} E_t$. □

THEOREM 4.9 If $m = n/2$ then B^f is decomposable. △
Proof. As before, B^f has the form (13).

$$B^f = \left(\underbrace{F^1}_{m} \overset{\tau}{\mid} \underbrace{G}_{m} \right)$$

and

$$d = \dim\left(\langle F^1 \rangle \cap \langle G \rangle\right) \leq m.$$

1). Suppose $m < t$. It is clear that $d = m$ if only $t = m = \operatorname{Rk} F^1 = \operatorname{Rk} G$. This contradicts the assumption $m < t$, hence $d < m$. By 4.4 B^f is decomposable.

2). Suppose $m = t$. If $d < m$ we apply 4.4. If $d = m$ then

$$B^f \xrightarrow{r} \left(\underbrace{E}_{m} \overset{\tau}{\mid} \underbrace{F}_{m} \right) \xrightarrow{c} \left(F \overset{\tau}{\mid} F \right) \xrightarrow{r} \left(E \overset{\tau}{\mid} E \right).$$

Hence, the group X_f is decomposable into summands of rank 2.

3). Suppose $m > t$, then B^f is decomposable by 4.7. □

COROLLARY 2 If $m \geq n/2$ then B^f is decomposable (4.5, 4.9).

Recalling that $\operatorname{Rk} B^f = t = \operatorname{rk} C$, we summarize all previous results in the following theorem which is the main result of this paper and gives necessary conditions for the decomposability of such groups.

THEOREM 4.10 Suppose that X_f, $f \in \operatorname{Mon}(C, \overline{A})$, is a group of rank n so that the maximal rank of a homogeneous summand of the completely decomposable group A is equal to m ($m \geq 2$) and C is an elementary p-group of rank t. Then, if X_f is indecomposable, necessarily $m < n/2$ and $m \leq t \leq n - m$. △

COROLLARY 3 There is not such indecomposable group of rank 3 or rank 4.

The following examples show that these conditions on the values t are best possible.

EXAMPLE 1 Let h, h_1, h_2, h_3 be distinct primes which are different from p. Define:

$$A = h^{-\infty} Za_1 \oplus h^{-\infty} Za_2 \oplus h_1^{-\infty} Za_3 \oplus h_2^{-\infty} Za_4 \oplus h_3^{-\infty} Za_5$$

and $X = \langle A; b_1, b_2, b_3 \rangle$ where

$$b_1 = \frac{a_1 + a_3}{p}, \qquad b_2 = \frac{a_2 + a_4}{p}, \qquad b_3 = \frac{a_1 + a_2 + a_5}{p}.$$

Then $X \cong X_f$, $f \in \operatorname{Mon}(C, \overline{A})$, where $\operatorname{rk} C = 3$. So, $n = 5$, $m = 2 < n/2$, $t = 3 = n - m$.

$$B^f = \left(\begin{array}{cc|c|c|c} 1 & 0 & 1 & 0 & 0 \\ 0 & 1 & 0 & 1 & 0 \\ 1 & 1 & 0 & 0 & 1 \end{array}\right) \xrightarrow{r} \left(\begin{array}{cc|c|c|c} 1 & 0 & 1 & 0 & 0 \\ 0 & 1 & 0 & 1 & 0 \\ 0 & 0 & -1 & -1 & 1 \end{array}\right)$$

(In the printed matrices each vertical line carries a τ mark above it.)

All vertical lines are type-dividing. It is clear that the condition of 4.3 can not be satisfied. Therefore X_f is indecomposable.

Since all necessary information of X_f can be obtained from the matrix B^f with type-dividing lines, from now on we restrict ourselves to displaying B^f.

It is almost evident that in general

$$B^f = (e_1 + e_{m+1}, \ldots, e_{m-1} + e_{m+1}, e_m + e_{m+1} + \ldots + e_t | e_1 | e_2 | \ldots | e_t)$$

is indecomposable for all $n \geq 5$, provided $m < n/2$, $t = n - m$ (it can be explicitly proved by induction on n).

Return to the case $n = 5$. For comparison let us suppose that $h_1 = h_2$. Then B^f has only two type-dividing lines and

$$B^f = \begin{pmatrix} 1 & 0 & \overset{\tau}{|} & 1 & 0 & \overset{\tau}{|} & 0 \\ 0 & 1 & | & 0 & 1 & | & 0 \\ 1 & 1 & | & 0 & 0 & | & 1 \end{pmatrix} \xrightarrow{r} \begin{pmatrix} 1 & 0 & \overset{\tau}{|} & 1 & 0 & \overset{\tau}{|} & 0 \\ 0 & 1 & | & 0 & 1 & | & 0 \\ 0 & 0 & | & -1 & -1 & | & 1 \end{pmatrix}$$
$$\xrightarrow{c} \begin{pmatrix} 1 & 0 & \overset{\tau}{|} & 1 & 0 & \overset{\tau}{|} & 0 \\ -1 & 1 & | & -1 & 1 & | & 0 \\ 0 & 0 & | & 0 & -1 & | & 1 \end{pmatrix} = B^{f_1}.$$

After suitable rearrangement of the columns of B^{f_1} we obtain the matrix $\widetilde{B^{f_1}}$:

$$\widetilde{B^{f_1}} = \begin{pmatrix} 1 & 1 & 0 & 0 & 0 \\ -1 & -1 & 1 & 1 & 0 \\ 0 & 0 & 0 & -1 & 1 \end{pmatrix} \xrightarrow{r} \begin{pmatrix} 1 & 1 & 0 & 0 & 0 \\ 0 & 0 & 1 & 1 & 0 \\ 0 & 0 & 0 & -1 & 1 \end{pmatrix}.$$

This means that X_f is decomposable into summands of ranks 2 and 3. □

EXAMPLE 2 Let $n = 5$, $m = t = 2$ and

$$B^f = \begin{pmatrix} 1 & 0 & \overset{\tau}{|} & 1 & \overset{\tau}{|} & 0 & \overset{\tau}{|} & 1 \\ 0 & 1 & | & 0 & | & 1 & | & 1 \end{pmatrix}.$$

Here $m < n/2$, $m = t < n - m$ and X_f is indecomposable by Lemma 4.3.

It is easy to see that in general

$$B^f = (e_1, \ldots, e_m | e_1 | e_2 | \ldots | e_t | e_t | \ldots | e_t | e_1 + \ldots + e_t)$$

is indecomposable for all $n \geq 5$, provided $m < n/2$, $m \leq t < n - m$. □

EXAMPLE 3 It is clear that

$$B^f = (e_1, \ldots, e_m | e_1 | e_2 | \ldots | e_t | e_t | \ldots | e_t)$$

is decomposable for all $n \geq 5, \quad m \leq t \leq n - m$. □

As seen in these Examples there are indecomposable groups as well decomposable groups for all $n \geq 5, \quad 2 \leq m < n/2, \quad m \leq t \leq n - m$.

Finally, we construct the following table containing the connection between values of $\text{rk}\, X/A = t$ and decomposability properties of X, introduced at the beginning of the paper, provided $m < n/2 \quad (n = \text{rk}\, X)$.

t	X is
0	completely decomposable
1 $\vdots$ $m-1$	decomposable
m $\vdots$ $- n/2 -$ $\vdots$ $n-m$	decomposable or indecomposable
$n-m+1$ $\vdots$ $n-1$	decomposable
n	completely decomposable

If $m \geq n/2$ then X is decomposable.

REFERENCES

1. D.Arnold, Finite Rank Torsion Free Abelian Groups and Rings, *Lecture Notes in Mathematics,* vol. 931, Springer Verlag (1982).
2. E.A.Blagoveshchenskaya, Direct decompositions of torsion-free abelian groups of finite rank, *Zap. Nauch. Sem. Leningrad. Otdel. Mat. Inst. V.A.Steklova,* vol. 132, pp. 17-25, (1983).

3. E.A.Blagoveshchenskaya, Decompositions of torsion-free abelian groups of finite rank into direct sums of indecomposable groups, *St. Petersburg Math. J.*, vol. 4, pp. 251-257, (1993).
4. E.A.Blagoveshchenskaya and A.Mader, Decompositions of almost completely decomposable abelian groups, *Contemporary Mathematics*, vol. 171, pp. 21-36, (1994).
5. E.A.Blagoveshchenskaya and A.V.Yakovlev, Direct decompositions of torsion-free abelian groups of finite rank, *Leningrad Math. J.*, vol. 1, pp. 117-136, (1990).
6. L.Fuchs, Infinite Abelian Groups, vol. 1, 2, Academic Press (1970, 1973).
7. A.Mader, Almost completely decomposable abelian groups, Montreal 1992, photocopied notes.

On the Group of Eventually Divisible Integer Sequences

Andreas R. Blass, University of Michigan, Ann Arbor, MI

John Irwin, Wayne State University, Detroit, MI

ABSTRACT. Let D be the group, under componentwise addition, of infinite sequences x of integers such that each positive integer divides all but finitely many terms of x. We study properties of D and related groups. We show, among other things, that all infinite-rank summands of D are isomorphic to D and that D has essentially indecomposable subgroups of all coranks from 1 to $2^{\aleph_0}$. Several of our theorems and proofs involve the notion of a basic subgroup, i.e., a pure, free subgroup with divisible quotient.

INTRODUCTION

This paper is about certain subgroups of the additive group $\Pi = \mathbb{Z}^{\aleph_0}$ of infinite sequences of integers. Our primary interest will be in the subgroup D consisting of those sequences $x \in \Pi$ that are *eventually divisible*, meaning that, for every positive integer d, all but finitely many terms of x are divisible by d. We write e_n for the element of Π whose n^{th} component is 1 while all other components are 0, and we write Σ for the subgroup of Π generated by these e_n's, i.e., the subgroup of sequences having only finitely many non-zero components. Then Σ is the free abelian group on the e_n's, and D is the $\mathbb{Z}$-adic closure of Σ in Π. In other words, D/Σ is the divisible part of Π/Σ.

All groups in this paper, except automorphism groups, are tacitly assumed to be abelian. By a *basic* subgroup [7, 8] of a group G, we mean a pure, free subgroup $H \subseteq G$ such that the quotient G/H is divisible. Thus, Σ is a basic subgroup of D. The group B of bounded sequences of integers is a basic subgroup of Π; the only difficult part of this is Nöbeling's theorem [11] that B is free. We remark for future reference that Nöbeling also showed that Σ is a direct summand of B. In Section 2 of this paper, we shall prove several results about basic subgroups of D and related groups, as well as some results which don't mention basic subgroups but whose proofs use them. In particular, we show that the automorphism group of D acts transitively on the set of basic subgroups of D, that every separable group with a countable basic subgroup can be embedded in D as a pure subgroup, and that Π is not the sum of fewer than $2^{\aleph_0}$ subgroups each isomorphic to D.

Partially supported by NSF grants DMS-9204276 and DMS-9505118.

In Section 3, we discuss direct sums and summands. In particular, we show that every infinite-rank direct summand of D is isomorphic to D. We also construct essentially indecomposable subgroups of D of every co-rank from 1 to $2^{\aleph_0}$.

Finally, in Section 4, we present some examples to clarify the connection between copies of Σ, D, and Π inside Π.

1. Preliminaries

In this section, we list some known results that we shall need later. We write G^* for the dual $\mathrm{Hom}(G,\mathbb{Z})$ of a group G, and we write $\eta : G \to G^{**}$ for the canonical embedding of G in its double-dual.

We use the symbol $\mathfrak{c}$ for the cardinality of the continuum, $\mathfrak{c} = 2^{\aleph_0}$.

Proposition 1.1 (Specker [13]). *The dual of* Π *is isomorphic to* Σ*, via the pairing* $\Sigma\times\Pi \to \mathbb{Z} : \langle x, y\rangle \mapsto \sum_n x(n)y(n)$.

Proposition 1.2 (Yen, cf. [9], vol. II, p. 163, and Blass, Irwin, and Schlitt [3]). *The dual of D is isomorphic to Σ, via the (restriction of the) same pairing as in Proposition 1.1. The same applies to any pure subgroup G that has corank $< \mathfrak{c}$ in D and includes Σ.*

Since the dual of Σ is clearly isomorphic to Π via the same pairing, it follows that Π is reflexive and that the double-dual D^{**} (or G^{**} for any G as in the proposition) is identified with Π in such a way that the canonical embedding η becomes the inclusion map $D \to \Pi$ (or $G \to \Pi$).

Corollary 1.3. *Every endomorphism of D extends uniquely to an endomorphism of* Π.

Proof. The existence of the extension is immediate because double-dualization is a functor. The uniqueness comes from the facts that a homomorphism $\Pi \to \mathbb{Z}$ is determined by its action on Σ (clear from Prop. 1.1) and that a homomorphism $\Pi \to \Pi$ is determined by its components $\Pi \to \mathbb{Z}$. □

In the next proposition, we need the notion of a dense subset of Π with respect to the product topology, i.e., the Tychonoff topology on $\Pi = \mathbb{Z}^{\aleph_0}$ induced by the discrete topology on $\mathbb{Z}$. To avoid confusion with the $\mathbb{Z}$-adic topology, which will also play a role in our work, we write "product-dense," and similarly for other topological notions. Note that a subset X of Π is product-dense if and only if every finite sequence of integers occurs as an initial segment of some sequence $x \in X$.

Proposition 1.4 (Chase [5]). *Every countable, pure, product-dense subgroup of* Π *can be mapped onto* Σ *by an automorphism of* Π*. Therefore, every countable subgroup of* Π *can be mapped into* Σ *by an automorphism of* Π.

Corollary 1.5 (Baer [1], Thm. 4.7 and Cor. 12.8). *Every countable subgroup of* Π *is free.*

Recall that a group G is *torsionless* if each non-zero element of it has a non-zero image under some homomorphism $G \to \mathbb{Z}$. Equivalently, G is isomorphic to a subgroup of $\mathbb{Z}^\kappa$ for some κ. If the subgroup here can be taken to be pure in $\mathbb{Z}^\kappa$ then G is *separable.*

Proposition 1.6 (Nunke [12]). *Every torsionless quotient of* Π *is isomorphic either to* Π *or to a finite power of* $\mathbb{Z}$.

Proposition 1.7 (Dugas and Irwin [7]). *A separable group with a basic subgroup of cardinality κ can be embedded as a pure subgroup of $\mathbb{Z}^\kappa$.*

Proposition 1.8 (Dugas and Irwin [8]). *Any two basic subgroups of the same group are isomorphic.*

Proposition 1.9 (Dugas and Irwin [8]). *If a separable group G has a countable basic subgroup, then so does every pure subgroup of G.*

Proposition 1.10 (Nöbeling [11]), see also [9], vol. 2, Thm. 97.3. *The group $B \subseteq \Pi$ of bounded sequences is free, and Σ is a direct summand of it.*

2. Basic Subgroups

We begin by showing that basic subgroups of D all "look alike," i.e., the automorphism group of D acts transitively on them.

Theorem 2.1. *If S_1 and S_2 are basic subgroups of D, then there is an automorphism h of Π that maps D onto itself and maps S_1 onto S_2.*

Remark. Once one has an automorphism of D sending S_1 onto S_2, the additional information in the theorem, namely that it extends to an automorphism of Π, could be obtained by applying Corollary 1.3 to the automorphism and its inverse. Our proof of the theorem will, however, give the automorphism of Π directly.

Proof of Theorem 2.1. It suffices to consider the case $S_2 = \Sigma$, for once this case is established we can obtain the general case by composing automorphisms. So let $S_1 = S$ be any basic subgroup of D. By definition, S is pure in D and therefore also in Π. Also, S is countable, by Prop. 1.8 since D has a countable basic subgroup Σ. We claim that S is product-dense in Π.

To prove this claim, suppose it were false, and fix a finite sequence u of integers that does not occur as an initial segment of any member of S. If n is the length of u, then the projection $p : \Pi \to \mathbb{Z}^n$ to the first n components sends S to a proper subgroup $p(S)$ of $\mathbb{Z}^n$, not containing u. So there is a non-trivial equation or congruence in n variables, satisfied by the first n components of any element of S but not satisfied by u. We may assume it's a congruence, since an equation could be replaced with a congruence having any sufficiently large modulus. Let m be the modulus of this congruence. Then, as S is basic in D, every element of D can be written as $s + mx$ for some $s \in S$ and some $x \in D$, and therefore every element of D satisfies our congruence. That is absurd, as u is an initial segment of some elements of D. This contradiction establishes the claim.

Being a countable, pure, product-dense subgroup of Π, S can be mapped to Σ by an automorphism h of Π, because of Prop. 1.4. It remains to prove that h maps D onto itself.

D is $\mathbb{Z}$-adically closed in Π, and each of Σ and S are $\mathbb{Z}$-adically dense in it since the quotients are divisible. So D is the $\mathbb{Z}$-adic closure of Σ and of S in Π. The automorphism h, being a $\mathbb{Z}$-adic homeomorphism and sending S to Σ, therefore sends D to D, as required. □

Remark. We cannot strengthen the theorem to say that every isomorphism between two basic subgroups of D extends to an automorphism of D. In fact, there are automorphisms of Σ that don't extend to automorphisms of D. See Example 1 in Section 4.

Corollary 2.2. *Every basic subgroup of D is a direct summand of a basic subgroup of* Π.

Proof. By Prop. 1.10, Σ is a direct summand of B, the group of bounded sequences, which is a basic subgroup of Π. So the corollary is true for Σ, and the theorem allows us to transfer it from Σ to any other basic subgroup of D. □

Theorem 2.3. *Every separable group with a countable basic subgroup can be embedded as a pure subgroup of D.*

Proof. Let G be a separable group with a countable basic subgroup H. By Prop. 1.7, G can be embedded as a pure subgroup in Π. By Prop. 1.4, we can arrange (by composing with an automorphism of Π) that the embedding sends H to a subgroup of Σ. Identifying (without loss of generality) G and H with their images in Π, we have that $(G+\Sigma)/\Sigma \cong G/(G\cap\Sigma)$ is a quotient of G/H, hence is divisible. So $G+\Sigma \subseteq D$ and, in particular, $G \subseteq D$. Finally, since G is pure in Π, it is pure in D as well. □

Remark. The proof establishes a stronger form of the theorem where a particular basic subgroup of G is specified and the embedding is required to send that basic subgroup into Σ.

Π is the union of its subgroups that are isomorphic to D. Indeed, every element of Π can be sent into Σ by an automorphism of Π. Pulling back D along that automorphism, we get a subgroup of Π isomorphic to D and containing the given element. The next theorem shows that we cannot represent Π as the union of a small number of copies of D. In fact, it shows more, replacing "union" with "sum."

Theorem 2.4. *Π is not the sum of fewer than $\mathfrak{c}$ subgroups isomorphic to D.*

Proof. Suppose Π were the sum of a family of fewer than $\mathfrak{c}$ copies of D. In each of these copies of D, fix a countable basic subgroup, and let H be their sum. Then $|H| < \mathfrak{c}$ and Π/H is divisible. This is impossible by the proof of Theorem 2 in [8]; that result would apply directly if H were free and pure in Π, hence basic, but the proof applies in any case, as follows.

Since Π/H is divisible, we have $\Pi = H + 2\Pi$. Therefore, $\Pi/2\Pi = (H + 2\Pi)/2\Pi \cong H/(H\cap 2\Pi)$. But the group $\Pi/2\Pi$ on the left here has the cardinality of the continuum, while the group on the right, a quotient of H, is smaller, a contradiction. □

The representation of Π as a union of $\mathfrak{c}$ copies of D can be tidied up a bit, as the following theorem shows; a corollary will give an even tidier representation when the continuum hypothesis holds.

Theorem 2.5. *Π is the union of a countably-directed family of pure subgroups isomorphic to D.*

Recall that a family is *countably directed* if every countable subfamily has an upper bound.

Proof. For each countable, pure, product-dense $S \subseteq \Pi$, let $D(S)$ be its $\mathbb{Z}$-adic closure in Π. Then each $D(S)$ is isomorphic to D because of Prop. 1.4. Each element of Π is in some such S and therefore in the corresponding $D(S)$. Finally, each countably many S's, say S_n, are in a single such S, say S^+, namely the purification of their sum. Then all the $D(S_n)$ are subgroups of $D(S^+)$. □

Corollary 2.6. *Assume the continuum hypothesis. Then Π is the union of a chain, of length $\aleph_1$, of pure subgroups isomorphic to D.*

Proof. The countably directed system in Theorem 2.5 clearly has the cardinality of the continuum, so under the continuum hypothesis it has cardinality $\aleph_1$. But any countably directed partial ordering of cardinality $\aleph_1$ with no maximal element has a cofinal well-ordered subset of length $\aleph_1$. □

3. Direct Sums

This section is about direct sum decompositions of D and certain subgroups. We show first that D has no "strange" summands. Notice that Nunke's Prop. 1.6 above implies the analogous result for Π.

Theorem 3.1. *Every direct summand of D of infinite rank is isomorphic to D.*

Proof. Suppose $D = D_1 \oplus D_2$. Since double-dualization is an additive functor, we have $\Pi = D^{**} = D_1^{**} \oplus D_2^{**}$. By Prop. 1.6, each D_i^{**} is isomorphic to Π or to a finite power of $\mathbb{Z}$. If D_i has infinite rank, then the latter possibility is excluded and we have $D_i^{**} \cong \Pi$.

The topology on D_i^{**} corresponding to the product topology on Π via such an isomorphism will be called the product topology on D_i^{**}. Notice that it doesn't depend on the choice of the isomorphism, since any two such isomorphisms differ by an automorphism of Π and any such automorphism is, thanks to Prop. 1.1, a homeomorphism of product spaces.

Since D has a countable basic subgroup Σ, Prop. 1.9 says that each D_i has a countable basic subgroup, say S_i. Then $S_1 \oplus S_2$ is clearly a basic subgroup of D. Applying an automorphism obtained from Theorem 2.1, we may assume without loss of generality that $S_1 \oplus S_2 = \Sigma$.

The divisible part of Π/Σ is, on the one hand,

$$\mathrm{Div}\left[\frac{\Pi}{\Sigma}\right] = \frac{D}{\Sigma} = \frac{D_1 \oplus D_2}{S_1 \oplus S_2} = \frac{D_1}{S_1} \oplus \frac{D_2}{S_2}$$

and, on the other hand

$$\begin{aligned}\mathrm{Div}\left[\frac{\Pi}{\Sigma}\right] &= \mathrm{Div}\left[\frac{D_1^{**} \oplus D_2^{**}}{S_1 \oplus S_2}\right] \\ &= \mathrm{Div}\left[\frac{D_1^{**}}{S_1} \oplus \frac{D_2^{**}}{S_2}\right] \\ &= \mathrm{Div}\left[\frac{D_1^{**}}{S_1}\right] \oplus \mathrm{Div}\left[\frac{D_2^{**}}{S_2}\right]\end{aligned}$$

After checking that all the identifications here are coherent, we conclude that

$$\mathrm{Div}\left[\frac{D_i^{**}}{S_i}\right] = \frac{D_i}{S_i}.$$

We claim that S_i is countable, pure, and product-dense in D_i^{**}. (Recall that the product topology on D_i^{**} was defined above by transferring the product topology of Π via an isomorphism. We assumed there that D_i has infinite rank; in the finite rank case, the product topology would be discrete.) Countability was already pointed out when S_i was introduced, and purity follows from the fact that D_i^{**}/S_i is a subgroup of the torsion-free group Π/Σ. As for product-density, notice that the projection of Π onto D_i^{**} (coming from $\Pi = D_1^{**} \oplus D_2^{**}$) sends $\Sigma = S_1 \oplus S_2$ onto S_i, that Σ is product-dense in Π, and that the projection is product-continuous (since its components are, by Prop. 1.1). But a continuous surjection sends dense sets to dense sets, so the claim is established.

Now assume that D_i has infinite rank, so there is an isomorphism $D_i^{**} \to \Pi$. Such an isomorphism, the claim just established, and Prop. 1.4 yield a new isomorphism $D_i^{**} \to \Pi$, sending S_i to Σ. But then this isomorphism also sends the pre-image in D_i^{**} of the divisible part of D_i^{**}/S_i onto the pre-image in Π of the divisible part of Π/Σ. That is, it sends D_i onto D. In particular, D_i is isomorphic to D. $\square$

It seems reasonable to conjecture for D the full analog of Prop. 1.6, namely that any torsionless quotient of D is isomorphic to either D or a finite power of $\mathbb{Z}$. We do not know how to prove this conjecture, but the following proposition reduces it, at least for separable groups, to the special case where the quotient in question is of a rather special form.

Proposition 3.2. *Every separable, infinite-rank, homomorphic image of D is isomorphic to $h(D)$ for some surjective endomorphism h of Π that maps Σ onto itself.*

Proof. Let G be a separable group of infinite rank with a surjective homomorphism $h : D \to G$. We shall modify h in a number of steps, composing it with isomorphisms either on the left or on the right, until we obtain the conclusion of the proposition.

In the first step, we arrange that $h(\Sigma)$ is pure in G. If it isn't, then we can take a countable Σ' with $\Sigma \subseteq \Sigma' \subseteq D$, with Σ' pure in D, and with $h(\Sigma')$ pure in G. Finding such a Σ' is easy, as we just need to close Σ under countably many functions. As both Σ and Σ' are basic in D, Theorem 2.1 gives us an automorphism of D sending Σ onto Σ'. Composing such an automorphism of D with h we get a new h, still mapping D onto G, and now mapping Σ onto a pure subgroup of G.

We claim that $h(\Sigma)$ is a countable, basic subgroup of G. It is clearly countable and therefore free, since G is separable. We've just arranged for it to be pure in G. And the quotient is divisible because, with K denoting the kernel of h, we have

$$\frac{G}{h(\Sigma)} \cong \frac{D}{K+\Sigma} \cong \frac{D/\Sigma}{(K+\Sigma)/\Sigma},$$

a quotient of the divisible group D/Σ.

Thus, we can apply Theorem 2.3 to assume, without loss of generality, that G is a pure subgroup of D and therefore of Π. Thus, we regard h as a homomorphism $D \to \Pi$. As such, it extends naturally to the double-duals, and we use the same symbol h for this extension, $h : \Pi \to \Pi$.

The range of this extended h is a torsionless (because it's $\subseteq \Pi$) homomorphic image of Π, and it has infinite rank because it includes G. By Prop. 1.6, it is isomorphic to Π. Composing h with such an isomorphism, we may assume that h maps Π onto Π. So we have a surjective endomorphism h of Π that maps D onto G.

To complete the proof, we want to arrange that h maps Σ onto itself. We do this by composing h with an automorphism of Π that sends $h(\Sigma)$ back to Σ. The existence of the required automorphism will follow from Prop. 1.4 once we check that $h(\Sigma)$ is a countable, pure, product-dense subgroup of Π. It is obviously countable, and it is pure in Π because it is pure in G which in turn is pure in Π. Finally, it is product-dense because Σ is product-dense in Π, h is product-continuous, and continuous surjections preserve density. $\square$

Remark. In view of the conjecture preceding the proposition, one might hope to prove that, if h is as in the proposition, then $h(D)$ is actually equal to D. We shall see an example in Section 4 where this is not so.

We turn now from direct summands (or separable quotients) of D to direct summands of certain subgroups of D. Dugas, Irwin, and Khabbaz [6] constructed subgroups of Π that are essentially indecomposable, i.e., when they are split into a direct sum of two subgroups then one of the summands must be free of finite rank. In fact, the groups they constructed are subgroups of D, and they have the property, stronger than essential indecomposability, that their only endomorphisms are (1) multiplication by integer scalars, (2) endomorphisms of finite rank, and (3) sums of (1) and (2). We shall construct such subgroups with any prescribed co-rank (except 0) in D.

To avoid interrupting the argument later, we first give a lemma combining some results from [2] and [4] that we shall need.

Lemma 3.3. *Suppose that h is an endomorphism of D, that α and β are integers with $\beta \neq 0$, and that the range of the endomorphism $\beta h - \alpha$ has cardinality $< \mathfrak{c}$. Then h is the sum of an (integer) scalar multiplication and an endomorphism of D of finite rank.*

Proof. It is shown in [4] that any torsionless quotient of D of cardinality $< \mathfrak{c}$ is in fact countable. In the present situation, this means that the range of $\beta h - \alpha$ is countable, hence free, and hence isomorphic to a direct summand of D. But D has no free summands of infinite rank (because its dual Σ is too small to have as a summand the dual $\mathbb{Z}^\kappa$ of a free group of infinite rank κ). So the range of $\beta h - \alpha$ has finite rank.

Passing to the divisible hull of D, where multiplication by rational scalars makes sense, we have that h and the rational scalar multiplication α/β differ by an endomorphism of finite rank. It is shown in Lemma 2 of [2] that, in this situation, the scalar α/β must be an integer. □

Theorem 3.4. *For every cardinal number in the range $1 \leq \kappa \leq \mathfrak{c}$, there exists a pure subgroup of co-rank κ in D, all of whose endomorphisms are sums of scalar multiplications and endomorphisms of finite rank.*

Proof. The case $\kappa = \mathfrak{c}$ is easy. First find inside D a pure copy of D of the desired co-rank, for example $\{x \in D \mid x(n) = 0 \text{ for all even } n\}$. Then apply the construction from [6] inside this copy of D.

We assume from now on that $1 \leq \kappa < \mathfrak{c}$. The idea of our proof is very similar to the idea in [6]; we shall list all possible endomorphisms h not of the desired form, and we shall get rid of each one by putting into the group G under construction some element x while keeping $h(x)$ out of G. The new twist is that, to keep the co-rank of G small, we mustn't keep too many elements out of G. To get around the difficulty, we put a few elements permanently out of G and then, when we want to keep some y out of G, we instead put *into* G the difference between y and one of the permanently out elements. Here are the details.

There are $\mathfrak{c}$ endomorphisms h of D, because each one is determined by where it sends the countably many elements of Σ. List these endomorphisms in a well-ordered sequence $(h_\xi)_{\xi<\mathfrak{c}}$ in which each element has $< \mathfrak{c}$ predecessors. We shall build, by a transfinite induction of this same length $\mathfrak{c}$, an increasing sequence of pure subgroups G_ξ of D. Each G_ξ will have cardinality smaller than $\mathfrak{c}$; in fact, $|G_\xi| \leq (1 + |\xi|)\aleph_0$.

The beginning of the construction and the limit stages are easy. $G_0 = \Sigma$ and, for limit ordinals λ, $G_\lambda = \bigcup_{\xi<\lambda} G_\xi$. Before describing the successor stages, we need some preliminary work.

Fix a pure subgroup $J \subseteq D$ of rank κ such that $J \cap \Sigma = 0$. One way to get such an J is to choose κ infinite subsets S of $\mathbb{N}$ that are almost disjoint, i.e., the intersection of any two is finite. Such sets exist; cf. Lemma 23.9 in [10]. Then associate to each chosen S the element $j(S) \in D$ defined by $j(S)_n = n!$ for $n \in S$ and $j(S)_n = 0$ for $n \notin S$. The purification of the subgroup generated by these $j(S)$'s will serve as J.

We shall choose the G_ξ so as to have zero intersection with J. Note that this is the case for $\xi = 0$ and will be preserved at limit stages, according to the description already given for these G_ξ.

We intend to carry out the successor steps of the inductive construction in such a way that, unless h_ξ is of the form scalar plus finite rank, there will be $x, y \in G_{\xi+1}$ such that $h_\xi(x) - y \in J - \{0\}$. Before showing how to do this, we show that it will complete the proof.

Suppose then that G_ξ has been defined for all $\xi < \mathfrak{c}$ in conformity with the preceding specification. Let $G_\mathfrak{c}$ be the union of all the G_ξ. Its intersection with J is 0. By Zorn's Lemma, enlarge $G_\mathfrak{c}$ to a maximal $G \subseteq D$ with zero intersection with J. Then this G is pure in D because of its maximality. It has co-rank κ in D. Indeed, the projection from D to D/G is one-to-one on J (as $G \cap J = 0$), and every element of D/G has a multiple in the image of J by maximality of G, so the rank of D/G equals the rank of J, namely κ. It remains to show that every endomorphism of G has the form scalar plus finite rank. Suppose h were a counterexample. It extends to an endomorphism, still called h, of the second-dual of G, which is Π by Prop. 1.2. And that extension maps D into itself, since it is $\mathbb{Z}$-adically continuous and D is the $\mathbb{Z}$-adic closure of G in Π (because $\Sigma \subseteq G \subseteq D$). So h occurs as some h_ξ in our list. If it has the form scalar plus finite rank (on D, therefore also on G), there is nothing to prove. Otherwise, let $x, y \in G_{\xi+1} \subseteq G$ be such that $h(x) - y \in J - \{0\}$. As h is an endomorphism of G, we'd have $h(x) - y \in G$, contradicting $G \cap J = 0$. That contradiction shows that G is as required.

All that remains, then, is to carry out the successor steps of the induction so as to satisfy the specifications above. Consider, therefore, the step from G_ξ to $G_{\xi+1}$. We shall let $G_{\xi+1}$ be the purification of the subgroup of D generated by G_ξ and some $x, y \in D$ chosen so that $h_\xi(x) - y \in J - \{0\}$. Of course, we must be careful lest this $G_{\xi+1}$ have a non-trivial intersection with J.

Let H be the purification of $G_\xi + J$ in D, and note that its cardinality is $< \mathfrak{c}$. We shall choose an appropriate $x \in D - H$ and $y \in D$. Let us temporarily consider an arbitrary $x \in D - H$ and consider under what circumstances we *cannot* find an appropriate y. That would mean that, for any element $j \in J - \{0\}$ (the intended $h_\xi(x) - y$), adjoining x and $y = h_\xi(x) - j$ to G_ξ and purifying the result leads to a non-trivial intersection with J. In other words, for any such j, there are integers α and β and there is an element $g \in G_\xi$ such that $g + \alpha x + \beta h_\xi(x) - \beta j \in J - \{0\}$. Then $\alpha x + \beta h_\xi(x) \in H$. Since we are considering only x's not in H, we have $\beta \neq 0$ here.

Let p be the projection map from D to D/H. We have seen that, if $x \in D - H$ and if no appropriate y can be found, to go with this x and produce $G_{\xi+1}$, then $\alpha p(x) + \beta p(h_\xi(x)) = 0$ for suitable integers α and $\beta \neq 0$. We now assume that *no* $x \in D - H$ admits an appropriate y, and we show that then h_ξ is of the form scalar plus finite rank. This will show that the construction of $G_{\xi+1}$ can be carried out whenever necessary, i.e., whenever h_ξ is not of that form.

By our assumption, we have, for each $x \in D - H$, integers α and $\beta \neq 0$ with $\alpha p(x) + \beta p(h_\xi(x)) = 0$. Let us work in the quotient D/H, which is a torsion-free, divisible group, because H is pure and includes $G_0 = \Sigma$. Writing ρ for $-\alpha/\beta$, we have, for each non-zero element $p(x) \in D/H$, a rational scalar ρ such that $p(h_\xi(x)) = \rho p(x)$. There could be several such ρ for one $p(x)$ as $p(x) = p(x')$ for many $x' \neq x$; in addition, it seems likely that ρ would be different for different $p(x)$'s. We shall show that a single ρ works for all non-zero $p(x) \in D/H$ simultaneously.

To this end, suppose ρ and ρ' work for (non-zero) $p(x)$ and $p(x')$, respectively. Multiplying x' by 2 if necessary, we can assume that $x + x' \notin H$, so some τ works for $x + x'$. (Note that the multiplication by 2 does not interfere with the assumption that ρ' works for x'.) Thus we have both

$$ph_\xi(x + x') = ph_\xi(x) + ph_\xi(x') = \rho p(x) + \rho' p(x')$$

and

$$ph_\xi(x + x') = \tau p(x + x') = \tau p(x) + \tau p(x').$$

If $\rho \neq \rho'$ then these two equations say that $p(x)$ and $p(x')$ are linearly dependent in D/H. In other words, if ρ works for $p(x)$, then no other ρ' can work for any $p(x')$ linearly independent of $p(x)$. So ρ must work for these $p(x')$ and therefore also, by the same argument, for all $p(x'')$ independent of any such $p(x')$. But that's all elements of D/H. (We've used that there are enough linearly independent elements in D/H. The argument would fail if the rank of D/H were 1. But the rank is $\mathfrak{c}$, since H has smaller cardinality.) So a single ρ works simultaneously for all non-zero $p(x) \in D/H$.

Returning to D, we see that we have a single pair of integers α and $\beta \neq 0$ such that, for all $x \in D - H$, $\alpha x + \beta h_\xi(x) \in H$. This actually holds for all $x \in D$, since every such x is the sum of two elements of $D - H$.

Since $|H| < \mathfrak{c}$, Lemma 3.3 now tells us that h is the sum of a scalar multiplication and a finite-rank endomorphism, as required. □

4. Examples

Many of the results in the previous sections are based on the rather tight connections that exist between Π, D, and Σ. The fact that Π is the double-dual of D allows us to extend endomorphisms from D to Π (Cor. 1.3). That D is the $\mathbb{Z}$-adic closure of Σ in Π ensures that automorphisms of Π that preserve Σ also preserve D. That Σ is $\mathbb{Z}$-adically dense in D and product-dense in Π means that homomorphisms are often determined by their restrictions to Σ.

In this section, we explore the limitations of the connections between Π, D, and Σ by giving examples where the three groups exhibit different behavior. These provide counterexamples to some natural extensions of our earlier results.

Example 1. Let δ be the endomorphism of Π defined by

$$(\delta(x))_n = x_n - x_{n+1}.$$

This is a surjective endomorphism of Π. It is almost but not quite one-to-one, for its kernel is the subgroup, isomorphic to $\mathbb{Z}$, generated by the sequence all of whose components are 1.

If we restrict δ to D or to Σ, it becomes one-to-one, since these groups have zero intersection with the kernel. In fact, the restriction of δ to Σ is easily seen to be an automorphism of Σ. Its inverse is the function σ defined by

$$(\sigma(x))_n = \sum_{k=n}^{\infty} x_k.$$

Notice that σ is well-defined on Σ (since the apparently infinite sum is finite for $x \in \Sigma$) but not on D or on Π. In fact, σ cannot be extended to an automorphism or even an endomorphism of D or of Π. This is because its first component, $\Sigma \to \mathbb{Z} : x \mapsto \sum_{k=0}^{\infty} x_k$, cannot be extended to a homomorphism $D \to \mathbb{Z}$, since Prop. 1.2 would require such an extension to depend on only finitely many components of its input. Thus we have the example claimed in the remark following the proof of Theorem 2.1.

If we restrict δ to D, the result is a one-to-one map of D into itself. That $\delta(D) \subseteq D$ follows from $\delta(\Sigma) = \Sigma \subseteq D$ and the fact that D is the $\mathbb{Z}$-adic closure of Σ in Π. Notice, however, that the $\mathbb{Z}$-adic closure relationship between Σ and D does not allow us to conclude that δ maps D onto itself just because it maps each of Π and Σ onto itself. In fact, $\delta(D)$ is, in a sense, a rather small part of D; it has co-rank $\mathfrak{c}$ in D. In another sense though, it is large; it is dense in D with respect to both the product and $\mathbb{Z}$-adic topologies. Thus, we have a dense isomorphic copy of D, between Σ and Π, and having co-rank $\mathfrak{c}$ in D.

These facts follow easily from an explicit description of $\delta(D)$. To give this description, we first notice that there is a "summation" homomorphism θ, from D onto the $\mathbb{Z}$-adic completion $\hat{\mathbb{Z}}$ of $\mathbb{Z}$, sending every sequence $x \in D$ to $\sum_{k=0}^{\infty} x_k$; this sum converges ($\mathbb{Z}$-adically) in $\hat{\mathbb{Z}}$ precisely because $x \in D$. Then $\delta(D)$ is the inverse image of $\mathbb{Z}$ under θ.

Finally, we observe that, when h is as in Theorem 3.2, we cannot conclude that $h(D) = D$, not even if h restricts to an automorphism of Σ and is one-to-one on D; a counterexample is given by $h = \delta$.

Example 2. This example looks very similar to Example 1, but some of its properties are quite different. We define $\delta : \Pi \to \Pi$ (recycling the notation δ) by

$$(\delta(x))_0 = x_0 \qquad (\delta(x))_{n+1} = x_{n+1} - x_n.$$

(The reversal of signs relative to Example 1 is convenient but inessential; it makes the diagonal entries of the matrix representing δ all 1's rather than -1's. The essential change is the 0-component.) This δ is (unlike the one in Example 1) an automorphism of Π. Its inverse is given by

$$(\sigma(x))_n = \sum_{k=0}^{n} x_k.$$

The restrictions of δ to D and to Σ are one-to-one endomorphisms of these groups, but they are not surjective. $\delta(\Sigma)$ is the subgroup of Σ consisting of those elements x for which $\sum_{k=0}^{\infty} x_k = 0$. (Notice that the apparently infinite sum is finite for $x \in \Sigma$.) This $\delta(\Sigma)$ is a co-rank 1 direct summand of Σ, a complementary summand being the subgroup generated by $e_0 = (1, 0, 0, \dots)$.

One might think that, since δ is an automorphism of Π and maps Σ onto something that is "almost" Σ, it ought to map D onto something that is "almost" D, but that is not the case. $\delta(D)$, the $\mathbb{Z}$-adic closure of $\delta(\Sigma)$ in Π, is a co-rank $\mathfrak{c}$ subgroup of D and not a direct summand. In fact, it is exactly the kernel of the summation map $\theta : D \to \hat{\mathbb{Z}}$ defined in Example 1.

This $\delta(D)$ is product-dense but not $\mathbb{Z}$-adically dense in D.

An easy way to get a pure subgroup of Π that is isomorphic to D but properly includes D is to take $\delta^{-1}(D)$. In fact, by iterating this, one gets a tower of such subgroups $\delta^{-n}(D)$. The process can be iterated transfinitely; at countable limit stages, take the $\mathbb{Z}$-adic closure of the union of the previous stages. (Each previous stage has a countable, $\mathbb{Z}$-adically dense subset; therefore so does the union; therefore the $\mathbb{Z}$-adic closure of the union is still isomorphic to D by Prop. 1.4.) $\delta^{-1}(D)$ admits an alternative description; it is the $\mathbb{Z}$-adic closure of the group generated by D and the all-ones sequence in Π.

We close our dicussion of this example by showing that the direct sum decomposition $\Sigma = \delta(\Sigma) \oplus \langle e_0 \rangle$ does not extend to a direct sum decomposition of D or of Π. Indeed, suppose $D = X \oplus Y$ were such an extension. (The proof with Π in place of D is the same.) So $X \cap \Sigma = \delta(\Sigma)$ and $Y \cap \Sigma = \langle e_0 \rangle$. Define a homomorphism $f : D \to \mathbb{Z}$ by composing the projection $p : D \to Y$ (with kernel X), the inclusion $Y \hookrightarrow \Pi$, and the first component map $\Pi \to \mathbb{Z} : x \mapsto x_0$. For each n, we have $e_n = (e_n - e_0) + e_0$ where $e_n - e_0 \in \delta(\Sigma) \subseteq X$ and $e_0 \in Y$. So $p(e_n) = e_0$ and $f(e_n) = 1$ for all n. That contradicts Prop. 1.2 (or 1.1 if D is replaced by Π).

If we take the $\mathbb{Z}$-adic closures (in Π) of the two summands $\delta(\Sigma)$ and $\langle e_0 \rangle$ of Σ, then the sum of the resulting groups is not the $\mathbb{Z}$-adic closure D of Σ but only the subgroup $\delta(\Sigma)$.

Example 3. This final example is less relevant to the study of D, but we include it to show how a minor change in the definition can influence the behavior of δ. We define

$$(\delta(x))_n = x_n - 2x_{n+1},$$

modifying Example 1 only by inserting a factor 2. As in Example 1, we have an automorphism of Σ, the inverse now being

$$(\sigma(x))_n = \sum_{k=n}^{\infty} x_k 2^{k-n}.$$

In contrast to Example 1, however, this δ, considered as an endomorphism of Π, is not surjective but rather injective. Its range can be described as the pre-image of $\mathbb{Z}$ under a "power-series" map ϕ from Π to the group of 2-adic integers, namely $\phi(x) = \sum_{k=0}^{\infty} x_k 2^k$. (The sum clearly converges 2-adically.) Since ϕ maps Π onto the group of 2-adic integers, whose cardinality is $\mathfrak{c}$, the co-rank of $\delta(\Pi)$ in Π is also $\mathfrak{c}$.

References

1. R. Baer, *Abelian groups without elements of finite order*, Duke Math. J. **3** (1937), 68–122.
2. A. Blass and R. Göbel, *Subgroups of the Baer-Specker group with few endomorphisms but large dual*, Fund. Math. (to appear).
3. A. Blass, J. Irwin, and G. Schlitt, *Some abelian groups with free duals*, Abelian Groups and Modules (A. Facchini and C. Menini, eds.), (Proceedings of 1994 conference in Padova), Kluwer (to appear).
4. A. Blass and J. Irwin, *Baer meets Baire: Applications of category arguments and descriptive set theory in $\mathbb{Z}^{\aleph_0}$* (to appear).
5. S. U. Chase, *Function topologies on abelian groups*, Illinois J. Math. **7** (1963), 593–608.
6. M. Dugas, J. Irwin, and S. Khabbaz, *Countable rings as endomorphism rings*, Quart. J. Math. Oxford **39** (1988), 201–211.
7. M. Dugas and J. Irwin, *On pure subgroups of cartesian products of integers*, Resultate Math. **15** (1989), 35–52.
8. M. Dugas and J. Irwin, *On basic subgroups of $\prod \mathbb{Z}$*, Comm. Alg. **19** (1991), 2907–2921.
9. L. Fuchs, *Infinite Abelian Groups, vols. I and II*, Academic Press, 1970 and 1973.
10. T. Jech, *Set Theory*, Academic Press, 1978.
11. G. Nöbeling, *Verallgemeinerung eines Satzes von Herrn E. Specker*, Invent. Math. **6** (1968), 41–55.
12. R. J. Nunke, *On direct products of infinite cyclic groups*, Proc. Amer. Math. Soc. **13** (1962), 66–71.

13. E. Specker, *Additive Gruppen von Folgen ganzer Zahlen*, Portugaliae Math. **9** (1950), 131–140.

MATHEMATICS DEPT., UNIVERSITY OF MICHIGAN, ANN ARBOR, MI 48109, U.S.A.
E-mail address: ablass@umich.edu

MATHEMATICS DEPT., WAYNE STATE UNIVERSITY, DETROIT, MI 48202, U.S.A.

Baer Meets Baire: Applications of Category Arguments and Descriptive Set Theory to $\mathbb{Z}^{\aleph_0}$

Andreas R. Blass, University of Michigan, Ann Arbor, MI

John Irwin, Wayne State University, Detroit, MI

ABSTRACT. We apply the Baire category theorem and other classical results of descriptive set theory to the study of the structure of the group $\mathbb{Z}^{\aleph_0}$ of infinite sequences of integers and some of its subgroups.

1. INTRODUCTION

Let $\Pi = \mathbb{Z}^{\aleph_0}$ be the Baer-Specker group of all infinite sequences of integers, the group operation being componentwise addition. Our purpose in this paper is to show how certain classical results of descriptive set theory can be combined with known facts about Π to produce new information about the structure of Π and some of its subgroups. Several of our theorems are about the possibility (or impossibility) of expressing these groups as unions of chains of subgroups with specified properties. Others are concerned with the possible quotients of these groups.

We begin by listing the definitions and classical results that we shall need from descriptive set theory. General references for this material are the books of Kuratowski [10], Moschovakis [14], and Kechris [9]; we shall give specific references to [14] for the results we use.

Definition. A set in a metric space is *meager* (or *of first (Baire) category*) if it can be covered by countably many closed sets with empty interiors. A set in a metric space has the *Baire property* if it differs from some open set by a meager set.

Baire Category Theorem [2], [14, 2H2]. *In a complete metric space, no non-empty open set is meager.*

The union of countably many meager sets is meager, hence cannot cover a non-empty open set in a complete metric space. Thus, it is reasonable to regard meager sets as very small. A set with the Baire property is then very nearly an open set. In particular, if

Partially supported by NSF grants DMS-9204276 and DMS-9505118.

such a set is not itself meager, then it contains most (i.e., all but a meager subset) of some non-empty open set.

Definition. A *Borel* set is a set obtainable from open sets by repeated (possibly transfinitely repeated) formation of complements and countable unions. An *analytic* set is the image, under a continuous function, of a Borel set in a complete separable metric space.

The definition of "analytic" in [14] (page 43) seems more restrictive, but it follows from [14, 1E2 and 1E5] that it is actually equivalent.

Analytic sets share much of the well-known good behavior of Borel sets. For example, in Euclidean spaces, they are Lebesgue measurable. We shall need the following two results of this sort.

Lusin-Sierpiński Theorem [12], [14, 2H5]. *Every analytic set in a complete separable metric space has the Baire property.*

Suslin's Perfect Set Theorem [11], [14, 2C3]. *Every uncountable analytic set in a complete separable metric space has cardinality* $\mathfrak{c}$.

Here and throughout this paper, $\mathfrak{c}$ denotes the cardinality of the continuum, $2^{\aleph_0}$.

Notice that this last theorem says that an analytic set can never be a counterexample to the continuum hypothesis. The name of the theorem derives from the fact that the harder part of its proof establishes that an uncountable analytic set has a perfect (i.e., closed, non-empty, and without isolated points) subset; the easier part then infers the cardinality result.

We shall apply these results to the complete separable metric space Π. Its metric is defined by

$$\rho(x, y) = 2^{-n} \quad \text{where } n \text{ is the first number with } x(n) \neq y(n),$$

whenever $x \neq y$. To verify completeness, notice that, if (x_k) is a Cauchy sequence in Π then for each fixed n the sequence $(x_k(n))$ of integers is eventually constant; call its final value $y(n)$ and check that x_k converges to y. To verify separability, observe that the elements of Π that have only finitely many non-zero components form a countable dense set. The topology induced by the metric ρ is the product topology on $\mathbb{Z}^{\aleph_0}$ obtained from the discrete topology on $\mathbb{Z}$. (Since the product of countably many complete metric spaces always admits a complete metric, we didn't really need to exhibit ρ.) A basic neighborhood (a metric ball) about 0 is a subgroup of the form

$$V_k = \{x \in \Pi \mid x(n) = 0 \text{ for all } n < k\},$$

and a basic neighborhood of any other point in Π is a coset of such a subgroup V_k.

To connect these topological aspects of Π with its algebraic aspects, we need a theorem of Specker ensuring that homomorphisms are continuous.

Specker's Theorem [16, Satz III]. *Every homomorphism* $h : \Pi \to \mathbb{Z}$ *has the form* $h(x) = \sum_{i=0}^{n} a_i x(i)$ *for some finite* n *and some coefficients* $a_i \in \mathbb{Z}$.

The finiteness of n in this theorem ensures that h is continuous with respect to the product topology on Π and the discrete topology (or any other topology) on $\mathbb{Z}$. Since a function into a product space is continuous if and only if all its components are, the following corollary is immediate.

Corollary. *Every homomorphism $h : \Pi \to \Pi$ is continuous.*

We shall also need the following result about subgroups of Π.

Baer's Theorem [1, Thm. 4.7 and Cor. 12.8], [16, Satz I]. *Every countable subgroup of Π is free.*

We finish this introduction by defining and discussing two particular subgroups of Π that will play a role in our results.

For each natural number i, let e_i be the element of Π whose ith component is 1 and whose other components are all 0; these elements e_i are sometimes called the *standard unit vectors*. The subgroup Σ that they generate consists of those elements of Π that have only finitely many non-zero components. It is the free abelian group on the countably many generators e_i.

The group $D \subseteq \Pi$ is defined to consist of those $x \in \Pi$ such that, for each positive integer q, all but finitely many terms of x are divisible by q. Equivalently, D/Σ is the divisible part of Π/Σ. Equivalently again, D is the $\mathbb{Z}$-adic closure of Σ in Π. We record for future reference that D is a Borel set in Π:

$$D = \bigcap_q \bigcup_k \bigcap_{n \geq k} \{x \in \Pi \mid q \text{ divides } x(n)\},$$

where q and k range over positive integers and the sets $\{x \in \Pi \mid q \text{ divides } x(n)\}$ are both open and closed.

There is an analog of Specker's theorem for D and for many other groups "near" Π and D. The result for D is on page 163 of [8], attributed to T. Yen. We quote here a more general form from [3].

Continuity Theorem. *Let G be a pure subgroup of index $< \mathfrak{c}$ in either Π or D, and assume that $\Sigma \subseteq G$. Then every homomorphism $h : G \to \mathbb{Z}$ has the form $h(x) = \sum_{i=0}^{n} a_i x(i)$ for some finite n and some coefficients $a_i \in \mathbb{Z}$. Therefore, every homomorphism $G \to \Pi$ is continuous.*

Here G is taken to be topologized as a subspace of Π.

2. Unions of Chains

In this section, we shall apply the results quoted in Section 1 to obtain restrictions on possible representations of Π, D and similar groups as unions of chains of subgroups.

Theorem 2.1. *The group Π is not the union of a countable chain of proper subgroups, each isomorphic to Π.*

Proof. Suppose we had a countable chain of subgroups $P_0 \subseteq P_1 \subseteq \dots$ of Π, each isomorphic to Π, with union Π. We shall show that some P_n is all of Π.

By the Baire Category Theorem, fix an m such that P_m is not meager. This P_m, being isomorphic to Π, is the image of a homomorphism $h : \Pi \to \Pi$. By the corollary to Specker's Theorem, h is continuous, so P_m is analytic and therefore has the Baire property. (In fact, because h is one-to-one, P_m is a Borel set, by [14, 2E7].) Since it isn't meager, it is comeager in some basic open set, say $a + V_k$ for some $a \in \Pi$. (Recall that $a + V_k$ is the coset consisting of those $x \in \Pi$ such that $x(n) = a(n)$ for all $n < k$.)

We claim that $P_m \supseteq V_k$. To see this, fix any $z \in V_k$. Then the map $x \mapsto x + z$ is a homeomorphism of Π to itself, and therefore, since $(a + V_k) \setminus P_m$ is meager, so is its translate $z + ((a + V_k) \setminus P_m)$. By the Baire Category Theorem again, these two meager sets cannot cover $a + V_k$, so fix $x \in a + V_k$ belonging to neither of them. Then $x \in P_m$ and $x - z \in P_m$ (where we used that $x - z \in a + V_k$ since $z \in V_k$ and V_k is a subgroup). As P_m is a subgroup, we conclude that $z \in P_m$, as required.

Since the chain of subgroups P_i covers Π, some P_n must contain the finitely many unit vectors $e_0, \ldots, e_{k-1}$. Then $P_{\max\{m,n\}}$ contains these unit vectors and all members of V_k. But that's enough to generate Π, so $P_{\max\{m,n\}} = \Pi$. □

In the statement of Theorem 2.1, both "chain" and "each isomorphic to Π" are essential. Π is the union of countably many subgroups, each isomorphic to Π, for example the subgroup $V_1 = \{x \in \Pi \mid x(0) = 0\}$ and the subgroups $\{x \in \Pi \mid x(1) = rx(0)\}$ for all rational numbers r. Also, Π is the union of a countable chain of proper subgroups (not isomorphic to Π). One way to see this is to represent Π/Σ as the direct sum of its divisible part D/Σ and a reduced part R; as a rational vector space, D/Σ can easily be expressed as a countable increasing union of proper, pure subgroups, and, taking direct sums with R and then pre-images in Π, we get a similar representation for Π.

We do not know whether "countable" is essential in Theorem 2.1. It might be provable that Π is not the union of a chain (of any length) of proper subgroups isomorphic to Π. On the other hand, it may be that the continuum hypothesis implies that Π is expressible as such a union, for a chain of length $\aleph_1$.

The proof of Theorem 2.1 actually establishes considerably more than the theorem asserts. In the first place, the assumption that the subgroups in the chain are isomorphic to Π was used only to show that they are analytic. Thus, the following corollary is already proved.

Corollary 2.2. *The group Π is not the union of a countable chain of analytic, proper subgroups.*

If G is an analytic subgroup of index $< \mathfrak{c}$ in Π or in D and if $\Sigma \subseteq G$, then, by the Continuity Theorem, every homomorphic image of G in Π is also analytic. Thus, Π cannot be covered by a countable chain of such homomorphic images. In particular, Π cannot be covered by a countable chain of subgroups isomorphic to D. We digress for a moment to improve this last special case with a totally different method.

Proposition 2.3. *The group Π is not the sum (hence a fortiori not the union) of fewer than $\mathfrak{c}$ subgroups, each isomorphic to D.*

Proof. Suppose it were such a sum. Each of the summands has a countable subgroup for which the quotient is divisible (because D has such a subgroup, namely Σ). Let H be the sum of those fewer than $\mathfrak{c}$ countable subgroups. Thus, $|H| < \mathfrak{c}$ and Π/H is divisible. This leads to a contradiction by the following argument, essentially the proof of Theorem 2 in [6]. Since Π/H is divisible, we have $\Pi = H + 2\Pi$. Therefore $\Pi/2\Pi = (H + 2\Pi)/2\Pi \cong H/(H \cap 2\Pi)$. Here the group on the left has cardinality $\mathfrak{c}$ while the group on the right, a quotient of H, has smaller cardinality. This contradiction completes the proof. □

Returning to Theorem 2.1, we need a definition in order to state our next improvement of it. For ease of reference, we combine this definition with another that we shall need later.

Definition. The cardinal $\mathbf{cov}(B)$ is the smallest κ such that some κ meager sets cover the real line. The cardinal $\mathbf{add}(B)$ is the smallest κ such that some κ meager sets of reals have a non-meager union.

Clearly, $\aleph_1 \leq \mathbf{add}(B) \leq \mathbf{cov}(B) \leq \mathfrak{c}$. It is known that each of these three inequalities is strict in some but not all models of set theory [7]. It is also known that $\mathbf{cov}(B)$ and $\mathbf{add}(B)$ are unchanged if we replace the real line in their definition by Π. Indeed, Π is homeomorphic to a comeager subset of $\mathbb{R}$, namely the set of irrational numbers; see [9, Thm. 7.7].

The only use, in the proof of Theorem 2.1, of the countability hypothesis was to ensure that the subgroups P_n cannot all be meager. For this purpose, that hypothesis can obviously be weakened to say only that the number of P_n's is smaller than $\mathbf{cov}(B)$. (Whether this is a genuine weakening, i.e., whether $\mathbf{cov}(B) > \aleph_1$, depends, as mentioned above, on what set-theoretic universe one is in.)

Finally, the hypothesis that the P_n form a chain can be weakened to say only that they form a directed family, i.e., that every two of them (and therefore every finitely many of them) are contained in a single one. This weakening would have made no difference if we were still requiring countability, since every countable directed set contains a cofinal chain, but, now that we have weakened the countability assumption, this additional weakening makes sense. The following corollary summarizes all the preceding improvements of Theorem 2.1.

Corollary 2.4. *The group Π is not the union of a directed family of fewer than $\mathbf{cov}(B)$ proper, analytic subgroups.*

We turn next to analogous results for D in place of Π. The proof of Theorem 2.1 cannot be applied to D because D, unlike Π, is not a complete metric space. Not only is D not complete with respect to the metric of Π (because it's not closed in Π), it is not complete with respect to any metric inducing the same topology (because it's not a G_δ set in Π; see [9 Thm. 3.11]).

In fact, D is the union of a countable chain of proper subgroups, each isomorphic to D; consider for example, the subgroups

$$D_k = \{x \in D \mid x(n) \text{ is even for all } n \geq k\}.$$

Nevertheless, we can get an analog of Theorem 2.1 by adding the hypothesis that the subgroups in the chain are pure.

Theorem 2.5. *The group D is not the union of a countable chain of pure subgroups, each isomorphic to D.*

Before starting the proof, we give a definition that will be used both in this proof and in some later results.

Definition. For any $f \in \Pi$, let

$$f\Pi = \{x \in \Pi \mid f(n) \text{ divides } x(n) \text{ for all } n\}.$$

Clearly, $f\Pi$ is a Borel (in fact, closed) subgroup of Π. Notice that, if no component of f is 0, then componentwise multiplication by f is an isomorphism from Π onto $f\Pi$. We shall write $f \cdot -$ for this isomorphism and $-/f$ for its inverse. Notice also that, if $f \in D$ then $f\Pi \subseteq D$.

Proof of Theorem 2.5. Suppose D were the union of a countable chain of pure subgroups $D_0 \subseteq D_1 \subseteq \ldots$. Temporarily fix some $f \in D$ with no zero components, and use the isomorphism $f \cdot -$ mentioned above to pull back the subgroups $D_n \cap f\Pi$ to subgroups $P_n = (D_n \cap f\Pi)/f$ of Π. These P_n form a countable chain with union Π.

Recall that D is a Borel set in Π. Each D_n, being isomorphic to D, is then an analytic set, because of the continuity theorem. Since $f\Pi$ is Borel, it follows that $D_n \cap f\Pi$ and P_n are analytic; see [14, Thm. 1E2].

By Corollary 2.2, we deduce that the P_n cannot all be proper subgroups of Π. So there is n with $P_n = \Pi$, which means that $D_n \cap f\Pi = f\Pi$ and thus $f\Pi \subseteq D_n$.

Un-fix f. We have seen that, for every $f \in D$ with no zero components, there is an n such that $f\Pi \subseteq D_n$. (Of course, n can depend on f; otherwise, we could immediately conclude that $D \subseteq D_n$ and the proof would be finished.)

Since each D_n is a proper subgroup of D, fix some $g_n \in D \setminus D_n$. For each natural number z, define $h_n(z)$ to be an integer so large that, for all $k \geq h_n(z)$, $g_n(k)$ is divisible by $z!$; such an integer exists because $g_n \in D$. There is an increasing function h, from natural numbers to natural numbers, that eventually majorizes each h_n. For example, we can define $h(z)$ to be z plus the maximum of $h_0(z), h_1(z), \ldots, h_{z-1}(z)$, so that h majorizes h_n beyond n. It will be convenient to have $h(0) = 0$.

Now define $f \in \Pi$ by letting $f(k) = z!$ for the largest z such that $h(z) \leq k$. Such a largest z exists, because h is increasing and $h(0) = 0$.

Notice that, for any fixed positive integer q, all sufficiently large k satisfy $h(q) \leq k$, which means that the z in the definition of $f(k)$ is $\geq q$ and therefore q divides $f(k)$. This proves that $f \in D$. Since no component of f is 0, the first part of this proof provides an n such that $f\Pi \subseteq P_n$. Fix this n for the rest of the proof.

Our choice of h ensures that $h(z) \geq h_n(z)$ for all sufficiently large z. By taking k large enough, we can ensure that $f(k) = z!$ for a sufficiently large z, where $k \geq h(z) \geq h_n(z)$. Then, by definition of h_n, we know that $g_n(k)$ is divisible by $z! = f(k)$.

We thus have $f(k)$ dividing $g_n(k)$ for all sufficiently large k. It follows that, for a suitable positive integer c, $f(k)$ divides $cg_n(k)$ for all k (not just all sufficiently large k). For example, we could take c to be the product of those finitely many $f(k)$ that do not divide the corresponding $g_n(k)$. Thus, we have $cg_n \in f\Pi$. By our choice of n, we have $cg_n \in D_n$. Using the assumption that D_n is pure, we infer that $g_n \in D_n$, which contradicts the definition of g_n. □

Theorem 2.5 admits improvements similar (but not quite identical) to those of Theorem 2.1. As before, we can replace the assumption that the subgroups are isomorphic to D by the assumption that they are analytic, and we can weaken "chain" to "directed family."

But a difference arises when we try to weaken the countability hypothesis, since this hypothesis was used twice in the proof of Theorem 2.5. The first use occurred when we invoked Corollary 2.2, and we have already seen that "countable" can be replaced with "$<$ **cov**(B)" there. The second use occurred when we defined the function h to eventually majorize all the (countably many) h_n. Can uncountably many functions be similarly eventually majorized by a single one? The answer depends on one's set-theoretic universe. More precisely, define $\mathfrak{b}$ to be the smallest cardinality of any family of functions $\mathbb{N} \to \mathbb{N}$ not eventually majorized by a single function. Clearly, $\aleph_1 \leq \mathfrak{b} \leq \mathfrak{c}$; any combination of equalities and strict inequalities here is consistent with the usual axioms of set theory. The second use of countability in the proof of Theorem 2.5 still works as long as the number of h_n's is smaller than $\mathfrak{b}$.

Thus, for the proof of Theorem 2.5 to apply to an uncountable directed family of D_n's, we need the cardinality of this family to be smaller than both $\mathbf{cov}(B)$ and $\mathfrak{b}$. This requirement can be simplified by using the known fact [7, 13] that the smaller of $\mathbf{cov}(B)$ and $\mathfrak{b}$ is exactly $\mathbf{add}(B)$.

Corollary 2.6. *The group D is not the union of a directed family of fewer than* $\mathbf{add}(B)$ *pure, proper, analytic subgroups.*

We close this section with some results about isomorphic copies of Π inside D. We considered a family of such copies, namely those of the form $f\Pi$, in the proof of Theorem 2.5. It turns out that every copy of Π in D is included in one of this special form.

Theorem 2.7. *If $h : \Pi \to D$ is any homomorphism, then there is an $f \in D$ (with no zero components) such that $h(\Pi) \subseteq f\Pi$.*

Proof. Temporarily fix a positive integer q. The group D is covered by the countable chain of Borel subgroups

$$D_k = \{x \in D \mid q \text{ divides } x(n) \text{ for all } n \geq k\}.$$

Since h is continuous, the groups $h^{-1}(D_k)$ are Borel sets in Π. By Corollary 2.2, there must be a k such that $h^{-1}(D_k) = \Pi$, which means $h(\Pi) \subseteq D_k$.

Now un-fix q. Of course, the k we found in the preceding discussion may depend on q, so we call it $k(q)$. It has the property that, for all $x \in \Pi$ and all $n \geq k(q)$, we have $h(x)(n)$ divisible by q. We may assume that $k(1) = 0$.

Define f by letting $f(m) = q!$ for the largest $q \leq m$ such that $k(q!) \leq m$. Then $f \in D$ since $f(m)$ is divisible by $q!$ for all $m \geq \max\{q, k(q!)\}$. Finally, we check that $h(\Pi) \subseteq f\Pi$. Consider any $x \in \Pi$ and any m; we must show that $f(m)$ divides $h(x)(m)$. But $f(m) = q!$ for a certain q satisfying $k(q!) \leq m$, which means, by definition of $k(q!)$, that $h(x)(m)$ is divisible by $q!$, as required. □

In the following corollaries, we deal with isomorphic copies of Π although Theorem 2.7 would allow us to deal with homomorphic images of Π just as easily. The extra generality is illusory, since Nunke [15, Thm. 5] showed that all homomorphic images of Π that are subgroups of Π are isomorphic to either Π or $\mathbb{Z}^n$ for some finite n.

Corollary 2.8. *No isomorphic copy of Π in D includes Σ.*

Proof. If an isomorphic copy G of Π in D included Σ, then Theorem 2.7 would give an $f \in D$ such that $\Sigma \subseteq G \subseteq f\Pi$. But the only functions f for which $\Sigma \subseteq f\Pi$ are those that take only the values ± 1, and these are not in D. □

Corollary 2.9. *The group D is not the union of countably many isomorphic copies of Π.*

Proof. If D were such a union, then by Theorem 2.7 we could assume that the countably many isomorphic copies of Π involved in the union are of the form $f_n\Pi$ for certain $f_n \in D$ with no zero components. As in the proof of Theorem 2.5, we can find a single $f \in D$ such that, for each n, all sufficiently large k have $f(k)$ dividing $f_n(k)/2$. In particular, for each n there exists some k (in fact any sufficiently large k will do) such that $f_n(k)$ does not divide $f(k)$. But then f is not in any of the $f_n\Pi$, contrary to assumption. □

Corollary 2.10. *The group D is not the union of a chain (of any length) of isomorphic copies of Π.*

Proof. Suppose it were such a union. For each of the countably many members s of Σ, fix a group in the chain that contains s. The union of the countably many chosen groups is either all of D (if these groups are cofinal in the chain) or included in a larger subgroup from the chain. The first possibility contradicts Corollary 2.9, and the second contradicts Corollary 2.8. □

3. Quotients of D

In this section, we prove two theorems saying that, under certain circumstances, a small quotient of D is necessarily free of finite rank. The "certain circumstances" are that the quotient is separable or that the kernel is analytic; the necessary "smallness" of the quotient depends on which of these two circumstances is assumed.

Recall that a group G is called *separable* if it can be embedded as a pure subgroup in $\mathbb{Z}^\kappa$ for some cardinal κ.

Theorem 3.1. *Every separable quotient of D of cardinality $< \mathfrak{c}$ is free of finite rank.*

Proof. Let G be such a quotient, with epimorphism $h : D \to G$. Let B be the purification in G of $h(\Sigma)$. Since D/Σ is divisible, so is G/B; furthermore, B is pure in G and, as a countable subgroup of some $\mathbb{Z}^\kappa$, B is free. Therefore, the hypotheses of Theorem 1 of [5] are satisfied, and we conclude that G may be identified with a pure subgroup of Π (rather than some larger $\mathbb{Z}^\kappa$).

Making this identification and applying the continuity theorem to h, we conclude that G is analytic. By Suslin's Perfect Set Theorem, the assumption that $|G| < \mathfrak{c}$ implies that G is countable. By Baer's Theorem, G is free and is therefore a direct summand of D. But it follows immediately from the continuity theorem that D has no free direct summands of infinite rank. (The continuity theorem says that D has only countably many homomorphisms to $\mathbb{Z}$, but such a summand would have at least $\mathfrak{c}$ such homomorphisms.) Therefore, G has finite rank. □

Theorem 3.2. *Suppose E is a pure, analytic subgroup of D of countable index. Then D/E is free of finite rank.*

Proof. Temporarily fix some $f \in D$ with no zero components. Recall that we have $f\Pi \subseteq D$ with $f\Pi$ isomorphic to Π via $-/f$ and $f \cdot -$. By hypothesis, E is analytic, and so are all its cosets $y + E$ because addition is continuous. Since $f \cdot -$ is clearly continuous, the inverse images in Π of these cosets, namely $((y + E) \cap f\Pi)/f$, are also analytic. They can't all be meager, as there are only countably many of them and they cover Π. By the Lusin-Sierpiński Theorem, one of them almost (i.e., except for a meager set) covers a basic neighborhood $x + V_k$. By a Baire category and subtraction argument, as in the proof of Theorem 2.1, it follows that $(E \cap f\Pi)/f$ includes V_k. This means that $V_k \cap f\Pi \subseteq E$.

Consider now some $z \in V_k$ such that $f(n)$ divides $z(n)$ for all but finitely many n. We say that such a z is *almost in* $f\Pi$. Then multiplying z by a suitable positive integer (for example the product of all the finitely many $f(n)$ that don't divide the corresponding $z(n)$) we obtain an element in $V_k \cap f\Pi$, hence in E. But E is a pure subgroup of D, so $z \in E$.

Now un-fix f. We have shown that, for every $f \in D$ with no zero components, there exists k such that E contains all elements of V_k that are almost in $f\Pi$. Let $k(f)$ be the smallest such k.

We claim that, as f varies, $k(f)$ remains bounded. To prove this claim, suppose it were false, and fix a sequence of f's, say f_m, such that the corresponding $k(f_m)$ increase without bound. As in the proof of Theorem 2.5, obtain a single $f \in D$ (with no zero components) such that, for each m, all sufficiently large n have $f(n)$ dividing $f_m(n)$. It follows that everything that is almost in $f_m\Pi$ is also almost in $f\Pi$. Therefore, $k(f_m) \leq k(f)$ for all m. This contradicts the choice of the f_m, so the claim is established.

By virtue of the claim just verified, we fix a single k so large that E contains all elements of $V_k \cap f\Pi$, for all $f \in D$ without zero components. In other words, $E \supseteq D \cap V_k$. Therefore, D/E is a quotient of $D/(D \cap V_k) \cong \mathbb{Z}^k$. As E is pure in D, D/E is torsion-free. But the only torsion-free quotients of $\mathbb{Z}^k$ are free abelian groups of finite rank, as required. □

In Theorem 3.2, the hypothesis that the index of E is countable was used only to apply the Baire category theorem to see that the countably many sets $((y+E) \cap f\Pi)/f$ cannot all be meager. So it suffices to assume that the index of E in D is $< \mathbf{cov}(B)$. The following corollary summarizes this improvement of Theorem 3.2 along with Theorem 3.1 and some known results.

Corollary 3.3. *For any subgroup E of D, the following are equivalent.*

(1) *D/E is a free abelian group of finite rank.*
(2) *E is a pure subgroup of D of countable index and $E \cong D$.*
(3) *E is a pure, analytic subgroup of D of index $< \mathbf{cov}(B)$.*
(4) *E is a subgroup of D of index $< \mathfrak{c}$ and D/E is separable.*

Proof. Theorem 3.1 says that (4) implies (1), and that (1) implies (4) is obvious. Theorem 3.2 and the discussion following its proof give that (3) implies (1). The only non-obvious part of the implication from (1) to (2) is that $E \cong D$, but this follows since the freeness of the quotient implies that E is a summand of D and it is shown in [4] that all infinite rank direct summands of D are isomorphic to D.

Finally, for the implication from (2) to (3), it suffices to recall that D is a Borel set and so, by the Continuity Theorem, every isomorphic copy of D in Π is analytic. □

References

1. R. Baer, *Abelian groups without elements of finite order*, Duke Math. J. **3** (1937), 68–122.
2. R. Baire, *Sur les fonctions de variables réelles*, Ann. Mat. Pura Appl. (3) **3** (1899), 1–122.
3. A. Blass, J. Irwin, and G. Schlitt, *Some abelian groups with free duals*, Abelian Groups and Modules (A. Facchini and C. Menini, eds.), (Proceedings of 1994 conference in Padova), Kluwer (to appear).
4. A. Blass and J. Irwin, *On the group of eventually divisible integer sequences* (to appear).
5. M. Dugas and J. Irwin, *On pure subgroups of cartesian products of integers*, Resultate Math. **15** (1989), 35–52.
6. M. Dugas and J. Irwin, *On basic subgroups of $\prod \mathbb{Z}$*, Resultate Math. **12** (1991), 2907–2921.
7. D. H. Fremlin, *Cichoń's diagram*, Séminaire Initiation à l'Analyse (G. Choquet, M. Rogalski, and J. Saint-Raymond, ed.), Univ. Pierre et Marie Curie, 1983/84, pp. (5-01)–(5-13).
8. L. Fuchs, *Infinite Abelian Groups, vol. II*, Academic Press, 1973.
9. A. S. Kechris, *Classical Descriptive Set Theory*, Graduate Texts in Math. 156, Springer-Verlag, 1995.
10. C. Kuratowski, *Topologie, vol. I*, 4th edition, Państwowe Wydawnictwo Naukowe, Warszawa, 1958.
11. N. Lusin, *Sur la classification de M. Baire*, C. R. Acad. Sci. Paris **164** (1917), 91–94.
12. N. Lusin and W. Sierpiński, *Sur un ensemble non mesurable B*, Journal des Mathématiques (9) **2** (1923), 53–72.
13. A. Miller, *Some properties of measure and category*, Trans. Amer. Math. Soc. **266** (1981), 93–114.
14. Y. N. Moschovakis, *Descriptive Set Theory*, Studies in Logic and Foundations of Math. 100, North-Holland, 1980.

15. R. J. Nunke, *On direct products of infinite cyclic groups*, Proc. Amer. Math. Soc. **13** (1962), 66–71.
16. E. Specker, *Additive Gruppen von Folgen ganzer Zahlen*, Portugaliae Math. **9** (1950), 131–140.

MATHEMATICS DEPT., UNIVERSITY OF MICHIGAN, ANN ARBOR, MI 48109, U.S.A.
E-mail address: ablass@umich.edu

MATHEMATICS DEPT., WAYNE STATE UNIVERSITY, DETROIT, MI 48202

Primes and Completions of Right Chain Rings

Hans-Heinrich Brungs* University of Alberta, Edmonton, Alberta, Canada

Günter Törner Gerhard-Mercator-Universität, Duisburg, Germany

1 Introduction

A commutative valuation ring V which has no proper extension with the same value group and the same residue field as V is called maximal (KRULL [Kr32]). Such valuation rings can also be characterized by the fact that every pseudo convergent sequence in V has a limit in V, that is, V is maximally complete (KAPLANSKY [Ka42]).

A *right chain domain* R is a not necessarily commutative integral domain such that $aR \subseteq bR$ or $bR \subset aR$ for any elements $a, b \in R$. The concepts of maximal and maximally complete extend in a natural way to right chain domains and a maximally complete right chain domain is maximal but we will discuss an example of a maximal right chain domain that is not maximally complete; see section 4.

Essential for the understanding of the ideal theory of right chain rings R is the understanding of the ideals between two neighbouring completely prime ideals $P_1 \supset P_2$; such a pair (P_1, P_2) with no further completely prime between them is called a *prime segment* of R. Prime segments fall in exactly one of three classes: they are either invariant, simple, or exceptional and the last two types do not occur if the skew field of quotients of R is finite dimensional over its center.

On the other hand we are concerned with completions of a right chain ring R and say that R is I-complete (or I-compact) for a right ideal $I \neq R$ if the natural mapping of R into the inverse limit of R/I_λ is onto where I_λ, $\lambda \in \Lambda, (I_\alpha \supseteq I_\beta$ if $\alpha \leq \beta$ in $\Lambda)$ is any family of right ideals of R with $\bigcap_{\lambda\in\Lambda} I_\lambda = I$; this can also be expressed in terms of the existence of limits of pseudo convergent sequences and the definitions clearly can be extended to the right R-modules of the form R/I for a right ideal I of R.

If a right chain domain R is I-complete, then R is necessarily I'-complete for a certain set $\mathcal{C}(I)$ of right ideals I' of R that contains all right ideals of the form aI for

*Supported in part by NSERC

$0 \neq a \in R$. More generally, if $s^{-1}B = t^{-1}A = \{r \in R \mid tr \in A\}$ for proper right ideals A and B of R and $s \notin B, t \notin A$ then A and B are called *related*, and it follows that R is A-complete if and only if R is B-complete. It is also proved that the union of right ideals related to I is equal to $P(I) = \{p \in R \mid Ip^{-1} \supset I\}$, the complete prime ideal of R associated with I.

As one of the consequences of this result it follows that $\mathcal{N}$, the family of right ideals I of R for which R/I is not maximally complete, is either empty or there exists a completely prime ideal P with $\mathcal{N} = \{I \mid I \subset P\}$ or $\mathcal{N} = \{I \mid I \subseteq P\}$. As another consequence one obtains the following extension of a theorem by (RIBENBOIM [Ri64]) for commutative valuation rings, see also (MAC LANE [McL38]). THEOREM: *Let R be a right chain domain with $P \neq P^2$ for all completely prime ideals $P \neq (0)$ of R. Then R is maximally complete if and only if R is P-complete for all completely prime ideals.*

Finally, we construct chain domains that are I-complete for a prescribed set of right ideals and are not B-complete for all right ideals outside this set by imposing certain conditions on the support of elements in a Malcev-Neumann power series ring.

2 Prime ideals

Let R be a right chain domain. We denote with $J = J(R)$ the unique *maximal right* and left *ideal* and it follows that $U = U(R) = R\backslash J$ is the *group of units* of R and R/J is the residue skew field. As usual, an ideal $P \neq R$ of R is called *prime* if $I_1 I_2 \subseteq P$ for ideals I_1, I_2 of R implies $I_1 \subseteq P$ or $I_2 \subseteq P$. If the same property holds for elements of R instead of ideals, P is called *completely prime*.

LEMMA 2.1 *Let R be a right chain domain. Then:*

(i) *An ideal $I \neq R$ is completely prime if $x^2 \in I$ implies $x \in I$ for $x \in R$.*

(ii) *Let $I \neq R$ be an ideal in R. Then $P = \bigcap_n I^n$ is a completely prime ideal.*

PROOF: (i) Assume that the condition holds for I and that $xy \in I$ for $x, y \in R$. If $x = yz$ for $z \in R$ then $x^2 = xyz \in I$ and $x \in I$. If $y = xz$ then $(yx)^2 \in I$ implies $yx \in I$ and with the previous argument $y \in I$.

(ii) If $I = I^2$, then $P = I$ and we assume $x^2 \in I$, $x \notin I$ for an element $x \in R$. Then $I \subseteq xJ$ and $I = I^2 \subseteq xI \subseteq x^2 J \subset x^2 R \subseteq I$, a contradiction that shows that $I = P$ is completely prime in this case. If $P = I^n$ for some $n \geq 2$, then $P = I^n = (I^n)^2$ is completely prime by the previous argument. Finally, if $P \subset I^n$ for all n, and $x \notin P$ then $I^n \subset xR$ for some n and $P \subset I^{2n} \subseteq xI^n \subseteq x^2 R$ and $x^2 \notin P$ follows. ■

We consider prime segments (P_1, P_2) of R, i.e. $P_1 \supset P_2$ are completely prime ideals of R with no further completely prime ideal of R between P_1 and P_2. By considering $L = \bigcup I$, the union of ideals of R properly contained in P_1 it can be proved that the following three possibilities arise (see [BBT90]):

THEOREM 2.2 *Let (P_1, P_2) be a prime segment of a right chain domain. Then one of the following alternatives occurs:*

(a) *$L = P_1$ if and only if $P_1 a \subseteq aP_1$ for all $a \in P_1 \setminus P_2$; the prime segment is called* invariant *in this case.*

(b) $L = P_2$; *the prime segment is* simple - *there are no further ideals between* P_1 *and* P_2.

(c) $P_1 \supset L = Q \supset P_2$. *In this case* Q *is a prime ideal that is not completely prime and* $\cap Q^n = P_2$. *The prime segment is called* exceptional *and there are no further ideals between* P_1 *and* L.

The set of prime segments of R is totally ordered by defining $(P_1, P_2) \leq (P_1', P_2')$ if and only if $P_1 \supseteq P_1'$. The order type of this set is called the *rank* of R. Hence, R is of rank one if $J \neq R$ is the only non-zero completely prime ideal of R.

If (P_1, P_2) is a prime segment of R, then the localization R_{P_1} of R at P_1 exists since $S = R \setminus P_1$ is a right Ore system; and $R_{P_1} = RS^{-1}$ contains $P_2R_{P_1}$ as a completely prime ideal. The ring $R_{P_1}/P_2R_{P_1} = \overline{R}(P_1, P_2) = \overline{R}$ is a rank-one right chain domain that is right invariant, i.e. all right ideals are two-sided if (P_1, P_2) is invariant; it is *nearly simple*, that is $J(\overline{R}(P_1, P_2))$ is its only ideal $\neq \overline{R}, (0)$ if (P_1, P_2) is simple, and it is exceptional, which means it contains an additional prime ideal, if (P_1, P_2) is exceptional. The non-zero principal right ideals of an invariant rank-one right chain domain form an ordered semigroup under multiplication which is isomorphic to a subsemigroup of $(\mathbb{R}, +)$.

The first examples of nearly simple rank one right chain rings were constructed by DUBROVIN and MATHIAK in 1980/81. DUBROVIN also announced the construction of an exceptional rank one chain domain in 1993; see [Du93].

Let $I \neq R$ be a right ideal of R. We define the *associated prime ideal* of I as follows: $P_r(I) = P(I) = \{p \in R \mid \exists t \in R\backslash I \text{ such that } tp \in I\}$. That $P(I)$ is indeed a completely prime ideal of R is shown in the next result:

PROPOSITION 2.3 *Let* R *be a right chain domain and* $I \neq R$ *a right ideal of* R. *Then* $P(I)$ *is a completely prime ideal of* R.

PROOF: If p_1 and p_2 are elements in $P(I)$ we can assume $p_1 = p_2 s$ for some $s \in R$. Hence, for $t_2 \in R\backslash I$ with $t_2p_2 \in I$ we have $t_2(p_1 - p_2) \in I$, which shows $p_1 - p_2 \in P(I)$. It follows that $P(I)$ is a right ideal. To see that it is also a left ideal we consider rp_1 for $r \in R$. Then either $rp_1 = p_1r' \in P(I)$ for some $r' \in R$ or $rp_1j = p_1$ for $j \in J(R)$. Hence, $rp_1(j-1) = rp_1j - rp_1 = p - rp_1 = (1-r)p_1$. Since $j-1$ is a unit in R, it follows that $rp_1 = up_r''$ for a unit $u \in R$ and an element r'' in R. ∎

3 I-completeness and related right ideals

Let R be a right chain domain. For A a right ideal and $s \in R$ we write $s^{-1}A = \{r \in R \mid sr \in A\}$. It follows that $s^{-1}A$ is again a right ideal and we say that two proper right ideals A and B of R are *related* if $s^{-1}A = t^{-1}B$ for some elements $s \in R\backslash A$ and $t \in R\backslash B$.

PROPOSITION 3.1 *Let* R *be a right chain domain. Then:*

(i) *If* A *and* B *are related right ideals of* R, *then* $P(A) = P(B)$ *and* R *is* A-*complete if and only if* R *is* B-*complete.*

(ii) *Let A be a proper right ideal in R. Then $P(A)$ is equal to the union of the right ideals of R related to A.*

PROOF: (i) The first part is proved in Lemma 3.1 in [BT92], the second in Theorem 4.15 in (see [BT89]).

(ii) If B is related to A then $t^{-1}B = s^{-1}A$ for some $t \notin B$, $s \notin A$. For $b \in B$ there exists $w \in t^{-1}B$ with $tw = b$ and $sw \in A$. Hence, $w \in P(A)$ and $b \in P(A)$, since $P(A)$ is an ideal. Conversely, let $z \in P(A)$; hence there exists $t \notin A$ with $tz \in A$ and $z \in t^{-1}A$ follows. This proves (ii). ∎

We can apply these results to *discrete right chain domains* R. This is a right chain domain such that $P^2 \neq P$ for every completely prime ideal $P \neq (0)$ of R.

THEOREM 3.2 *Let R be a discrete right chain domain, then R is maximally complete if and only if R is P-complete for every completely prime ideal P of R.*

PROOF: Let A be any right ideal of R and we want to show that R is A-complete. This is clear if $A = (0)$. Otherwise let $P = P(A)$ be the associated prime ideal of A and $P \neq P^2$ by assumption. It follows from Lemma 2.1(ii) that $P' = \bigcap_n P^n$ is a completely prime ideal and that (P, P') is an invariant segment using Theorem 2.2 and the fact that $P_1 = P_1^2$ if (P_1, P_2) is a prime segment of type (b) or (c). We are done by Proposition 3.1(i) if we can show that P is actually related to A. Since $P \supset P^2$ there exists by Proposition 3.1(ii) a right ideal B related to A with $P \supseteq B \supset P^2$ and $P(B) = P$. If $P \supset B \supset P^2$ then there exists $x \in P \setminus B$ and $b \in B \setminus P^2$ and $xt = b$ for some $t \in R \setminus P$ (because $b \notin P^2$). Since $b \in B$ and $x \in R \setminus B$, $xt = b$ implies $t \in P(B) = P$ by definition of $P(B)$. But we know that $t \notin P$. This is a contradiction that proves $B = P$, and R is A-complete since A is related to P. ∎

This result was proved in the commutative case by MAC LANE [McL38]. Valuation rings that are P-complete for every prime ideal P are closely related to the perfect valuation rings considered by KRULL [Kr32] and those that are 'complet par etages' as defined by RIBENBOIM [Ri64].

If R is a right chain ring and I a right ideal, then R/I is a right R-module whose lattice of submodules is totally ordered. The earlier introduced notion of maximally complete right chain domain carries over to right R-modules with the just mentioned property. One obtains the following result:

THEOREM 3.3 *Let $\mathcal{N}$ be the family of right ideals I of a right chain ring R such that R/I is not maximally complete. Then $\mathcal{N} = \emptyset$ or there exists a completely prime ideal P of R with $\mathcal{N} = \{I \mid I \subset P\}$ or $\mathcal{N} = \{I \mid I \subseteq P\}$.*

PROOF: Let I be a right ideal of R such that R/I is not maximally complete. Then R is not L-complete for some right ideal L of R with $I \leq L$. We consider the set $\mathcal{N}$ of right ideals L' of R with R not L'-complete. It follows by the above observation that the union V of right ideals $L' \in \mathcal{N}$ is equal to the union of right ideals I of R with R/I not maximally complete.

From Proposition 3.1 we conclude that the set $\mathcal{N}$ consists of classes of related right ideals and the union of elements in a class is a completely prime ideal. Hence, V itself is the union of completely prime ideals which implies $V = P$ for a completely prime ideal if $\mathcal{N}$ is not empty. We therefore have the three alternatives listed in the theorem: The ring R and hence R/I is maximally complete for every I, i.e. $\mathcal{N} = \emptyset$. The ring R/P is not maximally complete and $\mathcal{N} = \{I \mid I \subseteq P\}$. Finally, $P \notin \mathcal{N} = \{I \mid I \subset P\}$. ■

We close this section by showing that an almost maximally complete right chain ring with zero-divisors is actually complete. As was observed above the notion of I-complete and maximally complete extend immediately to right chain rings which are not necessarily domains. A right chain ring R is called *almost maximally complete* if it is I-complete for all right ideals I with the possible exception of $I = (0)$.

If $a \neq 0 \neq b$ in a right chain ring R with $ab = 0$ then $I = a^{-1}(0) \ni b$ is a nonzero right ideal of R related to (0). By Proposition 3.1(i) it follows that R is (0)-complete if it is I-complete. In the commutative case this was proved by GILL (see [Gi71]).

4 Examples and applications

We use the results in the previous sections to discuss the example of a maximal right chain domain R that is not maximally complete, and examples of right chain domains that are I-complete exactly for the right ideals in a predetermined set.

EXAMPLE 4.1 Let $\mathbb{Q}(t_1, t_2, \ldots)$ be the function field over the field $\mathbb{Q}$ of rational numbers in the variables $t_i, i = 1, 2, \ldots$ and $\mathbb{Q}(t_1, t_2, \ldots)[x]_{(x)}$ is the localization at the prime ideal (x) of the polynomial ring $\mathbb{Q}(t_1, t_2, \ldots)[x]$; it is a discrete valuation ring with a monomorphism σ defined by $\sigma(x) = t_1,\ \sigma(t_i) = t_{i+1}$.

Finally,

$$R = \mathbb{Q}(t_1, t_2, \ldots)\ [x]_{(x)}\ [[y, \sigma]] = \{\sum_{n=0}^{\infty} y^n f_n\,(t_i, x) \mid f_n\,(t_i, x) \in Q(t_1, t_2, \ldots)\ [x]_{(x)}\}$$

with $f_n(t_i, x)y = y\sigma(f_n(t_i, x))$, the skew power series ring in the variable y, is a discrete right invariant right chain domain with $\{y^n x^m R \mid n, m \geq 0\}$ the set of right ideals $\neq (0)$. This ring has rank 2 and $R \supset xR \supset yR \supset (0)$ is its chain of completely prime ideals; in addition, R is xR as well as (0)-complete. By Theorem 3.2 it follows that R is maximally complete if and only if R is (y)-complete. However, $R/yR \cong Q(t_1, t_2, \ldots)\ [x]_{(x)}$ is not complete and hence, R is not (y)-complete and not maximally complete. Before we indicate why R is a maximal right chain domain we recall the definition of an immediate extension.

Let R be a right chain domain. We say that a ring extension $R_1 \supseteq R$ is an *immediate extension* of R if the follwing conditions hold:

(i) Every element $a_1 \in R_1$ has the form $a_1 = a \cdot u_1$ for $a \in R$ and a unit $u_1 \in U(R_1)$.

(ii) $\varphi(r + J(R)) = r + J(R_1)$ defines an isomorphism between the skew fields $R/J(R)$ and $R_1/J(R_1)$.

It follows that for any right chain domain R there exist maximal immediate extensions and that every element $z \in R_1$ where R_1 is an immediate extension of R, is the limit of a pseudo convergent sequence in R that has no limit in R.

We return to our example R above and we know that R is not maximally complete. We want to show that on the other hand R has no proper immediate extension $R_1 \supset R$. Otherwise such a ring R_1 is again a right invariant right chain domain of rank two with $\{y^n x^m R \mid n, m \geq 0\}$ as its set of non-zero right ideals; however, $R_1/yR_1 \supset R/yR$. The existence of an element $f \in R_1/yR_1$ which is not in R/yR leads to the contradiction that x^k is a unit in R for some $k > 0$. Hence, no such proper immediate extension R_1 of R exists, i.e. R is maximal even though it is not maximally complete.
We don't know the answer to the following question.

PROBLEM 4.2 *Do there exist right and left chain domains which are maximal but not maximally complete?*

If R is a right chain domain with maximal ideal $J = J(R) \neq (0)$ and $0 \neq c \in R$, then $\bigcap_\lambda I_\lambda = cJ$ for a family of right ideals $I_\lambda \subset R$ implies $I_\lambda = cJ$ for an index λ since otherwise $I_\lambda \supseteq cR$ for all λ and $\bigcap_\lambda I_\lambda \supseteq cR \supset cJ$.The ring R is therefore trivially I-complete for every right ideal I of the form $cJ \neq (0)$.

A right ideal $I \neq R$ of R is either of the form $cJ \neq (0)$ or the set $L_I = \{aR \mid aR \supset I\}$ has no last element, i.e. L_I is a limit point of $W(R)$, the totally ordered set of principal ideals of R with $aR \leq bR$ if and only if $bR \subseteq aR$. Here we say that a convex subset L of a totally ordered set $(W, \geq)$ with first element $e \in W$ is a *limit point* of W if $e \in L$ and L has no last element.

The set $W(R)$ of a right invariant rank one right chain domain forms a semigroup with ideal multiplication as operation that can be identified with a subsemigroup D of $(\mathbb{R}^+, +)$ the non-negative real numbers under addition. With D' we denote the closure of D in $\mathbb{R}^+$; it follows that $D' = \mathbb{R}^+$ or $D' = \mathbb{N}_0$ in the discrete case. A limit point L of D is therefore given by a real number $0 \neq \rho \in D'$ with $L = \{b \in D \mid b < \rho\}$ or $L = D$.

We now consider a right ideal $I \neq cJ$ in a right chain ring R. Let $P_1 = P(I)$ be the prime ideal associated with I. We assume further that P_1 has a lower neighbour P_2 in the chain of completely prime ideals and that the rank one valuation ring $\overline{R}(P_1, P_2) = R_{P_1}/P_2 R_{P_1}$ associated with the prime segment (P_1, P_2) is right invariant with value semigroup D and completion $D' \subseteq \mathbb{R}^+$. Then either $I = dP_1$ for some non-zero $d \in R$ or there exists a right ideal I_1 of R related to I with $P_1 \supset I_1 \supset P_2$ and $\overline{I}_1 = I_1 \overline{R}(P_1, P_2)$ defines a limit point of D which in turn is defined by a real number $\rho \in D'$.

We summarize these results:

LEMMA 4.3 *Let R be a right chain domain, (P_1, P_2) a prime segment of R and I a right ideal of R with $P(I) = P_1$. If the associated rank one chain domain ring $\overline{R}(P_1, P_2)$ is right invariant, then either $I = dP_1$ for some $0 \neq d \in R$ or there exists a real number ρ and a right ideal I_1 of R related to I with $P_1 \supset I_1 \supset P_2$ such that $L_{I_1} = \{aR \mid a\overline{R}(P_1, P_2) < \rho\}$.*

This lemma provides a satisfactory description of the right ideals of a right chain domain with maximum condition for completely prime ideals and for which all the rank one chain domains corresponding to prime segments of R are right invariant, we say that R is locally invariant in this case.

We don't have similar results if completely prime ideals exist that are the union of completely prime ideals, or nearly simple or exceptional rank one chain domains correspond to some prime segments.

We return to the question about the description of the set of right ideals I' for which the right chain domain R must be I'-complete if it is known that R is I-complete for all right ideals in a set $\mathcal{Z}$. We restrict ourselves to locally invariant right chain domains R with maximum condition on completely prime ideals. Each right ideal I is then either of the form $cJ \neq (0)$ and R is trivially I-complete, or of the distinguished form dP for a completely prime ideal P for d in R, or is related to a right ideal I_1 that corresponds to a real number ρ in one of these 'local maps' $D' \subseteq \mathbb{R}^+$ that describe the limit points of the corresponding right invariant rank one right chain domains. It is possible to obtain some results on the closure properties of the set of right ideals I for which R must be I-complete under these circumstances. Finally, chain domains can be constructed that are I-complete exactly for right ideals I in a predesigned set $\mathcal{V}$. We illustrate this in the next example.

EXAMPLE 4.4 Let $D = \mathbb{Q}^+ = \{q \in \mathbb{Q} \mid q \geq 0\} \subset D' = \mathbb{R}^+$. Let T be a subset of D' with the following properties: T does not contain 0, but $s \in D$ and $r \in T$ implies $s + r \in T$; if $s + \mu = \rho \in T$ for $s \in D$ and $\mu \in D' \backslash \{0\}$, then $\mu \in T$. We then consider $\hat{R} = \mathbb{Q}\{\{D\}\}$, the Malcev-Neumann ring of all power series $\alpha = \sum a_i d_i$ with the coefficients a_i in $\mathbb{Q}$ and d_i in D such that $support(\alpha) = \{d_i \mid a_i \neq 0\}$ is a well ordered subset of D. We define R_T as the subset of $\hat{R}$ consisting of those elements α in $\hat{R}$ such that L a limit point of D and $L \cap support(\alpha)$ cofinal in L implies $L = D$ or $L = \{d \in D \mid d < \rho\}$ for some $\rho \in T$. By computing the set of limit points for $\alpha + \beta, \alpha\beta$ and $\sum \gamma^n$ for $\alpha, \beta \in \hat{R}, \gamma \in J(\hat{R})$ with given sets of limit points, one can prove that R_T is a chain domain of rank one with D as associated semigroup of values and that R_T is I-complete exactly for $I = (0)$, corresponding to $L = D$, for $I = I\rho = \cap zR_T, z < \rho$ for $\rho \in T$, and trivially for $cJ(R_T)$ for all $0 \neq c \in R_T$ and no other right ideals. We give a concrete example for T in the present situation:

$$T_1 = \{r = q + n\sqrt{2} \in \mathbb{R}^+ \mid r > 0, q \in \mathbb{Q}, n \in \mathbb{N}\}$$

satisfies the above condition as does

$$T_2 = \{r = q + n_1\sqrt{2} + n_2\sqrt{3} \in \mathbb{R}^+ \mid r > 0, q \in \mathbb{Q}, n_i \in \mathbb{N}\}.$$

Results similar to those illustrated by this example are obtained in a forthcoming paper in the case where D is replaced by certain semigroups, socalled right invariant right cancellative right holoids of finite rank n. The set $\mathcal{V}$ is then essentially determined by a collection of n subsets $T_i \subseteq \mathbb{R}^+$ satisfying appropriate properties.

References

[BBT90] C. BESSENRODT, H.H. BRUNGS, G. TÖRNER, Right chain rings. Part 1, *Fachbereich Mathematik, Universität Duisburg* **181**, (1990).

[Bra73] W. BRANDAL, Almost maximal integral domains and finitely generated modules; *Trans. Amer. Math. Soc.* **183** (1973), 203-222.

[BT89] H. H. BRUNGS; G. TÖRNER, Maximal immediate extensions are not necessarily maximally complete. *J. Austral. Math. Soc. (Series A)* **49** (1990), 196 - 211.

[BT92] H. H. BRUNGS; G. TÖRNER, *I*-compactness and prime ideals. *Publicationes Mathematicae (Debrecen)*, **40** (1992), 291-295.

[Du80] N.I. DUBROVIN, Chain domains. *Moscov Univ. Math. Bull. Ser. I* **35** (1980)2, 51 - 54.

[Du93] N.I. DUBROVIN, The rational closure of group rings of left-ordered groups; *Matemat. Sbornik.* **184**,7 (1994), 4-48.

[Gi71] D.T. GILL, Almost maximal valuation rings; *J. London Math. Soc.* **4** (1971), 140-146.

[Ka42] I. KAPLANSKY, Maximal fields with valuations. *Duke Math. J.* **9** (1942), 303 - 321.

[Kr32] W. KRULL, Allgemeine Bewertungstheorie. *J. Reine Angew. Math.* **167** (1932), 160 - 196.

[McL38] S. MACLANE, The uniqueness of power series representations of certain fields with valuations. *Annals of Math.* **39** (1938), 370 - 382.

[Ma81] K. MATHIAK, Zur Bewertungstheorie nicht kommutativer Körper. *J. Algebra* **73** (1981), 586-600.

[Ri64] P. RIBENBOIM, Théorie des valuations. *Les presses de l'université de Montreal*, Montreal 1964.

Hans-Heinrich Brungs
Department of Mathematics
University of Alberta
Edmonton
Canada T6G 2G1
hbrungs@vega.math.ualberta.ca

Günter Törner
Fachbereich Mathematik
Gerhard-Mercator-Universität
47048 Duisburg
Germany
toerner@math.uni-duisburg.de

Unsaturated Classes of Modules

JOHN DAUNS Department of Mathematics, Tulane University, New Orleans, LA 70118

Introduction

An unsaturated class Δ of right R-modules is one that is closed under (a) isomorphic copies, (b) arbitrary direct sums, (c) essential extensions, (d) essential submodules, (e) and is generated by any one of its elements. Let $\Xi(R)$ be the class of all such classes Δ. For any right R-module A, let $[A] \in \Xi(R)$ denote the unique smallest unsaturated class containing A. Then (e) says that for any $A \in \Delta$, $\Delta = [A]$. It is shown that for any finite or infinite pairwise disjoint subset $\Gamma(R) \subset \Xi(R)$ whose least upper bound $\sup \Gamma(R) = 1 \in \Xi(R)$, any module M contains submodules $M_\gamma < M$, $[M_\gamma] \leq \gamma \in \Xi(R)$, $\gamma \in \Gamma(R)$, such that $E(M) = E(\bigoplus_{\gamma \in \Gamma(R)} M_\gamma)$, where E denotes injective hulls, and the $E(M_\gamma)$ are unique up to not only isomorphim but superspectivity, a concept found in [MM; p.12, Def. 1.30].

In [D1], the above type of theory was developed for torsion free, that is nonsingular, modules only. In that case the $M_\gamma < M$ were completely unique fully invariant submodules of M, $[M_\gamma] \leq \gamma \in \Gamma(R) \subset \Xi(R)$. The torsion freeness was used explicity and implicitly in all proofs, which did not seem to generalize so as to include torsion (= singular) and mixed modules.

For example, "$\Xi(R)$" could still be defined, but for a long time it was not known whether it was a modular lattice in the presence of torsion, because the previous torsion free proofs based on traces simply failed in the now more general case. The objective of this article is to show that the theory initiated in [D1], and further developed in [D2-6], has an analogue and partial generalization to arbitrary modules, i.e. mixed modules. In view of the fact that the torsion free theory is simpler and easier than the general, and since the description of the torsion free case is not short ([D1-6]), the reader should not expect the whole complete general theory here in this one article (e.g. proof of 2.11, or the functorial aspects of Ξ). However some of the key new results are proved here, and the concepts needed for this are explained – such as saturated and unsaturated classes of modules.

Examples and applications to torsion modules are given in section three 3.9-3.12. Usually $\Gamma(R)$ as above is defined for one single specific ring R only. However, there are certain universal Γ's which are definable for every R, and this article develops and studies one such Γ, where $\Gamma(R) = \{\text{modules of local Goldie dimension } \aleph \mid \aleph = 1, \aleph_0, \aleph_1, \ldots\}$. In [DF1], the infinite Goldie dimension of a module was introduced, thus making it possible here to define and use the class of all modules of local Goldie dimension $\aleph$. (See Definitions 3.2 and 3.3, and Theorem 3.5.) Torsion free and also torsion examples of these are constructed using $\aleph$-products, a construct heavily used and developed in [DF2]. Moreover, several concrete examples are constructed to demonstrate that for any given cardinal $\aleph$, there exist both torsion free as well as torsion rings R where $\aleph \leq |\Gamma(R)|$.

Recently, much ring and module theory is done in the so called Wisbauer category $\sigma[M]$ ([W1,2]). Thus Page and Zhou ([PZ1-3], [Z]) study certain classes of modules $\mathcal{K} \subseteq \sigma[M]$ that they call M-natural classes. The overlap between their work and this article is minimal, even thought a saturated class of modules is a special case of an M-natural class.

Thus very recently a sizeable literature has appeared about various classes of modules which have the common characteristic that they are closed under (i) isomorphic copies, (ii) direct sums, (iii) certain limited kinds of essential extensions, and (iv) either essential submodules, or all submodules, and sometimes also closed under other possible operations, or possessing other additional properties depending on the author and article ([PZ1,2,3], [Z], [RT], [W2;p.118] and [D1-4]).

In descending order of generality, the Wisbauer category $\sigma[M]$ is the most general setting, then more concrete is the work of Page and Zhou in terms of the M-natural classes, and lastly the saturated classes used in this article are the very special case of M-natural classes if $M = R_R$. However, section 4 shows how the M-natural classes can be studied and understood by using the theory given in this article. (See Theorem 4.7). Also, some further problems and questions about $\sigma[M]$ and M-natural classes are listed in 4.8.

1. Preliminaries

Notation and some frequently used facts are explained.

1.1. Notation. Right unital modules over an arbitrary ring R are denoted by $K < M$, or $K \leq M$; essential or large ones by " $\leq\subset_E$ ". A complement submodule $P \leq M$ is one that does not have a proper essential extension inside M. For $x \in M$, $x^\perp = \{r \in R \mid xr = 0\} < R$, and $x^{-1}K = \{r \in R \mid xr \in K\} = (x+K)^\perp \leq R$. Thus $M^\perp \triangleleft R$, where " $\triangleleft$ " denotes ideals.

Right R injective hulls are denoted by both "$\widehat{\ }$" or "E", e.g. $M \leq\subset_E \widehat{M} = E(M) = EM$. The singular and second singular submodules are $Z(M) = ZM = \{x \in M \mid x^\perp \leq\subset_E R\} \leq\subset_E Z_2(M) = Z_2M \leq M$, where $Z[M/ZM] = Z_2M/ZM$. The *torsion submodule* of M is Z_2M, and M is *torsion* if $M = Z_2M$; M is *torsion free* – abbreviated "t.f." – if $Z_2M = 0$. Note that $ZM = 0 \Longleftrightarrow Z_2M = 0$. (See [D1; pp.52-55, 1.1-1.11], or [D5; p.748, 3.7]). The notation "$A \hookrightarrow B$" means that $A \cong A' \leq B$, some A'.

For a set X, $|X|$ denotes its cardinality, and $\mathcal{P}(X)$ the Boolean lattice and/or ring of all subsets $\langle \mathcal{P}(X), \cup, \cap, \setminus, \emptyset, X \rangle$. The category of right unital R-modules is denoted by $\mathcal{M}od_R$.

1.2. A decomposition $\widehat{M} = E(\oplus\{A_i \mid i \in I\})$ with certain properties $(\mathcal{P})$, is said to be *unique* up to *superspecitivity*, if for any other decomposition $M = E(\oplus\{B_j \mid j \in J\})$ also satisfying $(\mathcal{P})$, there is a bijection $\sigma : I \longrightarrow J$ such that $\widehat{A}_i \cong \widehat{B}_{\sigma(i)} \leq \widehat{M}$, $i \in I$, are superspective in the sense of [MM; p.12, Definition 1.30].

The following fact will be used repeatedly later on without mentioning it (see [D2; p.329, 1.4], [D3; p.101, 1.2], or [D4; p.43, 1.4]).

1.3. PROJECTION ARGUMENT. *For any indexed family W_γ, $\gamma \in \Gamma$ of modules and any $0 \neq \xi \in E(\bigoplus_{\gamma\in\Gamma} W_\gamma)$ there exist $a_0 \in R$, $\gamma \in \Gamma$, $0 \neq w \in W_\gamma$ such that $0 \neq \xi a_0 R \cong wR \leq W_\gamma$, and $\xi^\perp = w^\perp$.*

2. Module Classes

Unsaturated classes of modules will be the basic building blocks into which modules later on will be decomposed. In other words, all modules are formed by taking the injective hull of a direct sum of these basic primitive modules belonging to various unsaturated or saturated classes.

2.1. Definition. An *unsaturated class* of modules Δ is a nonempty class Δ closed under (a) isomorphic copies, (b) arbitrary direct sums, (c) essential extensions, (d) essential submodules, and satisfying the following minimality condition (e). (e) For any $A \in \Delta$, the (unique) smallest class of modules satisfying (a), (b), (c), (d) and containing A is all of Δ. Note that in the presence of (d), (c) is equivalent to closure under injective hulls.

A *saturated class* of modules is one that satisfies (a), (b), (c) and (d) is closed under arbitrary submodules.

For any class of modules $\mathcal{C}$ which is closed under 2.1 (a), (b) and also injective hulls, let $\langle \mathcal{C} \rangle$ denote the (unique smallest) saturated class generated by $\mathcal{C}$. If $\mathcal{C} = \{M\}$, write $\langle \{M\} \rangle = \langle M \rangle$.

2.2. Definition. The class of all unsaturated classes of R-modules will be denoted by $\Xi(R)$. For any R-module M, let $[M] \in \Xi(R)$ denote the unsaturated class of modules generated by M. Note that $[M] \subsetneq \langle M \rangle$ if $M \neq 0$. Thus $\Xi(R) = \{[A], [B], \ldots\}$ for (nonunique) $A, B \in \mathcal{M}od_R$. The class $\Xi(R)$ is partially ordered by $\Delta_1 \leq \Delta_2 \in \Xi(R)$ if some module of Δ_1 embeds into some module of Δ_2.

For a saturated or unsaturated class of modules Δ, define $t(\Delta) = t\Delta = \{Z_2A \mid A \in \Delta\}$, and $f(\Delta) = f\Delta = \{C \mid \exists\, A \in \Delta,\ Z_2A \oplus C \leq\subset_E A\}$. Since $t\Delta, f\Delta \subseteq \langle \Delta \rangle$, and since $t\langle \Delta \rangle = \{A \in \langle \Delta \rangle \mid A = Z_2A\}$ while $f\langle \Delta \rangle = \{B \in \langle \Delta \rangle \mid ZB = 0\}$, $t\Delta$ and $f\Delta$ are called the torsion and torsion free parts of Δ. Define $\Xi_t(R) = \{\Delta \in \Xi(R) \mid \forall\, A \in \Delta,\ A = Z_2A\}$ and $\Xi_f(R) = \{\Delta \in \Xi(R) \mid \forall\, B \in \Delta,\ ZB = 0\}$.

We now develop a concrete and readily applicable characterization of the unsaturated class of modules $[A]$ generated by any given module, as well as an alternate description of the relation $\leq$ on $\Xi(R)$.

2.3. Unsaturated classes. For modules $A, B, C, \ldots$ over a fixed ring R, define a relation " $\preceq$ " on $\mathcal{M}od_R \times \mathcal{M}od_R$ by $B \preceq A$ if $B \hookrightarrow E(\bigoplus_I A)$ for some finite or infinite index set I. Define $A \sim B$ if $A \preceq B$ and $B \preceq A$, i.e. if there exist I and J, and embeddings $A \hookrightarrow E(\bigoplus_J B)$ and $B \hookrightarrow E(\bigoplus_I A)$. We may take $I = J$. Then " $\sim$ " is an equivalence relation generated by the quasi-order " $\preceq$ ". It can be verified that any equivalence class modulo " $\sim$ " satisfies 2.1 (a), (b), (c), (d), and (e), and conversely for any Δ as in 2.1, and for any $A \in \Delta$, $\Delta = \{C \in \mathcal{M}od_R \mid A \sim C\}$. Hence, first of all $\Xi(R)$ is $\Xi(R) = \{[A], [B], [C], \ldots\}$, and secondly the quasi-order $\preceq$ induces a natural partial order on the equivalence classes $\mathcal{M}od_R/\sim = \Xi(R)$, where $[A] \leq [B]$ iff for some (or any) $A \in [A]$, $B \in [B]$ we have $A \preceq B$. It now can be verified that the latter partial order is the same as the one defined in 2.1. Thus $(\Xi(R), \leq, <)$ is a partially ordered class under $\leq$, which later will turn out to be a set, and with smallest element $0 = \{(0)\} \in \Xi(R)$. If least upper bounds or greatest lower bounds of finite or infinite subsets of $\Xi(R)$ exist, they are designated by the usual lattice symbols " $\vee$ " and " $\wedge$ ", e.g. $[A] \vee [B]$, $[A] \wedge [B] \in \Xi(R)$.

The next paragraph will clarify the connection of the present article with previous ones and will spell out how the present treatment is more general than previous ones.

2.4. All of the above can be carried out for torsion free modules only, and we would obtain a parallel Definition 2.1, analogous corresponding equivalence relation $\sim_f$ in 2.3, and equivalence classes ${}_f\Xi(R)$ modulo $\sim_f$. Indeed, this whole theory started with ${}_f\Xi(R)$ in [D1] and [D3], where ${}_f\Xi$ was then simply called "Ξ". And similarly we can form $\sim_t$ and ${}_t\Xi(R) = \text{tor } \mathcal{M}od_R/\sim_t$.

The next proposition begins to assemble a few facts about unsaturated classes.

2.5. PROPOSITION. *Let $\mathcal{S} = \{L \mid L \leq R\}$ be a set giving a set of representatives $\{E(R/L) \mid L \in \mathcal{S}\}$ of isomorphism classes of injective hulls of cyclic modules. Then*

(i) *every unsaturated class Δ is uniquely of the form $\Delta = [\oplus\{R/L \mid L \in \mathcal{T}\}]$ for a unique subset $\mathcal{T} \subseteq \mathcal{S}$. In particular,*

(ii) *$\Xi(R)$ is a set; $|\Xi(R)| \leq 2^{|\mathcal{P}(R)|}$.*

(iii) *$[\oplus\{R/L \mid L \in \mathcal{S}\}] = 1 \in \Xi(R)$ is the largest element.*

(iv) *$\Xi_t(R) = {}_t\Xi(R) = \{t\Delta \mid \Delta \in \Xi(R)\} = t(\Xi(R))$, and $\Xi_f(R) = {}_f\Xi(R) = \{f\Delta \mid \Delta \in \Xi(R)\} = f(\Xi(R))$.*

Proof. (i), (ii), and (iii) follow from the fact that any module contains an essential direct sum of cyclis together with the use of 2.3, and their proofs are omitted.

(iv) Since for any $M, ZEM = 0 \iff EZM = 0$, and since $EZ_2 = Z_2E$, it follows that for any $\Delta \in \Xi(R)$, $t\Delta \in \Xi_t(R) \subset \Xi(R)$ while $f\Delta \in \Xi_f(R) \subset \Xi(R)$. Hence $\sim_t \subset \sim$ and $\sim_f \subset \sim$. Conversely any equivalence class $\Delta = t\Delta \in \Xi(R)$ modulo $\sim$ which consists of torsion modules only, is also an equivalence class modulo $\sim_t$, and similarly for $\sim_f$. Thus (iv) follows. □

2.6. Definition and Notation. Let Δ be any class of modules which is closed under (a) isomorphic copies, (b) direct sums, and (c) injective hulls. For any module M, let either $M_0 = 0$, or let $M_0 \le M$, $M_0 \in \langle\Delta\rangle$. Upon taking the complement closure of a maximal direct sum of submodules of M belonging to $\langle\Delta\rangle$ and containing M_0, we obtain (in general a nonunique) submodule $N \le M$ satisfying (i) $M_0 \subseteq N$, (ii) $N \in \langle\Delta\rangle$, and (iii) $N \le M$ is a maximal submodule with properties (i) and (ii). I.e. if $N \subseteq L \le M$ and $\exists\ L \hookrightarrow W \in \Delta$, then necessarily $L = N$. If Δ is a saturated class to start out with, then $\Delta = \langle\Delta\rangle$. For any one such choice of $N \le M$ we write $M_\Delta = N$. Note that $M_\Delta = M_{\langle\Delta\rangle}$. Any $N \le M$ satisfying 2.6 (ii) and (iii) will be called a *maximal* Δ*-submodule* of M. Thus $M_\Delta \le M$ stands for any maximal Δ-submodule.

For Δ as above, define its *complementary saturated class* as $c(\Delta) = c\Delta = c(\langle\Delta\rangle) = c\langle\Delta\rangle = \{W \in \mathcal{M}od_R \mid \forall\, 0 \ne V \le W,\ V \notin \langle\Delta\rangle\}$. In terms of Δ only, $c\Delta = \{W_R \mid \forall\, 0 \ne V \le W,\ \forall\, A \in \Delta,\ \nexists\ V \hookrightarrow A\}$. It can be verified that $c\Delta$ is a saturated class. Upon replacing Δ with $c\Delta$, from the above we also get $M_{c\Delta} \le M$. Note that $M_{c\Delta} = M_{c\langle\Delta\rangle}$.

2.7. CONSEQUENCES. *For Δ and M as in 2.6, and for any choice of $M_\Delta \le M$, let $K \le M$ be maximal with respect to $M_\Delta \cap K = 0$. Then for any choice of $M_{c\Delta} \le M$,*

(i) $M_\Delta + M_{c\Delta} = M_\Delta \oplus M_{c\Delta} \le\subset_E M$;

(ii) $M_{c\Delta} = K$ *satisfies* 2.6 (i), (ii), *and* (iii) *with respect to* $c\Delta$.

2.8. SUPERSPECTIVITY. *For Δ as in* 2.6 (a), (b), (c) *and any module M assume that $N_1 \oplus C_1 \le\subset_E M$ and $N_2 \oplus C_2 \le\subset_E M$ where either N_i satisfies* 2.6 (i), (ii), (iii) *for Δ, or where C_i satisfies* 2.6 (i), (ii), (iii) *for $c\Delta$, $i = 1, 2$. Then we may take either*

(i) $M_\Delta = N_1,\ M_{c\Delta} = C_1$; *or* $M_\Delta = N_2,\ M_{c\Delta} = C_2$;

(ii) $\widehat{M} = \widehat{N}_1 \oplus \widehat{C}_1 = \widehat{N}_2 \oplus \widehat{C}_2$ *and*

(iii) $\widehat{N}_1 \cong \widehat{N}_2$ *and* $\widehat{C}_1 \cong \widehat{C}_2$;

(iv) $\forall\ M_\Delta \oplus M_{c\Delta} \le\subset_E M$, $\widehat{M_\Delta}$ *and* $\widehat{M_{c\Delta}} \le \widehat{M}$ *are unique up to superspectivity.*

Proof. (i) and (ii). Conclusion (i) follows from 2.7 (ii), while (ii) is a consequence of 2.7 (i).

(iii) By 2.7 (ii), with $M_\Delta = N_1$ and $M_{c\Delta} = C_2$, we get $\widehat{M} = \widehat{N}_1 \oplus \widehat{C}_2 = \widehat{N}_1 \oplus \widehat{C}_1$. Thus $\widehat{C}_1 \cong \widehat{C}_2$, and similary, $\widehat{N}_1 \cong \widehat{N}_2$.

(iv) If $\widehat{M} = \widehat{N}_1 \oplus D$, then take $M_\Delta = N_2$ and $M_{c\Delta} = M \cap D$. Use 2.7 (ii) to conclude that $\widehat{M} = \widehat{N}_1 \oplus D = \widehat{N}_2 \oplus D$. Thus $\widehat{M_\Delta}$, and similarly $\widehat{M_{c\Delta}}$ are unique up to superspectivity. □

Frequently both saturated and unsaturated classes have to be used simultaneously, and the next proposition provides an algorithm for going from one to the other.

2.9. PROPOSITION. *For any saturated class Υ of R-modules, the following hold.*

(1) (a) $\exists!\ \Delta \in \Xi(R)$, $\Delta \subseteq \Upsilon = \langle\Delta\rangle$.

(b) $\exists!\ \Delta \in \Xi(R)$ *such that* $\forall\ \Omega \in \Xi(R)$, $\Omega \subseteq \Upsilon \Longrightarrow \Omega \subseteq \Delta$.

(c) $\exists!\ \Delta \in \Xi(R)$ *such that* $\forall\ \Omega \in \Xi(R)$, $\Delta \subseteq \Omega \subseteq \Upsilon \Longrightarrow \Delta = \Omega$.

(d) *Either* (a), (b), *or* (c) *characterize* Δ.

(2) $\exists\ B \in \mathcal{M}od_R$, $\Upsilon = \langle B\rangle$.

(3) $\forall\ A \in \mathcal{M}od_R$, *if* $\Upsilon = \langle A\rangle$, *then* $\Longrightarrow \Delta = [A]$ *and*

(4) $\Upsilon = \{C \in \mathcal{M}od_R \mid \exists I,\ \exists C \hookrightarrow E(\bigoplus_I A)\} = \{C_R \mid [C] \le [A]\} = \bigcup\{x \mid x \in \Xi(R),\ x \le [A]\}$.

Proof. (2) For $\mathcal{S}$ as in 2.5, if $\mathcal{T} = \{L \in \mathcal{S} \mid \exists\ (x, W),\ x \in W \in \Upsilon,\ E(xR) \cong E(R/L)\}$, set $B = \oplus\{R/L \mid L \in \mathcal{T}\}$. Then $\Upsilon = \langle B \rangle$ because for any $C \in \Upsilon$, for some index set J, $C \hookrightarrow E(\bigoplus_J B)$. Note that for any R-module A

$$\Upsilon = \langle A \rangle \Longleftrightarrow \exists\ \Gamma,\ \exists\ B \hookrightarrow E(\bigoplus_\Gamma A) \Longleftrightarrow \langle A \rangle \text{ and}$$

Υ contain the same cyclics.

From the latter, (3) and (4) can be deduced.

(1) (a) Define $\Delta = [B]$ for B as in (2). Then by 2.9 (2), $\Delta \subseteq \Upsilon = \langle B \rangle$. Since $B \in \Delta \subset \Upsilon$, $\langle B \rangle \subseteq \langle \Delta \rangle \subseteq \langle \Upsilon \rangle = \Upsilon$. For the uniqueness of Δ, suppose $\Omega \in \Xi(R)$ with $\Omega \subseteq \Upsilon = \langle \Omega \rangle$. By an argument similar to 2.9 (2), there is a direct sum of cyclics D with $\Omega = [D]$. Since $B \in \Upsilon = \langle [D] \rangle = \langle D \rangle$. Now by 2.9 (4), $[B] = [D] = \Omega$. The proofs of (1) (b), (c), and (d) are omitted. $\square$

Because it includes torsion modules, the proof of the next theorem would be quite long, and beyond the scope of this proceedings article. Later on we will need to apply parts of this theorem to module direct sums.

2.10. THEOREM. *For any ring R with identity, let $\Xi(R)$ be the class of all unsaturated classes of R-modules (Definition 2.1) partially ordered as in Definition 2.2. Then*

(1) $\Xi(R)$ *is a set,* $|\Xi(R)| \le 2^{|\mathcal{P}(R)|}$.

(2) $\Xi(R)$ *is a complete Boolean lattice;*

(i) $\forall\ \Delta \in \Xi(R)$, $\exists!\ co(\Delta) = co\Delta \in \Xi(R)$ *with* $\Delta \wedge co\Delta = 0$, $\Delta \vee co\Delta = 1 \in \Xi(R)$.

(ii) $\forall\ \{[A_\gamma] \mid \gamma \in \Gamma\} \subseteq \Xi(R)$ *(see 2.3),*

$$\exists\ \sup_\gamma [A_\gamma] = \bigvee_\gamma [A_\gamma] = [\bigoplus_\gamma A_\gamma],$$
$$\exists\ \inf_\gamma [A_\gamma] = \bigwedge_\gamma [A_\gamma] \in \Xi(R).$$

(3) $\Xi(R) = \Xi_t(R) \oplus \Xi_f(R)$ *is a lattice direct sum of complete and convex sublattices, equivalently the canonical (associative) Boolean ring* $\Xi(R)$ *is a ring direct sum of convex (and complete) ideals* $\Xi_t(R)$, $\Xi_f(R) \triangleleft \Xi(R)$. *(See Definition 2.2).*

(4) $\forall\ \Delta \in \Xi(R)$, $t\Delta \in \Xi_t(R)$, $f\Delta \in \Xi_f(R)$, $t\Delta \wedge f\Delta = 0$, $t\Delta \vee f\Delta = \Delta$. *(See Definition 2.2).*

(5) $\forall\ M \in \mathcal{M}od_R$, $\forall\ \Delta \in \Xi(R)$, $[M_\Delta] \le \Delta$, $[M_{co\Delta}] \le co\Delta \in \Xi(R)$, $M_{t\Delta} \oplus M_{f\Delta} \le\subset_E M_\Delta$, $M_{t\,co\Delta} \oplus M_{f\,co\Delta} \le\subset_E M_{co\,\Delta}$, $M_{t\Delta} \oplus M_{t\,co\Delta} \le\subset_E Z_2 M$,

$$(M_{t\Delta} \oplus M_{f\Delta}) \oplus (M_{t\,co\Delta} \oplus M_{f\,co\Delta}) \le\subset_E M_\Delta \oplus M_{co\Delta} \le\subset_E M.$$

(6) $\forall\ \Delta \in \Xi(R)$, $co\Delta \subset c\Delta$ *is the unique largest unsaturated class contained in the saturated class* $c\Delta$. $\forall\ M \in \mathcal{M}od_R$, $M_{c\Delta} = M_{co\Delta} \le M$.

In view of 2.10 (6) above, and also since in general $M_{co\Delta} \notin co\Delta$, only $[M_{co\Delta}] \leq co\Delta$, whereas always $M_{c\Delta} \in c\Delta$, we write "$M_{c\Delta}$" even when we are working in the context of $\Xi(R)$ and should be writing "$M_{co\Delta}$".

3. Module Decomposition

Previously, in [D4; p.58] a direct sum module decomposition theorem was given for torsion free modules and where the direct summands belonged to saturated classes. Here this theorem will be generalized first by dropping the t.f. restriction, and secondly, it will be in terms of unsaturated classes. In the torsion free case $ZM = 0$ and for a saturated class Δ, $M_\Delta = \sum\{V \mid V \leq M,\ V \in \Delta\} \leq M$ is the unique fully invariant submodule of M maximal with respect to $M_\Delta \in \Delta$. Without the t.f. hypothesis, such a natural unique description of M_Δ, as well as the uniqueness of $M_\Delta \leq M$ no longer holds.

3.1. THEOREM II. *For any ring R $(1 \in R)$, suppose that $\Gamma = \{\alpha, \beta, \ldots\} \subset \Xi(R)$ is a finite or infinite pairwise disjoint $(\alpha \neq \beta,\ \alpha \wedge \beta = 0)$ subset whose supremum $\bigvee \Gamma =$* $\sup \Gamma = 1 \in \Xi(R)$. *Let M be any R-module, and for each $\alpha \in \Gamma$ select a submodule $M_\alpha \leq M$ maximal with respect to the property $[M_\alpha] \leq \alpha$, i.e. an maximal α-submodule of M. Let $\Omega \subset \Gamma$ be any subset and $K, L \leq M$ any two complement submodules such that*

$$\sum\{M_\alpha \mid \alpha \in \Omega\} \leq \subset_E K \quad and \quad \sum\{M_\beta \mid \beta \in \Gamma \backslash \Omega\} \leq \subset_E L.$$

Then the following hold:

(i) $\sum_{\alpha \in \Gamma} M_\alpha = \bigoplus_{\alpha \in \Gamma} M_\alpha \leq \subset_E M$;

(ii) $M_\alpha = M_{[M] \wedge \alpha} \in [M] \wedge \alpha = [M_\alpha]$.

(iii) *UNIQUE up to SUPERSPECTIVITY. If $N_\alpha \leq M$, $\alpha \in \Gamma$, is any other choice of maximal α-submodules, then*

$$\widehat{M} = E(\bigoplus_{\alpha \in \Gamma} N_\alpha) = E(\bigoplus_{\alpha \in \Gamma} M_\alpha) \quad and \quad \widehat{N}_\alpha \cong \widehat{M}_\alpha \quad are\ superspective\ in\ \widehat{M}.$$

(iv) $[K] \leq \bigvee \Omega \in \Xi(R)$, $[L] \leq co(\bigvee \Omega) = \bigvee(\Gamma \backslash \Omega) \in \Xi(R)$, *and* $(\bigoplus_{\alpha \in \Omega} M_\alpha) \oplus (\bigoplus_{\beta \in \Gamma \backslash \Omega} M_\alpha) \leq \subset_E K \oplus L$; *$K$ is $\bigvee \Omega$-maximal while L is $co(\bigvee \Omega)$-maximal in M, i.e.*

$$K = M_{\bigvee \Omega},\ L = M_{co(\bigvee \Omega)}, \quad and \quad K \oplus L \leq \subset_E M.$$

(v) *Suppose that P, $Q \leq M$ are any other $\bigvee \Omega$-maximal and $co \bigvee \Omega$-maximal submodules. In particular, they may be constructed as above K, L except by using the $\{N_\alpha \mid \alpha \in \Gamma\}$. Then*

$$\widehat{M} = \widehat{K} \oplus \widehat{L} = \widehat{P} \oplus \widehat{Q},\ \widehat{K} \cong \widehat{P} \quad and \quad \widehat{L} \cong \widehat{Q} \quad are\ superspective\ in\ \widehat{M}.$$

Proof. For $1 \leq n$ by induction suppose that for any $F \subset \Gamma$, $|F| = n$, $\sum\{M_\alpha \mid \alpha \in F\} = \bigoplus_{\alpha \in F} M_\alpha$. Assume that $\alpha(1), \ldots, \alpha(n), \alpha(n+1) \in \Gamma$ are $n+1$ distinct classes such that $M_{\alpha(n+1)} \cap (M_{\alpha(1)} \oplus \cdots \oplus M_{\alpha(n)}) \neq 0$. By the projection argument 1.3, there exists an $1 \leq i \leq n$ and submodules $0 \neq A \leq M_{\alpha(n+1)}$, $B \leq M_{\alpha(i)}$ with $A \cong B$. But then

$$[A] \leq \alpha(n+1),\ [A] = [B] \leq \alpha(i), \text{ and } [A] \leq \alpha(n+1) \wedge \alpha(i) = 0 = \{(0)\},$$

contradicting the fact that $0 \neq A \in [A] \neq 0$. Therefore, by induction $\sum_{\alpha\in\Gamma} M_\alpha = \bigoplus_{\alpha\in\Gamma} M_\alpha$.

Next, if $(\bigoplus_{\alpha\in\Gamma} M_\alpha) \oplus C \leq\subset_E M$, then by the limited infinite distributive law for complete Boolean lattice ([H; p.28, Lemma 4] or [B; p.118, Theorem 15])

$$[C] = [C] \wedge 1 = [C] \wedge (\bigvee_{\alpha\in\Gamma} \alpha) = \bigvee_{\alpha\in\Gamma} ([C] \wedge \alpha).$$

If $C \neq 0$, then for some $\alpha \in \Gamma$, there is a $0 \neq D \in [C] \wedge \alpha \neq 0$. Hence for some I, $D \hookrightarrow E(\bigoplus_I C)$. By 1.3, there exist $0 \neq V \leq D$, and $V \cong W \leq C \leq M$. Then $[M_\alpha] < [M_\alpha] \vee [W] = [M_\alpha \oplus W] \leq \alpha$. Hence $M_\alpha \oplus W$ violates the maximality of M_α. Thus $D = 0$, $C = 0$, and $\bigoplus_{\alpha\in\Gamma} M_\alpha \leq\subset_E M$.

(ii) Since $M_\alpha \leq M$, $[M_\alpha] \leq [M]$, and since $[M_\alpha] \leq \alpha$, trivially $[M_\alpha] \leq [M] \wedge \alpha$. A maximal $[M] \wedge \alpha$-submodule, automatically being a submodule of M, is simply a maximal α-submodule. Consequently, $M_\alpha = M_{[M]\wedge\alpha} \in [M_\alpha]$. It remains to prove that $[M_\alpha] = [M] \wedge \alpha$. Choose $M_\alpha \oplus M_{co(\alpha)} \leq\subset_E M$, where $\alpha \wedge [M_{co(\alpha)}] = 0$ because $[M_{co(\alpha)}] \leq co(\alpha)$. Then

$$\begin{aligned}\alpha \wedge [M] &= \alpha \wedge [M_\alpha \oplus M_{co(\alpha)}] = \alpha \wedge ([M_\alpha] \vee [M_{co(\alpha)}]) \\ &= (\alpha \wedge [M_\alpha]) \vee (\alpha \wedge [M_{co(\alpha)}]) = \alpha \wedge [M_\alpha] = [M_\alpha],\end{aligned}$$

since $[M_\alpha] \leq \alpha$. Thus $\alpha \wedge [M] = [M_\alpha]$.

(iii) By 2.7 (i) and (iii), first of all $\widehat{M_\alpha} \cong \widehat{N}_\alpha$. Now it suffices to show that for any $D < \widehat{M}$ such that $\widehat{M} = \widehat{M_\alpha} \oplus D$, also $\widehat{M} = \widehat{N}_\alpha \oplus D$. By 2.7 (ii) (with M, M_Δ, and K replaced by $\widehat{M}, \widehat{M_\alpha}$, and D respectively), $\widehat{M_{co(\alpha)}} = D$. Since $\widehat{N}_\alpha$ and D are maximal α and $co(\alpha)$-submodules of $\widehat{M}$, it follows from 2.7 (i) that $\widehat{M} = \widehat{N}_\alpha \oplus D$.

(iv) Use of 3.1 (ii) in 2.10 (2) (ii) followed by the limited infinite distributive law for complete Boolean lattices gives $[\bigoplus_{\alpha\in\Omega} M_\alpha] = \bigvee_{\alpha\in\Omega} [M_\alpha] = \bigvee_{\alpha\in\Omega} ([M] \wedge \alpha) = [M] \wedge (\bigvee \Omega)$. Thus $\bigoplus_{\alpha\in\Omega} M_\alpha \leq\subset_E K \in \bigvee \Omega$. Lengthy proofs requiring the semi-infinite distributive law and projection argument show that actually $K = M_{\bigvee\Omega}$, and similarly that $L = M_{co(\bigvee\Omega)}$ as well as (v), and are omitted. □

The generalized infinite Goldie dimension, studied in [DF1], is needed here to construct sets like Γ in 3.1.

3.2. Definition. For any R-module M, the *Goldie dimension* $\mathrm{Gd}\, M$ is the cardinal number

$$\mathrm{Gd}\, M = \sup\{|I| \mid \exists \bigoplus_{i\in I} V_i \leq M,\ 0 \neq V_i \leq M, \text{ all } i \in I\}.$$

Here, however, we need a new concept

3.3. Definition. For any cardinal $\aleph = 1, \aleph_0, \aleph_1, \ldots$ a module M is of *local Goldie dimension* $\aleph$, if

$$\forall\ 0 \neq V \leq M,\ \exists\ 0 \neq W \leq V, \text{ such that } \mathrm{Gd}\, W = \aleph.$$

In this case, write $\text{loc.Gd}\, M = \aleph$. Let $\Delta_\aleph(R)$ be the class of all R-modules of local Goldie dimension $\aleph$. Note that $\Delta_1(R)$ are just the discrete modules, i.e. those containing an essential direct sum of uniform submodules.

It can be shown that $\Delta_\aleph(R)$ is a saturated class of R-modules. However, in this article the emphasis is on unsaturated classes. We know by 2.9 (i) that each $\Delta_\aleph(R)$ determines, and in fact, contains a unique largest unsaturated class of local Goldie dimension $\aleph$. Here however, we will give a very explicit description of this unsaturated class.

3.4. Construction. *For any cardinal* $\aleph = 1, \aleph_0, \aleph_1, \ldots$ *and any* R, *let* $\mathcal{S} = \{L \mid L \leq R\}$ *as before give the isomorphism classes of injective hulls of cyclics. Then let* $\mathcal{S}_\aleph = \{L \in \mathcal{S} \mid \text{loc.Gd}\, R/L = \aleph\}$, *and set* $G_\aleph = \bigoplus_{L \in \mathcal{S}_\aleph} R/L$. *Define* $H_\aleph(R) = [G_\aleph] \in \Xi(R)$.

3.5. THEOREM. *For* $\aleph = 1, \aleph_0, \aleph_1, \ldots$ *and any ring* R *with identity, let* $H_\aleph(R) = [G_\aleph] \in \Xi(R)$ *and* $\mathcal{S}_\aleph \subseteq \mathcal{S}$ *as in 3.4. Then*

(1) $H_\aleph(R) \in \Xi(R)$ *is the unique largest unsaturated class such that for every* $N \in H_\aleph(R)$, $\text{loc.Gd}\, N = \aleph$. $\forall\, R$, $\exists!$ *cardinal* $\tau(R)$ *such that* $\forall\, \aleph \geq \tau(R)$, $H_\aleph(R) = 0$.

(2) *If* N *is any* R*-module with* $\text{loc.Gd}\, N = \aleph$, *then*

(i) $\exists\, N \hookrightarrow E(\bigoplus_\Gamma G_\aleph)$ *for some index set* Γ.

(ii) $[N] \leq H_\aleph(R) \in \Xi(R)$; *furthermore*

(iii) $[N] = [U]$, *where* $U \leq G_\aleph$, $U = \oplus\{R/L \mid L \in \mathcal{S}_\aleph,\ \exists\, x \in N,\ E(xR) \cong E(R/L)\}$.

(3) *For* $\kappa \neq \aleph$, $H_\kappa(R) \cap H_\aleph(R) = \emptyset$ *and* $H_\kappa(R) \wedge H_\aleph(R) = 0 \in \Xi(R)$; $\sup\{H_\aleph(R) \mid \aleph = 1, \aleph_0, \ldots\} = \bigvee_\aleph H_\aleph(R) = [\oplus\{R/L \mid L \in \mathcal{S}\}] = 1 \in \Xi(R)$.

(4) $\forall\, M \in \mathcal{M}\text{od}_R$, $\exists$ *a maximal* $H_\aleph(R)$*-submodule* $M_\aleph \leq M$, *i.e. (i)* $\text{loc.Gd}\, M_\aleph = \aleph$, *and (ii)* $M_\aleph \leq M$ *is maximal in* M *w.r.t. (i). Then each* $M_\aleph \leq M$ *is a complement and*

$$\sum_\aleph M_\aleph = \bigoplus_\aleph M_\aleph \leq\subset_E M, \quad E(M) = E(\bigoplus_\aleph M_\aleph).$$

(5) *If* $N_\aleph \leq M$, $\aleph = 1, \aleph_0, \ldots$ *is another choice of submodules of* M *maximal such that* $\text{loc.Gd}\, N_\aleph = \aleph$, *then*

$$E(M) = E(\bigoplus_\aleph M_\aleph) = E(\bigoplus_\aleph N_\aleph), \quad E(M_\aleph) \cong E(N_\aleph) \quad \text{all } \aleph,$$

and $\forall\, \aleph$, $\forall\, D \leq EM$,

$$E(M) = E(M_\aleph) \oplus D \iff E(M) = E(N_\aleph) \oplus D$$

i.e. $E(M_\aleph)$ *and* $E(N_\aleph)$ *are superspective.*

The reader should note that if $\text{Gd}\, M < \aleph_\omega$, that then all the above sums in (4) and (5) are finite sums.

3.6. COROLLARY 1. *Let* M *be t.f., i.e.* $ZM = 0$, *and let* $M_\aleph$, $N_\aleph \leq M$ *be as in 3.5 (4) and (5). Then for all* $\aleph$, $M_\aleph \leq M$ *is unique, i.e.* $M_\aleph = N_\aleph$.

3.7. COROLLARY 2. *For any* $\aleph$ *and* R, $H_\aleph(R) \subseteq \Delta_\aleph(R)$ *is the largest unique unsaturated class contained in the class* $\Delta_\aleph(R)$ *of all* R*-modules of local Goldie dimension* $\aleph$. *Furthermore,* $H_\aleph(R)$ *generates* $\Delta_\aleph(R) = \langle H_\aleph(R) \rangle$.

3.8. COROLLARY 3. *For any R, let $D(R)$ be the class of discrete R-modules, i.e. containing an essential direct sum of uniform submodules, and $C(R)$ the continuous modules which contain no uniform submodules. Let $\mathcal{T} = \{L \in \mathcal{S} \mid R/L \text{ is uniform}\}$. Then*

(i) $\Gamma = \{H_1(R), \bigvee_{\aleph_0 \leq \aleph} H_\aleph(R)\} \subset \Xi(R)$ *is pairwise disjoint with* $\sup\Gamma = \bigvee\Gamma = 1 \in \Xi(R)$.

(ii) $H_1(R) = [\oplus\{R/L \mid L \in \mathcal{T}\}] \subseteq D(R)$ *and* $\bigvee_{\aleph_0 \leq \aleph} H_\aleph(R) = [\oplus\{R/L \mid L \in \mathcal{S}\backslash\mathcal{T}\}] \subseteq C(R)$ *are the unique largest unsaturated classes contained in the saturated classes* $D(R), C(R)$.

(iii) $\forall\ M \in \mathcal{M}\mathrm{od}_R,\ \exists\ M_D,\ M_C \leq M,\ M_D \oplus M_C \leq\subset_E M,\ M_D$ *discrete,* M_C *continuous;* $\widehat{M} = \widehat{M}_D \oplus \widehat{M}_C$ *are unique up to superspectivity.*

Proof. The proof is omitted, except to say it is an illustration of how Theorem 3.1 (iv) can be used in more complicated circumstances. Here $\Omega = \{H_1(R)\}$ and thus $co(\Omega) = V(\Gamma\backslash\Omega) = \bigvee_{\aleph_0 \leq \aleph} H_\aleph(R),\ M_D = M_\Omega,\ M_C = M_{co(\Omega)}$. □

It is totally beyond the scope of this article to expound the theory of universal saturated or universal unsaturated classes of R-modules, or even to rigorously define the later concepts. Suffice it to say that since $H_\aleph(R)$ is defined for every R, $H_\aleph$ is a prototypical example of an unsaturated universal class. Prototypical, because essentially the last theorem holds for any set or class of such $H_\aleph(R)$, or $H_\aleph$'s provided they are pairwise disjoint and their supremum is 1. The simplest example of such is $\{D(R), C(R)\}$, or $\{D, C\}$ here as in 3.8. One of the more famous such pairwise disjoint classes with supremum 1 are the Murray-Von Neumann-Kaplansky-Goodearl-Boyle Type I, II, III modules. (For operator algebras, see [MN] and [Ka]; for torsion free modules [GB], [G], [D3,4], and [RT]). The author hopes to show in a future article that Types I, II, and III (and other such pairwise disjoint classes) can be generalized from the t.f. case to also include torsion and mixed modules. At one time ([D3,4]) only a finite number of these universal classes were know. However we will show that they do not even form a set by showing that the $\tau(R)$'s in 3.5 (1) are not bounded, i.e. for any $\aleph$, there exists a ring R with $H_\aleph(R) \neq 0$.

3.9. For any $\aleph_0 \leq \aleph$, and any ring F with identity with $|F| \leq \aleph$, and X any set of cardinality $|X| = \aleph$, let $R = F\langle X\rangle$ be the free noncommutative algebra on X over F with right hand F-coefficients, and $kx = xk$, for all $x \in X,\ k \in F$.

A *term* of R is any finite product of elements of X, the empty product being 1. A *monomial* is a term times a constant from F.

If F is a field, then R is torsion free with $\mathrm{loc.\,Gd}\, R_R = \aleph$. Hence $H_\aleph(R) \neq 0$.

We now give a similar torsion example.

3.10. EXAMPLE. For a prime p and integer $2 \leq m$, let F above in 3.9 be $F = \mathbf{Z}_{p^m} = \mathbf{Z}/p^m\mathbf{Z}$, and $R = \mathbf{Z}_{p^m}\langle X\rangle$. Every $r \in R$ is uniquely of the form $r = r_0 + pr_1 + \ldots + p^{m-1}r_{m-1}$, where the r_i are sums of monomials all of whose constant coefficients are not divisible by p, or they are zero. Moreover since F is local, all the nonzero monomials appearing in $r_0, \ldots, r_{m-1}$ may be taken to be jointly independent over $\mathbf{Z}_{p^m}$.

Any $0 \neq L \leq R_R$ contains a cyclic of the form $p^n rR$, $r = r_0 + \cdots + p^{m-1}r_{m-1}$ with $r_0 \in F$ a unit, and hence with $r^\perp = 0 < R$. Hence $\mathrm{Gd}\, p^n rR = \mathrm{Gd}\, rR = \mathrm{Gd}\, R = \aleph$, and hence $\mathrm{loc.\,Gd}\, R = \aleph$.

In order to show that the module R_R is torsion it suffices to show that $pR \leq\subset_E R$ and $pR \subseteq ZR$. If for the above r with $r_0, r_0^{-1} \in F$, $0 = rr' = (r_0 + \cdots + p^{m-1} r_{m+1})(r'_0 + \cdots + p^{m-1} r'_{m-1})$, $p \nmid r_i$, $p \nmid r'_j$, then

$$\begin{aligned} 0 = rr' =& r_0 r'_0 + p(r_0 r'_1 + r_1 r'_0) \\ &+ p^2(r_0 r'_2 + r_1 r'_1 + r_2 r'_0) + \cdots + p^{m-1}(r_{m-1} r'_0 + \cdots + r_0 r'_{m-1}) \end{aligned}$$

where the coefficients of the powers of the p's are units in F not divisible by p. Thus $r'_0 = 0$, $0 = r_0 r'_1 + r_1 r'_0 = r_0 r'_1$ and $r'_1 = 0$, and similarly all $r'_j = 0$. Hence we have shown that for any $r \in R$, either $r = r_0 + \cdots + p^{m-1} r_{m-1} \in R \backslash pR$, $0 \neq r_0$, $r_0^{-1} \in F$, in which case $r^{\perp} = 0$, and $(R \backslash pR) \cap ZR = \emptyset$, or $r \in pR$, $r = p^n(r_0 + pr_1 + \cdots + p^{m-1} r_{m-1})$, $r_0 \neq 0$, $p \nmid r_0$, and $r^{\perp} = p^{m-n} R \subseteq pR \leq\subset_E R$. Hence $ZR = pR \leq\subset_E R$, $Z_2 R = R$ is torsion. Thus $0 \neq [R] \leq tH_{\aleph}(R) = H_{\aleph}(R)$.

This raises the question whether a similar but much more general construction as in 3.10 works for any (not necessarily commutative) local ring F in the sense of [D7; p.133, Proposition 7-3.3] or [AF; p.170, Proposition 15.15] with the F-singular submodule of F being essential in F over F.

3.11. Example. For a set X with $\aleph_0 \leq |X| = \aleph$, let $J = \{Y \subset X \mid |Y| < \aleph\} \triangleleft \mathcal{P}(X)$ and set $R = \mathcal{P}(X)/J$. For any $A, A_1, \ldots, A_n \in \mathcal{P}(X)$, write $A + J = \overline{A} \in R$, and set $B = A_1 \cup \cdots \cup A_n$. Then

$$\begin{aligned} &\overline{A}_1 R + \cdots + \overline{A}_n R = (\mathcal{P}(B) + J)/J \quad \textit{and} \\ &\overline{A} R \cap (\overline{A}_1 R + \cdots + \overline{A}_n R) = (\mathcal{P}(A \cap B) + J)/J. \end{aligned}$$

As in [Ku; p.47, Def. 1.1] a subset $\mathcal{A} = \{Y, Z, W, \ldots\} \subset \mathcal{P}(X)$ is *almost disjoint* if for $W, Y \neq Z \in \mathcal{A}$ all $|W| = \aleph$ and $|Y \cap Z| < \aleph$. By Zorn's lemma there exists a maximal almost disjoint family $\mathcal{A} \subset \mathcal{P}(X)$. Then $\sum\{\overline{A}R \mid A \in \mathcal{A}\} = \oplus\{\overline{A}R \mid A \in \mathcal{A}\}$, because for any $A, A_1, \ldots, A_n \in \mathcal{A}$, it follows that also $B = A_1 \cup \cdots \cup A_n \in \mathcal{A}$, and hence $\overline{A}R \cap (\overline{A}_1 R + \cdots + \overline{A}_n R) = (\mathcal{P}(A \cap B) + J)/J = \overline{0}$. Hence $|\mathcal{A}| \leq \operatorname{Gd} R \leq 2^{\aleph}$. Also for any cyclic $0 \neq \overline{Y}R < R$, $\operatorname{Gd} \overline{Y}R = \operatorname{Gd} R = \operatorname{loc.Gd} R$.

Assume the G.C.H. By [Ku; p.48, Thm.1.3] there exists an $\mathcal{A}$ with $|\mathcal{A}| = 2^{\aleph}$, and hence $\operatorname{loc.Gd} R = \operatorname{Gd} R = 2^{\aleph}$. If we do not assume the G.C.H., then for $\aleph$ regular, by [Ku; p.48, Thm.1.2] we have $\aleph^+ \leq \operatorname{loc.Gd} R = \operatorname{Gd} R \leq 2^{\aleph}$. Note that either under the G.C.H., or with $\aleph$ regular, in either case, $\aleph = \operatorname{Gd} \mathcal{P}(X) < \operatorname{Gd} R$.

For any $\aleph_0 \leq \aleph$, the Boolean ring or lattice $R = \mathcal{P}(X)/J$ is not complete ([H; p.110, Lemma 1]). Thus since the injective hull of R is the completion of $R = \mathcal{P}(X)/J$ ([BL]), R is not self injective.

The next example uses $\aleph$-products, which were also extensively used in [DF2].

3.12. Example. For rings F_i with identity, $i < \aleph$, the $\aleph$-product of the F_i is the subring $\prod^{<\aleph}\{F_i \mid i < \aleph\} \subset \prod\{F_i \mid i < \aleph\}$ consisting of all vectors $f = (f_i)_{i<\aleph}$ whose support $\operatorname{supp} f = \{i < \aleph \mid f_i \neq 0\}$ has cardinality $|\operatorname{supp} f| < \aleph$. Define $R = \prod\{F_i \mid i < \aleph\}/\prod^{<\aleph}\{F_i \mid i < \aleph\}$.

First note that if all $F_i = \mathbf{Z}/2\mathbf{Z} = \mathbf{Z}_2$, $X = [0, \aleph) = \aleph$, and J is as in the last example, then $\prod_{\aleph} \mathbf{Z}_2 / \prod_{\aleph}^{<\aleph} \mathbf{Z}_2 \cong \mathcal{P}(X)/J$.

Next assume the G.C.H. and that all the F_i are fields with $|F_i| \leq \aleph$, so that in particular, R is t.f. and hence ER is a ring. We state without proof that then

$$\operatorname{loc.Gd} R = \operatorname{Gd} R = 2^{\aleph} = \operatorname{loc.Gd} ER = \operatorname{Gd} ER,$$

where $R \subsetneq ER$ properly.

4. THE CATEGORY $\sigma[M]$

For a ring R and right R-module M, $\sigma[M]$ is the full subcategory of $\mathcal{M}od_R$ subgenerated by M ([W2; p.118]). As shown in [W], almost all of the classical theorems of ring and module theory also carry over to their analogues in $\sigma[M]$. Thus $\sigma[M]$ has sums, products, and injective envelopes ([W2; p.118, 15.1(2), (6); p.141, 17.8; 17.9(1)]). In the special case $M = R$, $\sigma[R] = \mathcal{M}od_R$. For the latter reason, and for many other even more important reasons, some ring theorists have begun to do their module theory and to write their journal articles in the context of $\sigma[M]$ (including [PZ1-3], [Z], and [W1-2]). This section will define the analogues of Ξ and Σ in $\mathcal{M}od_R$ to corresponding objects in $\sigma[M]$, and then briefly build a bridge between the two. This section raises many more questions than it answers.

4.1. As before, let $\Sigma(R)$ be the set of all saturated classes of R-modules. By 2.9 (1), (2), and (3), the function $\Delta \longrightarrow \langle\Delta\rangle$, or $[A] \longrightarrow \langle A\rangle$, is a Boolean lattice isomorphism $\Xi(R) \cong \Xi(R)$. More is true, actually $\Xi \cong \Sigma$ as functors, but this is not proved here.

4.2. Definition. For any ring R and any fixed R-module M, the *Wisbauer category* $\sigma[M]$ ([W2; p.118]) is the smallest subcategory of $\mathcal{M}od_R$ which contains the category generated by M and also has kernels. Specifically, $\sigma[M] = \{A \in \mathcal{M}od_R \mid \exists\, I, \exists K, \exists A \hookrightarrow M^{(I)}/K\}$. Thus without loss of generality we may assume that $A \in \sigma[M]$ iff $A \leq_{CE} M^{(I)}/K$ for some I, and some $K < \bigoplus_I M$.

For R-modules M, A we write here $\mathrm{tr}(M, A)$ for the trace of M in A, i.e. $\mathrm{tr}(M, A) = \Sigma\{fM \mid f \in \mathrm{Hom}_R(M, A)\}$. The M-injective hull of A is $\mathrm{tr}(M, EA)$.

An *M-natural class* $\mathcal{K}$ in the sense of Page and Zhou ([PZ2; p.634]) is a subclass $\mathcal{K} \subseteq \sigma[M]$ closed under (i) isomorphic copies, (ii) arbitrary direct sums, (iii) M-injective hulls, and (iv) submodules.

Let $\mathcal{N}(R; M) \subseteq \mathcal{P}[\sigma[M]]$ be the set of all natural classes $\mathcal{K} \subseteq \sigma[M]$.

4.3. A subclass $\mathcal{C} \subset \sigma[M]$ is an *unsaturated M-natural class* if it is closed above in 4.2 under (i), (ii), (iii), (iv$'$) essential submodules, and (v) for any module $A \in \mathcal{C}$, $\mathcal{C}$ is the smallest subclass of $\sigma[M]$ containing A and still satisfying (i), (ii), (iii), and (iv$'$). Let $\mathcal{U}(R; M)$ be the set of all unsaturated classes $\mathcal{C}$ in $\mathcal{C} \subseteq \sigma[M]$.

Since Page and Zhou, so far at least, work with M-natural classes (which in the context of this article should be called saturated M-natural classes), and also for simplicity, we only discuss the connection between $\Sigma(R)$ and $\mathcal{N}(R; M)$.

4.4. REMARKS. For any ring R and R-module M the following hold.

(1) If $M = R_R$, or $M = \widehat{R}_R$, then $\sigma[M] = \sigma[R] = \mathcal{M}od_R$, in which case Δ is a saturated class of R-modules (Definition 2.1) if and only if Δ in an M-natural class (4.2).

(2) Hence $\mathcal{N}(R; R_R) = \Sigma(R) \cong \Xi(R)$.

(3) For any R-module A, $\mathrm{tr}(M, EA) \in \sigma[M]$. For any $N \in \sigma[M]$, $N \leq \mathrm{tr}(M, EN)$.

4.5. LEMMA. Let $\mathcal{T} = \{L < R \mid \exists\, 0 < m, \exists\, v_i \in M, L = \bigcap_1^m v_i^{\perp}\}$ and M^* be the direct sum $M^* = \bigoplus_n \bigoplus_K R^n/K_n$ of all modules of the form R^n/K_n where $O \leq n$, and

$L_1 \oplus \ldots \oplus L_n \leq K_n < R \oplus \ldots \oplus R$ for some $L_i \in \mathcal{T}$. Then $[M^*] \in \Xi(R)$ is the unique element (given by 2.9 (3)) such that

$$\langle [M^*] \rangle = \langle \sigma[M] \rangle = \langle M^* \rangle \in \Sigma(R).$$

Proof. By [W; p.118, 15 (4)], $\oplus\{R/L \mid L \in \mathcal{T}\}$ generates $\sigma[M]$. Consequently every module of $\sigma[M]$ is isomorphic to one of the form $Y = (\bigoplus_{L \in \mathcal{T}} R/L)^{(I)}/V$ for some index set I, and some $V < (\bigoplus_{L \in \mathcal{T}} R/L)^{(I)}$. Hence every module of $\sigma[M]$, and hence also of $\langle \sigma[M] \rangle$, can be embedded into the injective hull of a direct sum of cyclic submodules of Y of the form $(\overline{y} + V)R$, $\overline{y} = (y_1 + L_1, \ldots, y_n + L_n, 0\_0\_) \in (\bigoplus_{L \in \mathcal{T}} R/L), L_i \in \mathcal{T}$, $y_i \in R$. Let V_n be the projection of V into the above n-components (for which $y_i \neq 0$). Thus

$$(\overline{y} + V)R \cong (\overline{y} + V_n)R \cong [(y_1, \ldots, y_n) + K_n]R \leq R^n/K_n \hookrightarrow M^*,$$

where $K_n = \{(k_1, \ldots, k_n) \mid (k_1 + L_1, \ldots, k_n + L_n) \in V_n\}$.

At this point in time the present author does not know precisely when $\langle \sigma[M] \rangle$ is closed under quotient modules, even though $\sigma[M]$ is by definition.

4.6. Proposition. For any $\Delta \in \Sigma(R)$ Δ is closed under quotient modules if and only if for any $M \in \mathcal{M}od_R$, there is a unique choice of M_Δ.

Proof. $\Longrightarrow$: If not, then there exist $A \neq B$, $A, B < M$ each a maximal with submodule of M belonging to Δ. By maximality, $A + B \notin \Delta$. Let $C \leq A + B$ be maximal with respect to $C \in c(\Delta)$. If $C = 0$, then $A + B \in \Delta$. Since C contains no nonzero submodules of Δ, $A \cap C = 0$ and $B \cap C = 0$. But then $C \cong (C + A)/A \leq (A + B)/A \cong B/B \cap A \in \Delta$ is a contradiction.

$\Longleftarrow$: If $K < V \in \Delta$ with $V/K \notin \Delta$, then the above argument shows that there is a nonzero cyclic $0 \neq (v + K)R < V/K$, with $(v + K)R \in c(\Delta)$. $vR \cong (v, v + K)R \leq vR \oplus (v + K)R = M$. Since $vR \in \Delta$, there exists a submodule $B < M$ maximal with respect to $(v, v + K)R \subseteq B \in \Delta$. Similary $vR \subseteq A \in \Delta$ is maximal. By hypothesis $A = B$. But $v \notin K$. Then $(v, 0), (v, v + K) \in B$, hence $0 \neq (v + K)R \leq B \in \Delta$. But $(v + K)R \in c(\Delta)$, a contradiction.

4.7. THEOREM. *Let $\mathcal{N}(R; M)$ be the set of all M-natural classes in the category $\sigma[M]$ as in 4.2. Define $\Sigma_M = \Sigma(R) \wedge \langle \sigma[M] \rangle = \{\Upsilon \wedge \langle \sigma[M] \rangle \mid \Upsilon \in \Sigma(R)\}$, and $\Sigma_M^\perp$ as the orthogonal complement of Σ_M. For any $\Delta \in \Sigma(R)$ define $\varphi(\Delta) = \{V \cap \mathrm{tr}_{EV} M \mid V \in \Delta\}$. Then* (i) *$\Sigma(R) = \Sigma_M \oplus \Sigma_M^\perp$ is a lattice direct sum of convex sublattices; $\Sigma_M = \{\langle A \rangle \mid A \in \sigma[M]\}$.*

(ii) *$\varphi : \Sigma_M \longrightarrow \mathcal{N}(R; M)$ is a Boolean lattice isomorphism whose inverse is $\mathcal{K} \longrightarrow \langle \mathcal{K} \rangle$.*

(iii) *In particular, the set of M-natural classes forms a complete Boolean lattice; $\mathcal{K}^\perp = c(\langle \mathcal{K} \rangle) \wedge \langle \sigma[M] \rangle = \{N \in \sigma[M] \mid \forall\, 0 \neq V \leq N,\ V \notin \mathcal{K}\}$.*

(iv) *$\Sigma_M = \Sigma(R) \Longleftrightarrow \sigma[M] = \mathcal{M}od_R$; in particular $\mathcal{N}(R; R) = \Sigma(R)$.*

Proof. (i) follows from 2.9. (ii) It has to be shown first that $\varphi(\Delta) \in \mathcal{N}(R; M)$. But 4.2 (i), (iii), and (iv) are clear. For 4.2 (ii), if $V_i \in \Delta$, $i \in I$, then $\bigoplus_{i \in I} [V_i \cap \mathrm{tr}(M, EV_i)] =$

$\varphi(W) \in \varphi(\Delta)$, where $W = \bigoplus_{i\in I}[V_i \cap \mathrm{tr}(M, EV_i)] = \mathrm{tr}(M, EW) \cap W$. Note that $\varphi(\mathcal{C}) \subseteq \Delta$. Thus $\langle \varphi(\Delta) \rangle = \Delta$. That φ is monic follows upon combining 2.9 (3), (4) with 4.5.

4.8. Questions and Problems.

(1) Formulate and establish an appropriate and useful functoriality of $\mathcal{N}(\_;\_)$, and from $\mathcal{N}(R;M) \cong \Sigma_M$, that of $\Sigma\_$.

(2) What are necessary and sufficient conditions on M in order that $\langle \sigma[M] \rangle = \langle \oplus \{R/L \mid L \in \mathcal{T}\} \rangle$ for $\mathcal{T}$ as in 4.5?

(3) In the category $\sigma[M]$, a module $N \in \sigma[M]$ is defined to be M-singular if $N \cong L/K$ for some $K \leq\subset_E L \in \sigma[M]$. For $N = M$ a nonprojective simple module, always $Z(M) = M$ but $M \not\cong L/K$ for any $K \leq\subset_E L \in \sigma[M]$, i.e. N is not M-singular. (See [W1; p.4236, 4237] and [PZ2; p.634,2].) This raises at least two questions. First, formulate the analogues of 2.10 (4) and (5) for $\mathcal{N}(R;M) \cong \Sigma_M$ appropriate for the new concept of "M-singular" and "M-torsion free". And secondly, it is observed in [PZ2; p.641, 3(ii)] that the class of all modules with singular injective hulls is an R-natural class. Is the class of all modules in $\sigma[M]$ with M-singular M-injective hulls an M-natural class?

(4) The connection (i.e. the analogue of Theorem 4.7) between $\Xi(R)$ and $\mathcal{U}(R,M)$ should be clarified, as well as that of $\mathcal{N}(R,M)$ and $\mathcal{U}(R;M)$. Conjecture: $\mathcal{N}(R,M) \cong \mathcal{U}(R,M)$.

For the following informative suggestions, the author thanks the referee.

4.9. REMARKS. (1) In transferring module theory to $\sigma[M]$ (as for example in 4.8 (3)), all relevant notions should be considered 'relative to $\sigma[M]$'. Thus one might consider 'essential extensions in $\sigma[M]$', and the analogue of '$\langle\ \ \rangle$' (defined for $\mathcal{M}od_R$ in Definition 2.2), which would require transferring Definition 2.1 to $\sigma[M]$, and in particular 2.1 (c) and (d). Then of course $\sigma[M] =$ '$\langle \sigma[M] \rangle$'.

(2) However in 4.6, 4.7, and 4.8 we need to know how $\sigma[M]$ behaves in $\mathcal{M}od_R$. Consequently, "$\langle\ \ \rangle$" must refer to $\mathcal{M}od_R$, and $\sigma[M] \subsetneq \langle \sigma[M] \rangle$ in general there.

REFERENCES

[AF] F. Anderson and K. Fuller, *Rings and Categories of Modules*, Graduate Texts in Math. Vol. **13**, Springer Verlag, New York, 1974.

[B] G. Birkoff, *Lattice Theory*, Amer. Math. Soc. Colloquium Pub., **XXV**, Providence, R.I. (1967).

[BL] B. Brainerd and J. Lambek, *On the ring of quotients of a Boolean ring*, Can. Math. Bull. **2** (1959), 25-29.

[D1] J. Dauns, *Torsion free modules, Annali di Matematica Pura ed Applicata*, Vol. **CLIV** (1989), 49-81.

[D2] J. Dauns, *Classes of modules*, Forum Mathematicum **3** (1991), 327-338.

[D3] J. Dauns, *Torsion free types*, Fundamenta Mathematicae **139** (1991), 99-117.

[D4] J. Dauns, *Direct sums and subdirect products*, in: Methods in Module Theory, Proceedings, edited: G. Abrams, J. Haefner, K. M. Rangaswamy, Marcel Dekker, 1992, pp. 39-65.

[D5] J. Dauns, *Module classifying functors*, Czechoslovak Math. Jour. **42** (1992), 741-756.

[D6] J. Dauns, *Functors and σ-products*, pp. 149-171, in: Ring Theory, Proc. of Ohio State - Denison Conference 1992, Edited: S. K. Jain and S. Tariq Rizvi, World Scientific Publishers, Singapore, New Jersey, London, 1993.

[D7] J. Dauns, *Modules and Rings*, Cambridge University Press, Cambridge and New York, 1994.

[DF1] J. Dauns and L. Fuchs, *Infinite Goldie dimensions*, J. Algebra **115** (1988), 297-302.

[DF2] J. Dauns and L. Fuchs, *κ-Products of slender modules*, Acta Scient. Math. 56 (1992), 205-213.

[DHSW] N. V. Dung, D. V. Huynh, P. F. Smith, and R. Wisbauer, Extending Modules, Longman Scientific and Technical; John Wiley, New York, 1994.

[G] K. R. Goodearl, *Von Neumann Regular Rings*, Pitman, London, 1979.

[GB] K. R. Gooderal and A. K. Boyle, *Dimension theory for nonsingular injective modules*, Amer. Math. Soc. Memoir No. 177, Providence, R.I., 1976.

[H] P. Halmos, *Lectures on Boolean Algebras*, Studies No. 1, D. Van Nostrand Co., Princeton, New Jersey, 1963.

[Ka] I. Kaplansky, *Rings of Operators*, W. A. Benjamin, New York, 1968.

[Ku] K. Kunen, Set Theory, An Introduction to Independence Proofs, Studies in Logic and the Foundations of Math. Vol. 102, North-Holland Publishing Co., Amsterdam, New York, Oxford, 1980.

[MM] S. H. Mohamed and B. J. Müller, *Continuous and Discrete Modules*, London Math. Soc. Lecture Notes Series 147, Cambridge University Press, Cambridge, 1990.

[MN] F. J. Murray and J. Von Neumann, *On rings of operators*, Ann. of Math. **38** (1936), 116-229.

[PZ1] S. Page and Y. Zhou, *When direct sums of singular injectives are injective, in: Ring Theory*, Proceedings of the Ohio State - Denison Conference, 1992, World Scientific Publishing Co., Singapore, New Jersey, London, 1993.

[PZ2] S. Page and Y. Zhou, *On direct sums of injective modules and chain conditions*, Can. J. Math. **46** (1994), 634-642.

[PZ3] S. Page and Y. Zhou, *Direct sums of quasi-injective modules, injective covers, and natural classes*, Comm. Alg. **22** (1994), 2911-2923.

[RT] J. Rios-Montes and G. Tapia-Sánchez, *A general Theory of types for nonsingular injective modules*, Comm. Alg. **20** (1992), 2337-2364.

[W1] R. Wisbauer, *Generalized co-semisimple modules*, Comm. Alg. **18** (1990), 4235-4253.

[W2] R. Wisbauer, *Foundations of Module and Ring Theory, Algebra, Logic and Applications*, Vol. **3**, Gordon and Breach Science Publishers, Philadelphia, Reading U.K., Paris, Tokyo, 1991.

[Z] Y. Zhou, *Direct sums of M-injective modules and module classes*, Comm. Alg. **23** (1995), 927-940.

$\mathcal{B}^{(1)}$-Groups: Some Counterexamples

CLORINDA DE VIVO and CLAUDIA METELLI, Dipartimento di Matematica e Applicazioni, Universita' Federico II di Napoli, 80100 Napoli,Italy

1 INTRODUCTION

In this Note, 'group' stands for 'torsionfree Abelian group of finite rank'. In this setting a *Butler group* is a group that can be represented as the quotient of a completely decomposable group (of finite rank) over a pure subgroup. The class $\mathcal{B}$ of Butler groups is the union of the chain

$$\mathcal{B}^{(0)} \subset \mathcal{B}^{(1)} \subset \mathcal{B}^{(2)} \subset \ldots \subset \mathcal{B}^{(i)} \subset \ldots$$

where $\mathcal{B}^{(0)}$ denotes the class of completely decomposable groups, and for $i > 0$ $\mathcal{B}^{(i)}$ denotes the class of Butler groups that are quotients of a completely decomposable group over a pure subgroup of rank i.

We recall that in our setting a group is *strongly indecomposable* if it is indecomposable together with its subgroups of finite index; and two groups are *quasi-isomorphic* if one is isomorphic to a subgroup of finite index in the other.

The class $\mathcal{B}^{(1)}$ has been extensively investigated, but the question posed in [FM] remained open, whether strongly indecomposable $\mathcal{B}^{(1)}$-groups are classified up to quasi-isomorphism by their rank and typeset. There it was proved that two strongly indecomposable $\mathcal{B}^{(1)}$-groups with the same rank and typeset are quasi-isomorphic if and only if a certain matrix, associated to their representation as $\mathcal{B}^{(1)}$-groups, satisfies a linear condition called admissibility. We give here a negative answer to the question, by exhibiting (in Example 2) two such groups of rank 8, with the same typeset, whose associated matrix is not admissible.

We add three more examples (1, 3 and 4), which reassess conjectures about the relationship of a $\mathcal{B}^{(1)}$-group with its typeset.

In the References, besides the main reference [FM], we list some of the most recent papers on $\mathcal{B}^{(1)}$-groups.

2 PRELIMINARIES

Recall from [FM] that a strongly indecomposable $\mathcal{B}^{(1)}$-group G of rank m–1 (m≥2) can always be represented as

$$G = \sum_{i=1}^{m} R_i g_i,$$

where for all $i \in I = \{1, ..., m\}$ we have $\mathbb{Z} \leq R_i \leq \mathbb{Q}$ and $R_i g_i \leq_* G$; and where

$$\sum_{i=1}^{m} g_i = 0$$

is the only relation among the elements g_i of G.

Set $t_i = t(R_i)$ for all $i \in I$; and for each $E \subset I$, $E \neq \emptyset$, let

$$t_E = (\bigwedge_{i\in E} t_i) \vee (\bigwedge_{i\in I \setminus E} t_i).$$

Then (a): the *typeset of G* consists of all the types of the form

$$t_{\{A_1, ..., A_k\}} = t_{A_1} \wedge ... \wedge t_{A_k},$$

for each partition $\{A_1, ..., A_k\}$ of I into nonempty, proper subsets. Note that $t_{\{E, I\setminus E\}} = t_E = t_{I\setminus E}$.

(b): G is strongly indecomposable if and only if $t_{\{A_1, ..., A_k\}} \geq t_i$ for some $i \in I$ implies $\{A_1, ..., A_k\} = \{\{i\}, I\setminus\{i\}\}$: i.e. if and only if each t_i is *uniquely obtained* from $\{t_1, ..., t_m\}$.

(c): G is quasi-isomorphic to a similarly represented, strongly indecomposable

$$H = \sum_{i=1}^{m} S_i h_i$$

if and only if

i) for each $i \in I$, $u_i = t(S_i) = t_{E_i}$, for some $\emptyset \subset E_i \subset I$;

ii) the matrix $\mathfrak{E} = (E_1, ... , E_m)$, whose columns are the characteristic functions of the subsets E_i in I, is *admissible:* that is, for each $i \in I$ the matrix $\mathfrak{E}_i$ obtained by substituting the i'th column of $\mathfrak{E}$ with the column

$$I = \begin{pmatrix} 1 \\ 1 \\ \cdot \\ \cdot \\ 1 \end{pmatrix}$$

has nonzero determinant;

iii) the same conditions hold for G w.r.t. to H.

Stricter forms of the above statements, proofs and details can be found in [FM].

3 EXAMPLES

In the following examples, we denote types by substituting 1's for ∞'s, and by omitting infinite rows of zeroes: e.g. (0 1) denotes the type (0 ∞ 0 0 0 ...).

1. We start with the smallest of the four examples, where computations are easily checked. This example shows that strong indecomposability of a $\mathcal{B}^{(1)}$-group cannot be inferred from its rank and typeset. Consider the following rank 5 groups G and H (notations as above):

$t_1 = (1\ 1\ 0\ 0)$	$u_1 = (1\ 1\ 0\ 0)$
$t_2 = (0\ 1\ 1\ 0)$	$u_2 = (0\ 1\ 1\ 0)$
$t_3 = (0\ 0\ 1\ 1)$	$u_3 = (0\ 0\ 1\ 1)$
$t_4 = (1\ 0\ 0\ 1)$	$u_4 = (1\ 0\ 0\ 1)$
$t_5 = (1\ 0\ 1\ 0)$	$u_5 = (1\ 1\ 0\ 0)$
$t_6 = (0\ 1\ 0\ 1)$	$u_6 = (0\ 0\ 1\ 1)$

It is easy to check that the typeset of G consists of $t_1, ..., t_6$, and of the types (1 0 0 0) (= $t_{\{\{1\},\{4\},\{5\},\{2,3,6\}\}}$), (0 1 0 0), (0 0 1 0), (0 0 0 1), (0 0 0 0). The check shows that G is strongly indecomposable. The typeset of H contains, besides the u_j's, $t_5 = u_{\{1,4,5\}}$ (= $(u_1 \wedge u_4 \wedge u_5) \vee (u_2 \wedge u_3 \wedge u_6)$), $t_6 = u_{\{1,2,5\}}$, and the five smaller types listed above; hence G and H have equal rank and typeset. But H is not strongly indecomposable, since $u_5 = u_{\{1\}}$ and $\{\{1\},\mathbb{N}\setminus\{1\}\} \neq \{\{5\}, \mathbb{N}\setminus\{5\}\}$.

The computations are much more manageable if the types are represented as products of primes, with ∧ = g.c.d. and ∨ = l.c.m.: e.g. $t_1 = 2{\cdot}3$, $t_2 = 3{\cdot}5$, $t_3 = 5{\cdot}7$, etc. The rationale behind this method will be the subject of [9].

2. Consider the following $\mathcal{B}^{(1)}$-groups G and H of rank 8:

$t_1 = (100\ 111\ 000\ 011)$	$u_1 = (100\ 111\ 000\ 011)$
$t_2 = (010\ 111\ 000\ 101)$	$u_2 = (010\ 111\ 000\ 101)$
$t_3 = (001\ 111\ 000\ 110)$	$u_3 = (001\ 111\ 000\ 110)$
$t_4 = (111\ 100\ 011\ 000)$	$u_4 = (111\ 100\ 011\ 000)$
$t_5 = (111\ 010\ 101\ 000)$	$u_5 = (111\ 010\ 101\ 000)$
$t_6 = (111\ 001\ 110\ 000)$	$u_6 = (111\ 001\ 110\ 000)$
$t_7 = (100\ 111\ 000\ 100)$	$u_7 = (111\ 100\ 100\ 000)$
$t_8 = (010\ 111\ 000\ 010)$	$u_8 = (111\ 010\ 010\ 000)$
$t_9 = (001\ 111\ 000\ 001)$	$u_9 = (111\ 001\ 001\ 000)$

We have $u_1 = t_1, ..., u_6 = t_6$; $u_7 = t_{\{5,6\}}$, $u_8 = t_{\{4,6\}}$, $u_9 = t_{\{4,5\}}$; thus

$$\mathfrak{S} = \begin{pmatrix} 1&0&0&0&0&0&0&0&0\\ 0&1&0&0&0&0&0&0&0\\ 0&0&1&0&0&0&0&0&0\\ 0&0&0&1&0&0&0&1&1\\ 0&0&0&0&1&0&1&0&1\\ 0&0&0&0&0&1&1&1&0\\ 0&0&0&0&0&0&0&0&0\\ 0&0&0&0&0&0&0&0&0\\ 0&0&0&0&0&0&0&0&0 \end{pmatrix}; \text{ while } t_7 = u_{\{2,3\}},\ t_8 = u_{\{1,3\}},\ t_9 = u_{\{1,2\}}, \text{ hence}$$

$$\mathfrak{F} = \begin{pmatrix} 1&0&0&0&0&0&0&1&1\\ 0&1&0&0&0&0&1&0&1\\ 0&0&1&0&0&0&1&1&0\\ 0&0&0&1&0&0&0&0&0\\ 0&0&0&0&1&0&0&0&0\\ 0&0&0&0&0&1&0&0&0\\ 0&0&0&0&0&0&0&0&0\\ 0&0&0&0&0&0&0&0&0\\ 0&0&0&0&0&0&0&0&0 \end{pmatrix} \text{ is the matrix changing the u's into the t's.}$$

Clearly the two matrices are not admissible. Now a simple but lenghthy check shows that the typesets of G and H are the same, and that no type can be obtained equalling or exceeding any of the t_i's, nor of the u_j's. Thus G and H are strongly indecomposable, but not quasi-isomorphic. This example also answers in the negative questions I and II of [3], since G and H have the same maximal types, and all these types are uniquely obtained.

3. This example stems from the circumstance that the strongly indecomposable examples one happens onto seem to always have the type-matrices

$$\mathfrak{T} = \begin{pmatrix} t_1 \\ t_2 \\ \cdot \\ \cdot \\ t_m \end{pmatrix} \text{ and } \mathfrak{U} = \begin{pmatrix} u_1 \\ u_2 \\ \cdot \\ \cdot \\ u_m \end{pmatrix}$$

obtained from each other by row- and column-permutations. For instance, in the above Example 2, $\mathfrak{U}$ is obtained from $\mathfrak{T}$ by the column-permutation (1,4) (2,5) (3,6)(7,10)(8,11)(9,12) followed by the row-permutation (1,4)(2,5)(3,6) (this can be easily checked by looking at the 3×3 blocks evidenced for this purpose). In the following example of rank 11 this is not true, simply because the number of 1's in $\mathfrak{T}$ and $\mathfrak{U}$ is not the same.

t_1 = (100 111 000 011 100) u_1 = (100 111 000 011 100)
t_2 = (010 111 000 101 010) u_2 = (010 111 000 101 010)
t_3 = (001 111 000 110 001) u_3 = (001 111 000 110 001)
t_4 = (111 100 011 000 111) u_4 = (111 100 011 000 111)
t_5 = (111 010 101 000 111) u_5 = (111 010 101 000 111)
t_6 = (111 001 110 000 111) u_6 = (111 001 110 000 111)

$t_7 = (110\ 111\ 000\ 111\ 000)$ $u_7 = (110\ 111\ 000\ 111\ 000)$
$t_8 = (101\ 111\ 000\ 111\ 000)$ $u_8 = (101\ 111\ 000\ 111\ 000)$
$t_9 = (011\ 111\ 000\ 111\ 000)$ $u_9 = (011\ 111\ 000\ 111\ 000)$
$t_{10} = (100\ 111\ 000\ 100\ 100)$ $u_{10} = (111\ 100\ 100\ 000\ 111)$
$t_{11} = (010\ 111\ 000\ 010\ 010)$ $u_{11} = (111\ 010\ 010\ 000\ 111)$
$t_{12} = (001\ 111\ 000\ 001\ 001)$ $u_{12} = (111\ 001\ 001\ 000\ 111)$

We have $u_1 = t_1, \ldots, u_9 = t_9$; $u_{10} = t_{\{5,6\}}$, $u_{11} = t_{\{4,6\}}$, $u_{12} = t_{\{4,5\}}$; $t_{10} = u_{\{2,3,7,8,9\}}$, $t_{11} = u_{\{1,3,7,8,9\}}$, $t_{12} = u_{\{1,2,7,8,9\}}$. A check shows that G and H are strongly indecomposable, with the same typeset. The matrix here is not admissible; thus the problem remains open for quasi-isomorphic groups.

4. A small modification of the above example shows that, when the groups are not quasi-isomorphic, it is not sufficient to check that $\mathfrak{T}$ and $\mathfrak{U}$ are transformed into each other, for the typesets of G and H to coincide. Add to each t_i and u_j a 16th digit: a 1 to u_1, t_1, u_2, t_2, u_3, t_3, u_8, t_8, u_9, t_9; a 0 to all others. The above equalities remain true, but the element $u_{\{1,2,3,8,9\}} = (110\ 111\ 000\ 000\ 000\ 1)$ of the typeset of H does not belong to the typeset of G.

REFERENCES

[FM] L.Fuchs and C.Metelli, On a class of Butler groups, *manuscripta math. 71:* 1-28 (1991).

1. U. Albrecht, H.P. Goeters, C. Megibben, Zero-one matrices with an application to Abelian groups, *Rend. Sem. Mat. Univ. Padova 90:* (1993), 17-24.
2. D.M. Arnold and C. Vinsonhaler, Quasi-isomorphism invariants for two classes of finite rank Butler groups, *Proc. Amer. Math. Soc. 118:* (1993), 19-26.
3. D.M. Arnold and C. Vinsonhaler, Finite rank Butler groups: A survey of recent results, *Abelian Groups, Lecture Notes in Pure and Applied Math. 146* (Marcel Dekker, 1991), 17-41.
4. H.P. Goeters and W. Ullery, Quasi-summands of a certain class of Butler groups, *Abelian Groups, Lecture Notes in Pure and Applied Math. 146* (Marcel Dekker, 1991), 167-174.
5. H.P. Goeters, W. Ullery, C. Vinsonhaler, Numerical invariants for a class of Butler groups, *Contemporary Mathematics 171:* (1993),159-172.
6. P. Hill and C. Megibben, The classification of certain Butler groups, *J. Algebra 160:* (1993), 524-551.

7. B. Höfling, On direct summands of Butler $B^{(1)}$-groups, *Communications in Algebra 21:* (1993), 1849-56.

8. C. Metelli, On direct sums of $B^{(1)}$-groups, *Comment. Math. Univ. Carolinae 34:* (1993), 587-591.

9. C. De Vivo and C. Metelli, $B^{(1)}$-groups and finite partition lattices, (to appear).

Direct Decompositions of ACD Groups with Primary Regulating Index

THEODORE G. FATICONI Department of Mathematics, Fordham University, Bronx, New York, 10458, faticoni@murray.fordham.edu

PHILLIP SCHULTZ Department of Mathematics, University of Western Australia, Nedlands, Australia, 6907, schultz@maths.uwa.edu.au

1. INTRODUCTION AND NOTATION

Throughout this paper, G is an acd group with endomorphism ring $\mathcal{E} = \mathcal{E}(G)$. A ring E has Jacobson radical $\mathcal{J}(E)$ and nil radical $\mathcal{N}(E)$. For any prime p and for any group or ring X we denote the p–localisation $X \otimes \mathbb{Z}_p$ of X by X_p, and $\mathbb{Q}X = X \otimes \mathbb{Q}$.

Our object is to show that for all primes p, $\mathcal{E}_p$ is a semi–perfect ring, (Corollary 2.8). The main consequence of this result is an Azumaya–Krull–Schmidt Theorem for near–isomorphism classes of acd groups with primary regulating index. This is in complete contrast with the situation for acd groups whose regulating index has order divisible by several primes. For example, Blagoveshchenskaya and Mader [BM94] have given examples of block rigid acd groups G of rank n, with cyclic but not primary regulating quotient such that for any partitions $n = r_1 + r_2 + \cdots r_s = r'_1 + r'_2 + \cdots r'_t$ of n satisfying $r_i < n - t$ and $r'_j < n - s$, $G = \oplus_{i=1}^{s} G_i = \oplus_{j=1}^{t} G'_j$ where each G_i $[G'_j]$ has rank r_i $[r'_j]$ and no proper near–summands.

A G–projective group is a direct summand of G^n for some integer $n > 0$. We

The second author acknowledges the kind hospitality of Fordham University and the University of Connecticut during the period when this paper was written. The authors thank the referee for helpful comments which improved the paper and prevented the appearance of several errors.

show that if G has primary regulating index, then the category of G–projective groups has finite representation type. This is in sharp contrast to the situation for acd groups in general, ([AD93]).

We adopt the following conventions and notation. For each group or ring X, we identify X as a subset of $\mathbb{Q}X$ via the imbedding $x \mapsto x \otimes 1$. For any prime p, X_p is the group of elements v of $\mathbb{Q}X$ for which there is an integer m not divisible by p such that $mv \in X$. Given torsion–free groups G and H, we have

$$\mathrm{Hom}(G,H) \subseteq \mathrm{Hom}(G,H)_p \subseteq \mathbb{Q}\mathrm{Hom}(G,H) \subseteq \mathrm{Hom}(\mathbb{Q}G,\mathbb{Q}H),$$

where $\mathbb{Q}\mathrm{Hom}(G,H)$ is the set of linear transformations $f \in \mathrm{Hom}(\mathbb{Q}G,\mathbb{Q}H)$ such that $nf(G) \subseteq H$ for some integer $n \neq 0$, $\mathrm{Hom}(G,H)_p$ is the set of linear transformations f such that $mf(G) \subseteq H$ for some integer m not divisible by p, and $\mathrm{Hom}(G,H)$ is the set of linear transformations f such that $f(G) \subseteq H$. In particular, we have inclusions $\mathcal{E} \subseteq \mathcal{E}_p \subset \mathbb{Q}\mathcal{E}$.

The groups G and H are *quasi–isomorphic* if there are $f \in \mathbb{Q}\mathrm{Hom}(G,H)$ and $g \in \mathbb{Q}\mathrm{Hom}(H,G)$ such that $gf = 1_G$ and $fg = 1_H$ and *isomorphic at* p if there are $f \in \mathrm{Hom}(G,H)_p$ and $g \in \mathrm{Hom}(H,G)_p$ such that $gf = 1_G$ and $fg = 1_H$. It is well–known that if G is quasi–isomorphic to H then $\mathbb{Q}\mathcal{E}(G) \cong \mathbb{Q}\mathcal{E}(H)$, and if G is isomorphic to H at p then $\mathcal{E}_p(G) \cong \mathcal{E}_p(H)$.

A torsion–free group or ring X is *p–local* if it is isomorphic to its p–localisation. Equivalently, it is p–local if its additive structure is divisible by all primes other than p.

2. LIFTING IDEMPOTENTS

Let I be an ideal of a ring E. We say that *idempotents lift modulo* I *in* E if for any idempotent $\bar{e} \in E/I$ there is an idempotent $e \in E$ such that $e + I = \bar{e}$. For example, it is known that idempotents lift modulo nil ideals ([AF73, Proposition 27.1]), and hence modulo the nil radical.

We begin with a couple of ring theoretic results, the first of which at least is well–known. In both the following Lemmas, E is any ring with Jacobson radical $\mathcal{J}(E)$ and nil radical $\mathcal{N}(E)$. Recall that if $K \subseteq \mathcal{J}(E)$ is an ideal then $\mathcal{J}(E/K) = \mathcal{J}(E)/K$, [AF73, Proposition 9.15].

LEMMA 2.1 *Let $F = E/\mathcal{N}(E)$. If idempotents lift modulo $\mathcal{J}(F)$ in F, then idempotents lift modulo $\mathcal{J}(E)$ in E.*

Proof: Let $f + \mathcal{J}(E)$ be an idempotent of $E/\mathcal{J}(E)$. By the above comment, $\mathcal{J}(E)/\mathcal{N}(E) = \mathcal{J}(F)$, so $f - f^2 + \mathcal{N}(E) \in \mathcal{J}(F)$. Thus $(f + \mathcal{N}(E)) + \mathcal{J}(F)$ is an idempotent of $F/\mathcal{J}(F)$, so there is an idempotent $g + \mathcal{N}(E)$ of F such that $f - g + \mathcal{N}(E) \in \mathcal{J}(F)$. Thus there is an element n of $\mathcal{N}(E)$ such that $f - g + n \in \mathcal{J}(E)$. But idempotents lift modulo $\mathcal{N}(E)$ in E, so there is an idempotent e in E such that $g - e \in \mathcal{N}(E)$. Hence $f - e = (f - g + n) + (g - e) - n \in \mathcal{J}(E) + \mathcal{N}(E) = \mathcal{J}(E)$, as required.

Thus in order to show that idempotents lift modulo $\mathcal{J}(\mathcal{E}_p)$ in $\mathcal{E}_p$, it suffices to show that idempotents lift modulo $\mathcal{J}(\mathcal{E}_p)/\mathcal{N}(\mathcal{E}_p)$ in $\mathcal{E}_p/\mathcal{N}(\mathcal{E}_p)$. It is a consequence of a theorem of Faith, [Fa76, Theorem 18.23.5] that two idempotent liftings of a given idempotent in $E/\mathcal{J}(E)$ are conjugate. We provide a proof for the sake of completeness.

LEMMA 2.2 *Let E be a ring. Suppose $e, f \in E$ are such that $e^2 = e$, $f^2 = f$, and $e - f \in \mathcal{J}(E)$. There is a unit $w \in E$ such that $wew^{-1} = f$.*

Proof: Write $\mathcal{J}(E) = J$. Then $E = eE \oplus (1-e)E$ and $J = eJ \oplus (1-e)J$, so that $eE \cap J = eJ$. Then

$$e \cdot \frac{E}{J} = \frac{eE + J}{J} \cong \frac{eE}{eE \cap J} = \frac{eE}{eJ}$$

and similarly, $f \cdot \frac{E}{J} \cong \frac{fE}{fJ}$. Furthermore, since $e - f \in J$, we have $e \cdot \frac{E}{J} = f \cdot \frac{E}{J}$. Hence $\frac{eE}{eJ} \cong \frac{fE}{fJ}$ as right E-modules. By [AF73, Proposition 17.18] there are isomorphisms $u : eE \to fE$ and $v : (1-e)E \to (1-f)E$ of projective right E-modules.

By identifying the ring E with the ring of left multiplications acting on the right module E_E, we find that $u = fue$ and $v = (1-f)v(1-e)$. Thus $w = fue + (1-f)v(1-e)$ is an automorphism of E, and hence a unit of E. One verifies that $w^{-1} = eu^{-1}f + (1-e)v^{-1}(1-f)$ and that $wew^{-1}f = (fue)e(eu^{-1}f)f = f$. Consequently

$$wew^{-1} = wew^{-1}[(1-f) + f] = we(1-e)v^{-1}(1-f) + f = f.$$

Recall that a ring E with Jacobson radical $\mathcal{J}(E)$ is semi-perfect if $E/\mathcal{J}(E)$ is a semi-simple Artinian ring and idempotents lift modulo $\mathcal{J}(E)$ in E. We shall need the following property of semi-perfect rings.

LEMMA 2.3 *Let E be a semi–perfect ring and let K be an ideal of E contained in $\mathcal{J}(E)$. Then idempotents lift modulo K in E.*

Proof: Let $f + K$ be an idempotent in E/K. Since E is semi–perfect there is an idempotent $e \in E$ such that $e - f \in \mathcal{J}(E)$. Since $e + K$ and $f + K$ are idempotents in E/K, Lemma 2.2 produces a unit $\overline{w} \in E/K$ such that $\overline{w}(e+K)\overline{w}^{-1} = f + K$. Since $K \subseteq \mathcal{J}(E)$ and since units lift modulo $\mathcal{J}(E)$, there is a unit $w \in E$ such that $\overline{w} = w + K$. Then wew^{-1} is an idempotent in E such that $wew^{-1} + K = \overline{w}(e+K)\overline{w}^{-1} = f + K$.

We shall apply this result to a p–local finite product of matrix rings over subrings of $\mathbb{Q}$, so we need to prove that such products are semi–perfect.

LEMMA 2.4 *Let $E = \prod_{i=1}^{n} E_i$ where each E_i is a full matrix ring over $\mathbb{Z}_p$ or $\mathbb{Q}$. Then E is semi–perfect.*

Proof: Since each E_i is a matrix ring over a semi–perfect ring, it is semi–perfect by [AF73, Proposition 27.8]. Since the semi–perfect property is preserved by finite direct products, E is also semi–perfect.

The following result is easy to prove using maximal orders and their structure, (see e.g. [A82, Chapters 9-13]). However, we offer a more elementary proof.

LEMMA 2.5 *Let E be a p-local (not necessarily reduced) ring such that $\mathbb{Q}E$ is a finite product of matrix rings over $\mathbb{Q}$. Then E has finite index in a product of matrix rings over $\mathbb{Z}_p$ or $\mathbb{Q}$.*

Proof: Let $\mathbb{Q}E = \prod_{i=1}^{k} M_{n_i}(\mathbb{Q})$, and let $e_1, \cdots, e_k \in \mathbb{Q}E$ be the central idempotents such that $e_i\mathbb{Q}E = M_{n_i}(\mathbb{Q})$. Then $E \subseteq E' = e_1E \times \cdots \times e_kE$, a product of rings such that $e_i\mathbb{Q}E'$ is a matrix ring over $\mathbb{Q}$. There is an integer $m \neq 0$ such that $me_i \in E$ for each i, so $mE' \subset E$, and hence E'/E is finite. To complete the proof it suffices to show that E' has finite index in a product of matrix rings.

Without loss of generality, assume that $E' = e_1E'$ and that $\mathbb{Q}E' = M_n(\mathbb{Q})$. Let V be an n-dimensional $\mathbb{Q}$-vector space such that $\text{End}(V) = \mathbb{Q}E'$. Considered as a left $\mathbb{Q}E'$ module, $V = \mathbb{Q}E'x$ for some (indeed any) nonzero $x \in V$. Consider the left E'-module $L = E'x$. Since $\mathbb{Q}L = \mathbb{Q}E'x = V$, we have $\text{End}(L) \subseteq \text{End}(V) = \mathbb{Q}E'$.

Since $\mathbb{Q}E' = M_n(\mathbb{Q})$, the center Z of E', being a p–local subring of $\mathbb{Q}$, is either $\mathbb{Z}_p$ or $\mathbb{Q}$. By a theorem of Pierce, [A82, Theorem 9.9(b)], E', and hence L are finitely generated free Z-modules. Thus $\text{End}(L)$ is a matrix ring of rank n^2 over Z.

Since $L = E'x$ and $\mathbb{Q}L = V$, left multiplications by nonzero elements of E' are nonzero endomorphisms of L, so that $E' \subseteq \text{End}(L)$. Finally, $\text{End}(L)/E' \subset \mathbb{Q}E'/E'$ is finite because it is finitely generated and torsion.

LEMMA 2.6 *Let $C = \oplus_\tau C_\tau$ be completely decomposable, where the C_τ are maximal τ-homogeneous summands of C. Then $\mathcal{E}(C)/\mathcal{N}(\mathcal{E}(C)) \cong \prod_\tau \mathcal{E}(C_\tau)$.*

Proof: Since $\text{Hom}(C_\tau, C_\sigma) = 0$ unless $\tau \leq \sigma$, there is a direct sum of groups

$$\mathcal{E}(C) = \oplus_{\tau,\sigma}\text{Hom}(C_\tau, C_\sigma) = \oplus_\tau \mathcal{E}(C_\tau) \oplus N$$

where τ and σ range over the critical typeset of C, and where $N = \oplus_{\tau<\sigma}\text{Hom}(C_\tau, C_\sigma)$. One verifies that N is an ideal in the ring $\mathcal{E}(C)$, and hence $\mathcal{E}(C)/N \cong \prod_\tau \mathcal{E}(C_\tau)$ as rings.

We claim that $N = \mathcal{N}(\mathcal{E}(C))$. Since each C_τ is a homogeneous completely decomposable group, $\mathcal{E}(C_\tau)$ is a full matrix ring over a subring of $\mathbb{Q}$, and hence has no nonzero nilpotent ideals. Thus $\mathcal{N}(\mathcal{E}(C)) \subseteq N$.

Let n be the maximal length of a chain of critical types of C; we shall show that $N^n = 0$. Let $f_1, f_2, \ldots, f_n$ be a sequence of n elements of N. Each f_i can be expressed as a sum $\sum_{\tau<\sigma} f_i^{\tau,\sigma}$ where $f_i^{\tau,\sigma} \in \text{Hom}(C_\tau, C_\sigma)$. Thus if $f_1f_2\ldots f_n \neq 0$, it is a sum of non–zero terms of the form $\phi^{\tau_1,\tau_2}\phi^{\tau_2,\tau_3}\ldots\phi^{\tau_n,\tau_{n+1}}$, with $\phi^{\tau_i,\tau_{i+1}} \in \text{Hom}(C_{\tau_i}, C_{\tau_{i+1}})$ and $\tau_1 < \tau_2 < \cdots < \tau_{n+1}$. But this is impossible, since the length of a longest chain of critical types is n, so $f_1f_2\ldots f_n = 0$ and hence N is nilpotent.

Thus $N \subseteq \mathcal{N}(\mathcal{E}(C))$, and therefore $N = \mathcal{N}(\mathcal{E}(C))$ so $\mathcal{E}(C)/\mathcal{N}(\mathcal{E}(C)) \cong \prod_\tau \mathcal{E}_\tau$.

We now have the machinery to prove the main Theorem of this Section.

THEOREM 2.7 *Let G be an acd group and let p be a prime. Then idempotents lift modulo $\mathcal{J}(\mathcal{E}_p)$ in $\mathcal{E}_p$.*

Proof: Let $E = \mathcal{E}_p/\mathcal{N}(\mathcal{E}_p)$. By Lemma 2.1, it suffices to prove that idempotents lift modulo $\mathcal{J}(E)$ in E.

Let $C = \oplus_{\tau\in T}C_\tau$ be a completely decomposable subgroup of finite index in G, where C_τ is a maximal τ–homogeneous summand of C. By Lemma 2.6

$\mathcal{E}(C)/\mathcal{N}(\mathcal{E}(C)) \cong \prod_\tau \mathcal{E}(C_\tau)$ and clearly $\mathbb{Q}\mathcal{E}_p = \mathbb{Q}\mathcal{E}$, so

$$\mathbb{Q}E \cong \mathbb{Q}\mathcal{E}/\mathbb{Q}\mathcal{N}(\mathcal{E}) = \mathbb{Q}\mathcal{E}(C)/\mathbb{Q}\mathcal{N}(\mathcal{E}(C)) \cong \prod_\tau \mathbb{Q}\mathcal{E}(C_\tau)$$

is a product of matrix rings over $\mathbb{Q}$.

Let E_d be the divisible subgroup of E. Then E_d is an ideal of E so that $\mathbb{Q}E_d = E_d$ is an ideal of $\mathbb{Q}E$. By [Fa73, Proposition 3.66, page 177], $E_d = \epsilon\mathbb{Q}E$ for some central idempotent $\epsilon \in \mathbb{Q}E$. But then $\epsilon 1 \in E_d \subseteq E$ implies that ϵ is a central idempotent in E. Hence $E = (1-\epsilon)E \times \epsilon E = E_r \times E_d$, where $E_r = (1-\epsilon)E \cong E/E_d$ is an ideal of E whose additive structure is a reduced torsion-free p–local group. Evidently, $\mathcal{J}(E) = (1-\epsilon)\mathcal{J}(E) \times \epsilon\mathcal{J}(E) = \mathcal{J}(E_r) \times \mathcal{J}(E_d)$. Therefore, to prove that idempotents lift modulo $\mathcal{J}(E)$ in E, it suffices to show that idempotents lift modulo $\mathcal{J}(E_r)$ in E_r and modulo $\mathcal{J}(E_d)$ in E_d.

Since E_d is a factor of the semi-simple ring $\mathbb{Q}E$, E_d is also a semi-simple ring. Hence $\mathcal{J}(E_d) = 0$, so idempotents lift modulo $\mathcal{J}(E_d)$ in E_d trivially.

Now suppose $E_r \neq 0$. By Lemma 2.5 E_r has finite index in a product F of matrix rings over subrings of $\mathbb{Q}$. Since the additive structure of E_r is a reduced p-local group, F/E_r and $E_r/\mathcal{J}(E_r)$ are finite p-groups, [A82, Corollary 9.3(b)]. Thus we can choose $p^\ell \geq 1$ such that $K = p^\ell F \subseteq \mathcal{J}(E_r)$. Note that K is an ideal of F contained in $\mathcal{J}(F)$, so Lemma 2.3 applies. Note too that $E_r/K \subseteq F/K$.

Let $\bar{e}^2 = \bar{e} \in E_r/\mathcal{J}(E_r)$. Since $\mathcal{J}(E_r)/K$ is a nilpotent ideal in the finite ring E_r/K, there is an idempotent $f+K$ in E_r/K such that $f+\mathcal{J}(E_r) = \bar{e}$. Moreover, since $f+K$ is also an idempotent in F/K, Lemma 2.3 implies that there is an idempotent $e \in F$ such that $f+K = e+K$. Then $e = (e-f)+f \in K+E_r = E_r$, so that e is an idempotent lifting of $\bar{e}$ in E_r.

Thus idempotents lift modulo $\mathcal{J}(E)$ in E, and Lemma 2.1 implies that idempotents lift modulo $\mathcal{J}(\mathcal{E}_p)$ in $\mathcal{E}_p$.

COROLLARY 2.8 *Let G be an acd group with endomorphism ring $\mathcal{E}$. Then $\mathcal{E}_p$ is a semi-perfect ring for each prime p.*

Proof: As in Theorem 2.7, $\mathcal{E}_p/\mathcal{J}(\mathcal{E}_p)$ is a product of a finite ring $E_r/\mathcal{J}(E_r)$ and a semi-simple Artinian ring E_d, so $\mathcal{E}_p/\mathcal{J}(\mathcal{E}_p)$ is Artinian. Since $\mathcal{J}(\mathcal{E}_p/\mathcal{J}(\mathcal{E}_p)) = 0$, $\mathcal{E}_p/\mathcal{J}(\mathcal{E}_p)$ is a semi-simple Artinian ring. By Theorem 2.7 idempotents can be lifted modulo $\mathcal{J}(\mathcal{E}_p)$ in $\mathcal{E}_p$, so $\mathcal{E}_p$ is a semi-perfect ring.

3. AN AZUMAYA-KRULL-SCHMIDT THEOREM

Fix a prime p, and let $\mathcal{G}_p$ denote the category whose objects are the acd groups G such that G/C is a finite p-group for some completely decomposable subgroup $C \subseteq G$; for G, H in $\mathcal{G}_p$ define $\mathrm{Hom}_{\mathcal{G}_p}(G,H) = \mathrm{Hom}(G,H)_p$. Note that $\mathcal{G}_p$ is an additive category closed under near–isomorphism. By Theorem 1 of [L74], (see also [M95, Theorem 3.13]), the objects in $\mathcal{G}_p$ are those acd groups which have a p–power regulating index. Notice that $G \in \mathcal{G}_p$ has endomorphism ring $\mathcal{E}_p$. Theorem 3.4 of [S95] states that G, $H \in \mathcal{G}_p$ are isomorphic in $\mathcal{G}_p$ if and only if they are nearly isomorphic as groups.

Let $\mathcal{G}$ be an additive category. We say that *idempotents split in* $\mathcal{G}$ if given an object $G \in \mathcal{G}$ and an idempotent $e \in \mathcal{E}_{\mathcal{G}}(G)$, there is an object $H \in \mathcal{G}$ and maps $\pi \in \mathrm{Hom}_{\mathcal{G}}(G, H)$ and $\jmath \in \mathrm{Hom}_{\mathcal{G}}(H, G)$ such that $\jmath\pi = e$ and $\pi\jmath = 1_H$.

We shall show that idempotents split in $\mathcal{G}_p$. This implies by [A82, Proposition 7.2] that if $G \in \mathcal{G}_p$ and E_p has a non–trivial idempotent, then G is isomorphic in $\mathcal{G}_p$ to a non–trivial direct sum $H \oplus K$. An Azumaya-Krull-Schmidt Theorem and an exchange result follow.

LEMMA 3.1 *Idempotents split in* $\mathcal{G}_p$.

Proof: Let $G \in \mathcal{G}_p$ and let $e^2 = e \in \mathcal{E}_p$. Since $eG \subseteq G_p$ is a quasi-summand of the acd group G, eG is an acd group. Let $C \subseteq eG$ be a completely decomposable group that has finite index in eG, and let $C \subseteq H \subseteq eG$ be such that H/C is the p–primary component of eG/C. Evidently, $H \in \mathcal{G}_p$.

Let $\pi \in \mathrm{Hom}(G, eG)$ be the restriction of e to G. (Recall our conventions from Section 1.) The integer $m = |eG/H|$ is relatively prime to p by our choice of H, and $meG \subseteq H$, so $e|_G = \pi \in \mathrm{Hom}(G, H)_p$. Next, since $H \subseteq eG \subseteq G_p$, we let $\jmath : H \subseteq G_p$ be the inclusion map. There is an integer n relatively prime to p such that $ne \in \mathcal{E}$. Then $n\jmath H \subseteq neG \subseteq G$, so that $\jmath \in \mathrm{Hom}(H, G)_p$. Let $h = eg \in H$. Then $\pi\jmath(h) = e^2(g) = h$, so that $\pi\jmath = 1_H$ and clearly $\jmath\pi = e$.

The following result gives a short proof of [S95, Theorem 3.4].

LEMMA 3.2 *Let* G, $H \in \mathcal{G}_p$. *Then* G *is nearly isomorphic to* H *as groups if and only if* $G \cong H$ *in* $\mathcal{G}_p$.

Proof: It is easy to show that G and H are isomorphic in $\mathcal{G}_p$ if and only if they are isomorphic at p. So we show that G and H are nearly isomorphic groups if and only if they are isomorphic at p.

Assume that G and H are isomorphic at p, and let $C \subseteq G$, $D \subseteq H$ be completely decomposable subgroups such that G/C and H/D are finite p-groups. Then G and H are quasi-isomorphic acd groups, and so C and D are quasi-isomorphic, whence isomorphic, completely decomposable groups. Choose inverse isomorphisms $f : C \to D$ and $g : D \to C$. Then f and g lift to $f : G \to \mathbb{Q}D$ and $g : H \to \mathbb{Q}C$.

Because G/C is a finite p-group and because $f(C) = D$, $f(G)/D = f(G)/f(C)$ is a finite p-group. Similarly, $g(H)/C$ is a finite p-group. Then $p^k f(G) \subseteq H$ and $p^k g(H) \subseteq G$ for some p^k. We conclude that $f \in \mathrm{Hom}(G, H)_q$ and $g \in \mathrm{Hom}(H, G)_q$ for each prime $q \neq p$. Inasmuch as f and g are inverse functions, G and H are isomorphic at each prime $q \neq p$. Then by [A82, Theorem 7.16] G and H are nearly isomorphic groups.

The converse is also part of [A82, Theorem 7.16].

Our first application shows that every idempotent of $\mathcal{E}_p$ is conjugate to an idempotent of $\mathcal{E}$.

PROPOSITION 3.3 *Let* $G \in \mathcal{G}_p$. *If* $e^2 = e \in \mathcal{E}_p$, *there is a unit* w *of* $\mathcal{E}_p$ *such that* wew^{-1} *is an idempotent in* $\mathcal{E} = \mathcal{E}(G)$.

Proof: Let $e^2 = e \in \mathcal{E}_p$. Then $e\mathcal{E}_p$ is a direct summand of $\mathcal{E}_p$. By Lemma 3.1 $G \cong eG \oplus K$ in $\mathcal{G}_p$, so by Lemma 3.2 G is nearly isomorphic to $eG \oplus K$. By [A82,

Theorem 12.9] we can write $G = e'G \oplus K'$ as groups, where $e'^2 = e' \in \mathcal{E}$ and where eG is nearly isomorphic to $e'G$. But then eG and $e'G$ are isomorphic in $\mathcal{G}_p$, so that $e\mathcal{E}_p = \mathrm{Hom}(G, eG)_p \cong \mathrm{Hom}(G, e'G)_p = e'\mathcal{E}_p$ as right $\mathcal{E}_p$-modules. As in Lemma 2.2, we can lift an isomorphism $u : e\mathcal{E}_p \to e'\mathcal{E}_p$ to a unit $w \in \mathcal{E}_p$ such that $wew^{-1} = e' \in \mathcal{E}$.

We can use Lemmas 3.1 and 3.2 to relate direct decompositions in $\mathcal{G}_p$ to near direct decompositions of groups. For example, we now find a necessary and sufficient condition on $\mathcal{E}_p$ for the indecomposability of G. Recall that a ring E is local if $\mathcal{J}(E)$ is the unique maximal right ideal in E.

COROLLARY 3.4 *Let $G \in \mathcal{G}_p$. Then G is indecomposable as a group if and only if $\mathcal{E}_p$ is a local ring.*

Proof: By Corollary 2.8 and [AF73, Theorem 27.6], the semi-perfect ring $\mathcal{E}_p$ is a local ring if and only if $\mathcal{E}_p$ has no nontrivial idempotents. Hence by Proposition 3.3 G is an indecomposable group if and only if $\mathcal{E}_p$ is local.

In an additive category in which idempotents split, a finite direct sum of objects with local endomorphism rings satisfies the Azumaya-Krull-Schmidt Theorem, [A82, Theorem 7.4]. Thus we have shown for the category $\mathcal{G}_p$:

THEOREM 3.5 *Let $G \in \mathcal{G}_p$. Then $G = \oplus_{i=1}^{n} G_i$ as groups, where each G_i is an indecomposable acd group with local near endomorphism ring, $\mathcal{E}(G_i)_p$. If G is nearly isomorphic to $\oplus_{j=1}^{m} G'_j$ for some indecomposable acd groups G'_j then $n = m$, and after a suitable permutation of the indices, G_i and G'_i are nearly isomorphic groups.*

We say that a torsion-free group G of finite rank has the *near exchange property* if $G \cong \oplus_{i\in I} G_i \cong H \oplus K$ implies that H is nearly isomorphic to $\oplus_{j\in J} G_j$ for some $J \subseteq I$. R. Baer [Fu73, Proposition 86.1], showed that completely decomposable groups have the exchange property, while the examples of Blagoveshchenskaya and Mader cited in the introduction show that an acd group can fail to have the near exchange property. On the other hand, [A82, Theorem 7.4] implies that finite direct sums of objects with local endomorphism rings enjoy the exchange property. We can now show:

PROPOSITION 3.6 *Each group $G \in \mathcal{G}_p$ has the near exchange property.*

Proof: Suppose $G \cong \oplus_k G_k \cong H \oplus K$ as groups for some indecomposable groups G_k. Write $H \cong \oplus_i H_i$ and $K \cong \oplus_j K_j$ for some indecomposable H_i and K_j in $\mathcal{G}_p$. Then $G \cong (\oplus_i H_i) \oplus (\oplus_j K_j)$, so by Theorem 3.5 for each i there is a k_i such that H_i is nearly isomorphic to G_{k_i}, and the near exchange property follows.

4. REPRESENTATION TYPES IN $\mathcal{G}_p$

A group H is *G-projective* if there is an integer $n > 0$ and a group K such that $G^n \cong H \oplus K$. Let $\Sigma(G)$ denote the category of abelian groups whose objects are the G-projective groups.

LEMMA 4.1 *Let $G \in \mathcal{G}_p$. If G is isomorphic to $G_1 \oplus \cdots \oplus G_n$ for some indecomposable groups G_i then for each G-projective group H there are integers $a_1, \ldots, a_n \geq 0$ such that H is nearly isomorphic to $G_1^{a_1} \oplus \cdots \oplus G_n^{a_n}$.*

Proof: Suppose $H \oplus K \cong G^k$ as groups for some integer $k > 0$. Then $G^k \cong G_1^k \oplus \cdots \oplus G_n^k$ is a decomposition of G^k into indecomposables. Since G^k has the near exchange property by Proposition 3.6, $H = G_1^{a_1} \oplus \cdots \oplus G_n^{a_n}$ for some integers $a_i \geq 0$.

A category $\mathcal{G}$ has *finite representation type* if there are at most finitely many isomorphism classes of indecomposable objects in $\mathcal{G}$.

PROPOSITION 4.2 *Let $G \in \mathcal{G}_p$. The category $\Sigma(G)$ has finite representation type, and the indecomposable G-projective groups are the indecomposable near summands of G.*

Proof: Write $G = G_1 \oplus \cdots \oplus G_n$ for some indecomposable groups G_i. By Lemma 4.1 an indecomposable $H \in \Sigma(G)$ is nearly isomorphic to G_i for some indecomposable near summand G_i of G. Hence each indecomposable $H \in \Sigma(G)$ is a near summand of G, and there are at most n near isomorphism classes of indecomposable groups in $\Sigma(G)$. Furthermore, Mader [M95] showed that the near isomorphism class of an acd group contains at most finitely many isomorphism classes. Hence $\Sigma(G)$ has finite representation type.

Arnold and Dugas [AD93] showed that the category of uniform block rigid acd groups with 3 critical types and with exponent bounded by p^2 has finite representation type, whereas the category of uniform acd groups having more than 4 critical types and exponent bounded by p^2 has wild representation type. On the other hand, it is interesting to note that every such group is in an additive category closed under finite direct sums and direct summands that has finite representation type.

COROLLARY 4.3 *There is a collection $\mathcal{A}$ of subsets of $\mathcal{G}_p$ such that*

(1) *Each $\Sigma \in \mathcal{A}$ is closed under finite direct sums and direct summands;*
(2) *For any $G, H \in \mathcal{G}_p$ there is a $\Sigma \in \mathcal{A}$ such that $G, H \in \Sigma$;*
(3) *Each $\Sigma \in \mathcal{A}$ has finite representation type.*

Proof: Let $\mathcal{A} = \{\Sigma(G) | G \in \mathcal{G}_p\}$, where $\Sigma(G)$ is the set of G-projective groups. Clearly $\mathcal{A}$ satisfies condition (1). Condition (2) follows from part (1) and because $\Sigma(G \oplus H) \in \mathcal{A}$. Condition (3) follows from Proposition 4.2.

Finally, let $\Sigma_p(G)$ denote the full subcategory of $\mathcal{G}_p$ whose objects are those of $\Sigma(G)$. The next result explains to some extent the nice behaviour of groups in $\mathcal{G}_p$ with respect to decompositions.

PROPOSITION 4.4 *The functor $Hom(G, \cdot)_p$ induces a category equivalence $F : \Sigma_p(G) \to \mathcal{P}(\mathcal{E}_p)$.*

Proof: Define $F(\cdot) = \mathrm{Hom}(G, \cdot)_p$. The usual arguments show that F takes G–projective groups to finitely generated projective $\mathcal{E}_p$–modules and near homomorphisms to module homomorphisms. By Lemma 3.1 and [A82, Theorem 7.21], F defines a category equivalence.

REFERENCES

[A82] D M Arnold, *Finite Rank Torsion Free Abelian Groups and Rings, Lecture Notes in Mathematics* **931**, Springer Verlag, 1982.

[AD93] D M Arnold and M Dugas, *Block rigid almost completely decomposable groups and lattices over multiple pullback rings*, J. Pure and Appl. Alg. **87 (2)** (1993), 105–121.

[AF73] F W Anderson and K R Fuller, *Rings and categories of modules*, Springer Verlag, 1973.

[BM94] E A Blagoveshchenskaya and A Mader, *Decompositions of almost completely decomposable groups*, Abelian groups, Proceedings of the 1993 Oberwolfach Conference, Amer. Math. Soc., Contemporary Mathematics, Vol. 171, 1994, pp. 21–36.

[Fa73] C C Faith, *Algebra: Rings, Modules, and Categories I*, Springer Verlag, New York, 1973.

[Fa76] C C Faith, *Algebra II: Ring Theory*, Springer Verlag, 1976.

[Fu73] L Fuchs, *Infinite Abelian Groups*, Vol. 2, Academic Press, New York, 1973.

[L74] E L Lady, *Almost completely decomposable torsion–free abelian groups*, Proc. Amer. Math. Soc. **45** (1974), 41–47.

[M95] A Mader, *Almost completely decomposable groups*, Abelian groups and modules, Proceedings of the 1994 Padova Conference, Kluwer Academic Publishers, 1995.

[S95] P Schultz, *Near–endomorphism rings of almost completely decomposable groups*, Abelian groups and modules, Proceedings of the 1994 Padova Conference, Kluwer Academic Publishers, 1995.

On Transitive Mixed Abelian Groups

STEVE FILES Department of Mathematics, University of Essen, 45117 Essen, Germany

In this note, we wish to examine Kaplansky's notions of transitivity in the setting of mixed abelian groups. While historically it was unclear if an abelian p-group could ever fail to be transitive [14], the picture is markedly different for other kinds of p-local abelian groups, especially torsion-free ones. Nontransitive groups of the latter kind abound, and the rank preserving functors from [4, 7] can be used to embed these torsion-free groups in torsion-divisible mixed groups in a way that naturally passes on nontransitivity. It is worth noting that the torsion subgroups of the mixed groups obtained in this way are transitive and fully transitive (direct sums of cyclics), hence an obvious *necessary* condition for transitivity and full transitivity in mixed groups is fulfilled.

Within the smaller realm of p-local groups of rank 1, however, one might justifiably ask if (full) transitivity of the torsion subgroup actually implies (full) transitivity of the group itself. We will note that rank 1 groups do indeed enjoy both transitivity properties when their torsion subgroups are separable (Proposition 1) or totally projective (Corollary 1). In spite of that, the question has a negative answer. The example we give to this effect suggests that a more plausible conjecture is the following: a rank 1 group inherits (full) transitivity from its torsion subgroup provided it has no torsion-free elements of infinite height.

Of course, groups of greater rank may fail to be transitive even when their torsion subgroups are separable or totally projective. In the search for classes of transitive groups of greater rank than 1, we must impose further conditions to effectively rule out nontransitive torsion-free groups and the kinds of nontransitive mixed groups they can engender. In keeping, we focus in the latter part of the note on the important class of Axiom 3 $\mathbf{Z}_p$-modules, or local Warfield groups (see [6, 11 and 13] for detailed accounts of this class). In this setting we will obtain a positive result–Proposition 2–that reaches to rank 2, and see that there is a natural barrier to transitivity for greater rank. The severe rank limitation in Proposition 2 is slightly ameliorated by the fact that we consider groups for which $G/p^\sigma G$, but not

Supported by the Graduiertenkolleg *Theoretische und experimentelle Methoden der reinen Mathematik* of Essen University and project No. G-0294-081.06193 of the German-Israeli Foundation for Scientific Research and Development.

necessarily G, is of small rank. With this feature we are able to uncover classes of Axiom 3 modules of infinite rank which are transitive and fully transitive (Corollary 1).

The author thanks the referee for pointing out how a powerful theorem of Hill and Megibben [11, Theorem 3.4] can be used to simplify the proof of Proposition 2. (With extra work, the proposition can be deduced from Walker's version [22] of Hill's celebrated result [9, Theorem 5.1] for Axiom 3 p-groups.) As it stands here, Proposition 2 could be labeled a corollary of the Hill-Megibben theorem. We remark, finally, that Proposition 2 looks back to results of Hill [10, Theorems 3 and 4], extending those results as far as possible when the Axiom 3 p-groups in his theorems are replaced by the Axiom 3 mixed groups from [11].

From now on, we restrict attention to the category of p-local abelian groups, or $\mathbf{Z}_p$-modules. We refer to these objects simply as modules, and use "rank" to indicate torsion-free rank. For a module G, we write $|x|_G$ or $|x|$ for the p-height in G of an element x. Recall that the *height sequence* $U_G(x)$ of x in G is the sequence $(|x|_G, |px|_G, |p^2x|_G, \ldots)$ of ordinal numbers and perhaps ∞. We use a wedge to denote the minimum of two ordinals, and also the pointwise minimum of two height sequences. The module G is *transitive* (*fully transitive*) if x can be mapped to y by an automorphism (endomorphism) of G whenever the two elements of G satisfy $U_G(x) = U_G(y)$ ($U_G(x) \leq U_G(y)$, pointwise). Various accounts of transitivity and full transitivity–mainly for p-groups–are given in [1, 2, 3, 8, 10, 14, 15, 18, 19].

To prove the first result, we make use of the fact that finite subsets of a separable torsion module can be emedded in finitely generated direct summands.

PROPOSITION 1 *A reduced, rank 1 module is transitive and fully transitive if its torsion submodule is separable.*

To prove the proposition, let G be a reduced, rank 1 module with separable torsion T. We will assume $G/T \cong \mathbb{Q}$ and leave the easier case where G is split to the reader. We first establish full transitivity. Because G has rank 1, it follows that we can write $y = rx + t$ ($r \in \mathbf{Z}_p$, $t \in T$) for any two elements $x, y \in G$ that satisfy $U_G(x) \leq U_G(y)$. Note $U_G(x) \leq U_G(t)$, hence the proof reduces to the case where $y \in T$. If $x \in T$, the existence of an endomorphism mapping x to y follows easily from the separability of T and we omit the details. Therefore, assume $x \notin T$ and $y \in T$. Embed y in a finite direct summand S of T (hence of G), say $G = G_0 \bigoplus S$ and $p^n S = 0$. Write $x = (x_0, s)$ in this decomposition. Because G_0 is divisible modulo torsion, we can write $x_0 = p^n x_1 + t$ for some $x_1 \in G_0$ and $t \in tG_0$. We claim $U_G(t+s) \leq U_G(y)$. Since $|p^i(t+s)| = |p^it| \wedge |p^is|$, it suffices to check that $|p^it| \leq |p^iy|$ when $|p^is| > |p^iy|$ and $p^iy \neq 0$. In this case $|p^{i+n}x_1 + p^it| \wedge |p^is| = |p^ix| \leq |p^iy| < |p^is|$, hence $|p^it| = |p^ix| \leq |p^iy|$ because $|p^iy| < n$, as desired. Now embed t in S_0, where $G_0 = G_1 \bigoplus S_0$ and S_0 is finite. The claim shows $U_{S_0 \oplus S}(t+s) = U_G(t+s) \leq U_S(y)$, hence $t+s$ can be mapped to y by some $\varphi \in Hom(S_0 \bigoplus S, S)$ since $S_0 \bigoplus S$ is separable (finite). Extend φ to an endomomorphism of G by putting φ zero on G_1. Then $\varphi(x) = p^n\varphi(x_1) + \varphi(t+s) = y$ because $p^nS = 0$. Hence G is fully transitive.

We next demonstrate that G is transitive. Assume $U_G(x) = U_G(y)$. If $x \in T$, then both elements can be embedded in a finite direct summand S of G. There is an automorphism of S mapping x to y, and this map extends to an automorphism of G in an obvious way. Therefore, assume that x and y are torsion-free. As noted above, $y = rx + t$ for some $r \in \mathbf{Z}_p$ and $t \in T$. Because $U_G(x) = U_G(y)$ and G is reduced, it follows that r is a unit of $\mathbf{Z}_p$. We will induct of the order of t. If $t = 0$, then multiplication by r is an automorphism of G

mapping x to y. Assume $t \neq 0$ and note that $U_{pG}(px) = U_{pG}(py)$ with $py = rpx + pt$ and $o(pt) < o(t)$. By induction, we obtain $\varphi \in Aut(pG)$ with $\varphi(px) = py$. By [5, Proposition 113.3], φ is induced by some automorphism ψ of G. The proof is complete if we can find an automorphism of G mapping $x' = \psi(x)$ to y. Note $px' = py$, so $y = x' + s$ with $ps = 0$. Embed s in S, where $G = G_0 \bigoplus S$ and S is finite. Write $x = (x_0, s_1)$ in this decomposition, and note $y = (x_0, s_1 + s)$. If $|x'| = |x_0|$, then $U_G(x_0) \leq U_G(s)$ and there exists $\beta \in Hom(G_0, S)$ with $\beta(x_0) = s$ because G is fully transitive. In this case, the automorphism $\begin{pmatrix} 1 & 0 \\ \beta & 1 \end{pmatrix}$ of $G_0 \bigoplus S = G$ maps x' to y. To finish the proof, we may assume $|x'| = |s_1| < |x_0|$. Observe $|s_1| = |x'| = |y| = |x_0| \wedge |s_1 + s| < |x_0|$, hence $|s_1 + s| = |s_1|$ and therefore $U_S(s_1) = U_S(s_1 + s)$. Choose an automorphism γ of S mapping s_1 to $s_1 + s$. Then the automorphism $\begin{pmatrix} 1 & 0 \\ 0 & \gamma \end{pmatrix}$ of $G_0 \bigoplus S$ maps x' to y, as desired. □

As remarked earlier, Proposition 1 can fail easily if G has rank 2. An example of this failure for modules over the p-adic integers is harder to come by, though, so we give details in this case. Specifically, we construct a reduced $\mathbf{J}_p$-module M of rank 2 with separable torsion that is neither transitive nor fully transitive. Note that unlike the case of a $\mathbf{Z}_p$-module, M cannot be chosen torsion-free: the nontransitivity of M must be engendered by its torsion submodule. If T is a separable torsion module, let $\overline{T}$ denote its torsion-completion. Like on page 497 of [16], we can choose T so that $\overline{T}/T$ has p-rank at least 2, and $End(T)$ is the sum of $\mathbf{J}_p$ and the ideal of small endomorphisms of T. Let $T^\bullet$ be the cotorsion hull of T. Because $p^\omega T^\bullet \cong Hom(\mathbf{Z}(p^\infty), \overline{T}/T)$, it is possible to choose $\mathbf{J}_p$-independent elements $x, y \in T^\bullet$ of height ω. Let M be the purification of $T + \mathbf{J}_p x + \mathbf{J}_p y$ in $T^\bullet$. By [17, Lemma 1], endomorphisms of M which are small on its torsion submodule ($= T$) map $p^\omega M$ to zero. Because of the form of $End(T) = End(T^\bullet)$ it follows that $\mathbf{J}_p x$ is fully invariant in M, hence no endomorphism of M maps x to y. Since $U_M(x) = U_M(y) = (\omega, \omega+1, \omega+2, \dots)$, we have produced the desired $\mathbf{J}_p$-module M.

Now we return to $\mathbf{Z}_p$-modules and give an example to show that a reduced rank 1 module can fail to be transitive even if its torsion submodule is transitive and fully transitive. In fact, the torsion submodule T in our example will nearly be separable in the sense that $p^\omega T$ is cyclic of order p (by [2, Lemma 2.1], any such module T is transitive and fully transitive). Of the several constructions available to us at this point, we choose one which yields T as the direct sum of a separable and a totally projective module. By [20, Corollary 16.2], we may choose a separable torsion module T_1 with basic submodule B such that T_1 is not torsion-complete, and $Hom(T_1, B)$ consists of small homomorphisms. By standard methods, we obtain a second module T_2 such that $p^\omega T_2 = \langle t \rangle$ has order p and $T_2/p^\omega T_2$ is isomorphic to B. Put $G = G_1 \bigoplus T_2$, where G_1 is the purification of $T_1 + \langle x \rangle$ in $T_1^\bullet$ and $x \in T_1^\bullet$ has height ω. Then the torsion $T_1 \bigoplus T_2$ of G is of the desired form. Since $U_G(x) = U_G(x + t) = (\omega, \omega + 1, \omega + 2, \dots)$, it will follow that G is neither transitive nor fully transitive if $p^\omega G_1 = \langle x \rangle$ is fully invariant in G. Suppose φ maps G_1 into T_2. Then the restriction of φ to T_1, followed by canonical projection $T_2 \to T_2/p^\omega T_2 \cong B$, is necessarily small. Because $p^\omega T_2$ has order p, it follows easily that $p\varphi$, hence φ, is small on T_1. But like in the previous example, this forces φ to be zero on $p^\omega G_1$. Therefore the last submodule is fully invariant in G, as desired.

We desist with examples and turn to the task of proving that a module G is (fully) transitive when a height layer $p^\sigma G$ is (fully) transitive and $G/p^\sigma G$ is a Warfield module of small enough rank. A module is called an *Axiom 3 module* if it satisfies the third axiom of countability with respect to knice submodules; see [11] for full details. It turns out that Axiom 3 modules are precisely the well-known Warfield modules [11, Theorem 4.1], but we will stick with the former name from now on. If N is a submodule of G, we write the sets of relative Ulm invariants and relative Warfield invariants of G with respect to N as functions $f_{G,N}$ and $w_{G,N}$, respectively.

Part (2) of the following theorem is given in Hill and Megibben [11]; part (1) can be deduced from (2) using the trick in the proof of [5, Corollary 81.4].

THEOREM [11] *Let N_1 and N_2 be submodules of G and H, respectively. Assume G/N_1 is an Axiom 3 module and N_1 is knice in G.*

(1) Every homomorphism $N_1 \to H$ that does not decrease heights in G and H extends to a homomorphism $G \to H$.

(2) Assume $f_{G,N_1} = f_{H,N_2}$ and $w_{G,N_1} = w_{H,N_2}$. If N_2 is knice in H and H/N_2 is an Axiom 3 module, then every isomorphism $N_1 \to N_2$ that preserves heights in G and H extends to an isomorphism $G \to H$.

We require some additional concepts from [11]. A torsion-free element x of G is called *primitive* in G if its height sequence $U_G(x)$ necessarily agrees with one of the sequences $U_G(x_i)$ from some point on whenever $x = x_1 + \cdots + x_n$ in G and $U_G(x) \leq U_G(x_i)$ for $i = 1, \ldots, n$. Primitive elements of a module in which 0 is knice (in particular, an Axiom 3 module) generate knice submodules. The theorem can be applied to show that an automorphism of an Axiom 3 module G carrying x to y exists provided both elements are primitive in G and $U_G(x) = U_G(y)$. In fact, any Axiom 3 module whose torsion-free elements are all primitive must be transitive and fully transitive. Balanced projective modules meet this criterion, for example; see [15] for a more extensive result in this setting. For Axiom 3 modules in general (and many classes besides), the problem is that even a rank 2 module can possess nonprimitive torsion-free elements; for modules of rank 3, it is already implicit on page 716 of [11] that both kinds of transitivity will sometimes be lost. In the latter case, one can easily write down a reduced, simply presented module G of rank 3 that contains a primitive element y with the same height sequence as a nonprimitive element x of G; then no endomorphism of G can map x to y because y is primitive. Such an example indicates, incidentally, that the direct sum of two transitive and fully transitive mixed modules may not enjoy either property, even if the summands have isomorphic torsion submodules.

The following simple lemmas will aid us in the proof of Lemma 3. We omit the routine proof of the second lemma.

LEMMA 1 *Suppose $x = (a, b) \in G = A \bigoplus B$ and a is primitive in A. If $U_G(x)$ agrees with $U_G(a)$ from some point on, then x is primitive in G.*

To prove the lemma, suppose $x = (a_1, b_1) + \ldots (a_n, b_n)$ in G, where $U_G(x) \leq U_G((a_i, b_i))$ for all i. Then for some $k < \omega$, $U_A(p^k a) = U_G(x) \leq U_A(p^k a_i)$ for all i. Because the multiple $p^k a$ remains primitive in A, there is an index j such that $U_A(p^k a)$ agrees with $U_A(p^k a_j)$ from some point on. It follows that $U_G(x)$ and $U_G((a_j, b_j))$ agree from some point on as well. □

LEMMA 2 *Let A and B be submodules of G and H such that $f_{G,A} = f_{H,B}$. If $x \in G, y \in H$ and there is a height-preserving isomorphism $A' = A + \langle x \rangle \to B' = B + \langle y \rangle$ mapping A onto B and x to y, then $f_{G,A'} = f_{H,B'}$.*

If G is a module and σ an ordinal, we will use $\overline{(\ )} : G \to \overline{G}$ to denote the natural map. The final lemma deals with the case where $\overline{G}$ is completely decomposable of rank 2 (a direct sum of two rank 1 modules), but is equally valid when $\overline{G}$ has smaller rank; a direct application of the Hill-Megibben theorem proves the result in the latter case. To set up notation for the lemma, fix submodules $A, B \subset G$ containing $p^\sigma G$ such that $\overline{G} = \overline{A} \bigoplus \overline{B}$ and the summands have rank 1. Note that $\overline{G}$ is reduced. If $z \in G \backslash p^\sigma G$ and $g \in p^\sigma G$, then $|z+g| = |z| = |\overline{z}|_{\overline{G}}$ by the niceness of $p^\sigma G$ in G (if $\overline{G}$ is an Axiom 3 module, then $p^\sigma G$ is actually knice in G). It follows that if $a \in A$, $b \in B$ and at least one element has height less then σ, then $|a+b| = |a| \wedge |b|$ because $|a+b| \leq |\overline{a}+\overline{b}|_{\overline{G}} = |\overline{a}|_{\overline{G}} \wedge |\overline{b}|_{\overline{G}} = |a| \wedge |b| \leq |a+b|$. We will use these observations to compute heights in the next proof.

LEMMA 3 *If $p^\sigma G$ is (fully) transitive and $G/p^\sigma G$ is a completely decomposable Axiom 3 module of rank 2, then G is (fully) transitive.*

To prove Lemma 3, we first consider the fully transitive case. Assume $x, y \in G$ satisfy $U_G(x) \leq U_G(y)$. We begin by supposing $\langle \overline{x} \rangle$ to be knice in $\overline{G}$. Then $N_1 = p^\sigma G + \langle x \rangle$ is knice in G by [11, Proposition 2.7]. If $\overline{x}$ is torsion-free, we get a height-increasing homomorphism $\varphi : N_1 \to G$ by taking $\varphi(p^\sigma G) = 0$ and $\varphi(x) = y$ (heights are computed in G unless indicated differently). If $\overline{x}$ is torsion, we are guaranteed the existence of an endomorphism φ_0 of $p^\sigma G$ mapping $p^k x$ to $p^k y$, where $k < \omega$ is chosen minimal with respect to $p^k x \in p^\sigma G$. In this case we obtain a height-increasing homomorphism $\varphi : N_1 \to G$ by setting $\varphi = \varphi_0$ on $p^\sigma G$ and $\varphi(x) = y$. It is easy to verify that $\overline{G}$ factored by any finitely generated submodule is an Axiom 3 module, hence in both cases $G/N_1 \cong \overline{G}/\overline{N_1}$ is of the desired form. Part (1) of the theorem thus provides an endomorphism of G mapping x to y.

Now suppose $\langle \overline{x} \rangle$ is not knice in $\overline{G}$. Then $\overline{x}$ is torsion-free and nonprimitive. Write $\overline{G} = \overline{A} \bigoplus \overline{B}$ like in the discussion after Lemma 2, and fix elements $a, c \in A$, $b, d \in B$ with $x = a + b$ and $y = c + d$. Because torsion-free elements of rank 1 modules are primitive, Lemma 1 implies $\overline{a}$ and $\overline{b}$ are torsion-free and that there exist infinite sets $I_1, I_2 \subset \omega$ such that the following inequalites hold:

$$|p^i a| < |p^i b| \ (i \in I_1), \quad \text{and} \quad |p^i b| < |p^i a| \ (i \in I_2). \tag{*}$$

Since $\overline{A}$ and $\overline{B}$ have rank 1 we can write $p^m c = ra + g_1$ and $p^m d = sb + g_2$ for some $m < \omega$, where $r, s \in \mathbf{Z}_p$ and $g_1, g_2 \in p^\sigma G$. Choose i with $i+m \in I_1$ and observe $|p^{i+m}\overline{a}|_{\overline{G}} = |p^{i+m}a| = |p^{i+m}x| \leq |p^{i+m}y| \leq |p^{i+m}\overline{y}|_{\overline{G}} = |p^i r\overline{a}|_{\overline{G}} \wedge |p^i s\overline{b}|_{\overline{G}} \leq |p^i r\overline{a}|_{\overline{G}}$. This inequality implies that p^m divides r. Using I_2 in this fashion, we deduce that p^m divides s as well. Writing $r = p^m r_0$ and $s = p^m s_0$ in $\mathbf{Z}_p$, we get expressions $c = p^m r_0 a + g_3$ and $d = p^m s_0 c + g_4$ for elements $g_i \in G$ with $p^m g_i \in p^\sigma G$ $(i = 3, 4)$. Choose $k < \omega$ so that $p^i x$ and $p^i y$ are proper with respect to $\langle p^k b \rangle$ for all $i < m$. Put $N_1 = p^\sigma G \bigoplus \langle x \rangle \bigoplus \langle p^k b \rangle \subseteq G$. Then G/N_1 is totally projective, and N_1 is knice in G because the knice submodule $p^\sigma G \bigoplus \langle p^k a \rangle \bigoplus \langle p^k b \rangle$ of G has finite index in N_1. Define $\varphi : N_1 \to G$ by mapping $p^\sigma G$ onto 0, x to y and $p^k b$ to $p^k s_0 b$. Provided φ increases heights, the fully transitive case is complete by part (1) of the theorem. It suffices to check $|\varphi(z)| \geq |z|$ for $z \in N_1$ of the form $g + p^i x + rp^k b$, where $g \in p^\sigma G$ and $r \in \mathbf{Z}_p$. If $i < m$, we get $|z| = |p^i x + rp^k b| = |p^i x| \wedge |rp^k b| \leq |p^i y| \wedge |rp^k s_0 b| = |p^i y + rp^k s_0 b| = |\varphi(z)|$ by our choice of k. If $i \geq m$, then $p^i g_3$ and $p^i g_4$ lie in $p^\sigma G$ and hence $|\varphi(z)| = |p^i y + rp^k s_0 b| = |p^i(r_0 a + s_0 b) + rp^k s_0 b| = |p^i r_0 a| \wedge |s_0(p^i + rp^k)b| \geq |p^i a| \wedge |(p^i + rp^k)b| = |p^i x + rp^k b| = |z|$, as required.

Now we turn to the proof that G is transitive if $p^\sigma G$ is. Continue writing $x = a + b$ and $y = c + d$ as above, and suppose $U_G(x) = U_G(y)$. This time we begin by assuming that $\overline{x}$ is torsion-free and nonprimitive in $\overline{G}$. Like above, this implies $\overline{a}$ and $\overline{b}$ are torsion-free, and that there exist infinite sets $I_1, I_2 \subset \omega$ so that condition (*) above holds. We will see along the way that $\overline{y}$ is not primitive in $\overline{G}$ either. Like in the fully transitive case we get expressions $p^m c = ra + g_1$ and $p^m d = sb + g_2$, where $g_1, g_2 \in p^\sigma G$ and p^m divides both r and s. Choosing an i with $m+i \in I_1$, we compute $|p^{m+i}b| > |p^{m+i}a| = |p^{m+i}x| = |p^{m+i}y| = |p^{m+i}c| \wedge |p^{m+i}d| = |p^i ra| \wedge |p^i sb|$. Since $p^i r$ is already divisible by p^{i+m} we conclude $|p^{m+i}a| = |p^i ra|$, hence p^{m+1} does not divide r. Using I_2, we deduce the same about s. It follows that $c = ua + g_3$ and $d = vb + g_4$, where u and v are units in $\mathbf{Z}_p$ and $p^m g_i \in p^\sigma G$ $(i = 3, 4)$. Observe at this point that $\overline{c}$ and $\overline{d}$ are torsion-free, and that c and d can stand in (*) when i is taken greater than m. Therefore $\overline{y} = \overline{c} + \overline{d}$ is nonprimitive because $U_{\overline{G}}(\overline{y})$ fails to agree with either of the sequences $U_{\overline{G}}(\overline{c})$ or $U_{\overline{G}}(\overline{d})$ from some point on. Choose $k < \omega$ so that $p^i x$ and $p^i y$ are proper with respect to $\langle p^k b\rangle$ for $i < m$, and put $N_1 = p^\sigma G \bigoplus \langle x\rangle \bigoplus \langle p^k b\rangle \subseteq G$, $N_2 = p^\sigma G \bigoplus \langle y\rangle \bigoplus \langle p^k b\rangle \subseteq G$. Then N_1 and N_2 have totally projective quotients in G, and are knice because they are finite extensions of the knice submodule of G generated by $p^\sigma G$, $p^{k+m}a$ and $p^{k+m}b$. Define an isomorphism $\varphi : N_1 \to N_2$ by taking the identity on $p^\sigma G$ and mapping $x \mapsto y$, $p^k b \mapsto p^k vb$. Computation like in the fully transitive case above shows that φ preserves heights, and Lemma 2 shows $f_{G,N_1} = f_{G,N_2}$. The equality of relative Warfield invariants is automatic since the submodules are full rank, hence φ extends to an automorphism of G by part (2) of the theorem.

To finish the proof, we may assume that $\overline{x}$ is torsion or else is primitive in $\overline{G}$. By what was shown in the preceding case, it follows that $\overline{y}$ is also primitive in $\overline{G}$ if $\overline{x}$ is torsion-free. Note that $\overline{x}$ and $\overline{y}$ have equal height sequences in $\overline{G}$, hence by the remarks after the theorem we know that there already exists an automorphism of the Axiom 3 module $\overline{G}$ mapping $\overline{x}$ to $\overline{y}$. In particular, the knice submodules $N_1 = p^\sigma G + \langle x\rangle$ and $N_2 = p^\sigma G + \langle y\rangle$ of G must satisfy $w_{\overline{G},\overline{N_1}} = w_{\overline{G},\overline{N_2}}$. But it is easily verified that $w_{G,N_i} = w_{\overline{G},\overline{N_i}}$ $(i = 1, 2)$ for these submodules, hence the relative Warfield invariants of G with respect to N_1 and N_2 are equal as well. Now since $p^\sigma G$ is transitive, there is in any event a height-preserving isomorphism $N_1 \to N_2$ mapping $p^\sigma G$ onto itself, and x to y. By Lemma 2, $f_{G,N_1} = f_{G,N_2}$. Because all conditions needed to apply (2) in the theorem are met, the last isomorphism may be extended to an automorphism of G, completely establishing transitivity. □

If an Axiom 3 module G of rank 2 has a decomposition basis $\{x, y\}$ such that $\sup(U_G(x)) = \sup(U_G(y))$, it can be deduced from [13, Theorem 27] that G is completely decomposable. In case G is reduced and $\lambda = \sup(U_G(x)) < \sup(U_G(y))$, then both $p^\lambda G$ and $G/p^\lambda G$ are Axiom 3 modules of rank 1.

We now have all ingredients needed to prove the next proposition. As usual, G denotes a module and σ an ordinal number.

PROPOSITION 2 *Assume $G/p^\sigma G$ is an Axiom 3 module of rank less than 3. Then G is (fully) transitive if and only if $p^\sigma G$ is (fully) transitive.*

To prove necessity, suppose G is transitive and $x, y \in p^\sigma G$ satisfy $U_{p^\sigma G}(x) = U_{p^\sigma G}(y)$. Then $U_G(x) = U_G(y)$, hence $\varphi(x) = y$ for an automorphism φ of G. Since φ induces an automorphism of $p^\sigma G$, $p^\sigma G$ is transitive. Similarly, $p^\sigma G$ is fully transitive if G is fully transitive.

Now assume $\overline{G} = G/p^\sigma G$ is an Axiom 3 module of rank less than 3, and $p^\sigma G$ is (fully) transitive. The desired result follows immediately from Lemma 3 and the remarks after Lemma 2 if $\overline{G}$ is torsion or completely decomposable. By the preparatory remarks above, we may therefore assume $\text{rank}(\overline{G}) = 2$ and $\text{rank}(p^\lambda \overline{G}) = \text{rank}(\overline{G}/p^\lambda \overline{G}) = 1$ for an ordinal λ. Note that $\lambda < \sigma$, say $\sigma = \lambda + \nu$ for some ν. Since $p^\sigma G$ is nice in G, we have $G/p^\lambda G \cong (G/p^\sigma G)/(p^\lambda G/p^\sigma G) = \overline{G}/p^\lambda \overline{G}$, an Axiom 3 module of rank 1. By what we have already noted, G is (fully) transitive provided $H = p^\lambda G$ is (fully) transitive. Since $p^\nu H = p^\sigma G$ is (fully) transitive and $H/p^\nu H = p^\lambda \overline{G}$ is an Axiom 3 module of rank 1, the result follows for H. □

Proposition 2 shows that a reduced Axiom 3 module of rank less than 3 is both transitive and fully transitive (take σ to be the module's length). In order to single out classes of transitive and fully transitive Axiom 3 modules of larger rank, we need some simple invariants. If G is a module with decomposition basis X, let $\Lambda_G(\lambda)$ denote the cardinality of the set

$$\{x \in X : \sup(U_G(x)) = \lambda\},$$

where λ is any ordinal of cofinality ω.

It is an immediate consequence of the relationship between decomposition bases and Warfield invariants that $\Lambda_G(\lambda)$ is independent of the choice of decomposition basis X for G. If $\Lambda_G(\lambda) \leq 2$ for all λ of cofinality ω, we will say that G has *small Λ-values.*

COROLLARY 1 *Let G be a reduced Axiom 3 module with small Λ-values. Then G is fully transitive. If G is a direct sum of submodules of finite rank, then G is transitive.*

To prove the corollary, we may assume G is nontorsion. First suppose G has finite rank. Let $\lambda_1 < \lambda_2 < \cdots < \lambda_n$ be the ordinals where $\Lambda_G(\lambda_i) > 0$. If $n = 1$, then $\text{rank}(G) = \Lambda_G(\lambda_1) \leq 2$ and G is transitive and fully transitive by the remark after Proposition 2. Assume $n > 1$. Then $p^{\lambda_1} G$ is an Axiom 3 module with small Λ-values, only $n-1$ of which are positive. By induction, we may therefore assume $p^{\lambda_1} G$ is transitive and fully transitive. Since $\text{rank}(G/p^{\lambda_1} G) = \Lambda_G(\lambda_1) \leq 2$ and $G/p^{\lambda_1} G$ is an Axiom 3 module, Proposition 2 implies G is transitive and fully transitive.

In case G is a direct sum of submodules of finite rank, note that any given pair of elements of G is contained in a finite-rank direct summand H of G. By the preceding case, H is transitive and fully transitive. Depending on the elements in question, we get the desired automorphism or endomorphism of G by adding the appropriate map $H \to H$ to the identity map on a complement for H in G. Hence G is transitive and fully transitive in this case as well.

Finally, suppose G is any reduced Axiom 3 module with small Λ-values. There is a reduced torsion module T such that $G \bigoplus T$ is simply presented, hence completely decomposable. By the preceding case, $G \bigoplus T$ is fully transitive because it has the same Λ-values as G. It follows that the direct summand G is fully transitive as well, completing the proof. □

While the class of transitive modules is not closed under direct sums, it may be the case that the direct sum of copies of a single (fully) transitive $\mathbf{Z}_p$-module remains (fully) transitive, at least when $p \neq 2$. (This matter was raised for torsion modules by Carroll and Goldsmith in [2].) Such a closure property would enlarge many of the classes of transitive mixed modules discussed in this note.

References

[1] D. Carroll. Multiple transitivity in abelian groups, Arch. Math. **63** (1994), 9-16.

[2] D. Carroll and B. Goldsmith. On transitive and fully transitive abelian p-groups, Proc. Royal Irish Acad. (to appear).

[3] A.L.S. Corner, The independence of Kaplansky's notions of transitivity and full transitivity, Quart. J. Math. Oxford **27** (1976), 15-20.

[4] B. Franzen and B. Goldsmith, On endomorphism algebras of mixed modules, J. London Math. Soc. **31** (1985), 468-472.

[5] L. Fuchs, *Infinite Abelian Groups*, Vol. II, Academic Press, New York, 1973.

[6] L. Fuchs, *Abelian p-Groups and Mixed Groups*, University of Montreal Press, Montreal, 1980.

[7] R. Göbel and W. May, The construction of mixed modules from torsion-free modules, Archiv. Math. **48** (1987), 476-490.

[8] B. Goldsmith, On endomorphism rings of non-separable abelian p-groups, J. Algebra **127** (1989), 73-79.

[9] P. Hill, On the classification of abelian groups, photocopied manuscript, 1967.

[10] P. Hill, On transitive and fully transitive primary groups, Proc. Amer. Math. Soc. **22** (1969), 414-417.

[11] P. Hill and C. Megibben, Axiom 3 modules, Trans. Amer. Math. Soc. **295** (1986), 715-734.

[12] P. Hill and C. Megibben, The local equivalence theorem, Contemporary Math., Vol. 87, Gordon & Breach, New York, 1987, pp. 201-220.

[13] R. Hunter, F. Richman and E. Walker, Warfield modules, Lecture Notes in Math., Vol. 616, Springer-Verlag, New York, 1977, pp. 87-123.

[14] I. Kaplansky, *Infinite Abelian Groups*, Univ. of Michigan Press, Ann Arbor, 1954.

[15] M. Lane, Isotype subgroups of p-local balanced projective groups, Trans. Amer. Math. Soc. **301** (1987), 313-325.

[16] W. May, Isomorphism of endomorphism algebras over complete discrete valuation rings, Math. Zeit. **204** (1990), 485-499.

[17] W. May, The theorem of Baer and Kaplansky for mixed modules, J. Algebra **177** (1995), 255-263.

[18] C. Megibben, Large subgroups and small homomorphisms, Michigan J. Math. **13** (1966), 153-160.

[19] C. Megibben, A nontransitive, fully transitive primary group, J. Algebra **13** (1969), 571-574.

[20] R. Pierce, Homomorphisms of primary abelian groups, in *Topics in Abelian Groups*, Scott, Foresman & Co., Chicago, 1963, pp. 215-310.

[21] R. Stanton, Relative S-invariants, Proc. Amer. Math. Soc. **65** (1977), 221-224.

[22] E. Walker, Ulm's theorem for totally projective groups, Proc. Amer. Math. Soc. **37** (1973), 387-392.

[23] R. B. Warfield, Classification theorems for p-groups and modules over a discrete valuation ring, Bull. Amer. Math. Soc. **78** (1972), 88-92.

[24] R. B. Warfield, The structure of mixed abelian groups, Lecture Notes in Math., Vol. 616, Springer-Verlag, New York, 1977, pp. 1-38.

G.C.H. Implies Existence of Many Rigid Almost Free Abelian Groups

RÜDIGER GÖBEL Fachbereich 6 - Mathematik, Universität GH Essen, 45117 Essen, Germany
e-mail: R.Goebel@uni-essen.de

SAHARON SHELAH Department of Mathematics, Hebrew University, Jerusalem, Israel and Rutgers University New Brunswick, N.Y., U.S.A.
e-mail: Shelah@math.huji.ae.il

1 INTRODUCTION

More than thirty years ago A. L. S. Corner [2] proved a very remarkable theorem concerning the existence of "pathological" abelian groups; it is formulated in terms of endomorphism rings.

If A is a countable ring with reduced and torsion-free additive group $A_{\mathbb{Z}}$, then there exists a reduced and countable abelian group G with endomorphism ring $End_{\mathbb{Z}}G = A$.

This result has been generalized in various directions. e.g. replacing the ground ring $\mathbb{Z}$ by a more general ring and dropping the cardinal restriction (in Corner's result this is actually $|A| < 2^{\aleph_0}$), see [2], [20], [8], [9], [4], [23].

Other extensions include abelian groups which are not necessarily torsion- free, see Corner, Göbel [4] for references.

Here we are interested in a different strengthening of Corner's important result: As a byproduct of older constructions of abelian groups G with prescribed endomorphism ring A we have obtained (in a strong sense) almost-free groups provided $A_{\mathbb{Z}}$ is free and provided we are working in Gödel's universe $V = L$. Let us recall the appropriate definition. An R-module G (of cardinality λ) with $|R| < |G|$ is strongly λ-free if G has a λ-filtration (a continuous chain of submodules $\{G_\nu : \nu < \lambda\}$ with $|G_\nu| < \lambda$ and $\bigcup_{\nu<\lambda} G_\nu = G$) with two additional properties:

*This work is supported by GIF project No. G-0294-081.06/93 of the German-Israeli Foundation for Scientific Research & Development

(a) Any G_ν is a free R-module.

(b) If ν is a successor ordinal and $\nu < \rho < \lambda$, then G_ρ/G_ν is free as well.

We say that $G = \bigcup_{\nu<\lambda} G_\nu$ is a λ-filtration for strong λ-freeness.
First results on almost-free groups of cardinality $2^{\aleph_0}$ are in [1].

Because strong λ-freeness is a very natural definition, it can be reformulated in many ways, see Eklof, Mekler [15] for equivalent characterizations and many results on this class of modules. In order to illustrate such results we will restrict ourselves to abelian groups, i.e. $R = \mathbb{Z}$. In $V = L$ we have the following strengthening of Corner's result indicated above.

If A is a ring, $A_{\mathbb{Z}}$ is a free abelian group and $\lambda > |A|$ is any regular, not weakly compact cardinal, then we can find a strongly λ-free abelian group G with $|G| = \lambda$ and $End_{\mathbb{Z}}G = A$.

This older result [8] grew out of ideas of Eklof, Mekler [14], and Shelah [38]. It is closely related to the work of Jensen [30]. It is only natural to ask if such almost free groups exist without the vehicle of additional set theory. Extending a result of Shelah [36], Eda [11] showed that there is at least some hope.

There exists an $\aleph_1$-free abelian group G of cardinality $\aleph_1$ with trivial dual $G^* = 0$, see also Corner, Göbel [5]. Recall that $Hom(G, R) = G^*$ is the dual of an R-module, and G is κ-free for some cardinal κ if any submodule U with $|U| < \kappa$ is contained in a free submodule. In [22] we showed the following stronger result.

If A is a countable ring and $A_{\mathbb{Z}}$ is free, then we can find an $\aleph_1$-free abelian group G of cardinality $\aleph_1$ with endomorphism ring $End_{\mathbb{Z}}G = A$.

This result provides a satisfying answer in ZFC, but unfortunately (from the algebraist's point of view) it remains restricted to the world $\aleph_1$.

In fact, assuming ZFC + Martin's axiom, it was shown in [22] that $\aleph_2$-free groups of size $\aleph_2$ are already always separable (= pure subgroups of products $\mathbb{Z}^I$), hence they must split into various summands. Their endomorphism rings will never be $\mathbb{Z}$ for instance. This brings us to the starting point of this paper.

We want to derive a realization theorem of rings A with $A_{\mathbb{Z}}$ free as $A = End_{\mathbb{Z}}G$ for strongly $|G|$-free abelian groups G with $|G| \geq \aleph_1$, using as weak additional set theoretic conditions as possible. From the result mentioned above we know that Martin's axiom must be excluded. Our results discussed below are surprising in two ways. First of all, we will only need a special case of G.C.H., which ensures the existence of the desired modules. In another paper [43], Shelah and Spasojevic will show by proper forcing arguments that even G.C.H. will not be strong enough for $End\ G = \mathbb{Z}$, $|G| = \lambda$ and G λ-free if λ is too large, e.g. if λ is any strongly inaccessible cardinal. Hence we are limited to "small large cardinals". Secondly it turns out that the restriction to G.C.H. forces us to develop novel ideas to get the desired modules. A new Step-Lemma is needed! Here we will also apply combinatorial tools fortunately developed by Shelah [35], [41] a decade ago. The main results of this paper can be summarized as follows.

In order to construct pathological groups we begin with the existence of groups with trivial duals for cardinals $\aleph_n$ $(n \in \omega)$. Then we derive results about strongly $\aleph_n$-free abelian groups of cardinality $\aleph_n$ $(n \in \omega)$ with prescribed free, countable endomorphism ring. Finally we use combinatorial results [35], [41] to give similar answers for cardinals $> \aleph_\omega$. As in Magidor and Shelah [31], a paper concerned with the existence of κ-free, non-free abelian groups of cardinality κ, the induction argument breaks down at $\aleph_\omega$. Recall that $\aleph_\omega$ is the first singular cardinal and such groups of cardinality $\aleph_\omega$ do not exist by the well-known Singular Compactness Theorem [34], see also Hill [28] for cardinals cofinal to ω and, for a survey, Eklof and Mekler [15, p. 107].

We will fix a countable commutative ring R with multiplicatively closed subsets S of regular elements containing 1. An R-module G will be called torsion-free (with respect to S) if $sg = 0$ $(s \in S,\ g \in G)$ implies $g = 0$. Moreover G is reduced (with respect to S) if $\bigcap_{s \in S} sG = 0$. If A is an R-algebra, and the R-module A_R is R-free (torsion-free or reduced) we will say that the algebra A is free (torsion-free or reduced). We will also fix a countable, free R-algebra A and derive the following results.

THEOREM 5.1 Let (R, S) be as above. If H is a strongly μ-free R-module of regular cardinality μ with trivial dual $H^* = 0$ and $\lambda = 2^\mu = \mu^+$ (the successor cardinal of μ), then there exists a strongly λ-free R-module of cardinality λ with trivial dual.

As a consequence of (5.1) we obtain the existence of such R-modules of cardinality μ_n $(n \in \omega)$ for the n-th successor μ_n of μ provided $2^{\mu_i} = \mu_{i+1}$ for all $i < n$. The existence of H as in (5.1) for $|H| = \aleph_1$ under $2^{\aleph_0} = \aleph_1$ is well-known, see Dugas [7], Shelah [38] or Dugas, Göbel [8], cf. also Eklof, Mekler [15, p. 391]. Hence the existence of the derived strongly $\aleph_n$-free R-modules of cardinality $\aleph_n$ with trivial dual is immediate (see also (5.2)).

In §6 we will work with a stronger **algebraic** hypothesis, which again is satisfied automatically for $\aleph_1 = 2^{\aleph_0}$ by the papers [8],[38] above. We derive the

THEOREM 6.1 Let (R, S), $\lambda = 2^\mu = \mu^+$ and μ regular as above. If H is a strongly μ-free R-module of cardinality μ with $End_R H = A$ for some countable, free R-algebra A, then there exists a strongly λ-free R-module G of cardinality λ with $End_R G = A$.

Now the existence of strongly $\aleph_n$-free R-modules G of cardinality $\aleph_n$ with $End_R G = A$ follows for each countable, free R-algebra A, provided $\aleph_{i+1} = 2^{\aleph_i}$ for each $i < n$. As one of many consequences we see that under the mild restriction $\aleph_{i+1} = 2^{\aleph_i}$ $(i < n)$ there are counterexamples to Kaplansky's test problems for each cardinal $\aleph_n$ $(n \in \omega)$ even ones which are strongly $\aleph_n$-free (see [3] for the required rings). Similarly, we obtain indecomposable, strongly $\aleph_n$-free groups of cardinality $\aleph_n$. By the result of Shelah, Spasojevic [43] we know that a realization theorem similar to (6.1) and even the existence of modules with trivial dual (5.1) does not follow from G.C.H. for large enough cardinals, e.g. for the first strongly inaccessible cardinal ξ. What happens at smaller cardinals, for instance at $\aleph_{\omega+1}$? The Theorems 6.1 and (5.1) can be extended to $\aleph_{\omega+1}$, hence to $\aleph_{\omega+n}$ $(n \in \omega)$ by (5.1) and (6.1), and to certain larger cardinals $< \xi$. We must strengthen the methods from §§3,4 and 5, and this procedure is sketched in §7.

2 REPRESENTATION OF STRONGLY μ-FREE MODULES AND THE NOTION OF A TYPE OF FREENESS

We fix some notations throughout this paper.

(2.1) Let R be a commutative ring and $S \subseteq R$ a countable, multiplicatively closed subset of regular elements such that R_R (as R-modules) is reduced and torsion-free (with respect to S), compare §1. We will also assume $1 \in S$ and enumerate $S = \{s_n : n \in \omega\}$ and set $q_n = \prod_{i=1}^{n} s_i$.

Our results concern almost free R-modules (mainly for $|R| = \aleph_0$), hence we will refine the notion of almost free and strongly μ-free R-modules from §1:

We say that a strongly μ-free R-module H has freeness type $ftH = u$ if $u = \langle \mu_1, \cdots, \mu_n \rangle$ is a descending sequence of regular cardinals μ_i with $\mu_1 = \mu$, $\mu_n = \aleph_0$ and if $H = \bigcup_{\alpha<\mu_1} H_\alpha$ is a μ_1-filtration for strong μ_1-freeness with $H_{\alpha+1}/H_\alpha$ either free or of freeness type $\langle \mu_2, \cdots, \mu_n \rangle$. The definition is completed by induction if we say that countable groups which are not free are of freeness type $\langle \aleph_0 \rangle$. If H is of freeness type $u = \langle \mu_1, \cdots, \mu_n \rangle$, then we can choose a particular filtration of H, which we will call a u-filtration of H. We will write $H = \bigcup_{\alpha_1<\mu_1} H_{\alpha_1}$, such that $H_{\alpha_1+1}/H_{\alpha_1}$ is either free or of freeness type $\langle \mu_2, \cdots, \mu_n \rangle$ and set $\boldsymbol{S}_{\mu_1} = \{\alpha \in \mu_1 : H_{\alpha+1}/H_\alpha$ is not free $\}$. By inserting $< \mu_1$ members between H_α and $H_{\alpha+1}$ we may assume that $H_{\alpha+1}/H_\alpha \cong R$ if $\alpha \notin \boldsymbol{S}_{\mu_1}$. Moreover **we assume** $\boldsymbol{S}_{\mu_1} \subseteq E_{\mu_2} := \{\alpha \in \mu_1, \; cf\alpha = \mu_2\}$. If $\alpha \in \boldsymbol{S}_{\mu_1}$, then $H_{\alpha+1}/H_\alpha$ is not free and we proceed similarly. We can find a continuous chain

$$H_\alpha = H_{\alpha 0} \subseteq \ldots \subseteq H_{\alpha\beta} \subseteq \ldots \subseteq H_{\alpha\mu_2} = H_{\alpha+1} \quad (\beta < \mu_2)$$

with quotients $H_{\alpha\beta+1}/H_{\alpha\beta}$ either free or of type $\langle \mu_3, \cdots, \mu_n \rangle$. Recursively we unravel all of $\langle \mu_1, \cdots, \mu_n \rangle$ and let $u^i = \mu_1 \times \cdots \times \mu_i$ $(i = 1, \cdots, n)$. We obtain a refined, continuous chain $\{H_\alpha, \; \alpha \in u^i\}$ of free R-modules H_α with $H_\beta/H_{\alpha+1}$ free for $\beta \geq \alpha+1$. Moreover $H_{\alpha+1}/H_\alpha$ is either $\cong R$ or of type $\langle \mu_{i+1}, \cdots, \mu_n \rangle$. The set u^i is ordered lexicographically and each $\alpha \in u^i$ is of the form $\alpha = \alpha_1^\wedge \cdots {}^\wedge\alpha_i$ with $\alpha_j < \mu_j$. As usual, $\alpha + 1$ will denote the successor of α in u^i which is $\alpha + 1 = \alpha_1^\wedge \cdots {}^\wedge\alpha_i + 1$, and 0 denotes the initial element of u^i.

Moreover, again refining the chain $\{H_\alpha : \alpha \in u^n\}$, we may assume that $H_{\alpha+1}/H_\alpha \subseteq S^{-1}R$ for each $\alpha \in u^n$ and $H_0 = 0$. We will identify u^n with μ in §§3 - 6 and write for $\boldsymbol{S} = \{\alpha \in \mu : H_{\alpha+1}/H_\alpha$ is not free $\}$, hence

$$\textbf{(2.2)} \begin{cases} \text{either } H_{\alpha+1} = H_\alpha \oplus t_{\alpha 0}R, \text{ if } \alpha \notin \boldsymbol{S} \text{ or} \\ H_{\alpha+1} = \langle H_\alpha, t_{\alpha k} : k \in \omega \rangle \text{ is a free R-module with } q_{\alpha k} t_{\alpha k+1} = t_{\alpha k} + \sum_{j<k_\alpha} t_{\alpha_j k_j} \\ \text{where } q_{\alpha k} \in S, \; t_{\alpha_j k_j} \in H_{\alpha_j+1} \text{ for } \alpha_j < \alpha \text{ and } k < \omega. \end{cases}$$

REMARK The reader not interested in particular freeness-type or in modules of rank $\aleph_{\omega+1}$ may choose (2.2) immediately from $\{H_\alpha, \; \alpha \in u^1\}$. Recall that H is uniquely determined (up to isomorphism) by these generators and relations (as follows by a transfinite induction on $\alpha < \mu$). Inductively we will also use particular free resolutions.

DEFINITION 2.3 If $u = \langle \mu_1, \cdots, \mu_n \rangle$ is a (finite) descending sequence of regular cardinals, then we say that the R-module M has a u-resolution for $\kappa \in u$, if $\mu_1 < \mu$ and the following holds:

(1) M is strongly μ-free of cardinality μ.

(2) We have a free resolution $M \cong F/K$.

(3) $K = \bigoplus_{k\in\kappa} K_k$ and F, K_k are free R-modules of rank μ.

(4) K_k has a basis $\{x_{\alpha k} : \alpha < \mu\}$ and if
$K_{\beta k} = \langle x_{\alpha j} : j < k \text{ or } \alpha < \beta\rangle$ *for* $(k < \kappa)$ $(\beta < \mu)$, then $F/K_{\beta k}$ is free.

Sometimes R in (2.3) will be replaced by an R-algebra A, provided M is a strongly μ-free A-module.

3 A STEP-LEMMA AND A FREENESS-PROPOSITION

THE μ-$\aleph_0$-STEP-LEMMA 3.1 Let H be a strongly μ-free R-module of cardinality μ and $H^* = 0$. Moreover, let $F = \oplus_{n\in\omega} K_n$ be a direct sum of free modules K_n of rank μ. Choose a partial basis $x_{\alpha n}$ $(\alpha < \mu)$ for each K_n, let $K'_n = \langle x_{\alpha n} : \alpha < \mu\rangle$ and let $f \in K_0$ be any pure element with K_0/fR free. Then we can find two extensions $F^i \supset F$ $(i = 0, 1)$ with the following properties.

(1) If $K_{\beta n} = \langle x_{\alpha i} : i < n \text{ or } \alpha < \beta\rangle = \oplus_{i<n} K'_i \oplus \langle x_{\alpha i} : i \geq n,\ \alpha < \beta\rangle$ then $F^i/K_{\beta n}$ is free R-module for each $n \in \omega$, $\beta < \mu$.

(2) $F^i/F \cong H$ $(i = 0, 1)$.

(3) If $\varphi \in F^*$ extends to both $\varphi^i \in F^{i*}$ $(i = 0, 1)$, then $f\varphi = 0$.

REMARKS It follows from (1) that F^i is free, because $K_{\beta n}$ is free as well. The particular case $\beta = 0$ with $K_{0n} = \oplus_{i<n} K_i$ looks more familiar in connection with older Step-Lemmas, but we need more – that is stated in (1). Conditions (1) and (2) say that H has an $\langle\aleph_0\rangle$-resolution for $\aleph_0$ as defined in (2.3).

Later on we will need a more general Step-Lemma as well, in order to deal with abelian groups of cardinality $\geq \aleph_{\omega+1}$. This however is postponed to Section 7.

Proof: Recall our representation (2.2) of H. We have

$$(i) \left\{ \begin{array}{rcl} H_{\alpha+1} & = & \langle H_\alpha, t_{\alpha n} : n \in \omega\rangle \text{ with} \\ q_{\alpha n} t_{\alpha n+1} & = & t_{\alpha n} + \sum_{j<n_\alpha} t_{\alpha_j n_j} \text{ if } \alpha \in \boldsymbol{S} \text{ and} \\ H_{\alpha+1} & = & H_\alpha \oplus t_\alpha R \text{ if } \alpha \in \mu\backslash \boldsymbol{S} \end{array} \right.$$

where $q_{\alpha n} \in S$, $t_{\alpha_j n_j} \in H_{\alpha_j+1}$ for $\alpha_j < \alpha$. Let $C^i = \bigoplus_{\alpha\in \boldsymbol{S}} \bigoplus_{n\in\omega} s^i_{\alpha n} R \oplus \oplus_{\alpha\in\mu\backslash \boldsymbol{S}} s^i_\alpha R$ be freely generated by elements $s^i_{\alpha n}$ and s^i_α for $i = 0, 1$. Next we choose two submodules $N^i \subseteq F \oplus C^i$ by generators

$$N^0 = \langle q_{\alpha n} s^0_{\alpha n+1} - s^0_{\alpha n} - \sum_{j<n_\alpha} s^0_{\alpha_j n_j} - x_{\alpha n} : \alpha \in \boldsymbol{S},\ n \in \omega\rangle$$

and

$$N^1 = \langle q_{\alpha n}s^1_{\alpha n+1} - s^1_{\alpha n} - \sum_{j<n_\alpha} s^1_{\alpha_j n_j} - x_{\alpha n} - f : \alpha \in \boldsymbol{S},\ n \in \omega\rangle,$$

where $f \in K_0$ is taken from the hypothesis of the Lemma. If $F^i = F \oplus C^i/N^i$, then we identify $x \in F \oplus C^i$ with $x + N^i$ in F^i, hence $F \subset F^i$ $(i = 0, 1)$ and the following relations hold in F^i by choice of N^i:

$$(r^i) \quad q_{\alpha n}s^i_{\alpha n+1} = s^i_{\alpha n} + \sum_{j<n_\alpha} s^i_{\alpha_j n_j} + x_{\alpha n} + f \cdot \delta_{1i} \qquad (i = 0, 1)$$

where $\delta_{11} = 1$ and $\delta_{10} = 0$.

In order to show (2), we consider the maps $\psi_i : F^i \to H$ defined by $\psi_i|F = 0$ and $s^i_{\alpha n}\psi_i = t_{\alpha n}$ for all $\alpha \in \boldsymbol{S}$, $n \in \omega$, $s^i_\alpha\psi_i = t_\alpha$ for $\alpha \in \mu\backslash\boldsymbol{S}$ and $i = 0, 1$. These maps preserve the relations (r^i) and (i), hence ψ_i can be extended to an R-homomorphism $\psi_i : F^i \to H$. Its kernel is clearly F, ψ is epic and (2) follows.

We have a decomposition into free summands $K'_n \oplus L_n = K_n$ where $K'_n = \langle x_{\alpha n} : \alpha < \mu\rangle$, and L_n comes from the complement of the partial basis, hence

$$(ii)\begin{cases} F &= \oplus_{m<n}K_m \oplus \oplus_{m\geq n}L_m \oplus \langle x_{\alpha m} : \alpha < \mu,\ m \geq n\rangle \text{ and} \\ F^0 &= \oplus_{m<n}K_m \oplus \oplus_{m\geq n}L_m \oplus C \text{ where} \\ C &= \langle x_{\alpha m}, s^0_{\alpha k}, s^0_\beta : \alpha \in \boldsymbol{S},\ \beta \in \mu\backslash\boldsymbol{S},\ m \geq n,\ k \in \omega\rangle \end{cases}$$

Using (r^0) we see that the $x_{\alpha m}$'s can be expressed by $s^0_{\alpha k}$'s, hence C is also generated by

$$C = \langle s^0_{\alpha k}, s^0_\beta : \alpha \in \boldsymbol{S},\ \beta \in \mu\backslash\boldsymbol{S},\ k < \omega\rangle$$

If $i = 0$ is replaced by $i = 1$ we note that $f \in K_0$ and a similar submodule C is a new complement satisfying (ii). In order to show (1), we consider the case $i = 0$ only. Let $^- : F^0 \to F^0/K_{\beta n}$ denote the canonical epimorphism. The first step in proving that $\overline{F}^0$ is free is

$$(iii)\begin{cases} \overline{C}_{n\beta} := \langle \overline{s}^0_{\alpha k}, \overline{s}^0_{\alpha'} : \alpha' \in \mu\backslash\boldsymbol{S} \text{ and } \alpha \in \boldsymbol{S} \text{ such that } k < n \text{ or } k \geq n \\ \text{and } \alpha', \alpha < \beta\rangle \subseteq \overline{C} \text{ is freely generated.} \end{cases}$$

Note that $C_\gamma := \langle \overline{s}^0_{\alpha k}, \overline{s}^0_{\alpha'} : \alpha \in \boldsymbol{S},\ \alpha' \in \mu\backslash\boldsymbol{S}$ with $k < n,\ \alpha', \alpha < \gamma$ or $k \geq n,\ \alpha',\ \alpha < \beta\rangle$ $(\beta \leq \gamma < \mu)$ is an ascending, continuous chain of submodules of $\overline{C}_{n\beta}$ with $\overline{C}_{n\beta} = \bigcup_{\gamma<\mu} C_\gamma$. Because $|C_\gamma| \leq |\gamma| + |\beta| + \aleph_0 < \mu$ and $C_\gamma \subseteq \overline{F}^0$ is strongly μ-free by (2), as an extension of a free module by a strongly μ-free module, each C_γ is free as well. Now (iii) will follow if we show

$$C_{\gamma+1}/C_\gamma \cong \langle \overline{s}^0_{\gamma n-1} + C_\gamma\rangle \quad (\cong R)$$

for each $\gamma \geq \beta$. Observe that this is trivial if $\gamma \notin \boldsymbol{S}$, hence we assume $\gamma \in \boldsymbol{S}$. For simplicity we will ignore the elements s^i_α $(\alpha \notin \boldsymbol{S})$! The relation (r^0) under $^-$ become

$$q_{\gamma k}\overline{s}^0_{\gamma k+1} = \overline{s}^0_{\gamma k} + \sum_{j<n_\gamma} \overline{s}^0_{\gamma_j n_j},$$

as in (i), and $C_{\gamma+1}/C_\gamma = \langle \overline{s}^0_{\gamma k} + C_\gamma : k < n\rangle$. For large enough $q \in S$ we get $q\overline{s}^0_{\gamma n-1} = q'\overline{s}^0_{\gamma 0} + \sum_j q_j\overline{s}_{\beta_j 0}$ and $\sum_j q_j\overline{s}_{\beta_j 0}$ is in C_γ, $q'|q$. Hence the n generators above can be replaced by $C_{\gamma+1}/C_\gamma = \langle \overline{s}^0_{\gamma n-1} + C_\gamma\rangle$ and (iii) holds.

Next we show that

$$(iv)\begin{cases} \overline{C}_{n\beta} \oplus \overline{C}^{n\beta} = \overline{C} \text{ where } \overline{C}^{n\beta} = \langle \overline{s}^0_{\alpha k} :\ \beta \le \alpha < \mu,\ k \ge n \rangle \\ \text{is freely generated by the } \overline{s}^0_{\alpha k}s. \end{cases}$$

Clearly $\overline{C}$ is generated by $\overline{C}_{n\beta}$ and $\overline{C}^{n\beta}$

Any linear combination $s = \sum r_{\alpha k}\overline{s}^0_{\alpha k} \in \overline{C}^{n\beta}$ may be multiplied by a large enough $q \in S$ such that the summand $qr_{\alpha k}\overline{s}^0_{\alpha k}$ by (r^0) becomes a sum with summand $q'r_{\alpha k}x_{\alpha k}$. Now $s \in \overline{C}_{n\beta}$ is only possible if all the independent terms $q'r_{\alpha k}x_{\alpha k}$ $(\alpha \ge \beta,\ k \ge n)$ vanish, hence $r_{\alpha k} = 0$ and $s = 0$. The sum (iv) must be direct. The freeness of $\overline{C}^{n\beta}$ follows similarly, again using independence of the $x_{\alpha k}$ under $^-$, and (iv) follows.

The quotient $\overline{F}^0/\overline{C}_{n\beta} \cong \bigoplus_{m\ge n} L_m \oplus \overline{C}^{n\beta}$ is free by (iv) and $\overline{F}^0$ is free by (iii) and (1) is shown for $i = 0$.

In order to show (3) we consider any $\varphi \in F^*$ extending to $\varphi^i \in F^{i*}$ for $i = 0, 1$ and recall that $f \in F_0 \subseteq F$ is a prescribed basic element by hypothesis.

First we consider the pushout D of $F \subset F^i$, which is

$$D = F^0 \oplus F^1/\,\Delta \ \text{ with } \Delta = \{(g, -g) :\ g \in F\} \cong F.$$

We also have a pushout extension $\varphi^i \subset \Phi \in D^*$ $(i = 0, 1)$ by the pushout properties and $\varphi \subseteq \varphi^i$.

If $\overline{g} = (g, 0) + \Delta$, $\overline{g} = (0, g) + \Delta$ for $g \in F^0$ and $g \in F^1$ respectively, then we have canonical identifications $\overline{F^0} + \overline{F^1} = D$ with $\overline{F} = \overline{F^0} \cap \overline{F^1} = \{(g, 0) + \Delta :\ g \in F\} =$ $\{(0, g) + \Delta :\ g \in F\}$. We omit "$^-$" in the following. Let
$U = \langle F, d_{\alpha n}, d_{\alpha'} :\ \alpha \in \boldsymbol{S},\ n < \omega,\ \alpha' \in \mu\backslash\boldsymbol{S} \rangle \subseteq D$ for $d_{\alpha n} = s^0_{\alpha n} - s^1_{\alpha n}$, $d_\alpha = s^0_\alpha - s^1_\alpha$.
Observe that

$$U + F^i = F^0 + F^1 = D \text{ and } U \cap F^i = F^0 \cap F^1 = F.$$

The desired relations from (r^i) are

$$(v)\begin{cases} q_{\alpha n}d_{\alpha n+1} = d_{\alpha n} + \sum_{i<n_\alpha} d_{\alpha_i n_i} + f. \\ \text{The map } t_{\alpha n} \to d_{\alpha n} + \langle f \rangle \text{ induces } V/\langle f \rangle \cong H, \text{ with} \\ V = \langle f,\ d_{\alpha n},\ d_{\alpha'} :\ \alpha \in \boldsymbol{S},\ n \in \omega,\ \alpha' \in \mu\backslash\boldsymbol{S} \rangle. \end{cases}$$

In particular, $U/F \cong H$. Let $f\varphi = r \in R$ and consider the new homomorphism $\psi = \Phi - r :$ $D \to R$.

Now we have $f\psi = f(\Phi - r) = 0$ and the relations (v) become

$$q_{\alpha n}(d_{\alpha n+1}\psi) = (d_{\alpha n}\psi) + \sum_{i<n_\alpha} (d_{\alpha_i n_i}\psi)$$

and $t_{\alpha n} \xrightarrow{\eta} d_{\alpha n}\psi$, $f\psi = 0$, induces an epimorphism

$$\eta : H \to V\psi$$

and $H^* = 0$ forces $V\psi = 0$. Hence $\Phi|V$ is multiplication by r, which by cardinality reasons is only possible for $r = 0$. We conclude $f\varphi = 0$.

The following Freeness-Proposition is a special case of a more general Freeness-Proposition. In order to make the idea of the construction more transparent, we postpone again the extension of (3.2) to Section 7. The more general Freeness-Proposition 7.2 involves a few more technical difficulties, due to more complicated notation. The Proposition will ensure that continuous chains of free submodules in our construction remain free at limit ordinals below μ^+.

THE ω-FREENESS-PROPOSITION 3.2 Let μ be a regular cardinal and $\delta < \mu^+$ be an ordinal with $\{G_i : i \leq \delta\}$ an ascending, continuous chain of submodules with the following properties for all $i < \delta$.

(a) G_i is free of rank μ.

(b) G_{i+1}/G_i is strongly μ-free and let $\boldsymbol{S} = \{i < \delta : G_{i+1}/G_i \text{ is not free }\}$

(c) There exist $G_i \subseteq H_i \subseteq G_{i+1}$ and H_i^0 a free summand of G_i with free quotient such that $H_i = G_i \oplus_{H_i^0} H_i^1$ is a pushout with $H_i^0 \subseteq H_i^1$ free modules of rank μ.

(d) $G_{i+1} = H_i \oplus K_i$ with K_i freely generated by $x_{\alpha i}$ $(\alpha < \mu)$.

(e) If $i \notin \boldsymbol{S}$, then $G_i = H_i$ and $H_i^1 = H_i^0$ (hence $G_{i+1} = G_i \oplus K_i$).

(f) If $i \in \boldsymbol{S}$, then cf(i) $= \omega$. Moreover

 (i) there exists $a_i \subseteq i$ of order type ω with $a_i \cap \boldsymbol{S} = \emptyset$ and $\mathit{sup}\ a_i = i$.

 (ii) there exist $A_{ij} \subseteq \mu$ for all $j \in a_i$ such that $A_{ij} \cap A_{ij'} = \emptyset$ for $j \neq j'$ and $A_{ij} \cap A_{i'j'}$ is bounded in μ for $(ij) \neq (i'j')$.

 (iii) Let $K_j^\omega = \langle x_{\alpha j} : \alpha \in A_{ij}\rangle$, $H_i^0 = \bigoplus_{j\in a_i} K_j^\omega$. If $k \in a_i$, $\beta < \mu$ and $K_{\beta k} = \bigoplus_{j\in a_i}^{j<k} K_j^\omega \oplus \langle x_{\alpha j} : j \in a_i,\ j \geq k,\ \alpha < \beta\rangle$, then $H_i^1/K_{\beta k}$ is free.

It follows that G_δ is free as well.

Proof: Enumerate $\boldsymbol{S} = \{i_\epsilon : \epsilon < \sigma\}$ for some $\sigma \leq \mu$. By induction on $\epsilon < \sigma$ choose ordinals $\alpha_\epsilon < \mu$ and suppose $\alpha_0, \cdots, \alpha_\nu$ are defined for all $\nu < \epsilon$. Define α_ϵ subject to the following condition

$$(+)\ \text{For all } \nu < \epsilon,\ j \in a_{i_\nu} \cap a_{i_\epsilon} \text{ we have } A_{i_\epsilon j} \cap A_{i_\nu j} = \emptyset.$$

This can be arranged: we collect all troublemakers at stage ϵ, which is a set $M = \bigcup_{\nu<\epsilon} \bigcup_{j\in a_{i_\nu}\cap a_{i_\epsilon}} (A_{i_\nu j} \cap A_{i_\epsilon j}) \subseteq \mu$, and observe that $A_{i_\nu j} \cap A_{i_\epsilon j}$ is bounded below μ by (f)(ii), μ is regular and $|\epsilon| < \mu$. Hence M is bounded by some $\alpha_\epsilon < \mu$ and (+) follows. Rename $\alpha_\epsilon = \alpha_{i_\epsilon}$. Then (+) can be restated as

$$(+)\quad \text{For all } i \neq t \in \boldsymbol{S},\ j \in a_i \cap a_t \text{ we have } A_{tj}\backslash\alpha_t \cap A_{ij}\backslash\alpha_i = \emptyset.$$

If $i \in \boldsymbol{S}$, choose any fixed $\beta_i < \mu$ such that $\alpha_i < \beta_i$ and observe that $H_i^0 = K_i^* \oplus C_i$ by (f)(iii) for $k = \mathit{min}\ a_i$, $\beta = \beta_i$ and $K_{k\beta} = K_i^* := \langle x_{\alpha j} : j \in a_i,\ \alpha < \beta_i\rangle$.
Now we are able to collect a basis of G_δ: Choose all basis elements $x_{\alpha j}$ not involved in extension of G_i at $i \in \boldsymbol{S}$, and we have the set

$B = \{x_{\alpha j} : j < \delta$, there is no $i \in \boldsymbol{S}$ with $j \in a_i$, $\alpha \in A_{ij}$ and $\alpha > \beta_i\}$.

Moreover $C = \bigoplus_{i \in \boldsymbol{S}} C_i$ is direct because of $(+)$, and each C_i is free by (f)(iii). We may collect its basis B_C. Finally choose all basis elements $x_{\alpha j}$ $(\alpha < \beta_i)$ which are used for the distinct K_i^*, say B'. Then $B \cup B_C \cup B'$ is a basis for G_δ.

4 THE PREDICTION PRINCIPLES

We will use prediction principles that hold under G.C.H. First recall some well-known definitions, tailored for our applications. Let κ be a regular cardinal and $\theta = \{A_\nu : \nu < \kappa\}$ be a κ-filtration of $A = \bigcup_{\nu<\kappa} A_\nu$, i.e. θ is an ascending continuous chain with $A_0 = 0$ and $|A_\nu| < \kappa$ for all $\nu < \kappa$. Let $E \subseteq \kappa$. Then $\diamondsuit(E)$ holds if there is a family $\{S_\nu \subseteq A_\nu : \nu \in E\}$ such that for all $X \subseteq A$ the set $\{\nu \in E : X \cap A_\nu = S_\nu\}$ is stationary in κ. Besides this diamond principle, we also consider the weak diamond $\Phi_\kappa(E)$. If $P_\nu : \mathcal{P}(A_\nu) \to 2 = \{0, 1\}$ $(\nu \in E)$ is a given partition of subsets of A_ν, then $\Phi_\kappa(E)$ provides a prediction function $\varphi : E \to 2$ such that for all $X \subseteq A$ the set $\{\nu \in E : P_\nu(X \cap A_\nu) = \varphi(\nu)\}$ is stationary in κ. In this case E is called non-small. While $\diamondsuit(E)$ holds in $V = L$ for all stationary subsets E of regular, not weakly compact cardinals by Jensen [30], prediction principles are quite often valid under weaker assumptions, cf. Eklof, Mekler [15, pp.175-178], Shelah [35, p.376, Theorem 32]. See also Gregory [24].

PPROPOSITION 4.1 (Shelah [35]) Suppose $\lambda = 2^\nu = \mu^+$ and for some regular cardinal $\kappa < \mu$ either

(i) $\mu^\kappa = \mu$ or

(ii) μ is singular with $cf\mu \neq \kappa$ and $|\delta|^\kappa < \mu$ for all $\delta < \mu$

Then $\diamondsuit(E_\kappa)$ holds, where $E_\kappa = \{\alpha < \lambda : cf\alpha = \kappa\}$.

In particular we have a

COROLLARY 4.2 Assume G.C.H. and μ is some regular cardinal. Then $\diamondsuit_\kappa$ holds for any regular cardinal $\kappa < \mu$.

In the first application we will use $\Phi_\mu(E_{\aleph_0})$, a weak consequence of (4.2). Later we will replace $\kappa = \aleph_0$ by some regular $\kappa > \aleph_0$ (cf. §7). Moreover we will use

PROPOSITION 4.3 If $\lambda = \mu^+$ and $E \subseteq \lambda$ is non-small, then we can decompose $E = \coprod_{\beta<\kappa} E_\beta$ and each E_β is non-small as well.

Proof: See Eklof and Mekler [15, p.144].

5 CONSTRUCTION OF MODULES WITH TRIVIAL DUAL

We apply the $\mu - \aleph_0$-Step-Lemma 3.1, the ω-Freeness-Proposition 3.2 and combinatorial ideas from §4. The arguments are however partly standard (for working in L), so we keep them short. The reader may consult [15], [38] or [8] for details.

Let $\lambda = \mu^+ = 2^\mu$ for some regular cardinal μ, and let $E = \{\alpha \in \lambda,\ cf\alpha = \omega\}$. Recall that E is non-small by (4.1), for instance. Hence we assume $\Phi_\lambda(E)$, the weak diamond and decompose $E = \coprod_{\beta<\lambda} E_\beta$ into non-small sets E_β, cf. (4.3). For each $i \in E$ we also fix $a_i = \{i_n : n \in \omega\}$, a strictly increasing sequence $i_n < i$ of successor ordinals i_n with $sup\ a_i = i$, hence $E \cap a_i = \emptyset$. For each $j \in a_i$ we find $A_{ij} \subseteq \mu$ with:

(a) A_{ij} unbounded in μ and $A_{ij} \cap A_{i'j'}$ bounded in μ if $(ij) \neq (i'j')$ and

(b) $A_{ij} \cap A_{ij'} = \emptyset$ for $j \neq j' \in a_i$.

Choose a set of $2^\mu = \lambda$ subsets $X', X \subseteq \mu$ with $|X \cap X'| < \mu$ for distinct pairs X, X'. Decompose each X into ω subsets of cardinal μ to define the A_{ij}'s. Hence (b) follows and (a) holds because μ is regular. Suppose now that H is a given strongly μ-free R-module with $|H| = \mu$ and trivial dual $H^* = Hom\ (H, R) = 0$.

We want to derive the existence of a strongly λ-free R-module G of cardinality λ such that $G^* = 0$.

We have to fix the "prediction" and choose a λ-filtration $\{G_i : i < \lambda\}$ such that $G_0 = 0$ and $|G_i| = |G_{i+1} \backslash G_i| = \mu$ for all $i < \lambda$. Let $\{g_\nu : \nu < \lambda\} = G$ be an enumeration of $G = \bigcup_{i<\lambda} G_i$ such that $g_\beta \in G_i$ for all $i \in E_\beta$. We define partitions $P_i^\beta : \mathcal{P}(G_i \times R) \to 2$ for all $\beta < i$, $i \in E_\beta$ predicting homomorphisms in G_i^*.

Formally we also must predict an R-module structure on G_i, hence $G_i \times R$ must be replaced by $G_i \times G_i \times \cdots \times R$ to take care of this; however we will ignore this for simplicity; compare [8] or [19, p.284].

If $X \subseteq G_i \times R$, then define $P_i^\beta(X) = 0$ if the following holds.

(1) $G_i, G_j, G_i/G_j$ are free R -modules for $j \in a_i$ and $g_\beta \in G_{j_0}$, with $G_{j_0}/g_\beta R$ free, say $j_0 = min\ a_i$. Moreover $G_{j+1} = G_j \oplus K_j$ with K_j freely generated by some $x_{\alpha j}$ $(\alpha < \mu)$.

(2) X is (as a graph) a homomorphism $h_X \in G_i^*$. If $K_j^\omega = \langle x_{\alpha j} : \alpha \in A_{ij}\rangle (j \in a_i)$ and if we identify $(G_i,\ \oplus_{j<n} K_j^\omega,\ g_\beta,\ h_X : n \in \omega)$ with $(F,\ \oplus_{j<n} K_j',\ f,\ \varphi : n \in \omega)$ in (3.1), then we ***require*** that φ does not extend to F^{0*}.

From $\Phi_\lambda(E_\beta)$ we have prediction functions $\varphi_\beta : E_\beta \to 2$ such that

$$\mathcal{X}_\beta(X) := \{i \in E_\beta,\ P_i^\beta(X \cap G_i \times R) = \varphi_\beta(i)\}$$

is stationary in λ for all $X \subseteq G$, $\beta < \lambda$. The structure of the R-module $G = \bigcup_{i<\lambda} G_i$ is given inductively on the continuous chain $\{G_i : i < \lambda\}$ subject to the following conditions.

(i) G_i is a free R-module for any $i \in \lambda$.

$$(ii) \begin{cases} G_{i+1} = H_i \oplus K_i \text{ where } K_i \text{ is freely generated by elements } x_{\alpha i}\ (\alpha < \mu). \\ G_i \subseteq H_i \subseteq G_{i+1},\ G_{i+1}/G_i \text{ is strongly } \mu\text{-free and} \\ H_i = G_i \oplus_{H_i^0} H_i^1 \text{ is a pushout for some free summand } H_i^0 = G_i \cap H_i^1 \text{ of } G_i \\ \text{with free quotient and } H_i^0 \subseteq H_i^1 \text{ all free of rank } \mu. \end{cases}$$

$$(iii) \begin{cases} \text{If } i \in E_\beta \text{ and (1) holds, then we choose } H_i \text{ as an extension } F^{\varphi_\beta(i)} \\ \text{given by the identification } (F,\ \oplus_{j<n} K'_j,\ f,\ \varphi : n \in \omega) \text{ in (3.1)} \\ \text{with } (G_i,\ \oplus_{j<n} K_j^\omega,\ g_\beta,\ h_X : n \in \omega). \end{cases}$$

Let $H_i = G_i$ otherwise. Hence G is defined.

Note that $G_\delta = \bigcup_{i<\delta} G_i$ is a free R-module for limit ordinals $\delta < \lambda$ as follows from (3.2). We conclude that G is a strongly λ-free R-module of cardinality λ by (3.1) and the construction above.
If $\psi \in G^*$ and $\psi \neq 0$, then $C = \{i < \lambda : \psi|G_i \neq 0\}$ is a cub. Moreover $f\psi \neq 0$ for some $f \in G$ with G_i/fR free for large enough $i \in C$, and $f = g_\beta \in G_i$ for some $\beta < \lambda$. Choose an ordinal $i \in C \cap \mathcal{X}_\beta(\psi)$ and observe that $\psi|G_i$ extends to G_{i+1}, hence $P_i^\beta(\psi|G_i) = 1$ and $G_{i+1} = F^1$ as in the Step-Lemma 3.1. However $P_i^\beta(\psi|G_i) = 1$ and (2) tell us that $\psi|G_i$ extends to F^0 as in (3.1) as well. Therefore $f\psi = 0$ by (3.1), a contradiction. Hence $G^* = 0$ follows. We conclude the following results.

THEOREM 5.1 Let R be a countable commutative ring as in §2. If H is a strongly μ-free R-module of regular cardinality μ with trivial dual $H^* = 0$ and $\lambda = 2^\mu = \mu^+$, then there exists a strongly λ-free R-module of cardinality λ with trivial dual.

COROLLARY 5.2 If $2^{\aleph_i} = \aleph_{i+1}$ for $i = 0, \cdots, n-1$, then there exists a strongly $\aleph_n$-free abelian group of cardinality $\aleph_n$ with trivial dual.

The corollary follows by induction. If $i = 0$, then any subgroup of $\mathbb{Q}$ different from $\mathbb{Z}$ serves as H in (5.1). From (5.1) we obtain a strongly $\aleph_1$-free abelian group of cardinality $\aleph_1$ with trivial dual, and so on.

6 ENDO-RIGID MODULES

The following result establishes the existence of R-modules with endomorphism ring R, the so-called rigid R-modules. Such modules are known in ZFC, see e.g. Corner [2] or [20], [21], [40]. However we will also assume that the R-modules in question are almost free. A result in ZFC + MA (= Martin's axiom) shows that extra axioms are needed, because such modules turn out to be separable under ZFC + MA, see Göbel, Shelah [22]. Assuming $V = L$, almost free modules were constructed in Dugas [7], Shelah [38] and Dugas, Göbel [8] several years ago. Here we have to work harder to derive the existence under ZFC and a weak form of G.C.H.

THEOREM 6.1 Let R be a countable, commutative ring as in (2.1) and assume $\lambda = 2^\mu = \mu^+$ for some regular cardinal μ. Suppose there exists a strongly μ-free R-module H of cardinality μ with $End\ H = R$. Then we can find a strongly λ-free R-module G of cardinality λ with $End\ G = R$.

The proof is – as in (5.1) – divided into three parts, a Step-Lemma similar to (3.1), however with crucial, not-so-obvious changes. The construction of G and the actual proof of the claims in (6.1) are similar to §5. We will keep the proof short. However, the reader familiar with §§3-5 will follow the arguments easily. We begin with the Step-Lemma.

THE μ-$\aleph_0$-STEP-LEMMA 6.2 Let R be as in §2 and let H be a strongly μ-free R-module of cardinality μ with $End\ H = R$. Moreover, let $F = \oplus_{n\in\omega} K_n$ be a direct sum of free modules K_n of rank μ. Choose a partial basis $x_{\alpha n}$ ($\alpha < \mu$) for each K_n, let $f \in K_0$ and $h \in H$ such that K_0/fR is free and H/hR is strongly μ-free. Then we can find two extensions $F \subseteq F^i$ ($i = 0, 1$) with the following properties.

(1) If $K_{\beta n} = \langle x_{\alpha i} : i < n \text{ or } \alpha < \beta \rangle$, then $F^i/K_{\beta n}$ is a free R-module for each $n \in \omega$, $\beta < \mu$.

(2) $F^i/F \cong H/hR$ is strongly μ-free.

(3) If $\varphi \in End\ F$ extends to both $\varphi^i \in End\ F^i$ ($i = 0, 1$), then $f\varphi \in fR$.

Proof: Let $H' = H/hR$ and choose a μ-filtration $H = \bigcup_{i<\mu} H_i$ for strong μ-freeness with $h \in H_0$. Since H/hR is strongly μ-free, also $H'_i = H_i/hR$ is free and $\{H'_i : i < \mu\}$ constitutes a μ-filtration of H' for strong μ-freeness.

Let $\boldsymbol{S} = \{\alpha < \mu \text{ with } H_{\alpha+1}/H_\alpha \text{ not free }\}$. Next we apply the representation (2.2) to H' and obtain

$$(*)\begin{cases} \text{a continuous chain of free submodules } H_\alpha\ (\alpha < \mu) \text{ of } H = \bigcup_{\alpha<\mu} H_\alpha \\ \text{such that } H_0 = hR \text{ and } H_{\alpha+1} = H_\alpha \oplus t_\alpha R \text{ if } \alpha \notin \boldsymbol{S} \\ \text{or } H_{\alpha+1} = \langle H_\alpha, t_{\alpha n} : n < \omega\rangle \text{ with } q_{\alpha n} t_{\alpha n+1} = t_{\alpha n} + \sum_{j<n_\alpha} t_{\alpha_j n_j} + r_{\alpha n} h \\ \text{for } q_{\alpha n}, r_{\alpha n} \in S,\ t_{\alpha_j n_j} \in H_{\alpha_j+1} \text{ with } \alpha_j < \alpha,\ n < \omega. \end{cases}$$

Similar to (3.1), but identifying f and h, we define two extensions F^i of F for $i = 0, 1$. Let $C^i = \bigoplus_{\alpha\in\boldsymbol{S}} \bigoplus_{n<\omega} s^i_{\alpha n} R \oplus \bigoplus_{\alpha\in\mu\setminus\boldsymbol{S}} s^i_\alpha R$ be freely generated and $N^i \subseteq F \oplus C^i$ such that

$$\begin{aligned} N^0 &= \langle q_{\alpha n} s^0_{\alpha n+1} - s^0_{\alpha n} - \textstyle\sum_{j<n_\alpha} s^0_{\alpha_j n_j} - x_{\alpha n} : \alpha \in \boldsymbol{S},\ n \in \omega\rangle \\ N^1 &= \langle q_{\alpha n} s^1_{\alpha n+1} - s^1_{\alpha n} - \textstyle\sum_{j<n_\alpha} s^1_{\alpha_j n_j} - x_{\alpha n} - r_{\alpha n} f : \alpha \in \boldsymbol{S},\ n \in \omega\rangle \end{aligned}$$

Let $F^i = F \oplus C^i/N^i$ and identify $x \in F \oplus C^i$ with $x + N^i$ in F^i. The new relations are

$$(r^i) \quad q_{\alpha n} s^i_{\alpha n+1} = s^i_{\alpha n} + \sum_{j<n_\alpha} s^i_{\alpha_j n_j} + x_{\alpha n} - r_{\alpha n}\delta_{1i} f \quad (i = 0, 1)$$

The maps $\psi_i : F^i \to H/hR$ given by $\psi_i | F = 0$, $s^i_\alpha \psi_i = t_\alpha + hR$ and $s^i_{\alpha n}\psi_i = t_{\alpha n} + hR$ preserve the relation in $(*)$ modulo hR and are epic, hence (2) follows.

The proof of (1) is the same as in (3.1).

Suppose $\varphi \in End\ F$ extends to $\varphi^i \in End\ F^i$ ($i = 0, 1$). Again we consider the pushout

$F^0 + F^1 = D$ with $F = F^0 \cap F^1$ as in (3.1), and let
$U = \langle F, d_{\alpha n}, d_{\alpha'} : \alpha \in \boldsymbol{S}, n \in \omega, \alpha' \in \mu \backslash \boldsymbol{S} \rangle \subseteq D$
for $d_{\alpha n} = s^0_{\alpha n} - s^1_{\alpha n}$ and $d_\alpha = s^0_\alpha - s'_\alpha$. We also have $U + F^1 = D$ and $U \cap F^1 = F$, however the relations for $d_{\alpha n}$ from (r^i) are $q_{\alpha n} d_{\alpha n+1} = d_{\alpha n} + \sum_{j<n_\alpha} d_{\alpha_j n_j} + r_{\alpha n} f$. Hence

$$V = \langle f, d_{\alpha n}, d_{\alpha'} : \alpha \in \boldsymbol{S}, n \in \omega \textit{ and } \alpha' \in \mu \backslash \boldsymbol{S} \rangle \cong H$$

by $(*)$. We find a projection $\pi : D \to V$, because - as in (3.1) - fR has a complement L_{-1} in $K_0 \subseteq F$ (cf. 3.1) and we find complements L_n of each K_n in K_{n+1}, hence $D \subseteq V \bigoplus \bigoplus_{n \geq -1} L_n$ gives π. The pushout D provides $\Phi \in End\ D$ extending φ^i $(i = 0, 1)$ and $\Phi' = \Phi|V \cdot \pi$ is an endomorphism of V. Finally $End\ H = R$ forces $f\varphi = f\Phi \in fR$.

Proof (6.1): The proof is similar to §5, however (3.1) is replaced by (6.2). We only indicate the changes. Assuming $\lambda = \mu^+ = 2^\mu$ we prepare the application of $\Phi_\lambda(E)$ for $E = \{\alpha \in \lambda,\ cf\alpha = \omega\}$. Condition (2) in §5 must be replaced by a new condition

$$(2) \begin{cases} X \text{ is an endomorphism } h_X \in End\ G_i. \text{ If } K^\omega_j = \langle x_{\alpha j} : \alpha \in A_{ij} \rangle\ (j \in a_i) \\ \text{and if we identify } (G_i,\ \oplus_{j<n} K^\omega_j,\ g_\beta,\ h_X : n \in \omega) \text{ with} \\ (F,\ \oplus_{j<n} K'_j,\ f,\ \varphi : n \in \omega) \text{ in (6.2), then we } \textbf{require} \text{ that } \varphi \text{ does} \\ \text{not extend to } End\ F^0. \end{cases}$$

In the construction of G we must change condition (iii). Now we have

$$(iii) \begin{cases} \text{If } i \in E_\beta \text{ and (1) holds, then we choose } H_i \text{ as an extension } F^{\varphi_\beta(i)} \\ \text{given by the identification } (F,\ \oplus_{j<n} K'_j,\ f,\ \varphi : n \in \omega) \text{ in (6.2)} \\ \text{with } (G_i, \oplus_{j<n} K^\omega_j,\ g_\beta,\ h_X : n \in \omega). \end{cases}$$

As in §5, the final R-module G has size λ and is strongly λ-free. In particular $R \subseteq End\ G$ by scalar multiplication. If $\psi \in End\ G \backslash R$, then we can find a cub

$$C = \{i < \lambda : \psi|G_i = \psi_i \in End\ G_i \backslash R\}.$$

Moreover we can choose $f \in G_i$ with G_i/fR free such that $f\psi \notin fR$ and $f = g_\beta$ by enumeration. Choose an ordinal $i \in C \cap \mathcal{X}_\beta(\psi)$ and observe that ψ_i extends to an endomorphism $\psi' \in End\ G_{i+1}$. The latter follows from the fact that $G_{i+1}\psi \subseteq G_j$ for some $j < \lambda$ and G_{i+1} is a summand of G_j, so $\psi' = \psi|G_{i+1} \cdot \pi$ for some projection π. Hence $P^\beta_i(\psi_i) = 1$ and $G_{i+1} = F^1$ as in the Step-Lemma 6.2. However $P^\beta_i(\psi_i) = 1$ and (2) tell us that ψ_i extends to F^0 as in (6.2) as well. Therefore $f\psi \in fR$ by (6.1), a contradiction, and $End\ G = R$ follows.

Inductively we derive a

COROLLARY 6.3 If $2^{\aleph_i} = \aleph_{i+1}$ for $i = 0, \cdots, n-1$, then there exists a strongly $\aleph_n$-free abelian group G with $|G| = \aleph_n$ and $End\ G = R$.

We also have a corollary from the proof of (6.1) observing that the R-algebra R can be replaced by an R-algebra A.

THEOREM 6.1* Let R be a countable, commutative ring which is S-reduced and S-torsion-free for some multiplicatively closed set S of regular elements in R. Assume $\lambda = 2^{\mu} = \mu^{+}$ for some regular cardinal μ and suppose there exists a strongly μ-free A-module H of cardinality μ with $End_R H = A$ and $|A| < \mu$, A_R free. Then we can find a strongly λ-free A-module G of cardinality λ with $End_R G = A$.

Hence we have a corollary which ensures counterexamples for Kaplansky's test problems, cf. Corner [3] and Corner, Göbel [4]. The next corollary strengthens (6.3).

COROLLARY 6.3* Let R be as in (6.1*) and assume $2^{\aleph_i} = \aleph_{i+1}$ for $i = 0, \cdots, n-1$. If A is a countable, free R-algebra, then there exists a strongly $\aleph_n$-free R-module G with $|G| = \aleph_n$ and $End\ G = A$.

7 ALMOST FREE ABELIAN GROUPS WHICH ARE OF CARDINALITY $> \aleph_\omega$

In order to derive results like the existence of strongly $|G|$-free groups G with $|G| = \aleph_{\omega+1}$ and $End\ G = \mathbb{Z}$ under G.C.H., the induction used in §§3, 4 and 5 breaks down at $\aleph_\omega$ and we must extend the basic results of the earlier sections in order to overcome this difficulty. The changes will be sketched below. If μ is a regular cardinal, then let $\boldsymbol{r}(\mu)$ be the set of all decreasing sequences $u = \langle \mu_1, \cdots, \mu_n \rangle$ of regular cardinals $\mu \geq \mu_1 > \cdots > \mu_n$. Recall the notion of a u-resolution from (2.3) and of the freeness type of an R-module H from §2. In case $End\ H = A$ we assume that H is a strongly μ-free A-module and replace R by A in the above definitions. If H_0/hA is A-free for H_0 as in (2.2) and $h \in H_0$, then $ftH = ft(H/hA)$ follows. Observe that we showed in (3.1) (1) and (2) that a strongly μ-free R-module H of freeness type $u = \langle \mu, \aleph_0 \rangle$ has a $(\aleph_0)$-resolution. This implication does not depend on additional properties of H as $H^* = 0$ or $End\ H = A$.

Moreover, we must refine the use of ω-sequences converging to particular ordinals. If $u = \langle \mu_1, \cdots, \mu_n \rangle \in \boldsymbol{r}(\mu)$, then let $\times u = \mu_1 \times \cdots \times \mu_n$ be equipped with the lexicographical order. We consider strictly order preserving monomorphisms $\alpha : \times u \to \mu$ and let $a = Im(\alpha)$ and $i = sup\ a$ which is cofinal to μ_1.
We say that $X \subseteq a$ is bounded (with respect to α) if there is $\beta = (\beta_1, \cdots, \beta_n) \in \times u$ such that $X \subseteq Im(\alpha|\beta)$.
We may view α as an "n-ladder" at a. A bounded set X is bounded "on each ladder" in an obvious sense. If $n = 1$, then a is a μ_1-sequence representing $cf(i) = \mu_1$. The bounded subsets of a constitute an ideal of the set of subsets $\mathcal{P}(a)$ of a. The set $a \subseteq i$ will be used to find an extension on $G_i \subseteq G_{i+1}$ which will lead to the desired module $G = \bigcup_{i<\lambda} G_i$. If $u = \langle \aleph_0 \rangle$, as is the case considered in §3-6, then the 1-ladder system is an ω-sequence a_i at each ordinal i with $cf(i) = \omega$ which has an additional property (+) (§3) which ensures freeness of G_δ at limit ordinals $\delta < \mu^{+}$. A similar condition is required for a general u:

(+) $\begin{cases} \text{If } cf(i) = \mu_1, \text{ then there exists an } n\text{-ladder system } a_i \subseteq i \text{ such that} \\ \text{for each } \delta < \mu^{+} \text{ there exist bounded subsets } b_i \subseteq a_i \text{ (as defined above)} \\ \text{such that the sets } (a_i \setminus b_i)\ i \in S_{\mu_1},\ i < \delta \text{ are pairwise disjoint.} \end{cases}$

This is ensured by the following combinatorial result of Shelah, which follows from the main theorem of [35].

THEOREM 7.1 (ZFC + G.C.H.) If $S = \{i \in \aleph_{\omega+1}, cf(i) = \omega_1\}$ then $\diamondsuit_{\omega+1}(S)$ holds. Moreover, for any $i \in S$ we can find a 2-ladder system $\alpha^i : \omega_1 \times \omega \to i$ such that the sets $a_i = Im(\alpha^i)$ satisfy (+).

The μ-$\aleph_0$-Step-Lemma 3.1 and (6.2) can be refined to

THE μ-u-STEP-LEMMA 7.2 Let (R, S) be as in (2.1) and A be a countable, free R-algebra. We will distinguish two cases (a) and (b). In case (a) let $H = H'$ be a strongly μ-free R-module of cardinality μ of freeness type $ftH = u$. In case (b) let H be a strongly μ-free A-module of cardinality μ with $End_R H = A$ and $h \in H_0$ (in (2.2)) such that $ftH = ftH' = u$ for $H' = H/hA$.
Then we can find two u-resolutions of H' for μ_1 (with respect to R in case (a) and to A in case (b)) of the form F^j/K $(j = 0, 1)$ such that for any basic element $f \in K_0$ the following holds

(a) If $H^* = 0$ and $\varphi \in K^*$ extends to both $\varphi^j \in (F^j)^*$ $(j = 0, 1)$, then $f\varphi = 0$.

(b) If $\varphi \in End_R(K)$ extends to both $\varphi^j \in EndF^j$ $(j = 0, 1)$, then $f\varphi \in fA$.

Proof: We consider case (a) only and must find suitable extensions F^j of K. Recall that $K = \bigoplus_{i<\mu} K_i$ for $\mu = \mu_1$ is a direct sum of free R-modules K_i of rank μ, cf. (2.3). We will define F^j for $j = 0, 1$. We fix an n-ladder $a \subseteq \mu$ and write K as a continuous increasing chain $K = \bigcup_{i<\mu} F_i$ of free R-modules of rank μ such that $F_{i+1} = F_i \oplus K_i$ for all $i \in a$. The module H can also be represented by (2.2) using a in the form $H = \bigcup_{\alpha\in a} H_\alpha$. Using the ordering on a (induced by μ and also by $\times u$), equations (2.2) can be expressed as

$$q_\alpha t_{\alpha+1} = t_\alpha + \Sigma_{\alpha_j<\alpha} t_{\alpha_j}.$$

The free resolution is now similar to (3.1)(r^j). Defining C^j and N^j according to (3.1) we let F^j (j = 0, 1) be generated by K and new elements s^j_α $(\alpha \in a)$ subject to the relations

$$q_\alpha s^j_{\alpha+1} = s^j_\alpha + \Sigma_{\alpha_j<\alpha} s^j_{\alpha_j} + x_\alpha + \delta_{j1} f,$$

where $x_\alpha \in K_\alpha$ are fixed basic elements. Arguments as in (3.1) will give (a) and (b).

In order to ensure that unions at limit ordinals remain free in the construction of §5, the ω-Freeness-Proposition 3.2 was used. At limit ordinals not cofinal to ω, the construction was trivial. This is different here. If $u = \langle\mu_1, \cdots, \mu_n\rangle$ is a certain type of freeness, then the new construction will be non-trivial at limit ordinals cofinal to some μ_i. Hence we have to work at those limit ordinals to ensure freeness of the constructed μ^+ filtration. We need the generalized

u-FREENESS-PROPOSITION 7.3 Let μ be a regular cardinal, $u = \langle\mu_1, \cdots, \mu_n\rangle \in \boldsymbol{r}(\mu)$ some freeness type with $\mu_1 < \mu$ and $\delta < \mu^+$ an ordinal with $\{G_i : i \leq \delta\}$ an ascending, continuous chain of submodules with the following properties for all $i < \delta$.

(a) G_i is free of rank μ.

(b) G_{i+1}/G_i is strongly μ- free and let $\boldsymbol{S} = \{i < \delta : G_{i+1}/G_i \text{ is not free}\}$.

(c) There exist $G_i \subseteq H_i \subseteq G_{i+1}$ and H_i^0 a free summand of G_i with free quotient such that $H_i = G_i \oplus_{H_i^0} H_i^1$ is a pushout with $H_i^0 \subseteq H_i^1$ free modules of rank μ.

(d) $G_{i+1} = H_i \oplus \bigoplus_{\tau \in u} K_i^\tau$ with K_i^τ freely generated by $x_{\alpha i}^\tau$ $(\alpha < \mu)$.

(e) If $i \in \boldsymbol{S}$, then $G_i = H_i$ and $H_i^1 = H_i^0$ (hence $G_{i+1} = G_i \oplus \bigoplus_{\tau \in \boldsymbol{u}} K_i^\tau$).

(f) If $i \notin \boldsymbol{S}$, then cf(i)$= \kappa \in u$. Moreover

(i) there exists $a_i \subseteq i$ of order type κ with $a_i \cap \boldsymbol{S} = \emptyset$ and $sup\ a_i = i$.

(ii) there exist $A_{ij} \subseteq \mu$ for all $j \in a_i$ such that $A_{ij} \cap A_{ij'} = \emptyset$ for $j \neq j'$ and $A_{ij} \cap A_{i'j'}$ is bounded in μ for $(ij) \neq (i'j')$.

(iii) Let $K_j^\kappa = \langle x_{\alpha j}^\kappa : \alpha \in A_{ij} \rangle$, $H_i^0 = \bigoplus_{j \in a_i} K_j^\kappa$. If $k \in a_i$, $\beta < \mu$ and $K_{\beta k}^\kappa = \bigoplus_{j \in a_i}^{j<k} K_j^\kappa \oplus \langle x_{\alpha j}^\kappa : j \in a_i,\ j \geq k,\ \alpha < \beta \rangle$, then $H_i^1/K_{\beta k}^\kappa$ is free.

It follows G_δ is free as well.

The proof of (7.3) follows the proof of (3.2) up to minor changes.

Combining (7.1) and our next result (7.5) we can pass $\aleph_\omega$ and obtain the existence of certain modules of cardinal $\lambda = \aleph_{\omega+1}$.

COROLLARY 7.4 (ZFC + G.C.H.) If (R, S) is as in §2 and A is a countable, free R-algebra, then we can find strongly $\aleph_{\omega+1}$-free R-modules G of cardinality $\aleph_{\omega+1}$ with $End_R G = A$ and of freeness type $\langle \aleph_{\omega+1}, \aleph_1 \rangle$.

We would like to remark that a further extension of (7.5) would provide modules as in (7.4) for cardinals $\aleph_{\omega k+1}$ $(k \in \omega)$. Clearly (7.4) in conjunction with (5.1) and (6.1*), respectively, provides examples for all $\aleph_{\omega+n}$ $(n \geq 1)$. A similar result then follows for all $\aleph_{\omega k+n}$ $(k, n \geq 1)$. Magidor and Shelah [31] show, however, that there are models of ZFC + G.C.H. in which $\aleph_{\omega^2+1}$-free abelian groups of cardinality $\aleph_{\omega^2+1}$ are free. This shows that the results above cannot be extended any further.

THEOREM 7.5 (ZFC + G.C.H.) Suppose that the following conditions hold.

(1) Let (R, S) be as in §2 and A a countable, free R-algebra.

(2) There exists a strongly μ-free A-module H of cardinality μ with $End_R H = A$ of freeness type $u = \langle \mu_1, \cdots, \mu_n \rangle$.

(3) Let λ be a regular, uncountable cardinal with a stationary, non-small subset $\boldsymbol{S}$ of ordinals cofinal to some cardinal μ_1.

(4) There exists an n-ladder system $a_i \subseteq i$ $(i \in \boldsymbol{S})$ with (+) and $sup\ a_i = i,\ a_i \cap \boldsymbol{S} = \emptyset$.

If $\lambda = \mu^+$, then there exists a strongly λ-free R-module G with $|G| = \lambda$, $ft(G) = \lambda^\wedge u$ and $End_R G = A$.

REMARK A similar result holds for modules with trivial dual.

Proof: We indicate the changes in the construction of G, which are based on (2.2) using $u^n = \times u$ and the n-ladder $\alpha^i : \times u \to i$ with $Im(\alpha^i) = a_i$. Recall that $cf(i) = \mu$. From $a_i \cap S = \emptyset$ and the inductive construction of $F = \cup_{j<\lambda} F_j$ we have $F_{\alpha+1} = F_\alpha \oplus K_\alpha$ for $\alpha \in a_i$. We choose basic elements $x_\alpha \in K_\alpha$ (with respect to A in case of $End_R G = A$ and with respect to R for $G^* = 0$) which are used in (7.2) in order to find F^j_{i+1} $(j = 0, 1)$ with the desired properties. Moreover, (7.3) takes care of freeness at limit ordinals. The prediction principle (7.5)(c) picks the right branch in 2^λ which leads to $End_R G = A$ and $G^* = 0$, respectively.

References

[1] R. Baer, Abelian groups without elements of finite order, *Duke Math. J.*, *3:* 68–122 (1937).

[2] A.L.S. Corner, Every countable reduced torsion–free ring is an endomorphism ring, *Proc. London Math. Soc.*, *(3) 13:* 687–710 (1963).

[3] A.L.S. Corner, Additive categories and a theorem of W.G. Leavitt, *Bull. Amer. Math. Soc.*, *75:* 78–82 (1969).

[4] A.L.S. Corner and R. Göbel, Prescribing endomorphism algebras, a unified treatment, *Proc. London Math. Soc.* *(3) 50:* 447–479 (1985).

[5] A.L.S. Corner and R. Göbel, On the existence of an $\aleph_1$–free abelian group of cardinality $\aleph_1$ with no free summands (unpublished).

[6] K. Devlin and S. Shelah, A weak version of $\diamondsuit$ which follows from $2^{\aleph_0} < 2^{\aleph_1}$, *Israel J. Math.*, *29:* 239–247 (1978).

[7] M. Dugas, Fast Freie Abelsche Gruppen mit Endomorphismenring $\mathbb{Z}$, *Journal of Algebra*, *71:* 314–321 (1981).

[8] M. Dugas and R. Göbel, Every cotorsion–free ring is an endomorphism ring, *Proc. London Math. Soc*, *(3) 45:* 319–336 (1982).

[9] M. Dugas and R. Göbel, Every cotorsion–free algebra is an endomorphism algebra, *Math. Zeitschr.*, *181:* 451–470 (1982).

[10] M. Dugas and R. Göbel, On radicals and products, *Pacific J. Math.*, *18:* 70–104 (1985).

[11] K. Eda, Cardinal restrictions for preradicals, pp.277–283 in "Abelian Group Theory", *Contemporary Math.*, *87*, Providence (1989).

[12] P.C. Eklof, On the existence of κ–free abelian groups, *Proc. Amer. Math. Soc.*, *47:* 65–72 (1975).

[13] P.C. Eklof, Set theoretic methods in homological algebra and abelian groups. *Les Presses de l'Université de Montréal*, Montreal (1980).

[14] P.C. Eklof and A.H. Mekler. On constructing indecomposable groups in L. *Journal of Algebra. 49:* 96–103 (1977).

[15] P.C. Eklof and A.H. Mekler. Almost free modules, set-theoretic methods. *North-Holland*, Amsterdam (1990).

[16] P.C. Eklof and S. Shelah, On Whitehead modules, *Journal of Algebra, 142:* 492–510 (1991).

[17] L. Fuchs, Infinite abelian groups, Vol. I, *Academic Press*, New York (1970).

[18] L. Fuchs, Infinite abelian groups, Vol.II, *Academic Press*, New York (1973).

[19] R. Göbel, An easy topological construction for realizing endomorphism rings, *Proc. R. Irish Acad., 92 A:* 281–284 (1992).

[20] R. Göbel and W. May, Independence in completions and endomorphism algebras, *Forum Mathematicum, 1:* 215–226 (1989).

[21] R.Göbel and W. May, Four submodules suffice for realizing algebras over commutative rings, *J. Pure Appl. Algebra, 65:* 29–43 (1990).

[22] R. Göbel and S.Shelah, On the existence of rigid $\aleph_1$-free abelian groups of cardinality $\aleph_1$, to appear in Proceedings of the Padova Conference on Abelian Groups and Modules (1995).

[23] R. Göbel and B. Goldsmith, The Kaplansky Test problem — an approach via radicals, *Journ. Pure and Appl. Algebra, 99:* 331–344 (1995).

[24] J. Gregory, Higher Souslin trees and the generalized continuum hypothesis, *Journ. Symb. Logic, 41:* 663–667 (1976).

[25] Phillip Griffith, $\aleph_n$-free abelian groups, *Quart. J. Math., (2) 23:* 417–425 (1972).

[26] G. Higman, Almost free groups. *Proc. London Math. Soc., 1:* 184–190 (1951).

[27] G. Higman, Some countably free groups, pp. 129–150 in Proceedings "Group Theory", Singapore (1991).

[28] P. Hill, New criteria for freeness in abelian groups II, *Trans. Amer. Math. Soc., 196:* 191–201 (1974).

[29] T. Jech, Set Theory, *Academic Press*, New York (1978).

[30] R. Jensen, The fine structure of the constructible hierarchy, *Ann. of Math. Logic, 4:* 229–308 (1972).

[31] M. Magidor and S. Shelah, When does almost free imply free? (for groups, transversals, etc.), *Journ. Amer. Math. Soc.,(4) 7:* 769–830 (1994).

[32] A.H. Mekler, How to construct almost free groups, *Can. J. Math., 32:* 1206–1228 (1980).

[33] J. Rotman, Homological Algebra, *Academic Press*, New York (1979).

[34] S. Shelah, A compactness theorem for singular cardinals, free algebras, Whitehead problem and transversals, *Israel Journal of Math., 21:* 319–349 (1975).

[35] S. Shelah, On successors of singular cardinals, *Logic Colloquium '78 of Stud. Logic Foundations Math., 97:* 357–380, *North Holland*, Amsterdam-New York (1978).

[36] S. Shelah, On uncountable abelian groups, *Israel J. Math., 32:* 311–330 (1979).

[37] S. Shelah, Whitehead groups may not be free, even assuming H. II, *Israel J. Math., 35:* 257–285 (1980).

[38] S. Shelah, On endo–rigid strongly $\aleph_1$–free abelian groups in $\aleph_1$, *Israel J. Math., 40:* 291–295 (1981).

[39] S. Shelah, A combinatorial principle and endomorphism rings. I: On p–groups, *Israel J. Math., 49:* 239–257 (1984).

[40] S. Shelah, A combinatorial theorem and endomorphism rings of abelian groups. II, pp. 37 - 86 in Abelian groups and modules, *CISM Lecture Notes, Vol. 287, Springer*, Wien-New York (1984).

[41] S. Shelah, Incompactness in regular cardinals, *Notre Dame Journal of Formal Logic, 26:* 195–228 (1985).

[42] S. Shelah, A note on κ-freeness of abelian groups, pp. 260 - 268 in Around classification theory of models, *Lecture Notes in Math., Vol. 1182, Springer*, Berlin (1986).

[43] S. Shelah and Z. Spasojevic, A forcing axiom on strongly inaccessible cardinals making all uniformizations for all ladder systems lying on a fixed stationary set and applications for abelian groups, manuscript No. 587.

Extensions of Finitely Faithful S-Groups

H. Pat Goeters
Mathematics Department
Auburn University
Auburn, AL

Two papers of R. B. Warfield Jr. come to the forefront in the study of torsion-free abelian group theory, and have been instrumental in shaping the course of research in this area. In [W], he initiated the study of the structure of the divisible group $Ext(A, B)$ when A and B are torsion-free of finite rank. Recall that the *p-rank* of a torsion-free group G is $r_p(G) = dim_{Z(p)} G/pG$. Warfield noted that the dimension of the p-socle of the torsion subgroup of $Ext(A, B)$ is $r_p(A)r_p(B) - r_p(Hom(A, B))$, a formula which holds even if A and B are only assumed to have finite p-ranks. Confronted with the apparent difficulty of determining $r_p(Hom(A, B))$ for general A and B, Warfield sought conditions in order for $r_p(Hom(A, B)) = r_p(A)r_p(B)$. The main result from [W] is a complete characterization of when $Ext(A, B)$ is torsion-free, if A and B both have finite rank, and one of A or B is rank-1.

The finite rank torsion-free groups A for which $Ext(A, A)$ is torsion-free were given a surprising characterization by D. M. Arnold ([Ar]) in terms of group properties of A and relational properties with its endomorphism ring $End(A)$. Following the terminology of [Ar], we say that a torsion-free group B is *finitely faithful*, if for every right ideal I of finite index in $End(B)$, $IB \neq B$, and we say that B is an *S-group* if $Hom(B, C)B = C$ for all subgroups C of finite index in B (i.e. B generates each subgroup of finite index). Arnold showed that when A is torsion-free of finite rank, $Ext(A, A)$ is torsion-free precisely when A is a finitely faithful S-group. In [FG] it was shown that given any torsion-free group B, $Ext(B, B)$ is torsion-free and $r_p(B)$ is finite for all p, if and only if B is a finitely faithful S-group. It follows that if B_p is any pure subgroup of the p-adic integers, then $B = \Pi_p B_p$ is a finitely faithful S-group, providing a source of examples.

The theorem 3 of [W] mentioned above was generalized in [Goe] to determine exactly when $Ext(A, B)$ is torsion-free when both A and B have finite rank, however for technical reasons, we needed the added stipulation that for a particular prime p, $r_p(B) = rank\ B$. In section 1 we generalize this result almost verbatim and specify precisely when $Ext(A, B)$ is torsion-free in case one of the groups A or B is a finitely faithful S-group. We encounter a difficulty in the second situation, when B is a finitely faithful S-group, and must restrict the p-ranks of A to be finite in that case. Section 1 culminates in

MAIN THEOREM. *Let A and B be torsion-free groups. Then*

(1) *If A is a finitely faithful S-group, then $Ext(A, B)$ is torsion-free if and only if B/C is divisible by all primes p not dividing A, where $C = Hom(A, B)A$.*

(2) *If B is a finitely faithful S-group and $r_p(A) < \infty$ for all p, then $Ext(A, B)$ is torsion-free if and only if $K = \cap\{ker\ f | f : A \to B\}$ is divisible by all primes not dividing B, and A/K embeds as a subgroup in B^J (for some J) which is p-pure at all primes p not dividing B.*

In the foundational article [Wa], Warfield examined what is now commonly called *Warfield duality.* Specifically, given a rank one group A, he determined the largest class $\mathcal{C}$ of torsion-free finite rank groups such that $Hom(-, A) : \mathcal{C} \to \mathcal{C}$ defines a duality. Many authors have endeavored to extend this duality result. The approach taken by Huber and Warfield in [HW] was to consider the pair of functors $Hom(-, B)$ and $Hom_{End(B)}(-, B)$ and categories of groups $\mathcal{G}$ and left $End(B)$-modules $\mathcal{M}$ on which the pair of functors defines a duality. In [A] a general description of the abelian groups M for which $M \cong_{nat} Hom_{End(B)}(Hom(M, B), B)$ was given, for an arbitrary group B.

In section 2, given a finitely faithful S-group B with $QEnd(B)$ left artinian, we show that with $\mathcal{F}_B$ equal to the class of groups quasi-isomorphic to B, and $\mathcal{I}_{End(B)}$ the class of left $End(B)$-modules which are isomorphic to submodules of $QEnd(B)$ and quai-equal to $End(B)$ as groups, $Hom(-, B) : \mathcal{F}_B \to \mathcal{I}_{End(B)}$ and $Hom_{End(B)}(-, B) : \mathcal{I}_{End(B)} \to \mathcal{F}_B$ defines a duality. We use this fact to give a new description of the finitely faithful S-groups with left artinian quasi-endormorphism ring and apply our results to investigate which groups A are quasi-isomorphic to but not isomorphic to $Hom_{End(B)}(Hom(A, B), B)$.

In the text below all groups are torsion-free and abelian. We will adopt the conventional notation of using $E(B)$ when referring to $End(B)$, and the notation $\pi(B)$ is used to denote the set $\{p | pB \neq B\}$. When S is a set of primes, $A \triangleleft_S B$ signifies that A is a subgroup of B which is p-pure for all $p \in S$ (or $A \triangleleft B$ if S is the set of all primes), and Z_S represents the subring of Q generated by 1 and $\frac{1}{p}$ when $p \notin S$.

1. WARFIELD'S EXTENSION THEOREM

Given torsion-free groups A and B, a general characterization of when $Ext(A, B)$ is torsion-free is that A is projective with respect to any epimorphism $B \to C \to 0$ with C bounded ([FG]). Our intent is to bring out the full meaning of this property in case B is a finitely faithful S-group. Recall the key observation from [W].

LEMMA 1. *Let G and B denote torsion-free groups. Then $Ext(G, B)$ is torsion-free if and only if, for each prime p, G is projective with respect to $0 \to B \xrightarrow{p} B \to B/pB \to 0$.*

Proof. For any prime p, the exact sequence

$$(*) \qquad 0 \to B \xrightarrow{p} B \to B/pB \to 0,$$

induces the exact sequence

$$(**) \qquad 0 \to Hom(G,B) \xrightarrow{p} Hom(G,B) \to Hom(G,B/pB) \to Ext(G,B)[p] \to 0.$$

From $(**)$ it is clear that $Ext(G,B)[p] = 0$ if and only if G is projective with respect to $(*)$. $\square$

Throughout the text, we will frequently refer to the sequences $(*)$ and $(**)$ appearing in the proof of lemma 1. A standard functorial subgroup of A is, $A[B] = \bigcap\{\ker f | f : A \to B\}$. We often refer to $A[B]$ as the *B-radical* of A, and as the reader will observe, $A[B] = 0$ if and only if A embeds in a product of copies of B. In particular $(A/A[B])[B] = 0$.

THEOREM 1. *Let B be a finitely faithful S-group with $S = \pi(B)$ and assume that the p-rank of A is finite for each $p \in S$. Then $Ext(A,B)$ is torsion-free if and only if $Z_S \otimes_Z A \cong C \oplus D$ with D divisible, and $C \triangleleft B^I$ for some I.*

Proof. First assume that $Ext(A,B)$ is torsion-free. In order to show that the B-radical of A is divisble by each prime from $\pi(B) = S$ let us fix $p \in S$. If $p(A[B]) \neq A[B]$, then there must exist some $x \in A[B]$ having p-height zero. Since $A[B]$ is a pure subgroup of A, $\overline{x} = x + pA$ is non-zero in A/pA, so $\langle \overline{x} \rangle$ is a direct factor of A/pA. We can then construct a map $f : A \to B/pB$ with $f(x) \neq 0$. As covered in lemma 1, since $Ext(A,B)$ is torsion-free, A is projective with respect to $(*)$ (the sequence appearing in the proof). So, there must exist a map $\overline{f} : A \to B$ such that $\overline{f}(x) + pB = f(x) \neq 0$. However $\overline{f}(x) \neq 0$ contradicts $x \in A[B]$. Therefore, $Z_S \otimes A[B]$ is divisible which implies that $Z_S \otimes A \to Z_S \otimes (A/A[B]) \to 0$ is split exact. Write $Z_S \otimes A = D \oplus C$ with D divisible and $C \cong Z_S \otimes (A/A[B]) \cong (Z_S \otimes A)/(Z_S \otimes A)[B]$. We must show that C is isomorphic to a pure subgroup of B^J for some J. Due to $C[B] = 0$, C embeds in B^I for some I.

By lemma 1 again, A is projective with respect to the sequence $(*)$. Since B is naturally a Z_S-module, $Z_S \otimes A$ is projective relative to the sequence $(*)$, and consequently so is C. Fix an embedding $f : C \to B^I$ and set $U_p = \{x \in C | p\text{-height } x = 0\}$. For each $p \in S$ and $x \in U_p$, there is a map $f_{p,x} : C \to B/pB$ such that $f_{p,x}(x) \neq 0$. By the projectivity of C relative to $(*)$, there is a lifting $f^*_{p,x} : C \to B$ of $f_{p,x}$. Define $f^* : C \to B^I \oplus \prod_{p \in S} B^{U_p}$ by $f^*(x) = (f(x), \prod_{p\in S} \prod_{y \in U_p} f^*_{p,y}(x))$. Then $ker\ f^* \subseteq ker\ f = 0$. Also, if $x \in C$ has p-height zero, then $p \in S$, and $x \in U_p$, and consequently the p-height of $f^*_{p,x}(x)$ in B is zero. This implies that f^* is a pure embedding.

For the converse, the cokernel T of the embedding of A into the localization $A_S = Z_S \otimes A$ is a torsion group with zero p-component whenever $p \in S$. This supports the existence of $0 = Hom(T, QB/B) \cong Ext(T,B) \to Ext(A_S, B) \to Ext(A,B) \to 0$, and it is enough to show that $Ext(A_S, B)$ is torsion-free. With $A_S = D \oplus C$ as prescribed, we clearly have that $Ext(D,B)$ is torsion-free.

Let $C \triangleleft B^I$. To show that $Ext(C,B)$ is p-torsion-free for each p, let us again fix $p \in S$. By hypothesis, $r_p(C) = r_p(A)$ is finite, and we may select a complete set of representatives of the nonzero cosets in C/pC: call them $c_1, \ldots, c_k$. Because C is a pure subgroup of B^I, for each index $1 \leq i \leq k$, there is an appropriately restricted coordinate map $\pi_i : C \to B$ such that the p-height of $\pi_i(c_i)$ in B is zero.

We aim to show that $\pi : C \to B^k$ defined as $\pi(c) = (\pi_1(c), \ldots, \pi_k(c))$, is a map whose image is p-pure and kernel is p-divisible. Given $c \in C \backslash pC$, write $c = c_i + pc'$ for some i

and $c' \in C$. Then $\pi_i(c) = \pi_i(c_i) + p\pi_i(c')$ and the p-height of $\pi_i(c)$ is zero in B. Thus, for any $c \in C$ having finite p-height, the p-height of $\pi(c) \in B$ matches the p-height of $c \in C$. Consequently, $G = im\ \pi$ is p-pure in B^k and $K = ker\pi \subseteq p^\omega C$.

Now, the sequence $0 \to G \to B^k \to V \to 0$ where $V = B^k/G$, is p-pure and exact so that V has no p-torsion. Therefore, referring to property (I) on page 223 in [F] volume 1, $Ext(V, B)$ is p-divisible. Reading the sequence $Ext(V, B) \xrightarrow{\phi} Ext(B^k, B) \to Ext(G, B) \to 0$, we see that $Ext(G, B)$ is p-torsion-free since $im\ \phi$ is p-pure in the torsion-free group $Ext(B^k, B) \cong \oplus^k Ext(B, B)$ (B is a finitely faithful S-group). Consider the exact sequence $Hom(K, B) \xrightarrow{\alpha} Ext(G, B) \xrightarrow{\beta} Ext(C, B) \xrightarrow{\gamma} Ext(K, B) \to 0$, where $K = ker\ \pi$ is a pure subgroup of $p^\omega C$.

The p-divisibility of K implies the p-purity of $im\ \alpha$ in $Ext(G, B)$. From above, $Ext(G, B)$ is p-torsion-free, so $im\ \beta$ is p-torsion-free. Furthermore, by lemma 52.1 in [F], $Ext(K, B)$ is p-torsion-free because K is p-divisible. Therefore, $Ext(C, B)$ is an extension of two p-torsion-free groups and must be p-torsion-free. When $p \notin S$, C is p-divisible and $Ext(C, B)$ is necessarily p-torsion-free ([F]). Thus, we have reached our desired conclusion: $Ext(C, B)$ is torsion-free. □

As might be expected, some restrictions on the p-ranks of A are necessary in theorem 1. In particular, $Ext(Z_p^\omega, Z_p)$ is not torsion-free ([Go]). We feel a comparison between the statement of theorem 1 and that of theorem 3 in [W] is in order.

COROLLARY 1. *(Warfield [W]) If B is rank-1 and A has finite rank, then $Ext(A, B)$ is torsion-free if and only if $A[B]$ is divisible by all primes p not dividing B.*

Proof. It was shown in [Go] that any countable group G which is divisible by all primes $p \notin \pi(B)$, is isomorphic to a subgroup of a product of copies of B if and only if G is isomorphic to a pure subgroup of a product of copies of B. With this in mind, the corollary is just a restatement of theorem 1 under these special circumstances. □

As offered in the proof of the last corollary, in the case that B has rank one and A has countable rank, $A/A[B] \lhd_{\pi(B)} B^I$. So in this case, if $A[B] = 0$, then $Ext(A, B)$ is torsion-free by theorem 1. However, as the example to follow shows, it is not the case for all finitely faithful S-groups B that $A[B] = 0$ implies $Ext(A, B)$ is torsion-free. After a brief inspection of equation (**) from the proof of lemma 1 with $G = B$, we observe that if B is a Murley group (that is, $r_p(B) \leq 1$ for each prime p), then B is a finitely faithful S-group. This provides a wealth of examples of finitely faithful S-groups (other examples are given in [Ar]).

EXAMPLE. *There is a rank-2 finitely faithful S-group B and a rank-2 subgroup $A \leq B$ such that A is not a pure subgroup of B^I for any I.*

Proof. Let X and Y be subgroups of Q containing Z and having the respective height sequences of 1 equal to $(0, 1, 0, 1, 0 \dots)$ and $(0, 0, 1, 0, 1, \dots)$. Fix an isomorphism $\theta : Q/X \cong Q/Y$ and set $B = \{(a, b) \in Q \oplus Q | \theta(a + X) = b + Y\}$ (this follows the Beaumont-Wisner construction [BW]). Then $\overline{X} = X \oplus 0$, and $\overline{Y} = 0 \oplus Y$ are pure subgroups of B, with $B/\overline{X} \cong B/\overline{Y} \cong Q$. Moreover, $B/(X \oplus Y) \cong \oplus_p Z(p^\infty)$. Let $X \oplus Y \leq A \leq B$ satisfy $B/A \cong Z(2^\infty)$.

Note that $r_p(B) = 1$ for all p, so B is a finitely faithful S-group as mentioned prior to this example. Since type X and type Y are in both typeset B and typeset A, and finite typesets are closed under finite infimums (lemma 1 in [Arn]), $E(A)$ and $E(B)$ are rank-1 by theorem 3.3 in [A]. Also $r_2(A) = 2$, so every image of A in $Z(2^\infty)$ must be bounded. Therefore, $Hom(A, Z(2^\infty))$ is a torsion group, and the exactness of $0 \to Hom(A, A) \to Hom(A, B) \to Hom(A, Z(2^\infty))$ implies that $Hom(A, B)$ is rank-1 too. But, with reference to (**) from the proof of lemma 1 with $p = 2$ and $G = A$, we observe that $dim_{Z(2)}(Hom(A, B/2B)) = dim_{Z_{(2)}}(Hom(A/2A, B/2B)) = 2$ while $dim_{Z_{(2)}}(Hom(A, B)/2\,Hom(A, B)) = 1$. Thus, $Ext(A, B)[2] \neq 0$ so by theorem 1, A is not a pure subgroup of B^I for any I. □

We are led to consider the generalization of a duality first investigated by Warfield. Using the notation presented in [Al]: Set A^* to be the left $E(B)$-module $Hom(A, B)$ for an abelian group A, and M^* to be the abelian group $Hom_{E(B)}(M, B)$ for a left $E(B)$-module M; A is called B-*reflexive*, if $A \cong_{nat} A^{**} = Hom_{E(B)}(A^*, B)$. Likewise, a left $E(B)$-module M is called B-*reflexive*, if $M \cong M^{**} = Hom(Hom_{E(B)}(M, B), B)$ naturally as $E(B)$-modules. Here, the natural map sends $a \in A$ to the map $\varepsilon_a : A^* \to B$ which is evaluation at a ($\varepsilon_a(f) = f(a)$ for any $f \in A^*$). We like this notation although possible confusion may arise from the terseness of the symbol *. The contravariant pair $Hom(-, B)$ and $Hom_{E(B)}(-, B)$ will be considered more fully in the next section.

COROLLARY 2. *Let B be a finitely faithful S-group and suppose A is a group with $A[B] = 0$ and $r_p(A) < \infty$ for each $p \in S$. Then with $S = \pi(B)$, $Ext(A, B)$ is torsion-free if and only if $Z_S \otimes A \lhd A^{**}$.*

Proof. Set $A_S = Z_S \otimes A$. It really is easy to see that $A_S^* \cong_{nat} A^*$. So, $A_S^{**} \cong_{nat} A^{**}$. From this and theorem 1, we may assume that $A = A_S$ here. If $Ext(A, B)$ is torsion-free and $p\alpha = \varepsilon_a$ for some $\alpha \in A^{**}$ and $a \in A$, then either $a = pa' \in pA$ or $a \in A \backslash pA$. In the first situation, it follows that $\varepsilon_{pa'} = p\varepsilon_{a'}$ and so $\alpha = \varepsilon_{a'} \in im\ (A \to A^{**})$. In the latter case, there is a map $\pi : A \to B$ such that the p-height of $\pi(a)$ in B is zero because $A \lhd B^I$ for some I (theorem 1). But in this case, $p\alpha(\pi) = \varepsilon_a(\pi) = \pi(a)$ is impossible.

Conversely, the mapping $\theta : A^{**} \to B^I$ for $I = A^*$ given by $\theta(\alpha) = (\alpha(f)|f \in A^*)$ is a pure embedding. To see this suppose that $\alpha \in A^{**}$ has p-height zero. Since $\alpha : A^* \to B$, for some $f \in A^*, \alpha(f) \notin pB$. The component of the tuple $(\alpha(g)|g \in A^*)$ from B^I associated with the index $f \in A^*$, must have zero p-height. Hence so does $\theta(\alpha)$. Thus, A embeds as a pure subgroup in a product of copies of B, and $Ext(A, B)$ is torsion-free by theorem 1. □

COROLLARY 3. *Assume that B is a finitely faithful S-group, and that A has finite p-rank for each p. If A is quasi-isomorphic to a B-reflexive group then A is B-reflexive.*

Proof. Assume that A is of bounded index in a B-reflexive group C. Set $S = \pi(B)$, and let T denote the bounded group C/A. It follows from theorem 0.2 in [A] that C and A have the same p-ranks, and therefore A is a Z_S-module because $C \cong C^{**}$ is one. From the sequence $0 \to A \to C \to T \to 0$ we obtain the exact sequence $0 \to C^* \to A^* \to Ext(T, B) \to Ext(C, B) \to Ext(A, B) \to 0$. By corollary 2 $Ext(C, B)$ is torsion-free, and since $T' = Ext(T, B)$ is bounded, $Ext(A, B) \cong Ext(C, B)$ is torsion-free as well. Hence $A \lhd A^{**}$ by corollary 2.

We obtain a similar sequence, $0 \to A^{**} \to C^{**} \to Ext_{E(B)}(T', B)$, and with $T'' = im\ (C^{**} \to Ext_{E(B)}(T', B))$, there is a commutative diagram

$$\begin{array}{ccccccccc} 0 & \longrightarrow & A & \longrightarrow & C & \longrightarrow & T & \longrightarrow & 0 \\ & & \downarrow & & \wr & & \downarrow \gamma & & \\ 0 & \longrightarrow & A^{**} & \longrightarrow & C^{**} & \longrightarrow & T'' & \longrightarrow & 0. \end{array}$$

By the Snake Lemma ([R]), $A^{**}/A \cong ker\ \gamma$, a torsion group. But $A \triangleleft A^{**}$, so $A \cong A^{**}$ in the natural way. □

We emphasize the need for a restriction on the p-ranks of the group A in corollary 3. In [Go] it was shown that if B is rank-1, and $0 \to B \to G \to B^{\omega} \to 0$ represents an element of finite nonzero order in $Ext(B^{\omega}, B)$ then G is not B-reflexive. However, G is quasi-isomorphic to B^{ω}, a B-reflexive group.

Our last result of this section generalizes theorem 3 of [W] by removing the finite rank assumptions and replacing the conditions that either rank A or rank B is one, by the stipulation that either A or B is a finitely faithful S-group. As in the literature, the *A-socle* of a group B, is $S_A(B) = Hom(A, B)A$.

THEOREM 2. *Let A and B denote torsion-free groups.*

(1) *If A is a finitely faithful S-group, then $Ext(A, B)$ is torsion-free if and only if $B/S_A(B)$ is divisible by all primes not dividing A.*

(2) *Let B denote a finitely faithful S-group, and suppose $r_p(A) < \infty$ for all p. Then $Ext(A, B)$ is torsion-free if and only if $A[B]$ is p-divisible for all $p \in \pi(B)$, and $A/A[B]$ is isomorphic to a $\pi(B)$-pure subgroup of B^I for some I.*

Proof. (1). Assuming that A is a finitely faithful S-group, we will first prove the general statement that $Ext(A, C)$ is torsion-free for any torsion-free C with $S_A(C) = C$. To see this, we must check that A is projective with respect to the natural map $\pi : C \to C/pC$ for each p, as pointed out in lemma 1. Let $f : A \to C/pC$. Since A has finite p-rank and $pA \subseteq ker\ f$, there are $x_1, \ldots, x_n \in C$ such that $\sum \langle x_i + pC \rangle$ contains $im\ f$. We will now obtain a finitely A-generated subgroup G of C such that $(G + pC)/pC$ contains the image of f.

Because $S_A(C) = C$, for each $i = 1, \ldots, n$, there are $g_{ij} \in Hom(A, C)$ and $a_{ij} \in A$ with $j \leq m_i$ such that $\sum_j g_{ij}(a_{ij}) = x_i$. Set $m = \sum_i m_i$, and define $\phi : A^m = A^{m_1} \oplus \cdots \oplus A^{m_n} \to C$ by $\phi((b^1_1, \ldots, b^1_{m_1}), (b^2_1, \ldots, b^2_{m_2}), \ldots, (b^n_1, \ldots, b^n_{m_n})) = \sum g_{ij}(b^i_j)$. For $G = im\ \phi$, $x_i \in G$ for $i = 1, 2, \ldots, n$, so that $im\ f \subseteq (G + pC)/pC$. Also with $K = ker\ \phi$, the exactness of $Ext(A, K) \to Ext(A, A^m) \to Ext(A, G) \to 0$ and the divisibility of $Ext(A, K)$ imply that $Ext(A, G)$ is isomorphic to a summand of $Ext(A, A^m) \cong Ext(A, A)^m$. Hence $Ext(A, G)$ is torsion-free since A is a finitely faithful S-group.

Let π be the natural map from G onto $T = (G + pC)/pC$, and set $H = ker\ \pi$. We have an exact sequence $0 \to Hom(A, H) \to Hom(A, G) \to Hom(A, T) \xrightarrow{\alpha} Ext(A, H) \to Ext(A, G)$. From above, $Ext(A, G)$ is torsion-free, which implies that $Ext(A, H)/im\ \alpha$ is torsion-free. Consequetly, $im\ \alpha$ is pure in $Ext(A, H)$. Moreover, since $Ext(A, H)$ is divisible, the pure subgroup $im\ \alpha$ must be divisible as well. But $Hom(A, T)$ is bounded, so we conclude that $im\ \alpha = 0$. Therefore, there is a map $\bar{f} : A \to G$ which satisfies $\pi\bar{f} = f$. Once we regard

$\bar{f}: A \to C$, we have shown that A is projective with respect to the sequence $C \xrightarrow{\pi} C/pC$ with p arbitrary, so from lemma 1 $Ext(A,C)$ is torsion-free.

For a general B, set $C = S_A(B)$. We will now observe that $C \triangleleft B$. Suppose $C \subseteq H \subseteq B$ while $pH \subseteq C$ for some p with $pA \neq A$, and then consider the exact sequence $0 \to Hom(A,C) \to Hom(A,H) \xrightarrow{\alpha} Hom(A,H/C) \to Ext(A,C)$. We surmise that $im\ \alpha = 0$ since $S_A(H) = C$. But $Ext(A,C)$ is torsion-free from the arguments above, so $Hom(A,H/C) = 0$, forcing $H = C$. Thus, $C \triangleleft B$.

We now invoke theorem 1.2 in [Goe], which states that (for any torsion-free groups A and B), $Ext(A,B)$ is torsion-free if and only if with $S_* = \langle S_A(B)\rangle_*$, $Ext(A,S_*)$ is torsion-free and B/S_* is p-divisible whenever $pA \neq A$. We have just shown that $S_A(B) \triangleleft B$ and the property that $Ext(A, S_A(B))$ is torsion-free is automatic when A is a finitely faithful S-group. The confluence of these remarks yields (1).

(2). This is theorem 1. □

2. Warfield Duality

The endomorphism ring of B, $E(B)$, will be represented by E in this section. A further investigation of the finitely faithful S-groups B is undertaken by considering the pair of contravariant functors mentioned previously. We define the following two classes:

$$\mathcal{F}_B = \{A | A \text{ is quasi-isomorphic to } B\}, \text{ and}$$

$$\mathcal{I}_E = \{I | I \text{ is isomorphic to a left ideal of finite index in } E\}.$$

If B has $r_p(B) < \infty$ for each prime p, then from theorem 0.2 in [A] we observe that any $A \in \mathcal{F}_B$ is isomorphic to a subgroup of QB containing B as a subgroup of finite index. Then $Hom(-,B) : \mathcal{F}_B \to \mathcal{I}_E$ since if $0 \to B \to A \to T \to 0$ is exact with T finite, then $0 \to Hom(A,B) \to Hom(B,B) \to Ext(T,B)$ is exact. The fact that $Ext(T,B) \cong Hom(T, QB/B)$ is finite is also due to $r_p(B) < \infty$ for all p, so $Hom(A,B)$ belongs to $\mathcal{I}_E$. When $r_p(B) < \infty$, $r_p(E) \leq r_p(B)^2$ (this follows from (**) after setting $G = B$). Therefore, using similar arguments we find that $Hom_E(-,B) : \mathcal{I}_E \to \mathcal{F}_B$.

For future reference we make note of the following property of finitely faithful S-groups: For any p such that $pB \neq B$, $E/pE \cong Hom(B/pB, B/pB)$ as rings. One can see this by replacing G with B in (**).

LEMMA 2. *Let B be a reduced finitely faithful S-group with QE left artinian. Then B decomposes as $B_1 \oplus \cdots \oplus B_n$ such that:*

(a) *Each $QE(B_j)$ is simple artinian,*
(b) *Given a prime p, $pB_i \neq B_i$ holds for at most one of the B_i's, and*
(c) *For each i, when nonzero, any localization $E(B_i)_p$ at an integral prime p is a principal left (right) ideal ring.*

Proof. Since QE is left artinian, Reid's decomposition theorem ([Rd]) states that B is quasi-equal to $B_1 \oplus \cdots \oplus B_n$ with each B_j a pure subgroup of B and B_j is quasi-isomorphic to $A_j^{n_j}$ for some strongly indecomposable group A_j. Corollary III.3 (b) from [FG] asserts that the Jacobson radical of QE, $\mathcal{J}(QE)$, is zero. Thus, the nilradical of E, $\mathcal{N}(E) = \mathcal{J}(QE) \cap E = 0$

as well. Since $Hom(B_i, B_j) \subseteq \mathcal{N}(E)$, there must be no nonzero homomorphisms between distinct components in $B_1 \oplus \cdots \oplus B_n$.

Since $B_1 \oplus \cdots \oplus B_n$ is of finite index in B, it follows that $Ext(\oplus_1^n B_i, \oplus_1^n B_i)$ is torsion-free too. With $G = \oplus_{i \neq j} B_i$, since $Ext(B_j, G)$ is torsion- free and $Hom(B_j, G) = 0$, a careful look at $(**)$ from the proof of lemma 1 reveals that whenever $pB_j \neq B_j$, $pG = G$ (this also appears explicitly in [Goe]). Then (b) holds from which it is easy to see that $B = B_1 \oplus \cdots \oplus B_n$. Furthermore, $\mathcal{N}(E(B_j)) \subseteq \mathcal{N}(E) = 0$ so $\mathcal{J}(QE(B_j)) = 0$ and (a) holds.

Fix i and let $R = E(B_i)$ and $S = \mathcal{C}(R)$, the center of R. First let I be a nonzero ideal of R. Then QI is an ideal of the simple ring QR, so $QI = QR$. From this, R/I is torsion, so I contains a smallest positive integer m. This m is the order of the finite group R/I. We will show by induction on m that $I = mR$ (this does not support (c) but is helpful).

Let p denote a prime divisor of m. As mentioned prior to this lemma, R/pR, as a matrix ring over Z/pZ, is simple. Therefore, $I + pR = pR$ or $I + pR = R$. In case $I + pR = pR$, $I \subseteq pR$ and so $\frac{m}{p}R \subseteq \frac{1}{p}I \subseteq R$. By induction, $\frac{1}{p}I = \frac{m}{p}R$ and $I = mR$. Now assume $I + pR = R$. Then $p(R/I) = (pR + I)/I = R/I$ indicating that R/I has no elements of order p. This contradicts the choices of m and p.

We now will show that S is a *pid* and R is a finite rank free S-module. Let $s \in S$. The ideal $I = sR$ is generated by an integer m as above. So $s = mu$ and $m = ts$ for some $u, t \in R$. Then $m = mtu$ implying that t and u are units in R, and since S is a pure subring of R, $u \in S$. Thus, every element of S is an integral multiple of a unit of S so S is a *pid* with QS is a field. As a point to be taken later, note that if T is a pure subring of S for which QT is a field, then T has finite p-ranks for all primes p. Also if J is an ideal of T, then $QT = QJ$ from which we conclude that T/J is finite. It follows that T is a noetherian domain.

From the Jacobson Density Theorem, $QR = End_D(QR)$ where $D = End_{QR}(QR)$ is a division ring, and QR is finite dimensional over D. The proof of Pierce's theorem as conveyed in theorem 9.9 from [Ar] now goes through to show in sequence that (i) D is a subfield of QS; (ii) $D \cap R$ is a pure subring of S and is therefore noetherian (substantiated above); and (iii) R is a finitely generated torsion-free $D \cap R$-module. From (iii) we deduce that R is a finitely generated torsion-free S-module. In particular, R is left and right noetherian.

Fix a prime p with $pR \neq R$, and let I represent any left ideal of R_p. By (a), $QI \oplus QM = QR$ for some ideal M of R_p, so there is an element $e \in I$ for which $QI = QRe$. Note that $I \oplus M$ is of finite index in R_p so I has finite p-rank. We must show that I is principal. Since B_j is a finitely faithful S-group, $R_{(p)} = R_p/pR_p = R/pR$ is a matrix ring over Z/pZ and so R_p has Jacobson radical $\mathcal{J} = pR_p$.

The ideal R_pe is of torsion index in I, so $kI \leq R_pe$ for some $k \neq 0$. It follows that I/R_pe is a finite R_p-module. Also, R_p and therefore I and R_pe are torsion-free S_p-modules of finite rank, from which we deduce that I and R_pe are isomorphic as S_p-modules. Consider the natural map from $R_pe/\mathcal{J}e$ into $I/\mathcal{J}I$. Both $I/\mathcal{J}I$ and $R_pe/\mathcal{J}e$ are finite $R_{(p)}$-modules so by the simplicity of $R_{(p)}$, $I/\mathcal{J}I \cong L^{k_1}$ and $R_pe/\mathcal{J}e \cong L^{k_2}$ for some positive integers k_1 and k_2, where L is a minimal left ideal in $R_{(p)}$. But $I/\mathcal{J}I$ and $R_pe/\mathcal{J}e$ are isomorphic as S/pS-modules. Every element of S is an integral multiple of a unit of S from which it follows that pS is a maximal ideal of S. Therefore, $k_1 = k_2$ since S/pS is a field. Thus, there is an R_p-module epimorphism $\theta : R_pe \to I/\mathcal{J}I$. By Nakayama's lemma, $I = R_px$ where $x = \theta(e)$, and I is principal. The proof that R is a principal right ideal ring is similar. $\square$

To help alleviate ambiguities we will follow the notation from [HW] and use H_B and H_E to denote $Hom(-, B)$ and $Hom_E(-, B)$ respectively. However, when no confusion seems likely, we will revert to the notation of section 1 and use * to represent the appropriate functor. As usual, A is called *nearly isomorphic* to B if for every prime p, there exists $n > 0$ and maps $f : A \to B$ and $g : B \to A$ satisfying $fg = n1_B$ and $gf = n1_A$ ([A]). We will call B a *Q-group*, if whenever B is quasi-isomorphic to A, then B is nearly isomorphic to A. With this notation and terminology we have the main result of this section.

THEOREM 3. *Assume that B is reduced and that QE is left Artinian with $r_p(B)$ finite for all p. The following are equivalent:*

(1) *B is a finitely faithful S-group.*
(2) *The functors $H_B : \mathcal{F}_B \to \mathcal{I}_E$ and $H_E : \mathcal{I}_B \to \mathcal{F}_E$ define a duality.*
(3) *E is (left and right) hereditary, and B is a Q-group.*

Proof. (1) $\to$ (2). We must show that $H_B \circ H_E$ and $H_E \circ H_B$ are naturally equivalent to the identity functors on their respective categories (for the formal definition of duality we refer to [AF]). Let $A \in \mathcal{F}_B$. Then $r_p(A) = r_p(B)$ for each p, as mentioned in corollary 3. Furthermore, it the follows from corollary 3 that $A \cong_{nat} A^{**} = H_E \circ H_B(A)$. We now consider the composite functor $H_B \circ H_E$.

Let $I \in \mathcal{I}_E$. Following lemma 2, decompose $B = B_1 \oplus \cdots \oplus B_n$ so that $E = R_1 \times \cdots \times R_n$ with $R_j = E(B_j)$. In order to show that $I \cong I^{**}$ it will be enough to check that $R_jI \cong Hom(Hom_{R_j}(R_jI, B_j), B_j)$, and therefore we may assume that $B = B_j$.

If $J \in \mathcal{I}_E$ with $J \leq E$ and $T = E/J$, then we obtain

$$0 \to E^* \to J^* \to T^* \to 0,$$

with $T^* = Ext_E(T, B)$. Since T^* is a bounded group, J^* is quasi-equal to $E^* = B$ so by corollaries 2 and 3, $Ext(J^*, B)$ is torsion-free. Therefore, in setting $T^{**} = Ext(T^*, B)$ we obtain the commutative rectangle

$$\begin{array}{ccccccccc} 0 & \longrightarrow & J & \longrightarrow & E & \longrightarrow & T & \longrightarrow & 0 \\ & & \downarrow & & \wr & & \downarrow & & \\ 0 & \longrightarrow & J^{**} & \longrightarrow & E^{**} & \longrightarrow & T^{**} & \longrightarrow & 0. \end{array}$$

It follows from the snake lemma that J^{**}/J is isomorphic to a submodule of T. Now refer to lemma 2 (c) with $I \in \mathcal{I}_E$. For any prime p, there is an E-module embedding $f : I \to E$ such that $E/f(I)$ is a torsion group with zero p-component (i.e. $I_p \cong E_p$ as E-modules). Then from the diagram above (with $f(I)$ in place of J) $(f(I))^{**}/f(I) \cong I^{**}/I$ has zero p-component. Since p was arbitrary, $I = I^{**}$.

(2) $\to$ (3). We remark that each of the functors defines a lattice anti-isomorphism between the appropriate categories. To show that E is hereditary, we will first consider the proper left ideals $I \in \mathcal{I}_R$ containing but not equal to pE, and note that I/pE is a nonzero left ideal of E/pE. Then by (2), $I = A^*$ for some $A \in \mathcal{F}_B$ containing B with $pA \subseteq B$ (we are identifying I with $Hom(A, B)$). Write $(\frac{1}{p}B)/B = A/B \oplus C/B$. We now have $(\frac{1}{p}B)^* = pE$

and $A + C = \frac{1}{p}B$, together implying $A^* \cap C^* = pE$. Furthermore, $A \cap C = B$, implying $A^* + C^* = E$.

For $J = C^*$, consider the sequence $0 \to pE \to I \oplus J \to E \to 0$ obtained above. This is split exact so that $I \oplus J = E \oplus pE$ and I is projective. For arbitrary $I \in \mathcal{I}_E$ there is composition series $I = I_0 \subseteq I_1 \subseteq \cdots \subseteq I_n = E$ with I_j/I_{j-1} a finite E-module bounded by a prime p_j. Assuming that I_j is projective, let A_j and $A_{j-1} \in \mathcal{F}_B$ correspond to I_j and I_{j-1} respectively. Then $(\frac{1}{p_j}A_j)/A_j = A_{j-1}/A_j \oplus C_{j-1}/A_j$ and as above $A^*_{j-1} \oplus C^*_{j-1} \cong A^*_j \oplus (\frac{1}{p_j}A_j)^* = I_j \oplus p_i I_j$. Hence I_{j-1} is projective. This proves that any I belonging to $\mathcal{I}_E$ is projective. To complete the proof that E is left hereditary, we would like to know that QE is semi-simple.

Given a left ideal I of E containing pE, from (2), $I = A^*$ for some $B \subseteq A \subseteq \frac{1}{p}B$. As above, if $(\frac{1}{p}B)/B = A/B \oplus C/B$, then $J = C^*$ satisfies $I/pE \oplus J/pE = E/pE$. Consequently ideals of E/pE are complemented, and E/pE is a (finite) semi-simple ring. Set $\mathcal{N} = \mathcal{J}(QE) \cap E$, the nil-radical of E; an ideal of E which is pure as an additive subgroup. Because QE is left Artinian, $\mathcal{N}$ is nilpotent. If p be any prime such that $pE \neq E$, then $(\mathcal{N} + pE)/pE$ is nilpotent as well, implying $(\mathcal{N} + pE)/pE \subseteq \mathcal{J}(R/pR) =$ Jacobson radical of E/pE. Due to the semi-simplicity of the ring E/pE, $\mathcal{N}$ must be contained in pE. But $\mathcal{N}$ is pure in E so that $p\mathcal{N} = pE \cap \mathcal{N} = \mathcal{N}$ and consequently $\mathcal{N}$ is divisible. Since B is reduced $\mathcal{N}$ must be 0 and therefore QE is a semi-simple ring.

Since the lattice of left ideals of QE is complemented, if I is any left ideal of E, then $QI \oplus L = QR$ for some left ideal L of QE. So, $I \oplus M$ is a left ideal in E of torsion index in E (as a subgroup), where $M = L \cap E$. This implies that $n \in I \oplus M$ for some $n \neq 0$. Consequently, $nE \subseteq I \oplus M$, and $I \oplus M \in \mathcal{I}_E$. From what we have shown above, $I \oplus M$ and hence I are both projective . Therefore E is left hereditary.

We next show that E is a noetherian ring. Having done so, we will have established that E is a left hereditary ring which is right noetherian, and therefore by a theorem of Small ([S]), E is also right hereditary. Let I be a right ideal of E. Because QE is semi-simple, as argued above, there is a right ideal M of E such that $I \oplus M$ is of finite index in E. So, if $\{I_j | j \geq 1\}$ denotes a chain of right ideals of E, then for some positive integer m, $QI_j = QI_m$ for all $j \geq m$. If M is a right ideal such that $I_m \oplus M$ is of finite index in E, then $I_j \oplus M$ is also of finite index in E. Clearly this implies that the right ideals I_j eventually stablize and E is right noetherian (a similar argument will show that E is left noetherian). It remains to show that B is a Q-group.

Let $A \in \mathcal{F}_B$ contain B as a subgroup of index p. With $A/B \oplus C/B = \frac{1}{p}B/B$, as above $A^* \oplus C^* \cong E \oplus E$ and $A^{**} \oplus C^{**} \cong B \oplus B$. It follows that A is B-generated. A direct induction argument then shows that any element of $\mathcal{F}_B$ is B-generated, so B is an S-group. Since we have shown that E is hereditary, the proof of theorem 5.9 in [A] shows that $IB \neq B$ for any right ideal I of E (i.e. B is (finitely) faithful).

Lemma 2 applies: Write $B = B_1 \oplus \cdots \oplus B_n$ as presented there. Due to condition (b) of that lemma, given $A \leq QB$ containing B as a subgroup of finite index, $A = A_1 \oplus \cdots \oplus A_n$ with A_j containing B_j as a subgroup of finite index. It is then enough to show that B_j is nearly isomorphic to A_j. Let p be given. Then $I = A^*_j$ is an ideal of finite index in $R = E(B_j)$. It then follows from lemma 2 (c) that $I_p \cong R_p$ as R_p-modules. In particular, since I is a finitely generated R-module, there are R-module maps $\theta : I \to R$ and $\gamma : R \to I$ and a nonzero integer t prime to p such that $\theta\gamma = t1_R$ and $\gamma\theta = t1_I$. The induced maps $f = \theta^*$ and $g = \gamma^*$ satisfy $fg = t1_{A_j}$ and $gf = t1_{B_j}$ so A_j is nearly isomorphic to B_j.

($3 \to 1$). As mentioned above, since E is hereditary, B is finitely faithful. If A is a subgroup of finite index in B then B is nearly isomorphic to B since B is a Q-group. Set $C = Hom(B, A)B$. Given a prime p, there is an integer k prime to p and maps $f : B \to A$ and $g : A \to B$ so that $fg = k1_A$ and $gf = k1_B$. In particular, $A/f(B)$ has index bounded by k implying that A/C has no p-torsion. Since p was arbitrary, $A = C$ and B is an S-group. □

We will say that B is *finitely E-reflexive* if any $I \in \mathcal{I}_E$ satisfies $I \cong_{nat} Hom(Hom_E(I, B), B)$ as left E-modules. Our final result follows immediately from corollary 3 and theorem 3.

COROLLARY 4. *Assume that B is a reduced group for which QE is left Artinian and $r_p(B)$ is finite for all p. The following are equivalent:*

(1) *B is a finitely faithful S-group.*
(2) *E is hereditary and any group having finite p-ranks for all primes p, and which is quasi-isomorphic to a B-reflexive group, is B-reflexive.*
(3) *B is finitely E-reflexive and any group quasi-isomorphic to B is B-reflexive.*

As an example of how corollary 4 might be applied, suppose B is a reduced group with $r_p(B) < \infty$ for all primes p. If $E(B)$ is hereditary, then either B is a finitely faithful S-group, or, some $A \in \mathcal{F}_B$ is quasi-isomorphic to $A^{**} = Hom_{E(B)}(Hom(A, B), B)$ but not isomorphic to A^{**}.

REFERENCES

[Al] Albrecht, Ulrich, A-Reflexive abelian groups, Houston Journal of Mathematics, 15 (4), (1989), 459-480.

[AF] Anderson, Frank W.; Fuller, Kent R.; *Rings and Categories of Modules*, 2nd-Edition, Springer-Verlag, 1992.

[A] Arnold, D. M., *Finite Rank Torsion-Free Abelian Groups and Rings*, Lecture Notes in Mathematics 931, Springer-Verlag (1982).

[Ar] ______________ Endomorphism rings and subgroups of finite rank torsion-free abelian groups, Rocky Mountain Journal of Mathematics, 12 (2), (1982), 241-256.

[Arn] ______________ Pure subgroups of finite rank completely decomposable groups, *Abelian Group Theory Proceedings, Oberwolfach 1981*, Lecture Notes in Mathematics 874, Springer-Verlag (1981).

[BW] Beaumont, Ross, Wisner, R., Rings with additive group which is a torsion-free group of rank two, Acta. Math. Sci. Hungar. 20 (1959), 105-116.

[FG] Faticoni, Theodore, Goeters, H. Pat, On torsion-free *Ext*, Communications in Algebra, 16 (9), (1988), 1853-1876.

[F] Fuchs, Laszlo, *Infinite Abelian Groups, Volumes I and II*, Academic Press, 1970 and 1973.

[G] Goeters, H. Pat, Fuchs' problem 43, Rocky Mountain Journal of Mathematics, 22 (1), (1992), 197-201.

[Go] ______________ Cobalanced torsion-free groups, Communications in Algebra, 21 (8), (1993), 2715-2726.

[Goe] ______________ When is $Ext(A, B)$ torsion-free?, and related problems, Communications in Algebra, 16 (8), 1605-1619, (1988).

[H] Huber, Martin, On reflexive modules and abelian groups, Journal of Algebra, 82, (1983), 469-487.

[HW] Huber, M., Warfield, R. B. Jr., Homomorphisms between cartesian powers of an abelian group, *A-belian Group Theory Proceedings, Oberwolfach 1981*, Lecture Notes in Mathematics 874, Springer-Verlag (1981).

[Rd] Reid, J. D.; On the ring of quasi-endomorphisms of a torsion-free abelian group, *Topics in Abelian Groups*, Scott, Foresman, and Co., 1962.

[R] Rotman, Joseph, *Homological Algebra*, Academic Press, (1979).

[S] Small, L., Semihereditary rings, Bulletin of the American Mathematical Society, 73, (1967), 656-658.

[W] Warfield, R. B. Jr., Extensions of torsion-free abelian groups of finite rank, Archiv der Mathematik, 23, (1972), 145-150.

[Wa] Warfield, R. B. Jr., Homomorphisms and duality for torsion-free groups, Math Z., 107, (1968), 189-200.

Units of Modular Group Algebras of Isotype Subgroups of D.S.C.'s

PAUL HILL[1] AND WILLIAM ULLERY
Auburn University, Auburn, AL

ABSTRACT. Suppose F is a perfect field of characteristic $p \neq 0$ and H is an abelian p-group. If H is an isotype subgroup of a direct sum of countable reduced p-groups, it is shown that the group of units of the group algebra $F(H)$ is of the form $F^* \times H \times T$, where F^* is the multiplicative group of F and T is a totally projective p-group. More particularly, we show that H is a direct factor of the p-group $V(H)$ of normalized units and that $V(H)/H$ is totally projective.

Suppose F is a perfect field of characteristic $p \neq 0$. If H is an abelian p-group written multipicatively, let $U(H)$ denote the group of units of the group algebra $F(H)$, and write $V(H)$ for the group of normalized units (that is, those elements of $F(H)$ with coefficients summing to 1). It is completely elementary that $V(H)$ is a p-subgroup of $U(H)$ and $U(H) = F^* \times V(H)$, where F^* is the multiplicative group of F. This essentially reduces the structure theory of $U(H)$ to that of $V(H)$.

It is an open question whether or not $V(H)/H$ is totally projective for an arbitrary p-group H. However, an affirmative answer is known in the following cases.

Theorem 1 ([M1]). *If H is totally projective, then $V(H)/H$ is totally projective and H is a direct factor of $V(H)$.*

Theorem 2 ([HU]). *If $|H| \leq \aleph_1$, then $V(H)/H$ is totally projective and H is a direct factor of $V(H)$.*

We remark that in Theorem 2 it is perhaps surprising that the complementary factor of H in $V(H)$ is totally projective, no matter what the structure of H is (as long as $|H| \leq \aleph_1$).

It is our purpose here to extend the preceding results to the case where H is an arbitrary isotype subgroup of a d.s.c. As usual, by a d.s.c. we mean a direct sum of countable, reduced p-groups. Since our groups will be written multiplicatively, we should use the term "coproduct" instead of "direct sum", but nevertheless we shall continue to use the term "d.s.c." even in the multiplicative context. In view of the fact that Hill and Megibben have shown in [HM] that, in a certain sense, the structure of isotype subgroups of d.s.c.'s is no less complicated than the structure of arbitrary p-groups, our theorem should give credence to the general conjecture that $V(H)/H$ is totally projective for any p-group H.

Before proceeding to some preliminary results, we remind the reader of several definitions, some of which may be unfamiliar to nonspecialists. In the sequel, all

[1]Supported by NSF grant DMS 92-08199

groups are assumed to be abelian, but we deviate from the usual convention and write our groups multiplicatively. Thus, if G is a p-group, $G^p = \{g^p : g \in G\}$, and for each ordinal α we define G^α as follows.

$$G^\alpha = \begin{cases} G & \text{if } \alpha = 0 \\ (G^{\alpha-1})^p & \text{if } \alpha - 1 \text{ exists} \\ \bigcap_{\beta<\alpha} G^\beta & \text{if } \alpha \text{ is a limit.} \end{cases}$$

Since the prime p is fixed, there is no ambiguity in the abbreviated notation. The smallest ordinal α with $G^\alpha = G^{\alpha+1}$ is called the *length* of G. Recall that a reduced p-group is a d.s.c. if and only if it is totally projective of length not exceeding ω_1. A subgroup H of G is *isotype* if $G^\alpha \cap H = H^\alpha$ for all α and is *nice* if $(G/H)^\alpha = G^\alpha H/H$. A subgroup is *balanced* if it is both isotype and nice. For $g \in G$ we donote the *height* of g in G by $|g|_G$. Thus, $|g|_G = \alpha$ means $g \in G^\alpha \setminus G^{\alpha+1}$. If $g \in G^\alpha$ for all α, we set $|g|_G = \infty$ with the understanding that $\alpha < \infty$ for every ordinal α. If $H \subseteq G$, the *height spectrum* of H in G is the set $\{\alpha : |h|_G = \alpha \text{ for some } h \in H\}$. H has *countable height spectrum* if the height spectrum of H has cardinality not exceeding $\aleph_0$. Two subgroups A and B are *compatible* in G, written $A||B$, if whenever $a \in A$ and $b \in B$, there exists $z \in A \cap B$ such that $|ab|_G \leq |az|_G$. Observe that compatibility is a symmetric relation. Finally, a subgroup H of G is *separable* in G if for each $g \in G$ there is a corresponding countable subgroup C of H such that

$$\sup\{|gh|_G : h \in H\} = \sup\{|gc|_G : c \in C\}.$$

Clearly every nice subgroup is separable.

Lemma 3. *Suppose H is a subgroup of a p-group G. If for each $g \in G$ there exists a subgroup C of G such that $g \in C$, C has countable height spectrum and $C||H$, then H is separable in G.*

Proof. Since $C||H$, for each $h \in H$ there exists $z \in C \cap H$ such that $|gh|_G \leq |gz|_G$. Therefore, $\sup\{|gh|_G : h \in H\} = \sup\{|gz|_G : z \in C \cap H\}$. But C has countable height spectrum, so there exists a countable subset $\{z_n : n < \omega_0\} \subseteq C \cap H$ with $\{|gz|_G : z \in C \cap H\} = \{|gz_n|_G : n < \omega_0\}$. The result follows. □

A partial converse of the next lemma appears in [FH].

Lemma 4. *Suppose N, M and H are subgroups of a p-group G with $N \subseteq M$ and N nice in G. If $H||M$ in G, then $(HN/N)||(M/N)$ in G/N.*

Proof. If $hN \in HN/N$ and $mN \in M/N$ for some $h \in H$ and $m \in M$, then $|(hN)(mN)|_{G/N} = |hmn|_G$ for some $n \in N$ since N is nice. But $H||M$ and $mn \in M$ imply that there exists $z \in H \cap M$ such that $|hmn|_G \leq |hz|_G$. Therefore, $|(hN)(mN)|_{G/N} \leq |(hN)(zN)|_{G/N}$ with $zN \in (H \cap M)N/N = (HN/N) \cap (M/N)$. □

The next result is implicit in [H]. Nevertheless, we include its short proof below.

Lemma 5. *Suppose $N \subseteq M$ are subgroups of a p-group G with N nice in G. If M/N is a separable subgroup of G/N, then M is a separable subgroup of G.*

Proof. If $g \in G$, there is a countable subset $\{m_k : k < \omega_0\} \subseteq M$ with the property that for every $m \in M$ there exists $k < \omega_0$ such that $|gmN|_{G/N} \leq |gm_kN|_{G/N}$. For each k, select $n_k \in N$ with $|gm_kN|_{G/N} = |gm_kn_k|_G$. Then, for every $m \in M$, there exists $k < \omega_0$ such that $|gm|_G \leq |gmN|_{G/N} \leq |gm_kN|_{G/N} = |gm_kn_k|_G$. Since $m_kn_k \in M$ for every k, the proof is complete. □

In order to apply the previous lemmas to the group algebra setting, we need methods for recognizing compatible pairs of subgroups and nice subgroups inside $V(G)$, G a p-group. The following result provides the information we require. Its proof can be found in the sources cited and will be omitted here.

Lemma 6 ([M1], [M2], [HU]). *Suppose H is a subgroup of a p-group G.*

(a) *$GV(H) \| V(A)$ in $V(G)$ for every subgroup A of G.*
(b) *$GV(H)$ is a nice subgroup of $V(G)$.*
(c) *$V(H)$ is nice (respectively isotype) in $V(G)$ if and only if H is nice (respectively isotype) in G.*
(d) *If H is isotype in G, $GV(H)$ is balanced in $V(G)$. In particular, G is balanced in $V(G)$.*

Suppose C is a subgroup of a p-group G. The natural quotient map $G \twoheadrightarrow G/C$ induces a group-epimorphism $V(G) \twoheadrightarrow V(G/C)$ with kernel we denote by $K_G(C)$. In our next two lemmas we collect some pertinent facts regarding $K_G(C)$.

Lemma 7. *Suppose A is a subgroup of a p-group G and $A = B \times C$ for subgroups B and C. Then, $V(A) = V(B) \times K_A(C)$ and $K_A(C) = K_G(C) \cap V(A)$.*

Proof. The natural projection $A \twoheadrightarrow B$ induces a group-epimorphism $\pi : V(A) \twoheadrightarrow V(B)$ which is split by the inclusion map $V(B) \rightarrowtail V(A)$. Since $\operatorname{Ker} \pi = K_G(C) \cap V(A)$, the result follows. □

If $v \in F(G)$, write $\operatorname{supp} v$ for the *support* of v ; that is, those elements of G appearing nontrivially in v. It is clear that $C \cap \operatorname{supp} v \neq \emptyset$ whenever $v \in K_G(C)$.

Lemma 8. *Suppose C is a subgroup of a p-group G. If $1 \neq v \in K_G(C)$, there exists $c \in C$ such that $c \neq 1$ and $|v|_{V(G)} \leq |c|_G$.*

Proof. Since the field F is perfect, it follows easily that $|v|_{V(G)} = \min\{|g|_G : g \in \operatorname{supp} v\}$. Therefore, if $C \cap \operatorname{supp} v \neq \{1\}$, the result follows. So, it remains to consider the case $C \cap \operatorname{supp} v = \{1\}$. Since $v \neq 1$, we have $v = d_1 1 + d_2 g_2 + \cdots + d_n g_n$ for nonzero $d_1, d_2, \ldots, d_n \in F$ with $n \geq 2$, and distinct $g_2, \ldots, g_n \in G \setminus C$. From $v \in K_G(C)$ and $g_i \notin C$ for all i, we conclude that $n \geq 3$ and, after reindexing, $|v|_{V(G)} = |g_2|_G$ and $g_2C = g_3C$. Thus, there exists a nonidentity $c \in C$ such that $g_3 = g_2c$. Therefore, $|g_2|_G = |v|_{V(G)} \leq |g_2c|_G$ and we have $|c|_G \geq |g_2|_G$. □

The proof of our main result makes use of the result of W. May [M1] stated above as Theorem 1, and a result of P. Hill [H] which gives an Axiom 3 characterization of the isotype subgroups of totally projective groups which are themselves totally projective. The latter result has the following equivalent formulation.

Theorem 9 ([H]). *Suppose H is an isotype subgroup of a totally projective p-group G. Then H is totally projective if and only if there is a smooth chain*

$$H = A_0 \subseteq A_1 \subseteq \cdots \subseteq A_\alpha \subseteq \ldots \qquad (\alpha < \mu)$$

of subgroups of $G = \bigcup_{\alpha<\mu} A_\alpha$ such that, for each $\alpha < \mu$, A_α is a separable subgroup of G and $|A_{\alpha+1}/A_\alpha| \leq \aleph_0$ whenever $\alpha + 1 < \mu$.

We are now ready for the statement and proof of our main result. As mentioned previously, recall that a p-group is a d.s.c. if and only if it is reduced and totally projective of length not exceeding ω_1.

Theorem 10. *Suppose F is a perfect field of characteristic $p \neq 0$ and H is an isotype subgroup of a d.s.c. p-group. Then $V(H)/H$ is a d.s.c. and H is a direct factor of $V(H)$.*

Proof. Suppose G is a d.s.c. p-group which contains H as an isotype subgroup. In view of Theorem 2, we may assume without loss that $|G| \geq \aleph_2$. Thus, there is a limit ordinal μ such that $G = \coprod_{i<\mu} C_i$ with $|C_i| \leq \aleph_0$ for all i. Set $P_0 = 1$ and if $1 \leq \alpha < \mu$ define $P_\alpha = \coprod_{i<\alpha} C_i$. Therefore,

$$1 = P_0 \subseteq P_1 \subseteq \cdots \subseteq P_\alpha \subseteq \ldots \qquad (\alpha < \mu)$$

is a smooth chain of direct factors of $G = \bigcup_{\alpha<\mu} P_\alpha$, with $|P_{\alpha+1}/P_\alpha| \leq \aleph_0$ holding for all $\alpha < \mu$. As a consequence, there is a smooth chain

$$GV(H) \subseteq GV(H)V(P_1) \subseteq \cdots \subseteq GV(H)V(P_\alpha) \subseteq \ldots \qquad (\alpha < \mu)$$

from $GV(H)$ to $V(G) = \bigcup_{\alpha<\mu} GV(H)V(P_\alpha)$.

We claim that each $GV(H)V(P_\alpha)$ is a separable subgroup of $V(G)$. In fact, we show more generally that if W is a nice subgroup of $V(G)$ with $V(P_\alpha) \subseteq W \subseteq V(P_{\alpha+1})$, then $GV(H)W$ is a separable subgroup of $V(G)$. To obtain this result, Lemma 5 implies it is enough to demonstrate that $GV(H)W/W$ is separable in $V(G)/W$. Suppose vW is an arbitrary (but fixed) element of $V(G)/W$ with $v \in V(G) \setminus W$. Since $\operatorname{supp} v$ is finite, there exists a direct factor P of G such that $v \in V(P)$, $P_{\alpha+1} \subseteq P$ and $|P/P_{\alpha+1}| \leq \aleph_0$. By Lemma 6(a), $GV(H) \parallel V(P)$; so, from Lemma 4 we obtain $(GV(H)W/W) \parallel (V(P)/W)$ in $V(G)/W$. Therefore, to conclude from Lemma 3 that $GV(H)W/W$ is separable in $V(G)/W$, it suffices to show that $V(P)/W$ has countable height spectrum in $V(G)/W$. Consider the natural map $V(P)/V(P_\alpha) \twoheadrightarrow V(P)/W$ with kernel $W/V(P_\alpha)$ nice in $V(G)/V(P_\alpha)$. Recall from Lemma 6(c) that $V(P)$ is balanced in $V(G)$. Hence, $W/V(P_\alpha)$ is nice in $V(P)/V(P_\alpha)$, and both $V(P)/V(P_\alpha)$ and $V(P)/W$ are isotype in the respective quotient groups $V(G)/V(P_\alpha)$ and $V(G)/W$. We conclude that the height spectrum of $V(P)/W$ in $V(G)/W$ is a subset of the height spectrum of $V(P)/V(P_\alpha)$ in $V(G)/V(P_\alpha)$. Therefore, to obtain the conclusion that $GV(H)W$ is separable in $V(G)$, it is sufficient to show that $V(P)/V(P_\alpha)$ has countable height spectrum in $V(G)/V(P_\alpha)$.

Suppose $wV(P_\alpha) \in V(P)/V(P_\alpha)$ for some $w \in V(P) \setminus V(P_\alpha)$. By our choice of P and the construction of the P_α's, $P = P_\alpha \times C$ for some countable direct factor C of G. Since $V(P) = V(P_\alpha) \times (K_G(C) \cap V(P))$ by Lemma 7, we may assume that $w \in K_G(C) \cap V(P)$. Thus, noting that $V(P)$, $V(P_\alpha)$ and $K_G(C) \cap V(P)$ are all direct factors of $V(G)$, $|wV(P_\alpha)|_{V(G)/V(P_\alpha)} = |w|_{V(G)}$. Because C is countable, $G^{\omega_1} = 1$ and ω_1 is not cofinal with ω_0, the heights in G of the nonidentity elements of C are bounded by some countable ordinal. From Lemma 8 we now have that $K_G(C)$, and hence $K_G(C) \cap V(P)$, has countable height spectrum in $V(G)$. Therefore, $V(P)/V(P_\alpha)$ has countable height spectrum in $V(G)/V(P_\alpha)$ and the desired conclusion is obtained; that is, for all $\alpha < \mu$, $GV(H)W$ is a separable subgroup of $V(G)$ whenever W is a nice subgroup of $V(G)$ with $V(P_\alpha) \subseteq W \subseteq V(P_{\alpha+1})$.

By Theorem 1 and Lemma 7, $V(P_{\alpha+1})/V(P_\alpha)$ is a d.s.c. for all $\alpha < \mu$. Therefore, since $V(P_\alpha)$ is nice in $V(P_{\alpha+1})$, we obtain a smooth chain

$$V(P_\alpha) = W_{\alpha,0} \subseteq W_{\alpha,1} \subseteq \cdots \subseteq W_{\alpha,\beta} \subseteq \cdots \qquad (\beta < \gamma)$$

of nice subgroups of $V(P_{\alpha+1}) = \bigcup_{\beta<\gamma} W_{\alpha,\beta}$, where γ is an ordinal depending on the choice of α and $|W_{\alpha,\beta+1}/W_{\alpha,\beta}| \leq \aleph_0$ whenever $\beta+1 < \gamma$. Observe that $V(P_{\alpha+1})$ is balanced in $V(G)$ by Lemma 6(c). Consequently, each $W_{\alpha,\beta}$ is also a nice subgroup of $V(G)$. By what we have shown above, the subgroups $GV(H)W_{\alpha,\beta}$ form a smooth chain of separable subgroups of $V(G)$ from $GV(H)V(P_\alpha)$ to $GV(H)V(P_{\alpha+1})$ with each link of the chain having countable index in the next. Putting these chains together, we obtain a smooth chain of separable subgroups from $GV(H)$ to $V(G)$ with consecutive countable factors. Since $GV(H)$ is isotype in $V(G)$ by Lemma 6(d) and $V(G)$ is totally projective by Theorem 1, it follows from Theorem 9 that $GV(H)$ is totally projective. Also, by Theorem 1, G is a direct factor of $V(G)$ and hence one of $GV(H)$ as well. From this we conclude that $V(H)/H \cong GV(H)/G$ is totally projective and, in fact, a d.s.c. by the niceness of H in $V(H)$. Finally, in view of the facts that totally projective groups are balanced-projectives (see [F]) and that H is balanced in $V(H)$ by Lemma 6(d), H is a direct factor of $V(H)$. □

In conclusion, we present two corollaries of Theorem 10. Our first is simply a consequence of the identity $U(H) = F^* \times V(H)$ mentioned previously.

Corollary 11. *Let $U(H)$ denote the full group of units of the group algebra $F(H)$. If H is an isotype subgroup of a d.s.c., then*

$$U(H) = F^* \times H \times T$$

for some d.s.c. T.

Recall that two p-groups H and H' are *similar* if $H \times T \cong H' \times T$ for some totally projective T. For an indication of the importance of the similarity classes of isotype subgroups of d.s.c. p-groups, we refer the reader to Theorem 6.2 of [HM]. This theorem states that to each p-group A there corresponds an isotype subgroup $A^\star$ of a d.s.c. such that arbitrary p-groups A and B are isomorphic if and only if $A^\star$ and $B^\star$ are similar. For our final result it is not necessary to assume that the field of coefficients is perfect.

Corollary 12. *Suppose k is a field of characteristic $p \neq 0$ and H is an isotype subgroup of a d.s.c. p-group. If H' is a group such that $k(H) \cong k(H')$ as k-algebras, then H' is also an isotype subgroup of a d.s.c. p-group and H and H' are similar.*

Proof. If F is a perfect extension of k, from

$$F(H) \cong F \otimes_k k(H) \cong F \otimes_k k(H') \cong F(H')$$

we conclude that there is an F-algebra isomorphism $F(H) \cong F(H')$ which may be assumed to preserve augmentation. This algebra-isomorphism restricts to a group-isomorphism $V(H) \cong V(H')$, where the groups of normalized units are from the respective group algebras with coefficients in F. If G is a d.s.c. p-group that contains H as an isotype subgroup, then $V(H)$ is an isotype subgroup of $V(G)$, a d.s.c. by Theorem 1. Thus, $V(H')$ is an isotype subgroup of a d.s.c.; moreover, since H' is balanced in $V(H')$, H' is an isotype subgroup of a d.s.c. as well. By Theorem 10,

$$H \times T_1 = V(H) \cong V(H') = H' \times T_2$$

for d.s.c.'s T_1 and T_2. By selecting a d.s.c. T such that $T \cong T_1 \times T \cong T_2 \times T$, we obtain $H \times T \cong H' \times T$. □

References

[F] L. Fuchs, *Infinite Abelian Groups*, vol. II, Academic Press, New York, 1973.

[FH] L. Fuchs and P.Hill, *The balanced-projective dimension of abelian p-groups*, Trans. Amer. Math. Soc. **293** (1986), 90–112.

[H] P. Hill, *Isotype subgroups of totally projective groups*, Abelian Group Theory, Lecture Notes in Math. 874, Springer-Verlag, New York, 1981, pp. 305–321.

[HM] P. Hill and C. Megibben, *On the theory and classification of abelian p-groups*, Math. Zeit. **190** (1985), 17–38.

[HU] P. Hill and W. Ullery, *A note on a theorem of May concerning commutative group algebras*, Proc. Amer. Math. Soc. **110** (1990), 59–63.

[M1] W. May, *Modular group algebras of simply presented abelian groups*, Proc. Amer. Math. Soc. **104** (1988), 403–409.

[M2] W. May, *The direct factor problem for modular group algebras*, Contemporary Math. **93** (1989), 303–308.

Department of Mathematics, Auburn University, Auburn, Alabama 36849–5310
E-mail address: hillpad@mail.auburn.edu

Department of Mathematics, Auburn University, Auburn, Alabama 36849–5310
E-mail address: ullery@mail.auburn.edu

Calculating Indicators in a Class of Mixed Modules

Ralf Jarisch, Universitat Würzburg, Würzburg, Germany

Otto Mutzbauer, Universitat Würzburg, Würzburg, Germany

Elias Toubassi, University of Arizona, Tucson, AZ

1. Introduction

This paper deals with the indicators of mixed modules over discrete valuation domains. In particular the results hold for local mixed abelian groups.

We study a class of mixed modules $\mathcal{H}$ with the property that the torsion submodule is a direct sum of cyclics and the quotient modulo torsion is divisible of arbitrary rank. In this paper we work with representations of modules as extensions of torsion by torsion-free and make use of the ideas established in [4] particularly that of a relation system.

We develop methods to calculate with relation systems of rank 1 and show that for this class the p-indicators are either of finite type, ω-type or ∞-type. Finally we show that this characterization of the class $\mathcal{H}$ leads to a natural mapping into $\operatorname{Ext}(K, T)$, with the kernel consisting of "small" relation systems. Moreover, the image provides information on $p^\omega G$ and the divisible part of G.

2. Preliminaries

Let $\mathbb{N}$ denote the set of natural numbers and $\mathbb{N}_0 = \mathbb{N} \cup \{0\}$. Let R denote a *discrete valuation domain*, i. e. a local principal ideal domain with prime p and quotient field K.

All modules are understood to be R-modules. Let us quickly recall the main definitions and conventions. We adopt the generators and relations description of mixed modules G in the class $\mathcal{H}$ introduced in [4]. Let G be such a mixed module of torsion-free rank d with torsion submodule $\mathbf{t}\,G = \bigoplus_{i\in\mathbb{N}} \bigoplus_{v\in I_i} Rx_i^v$ of isomorphism type $\lambda = (s_i \mid i \in \mathbb{N})$, where $s_i = |I_i|$ and $\operatorname{ann} x_i^v = p^i R$ for $i \in \mathbb{N}, v \in I_i$. Then the quotient $G/\mathbf{t}\,G$ is a vector space over K of dimension d. Let I be a set of cardinality d, then the subset

$$B = \{x_i^v, a_{i-1}^k \mid i \in \mathbb{N}, v \in I_i, k \in I\} \subset G$$

is called a *basic generating system of* G, if

(1) $\{x_i^v \mid i \in \mathbb{N}, v \in I_i\}$ is a basis of $\mathbf{t}\,G$ with $\operatorname{ann} x_i^v = p^i R$ for all $i \in \mathbb{N}, v \in I_i$,
(2) $G/\mathbf{t}\,G = \bigoplus_{k\in I} K\bar{a}_0^k$,

where $\bar{a}_{i-1}^k = a_{i-1}^k + \mathbf{t}\,G$ and $p\bar{a}_i^k = \bar{a}_{i-1}^k$ for all $i \in \mathbb{N}, k \in I$. The definition implies that the elements a_0^k, $k \in I$, are independent. It is easy to see that

$$G = \langle x_i^v, a_{i-1}^k \mid i \in \mathbb{N}, v \in I_i, k \in I \rangle.$$

A basic generating system intrinsically defines a series of equations with coefficients in R that describe the relations among the generators. Modeling upon these relations we introduce the concept of an abstract array $(\alpha_{i-1,j}^{k,u})$ with $i, j \in \mathbb{N}, u \in I_j, k \in I$ having entries in R and which is *row finite in j and u*, i. e. for a fixed pair (k, i), $\alpha_{i,j}^{k,u} \in p^j R$, for almost all pairs (j, u). Such arrays $(\alpha_{i-1,j}^{k,u})$ are called *relation arrays*. If a torsion module $\mathbf{t}$ is a direct sum of cyclics and of isomorphism type $\lambda = (|I_i| \mid i \in \mathbb{N})$ and $d = |I|$ then we call the relation array $(\alpha_{i-1,j}^{k,u})$ of *format* (λ, d). Note that all relation arrays of a mixed module in the class $\mathcal{H}$ have the same format.

The following result gives a description of mixed modules in the class $\mathcal{H}$ by generators and relations.

Proposition 2.1. ([4, Proposition 3]) *Every mixed module G in the class $\mathcal{H}$ has a basic generating system $B = \{x_i^v, a_{i-1}^k \mid i \in \mathbb{N}, v \in I_i, k \in I\}$ and a relation array $(\alpha_{i-1,j}^{k,u})$ relative to B given by*

$$(1) \qquad pa_i^k = a_{i-1}^k + \sum_{j\in\mathbb{N}} \sum_{u\in I_j} \alpha_{i-1,j}^{k,u} x_j^u, \qquad i \in \mathbb{N}, k \in I.$$

The entries $\alpha_{i-1,j}^{k,u}$ are uniquely determined modulo $p^j R$ by the choice of a_i^k.

By [4, Theorem 4] all relation arrays are realized, i. e. there is a module in $\mathcal{H}$ having a basic generating system with a prescribed relation array.

There is a compact way of discussing relations namely, the so called relation systems. For this we need some notation. Let $R[x]$ denote the ring of polynomials over R with the usual degree function. Let $R[[x]]$ be the ring of power series with coefficients in R, cf. Jacobson [2, II.7.9]. The order, $\mathfrak{o}(\alpha)$, of a non-zero $\alpha = \sum_{i\geq 0} \alpha_i x^i \in R[[x]]$ is h if $0 = \alpha_0 = \ldots = \alpha_{h-1} \neq \alpha_h$ and define $\mathfrak{o}(0)$ to be ∞. For a power series α let $\alpha^{(s)} = \sum_{i=0}^{s-1} \alpha_i x^i$ denote the polynomial given by the s-th partial sum.

A matrix $A = (\alpha_{lh})$ with entries in $R[[x]]$ is called *order finite over $R[[x]]$ in h* if for each l the number $|\{h \mid \mathfrak{o}(\alpha_{lh}) \leq m\}|$ is finite for all $m \in \mathbb{N}_0$. If $\mathbb{N}$ is the index set of the columns, a matrix A over $R[[x]]$ is order finite if and only if for each row l and any natural number m there is a column index $h_0 = h_0(l)$ such that $\mathfrak{o}(\alpha_{lh}) > m$ for all $h \geq h_0$. We extend in the obvious way the formation of partial sums to matrices $A = (\alpha_{i,j})$ over $R[[x]]$, namely, define $A^{(s)} = \left(\alpha_{i,j}^{(s)}\right)$.

Lemma 2.2. [5, Lemma 3] *Let $(\alpha_{i-1,j}^{k,u})$ be a relation array of format (λ, d). Then the system*

$$\alpha = (\alpha_j \mid j) = \Big(\sum_{i=0}^{\infty} (\alpha_{i,j}^{k,u} + p^j R) x^i \mid j \in \mathbb{N}\Big),$$

has order finite $d \times s_j$-matrices $\alpha_j = \left(\alpha_j^{k,u}\right)$ with entries in the ring $\left(R/p^j R\right)[[x]]$ and

$$\lim_{j\to\infty} \left(\mathfrak{o}^k(\alpha_j)\right) = \infty$$

for each $k \in I$ and $j \in \mathbb{N}$, where $\mathfrak{o}^k(\alpha_j) = \min\{\mathfrak{o}(\alpha_j^{k,u}) \mid u \in I_j\}$.

Conversely an array $(\alpha^{k,u}_{i-1,j})$ *such that the corresponding* α *satisfies these two conditions is a relation array.*

A sequence $\alpha = (\alpha_j \mid j)$ of order finite matrices α_j is called a *relation system of format* (λ, d), corresponding to the relation array $(\alpha^{k,u}_{i-1,j})$ if $\lim_{j\to\infty}\big(\mathfrak{o}^k(\alpha_j)\big) = \infty$ for each $k \in I$. If $(\alpha^{k,u}_{i-1,j})$ is a relation array of a module G for a basic generating system B then the relation system $\alpha = (\alpha_j \mid j)$ corresponding to $(\alpha^{k,u}_{i-1,j})$ is called a *relation system of* G *relative to* B and *to* $(\alpha^{k,u}_{i-1,j})$. If a relation system corresponds to an array $(\alpha^{k,u}_{i-1,j})$ then the array is called an *expansion of the relation system.* Note that expansions of relation systems are relation arrays. Different expansions of a relation system result in the same relations given by formula (1) in Proposition 2.1.

A relation array $(\alpha^{k,u}_{i-1,j})$ is called *small* if $\left(\left(\sum_{i=0}^{j-1}\alpha^{k,u}_{s+i,j}p^i\right)^{k,u}_{s,j}\right)$ is also a relation array. The relation system α is said to be *small* if it has one small expansion. By [4] this implies all expansions are small. A relation system $\alpha = (\alpha_j \mid j)$ is called *restricted*, if the coefficients of the power series $\alpha^{k,u}_j \in R[[x]]/p^jR[[x]]$ are either 0 or units in R/p^jR, i. e. if $\alpha^{k,u}_{i,j} \in (R\setminus pR)\cup\{0\}$ for all $i \in \mathbb{N}_0$.

Let G be an arbitrary R-module with $g \in G$. Let $\mathrm{h}^*(g)$ denote the generalized p-height of g in G, cf. [1, Section 37]. Then

$$\mathbb{H}(g) = \big(\mathrm{h}^*(g), \mathrm{h}^*(pg), \ldots, \mathrm{h}^*(p^n g), \ldots\big)$$

is the *p-indicator of g*. We use the terms gap and equivalence of indicators as in [1, Section 103]. For a module G of torsion-free rank 1 we have a unique equivalence class of indicators of its elements, $\mathbb{H}(G)$, the *indicator of G*.

3. Relation Systems of Rank 1

Let $\mathcal{H}^1$ be the class of all modules of torsion-free rank 1 in the class $\mathcal{H}$. In this section we consider modules in the class $\mathcal{H}^1$. Thus the corresponding relation systems are of rank 1, too. Therefore we omit the index k and the relation arrays have the form $(\alpha^u_{i-1,j})$. Two relation systems of the same format are called *similar*, $\alpha \sim \beta$, if their difference is small. It is clear that smallness is an equivalence relation.

Lemma 3.1. *Let* G *be a module in the class* $\mathcal{H}^1$ *with relation system* α *(of rank 1). Then* G *has a restricted relation system which is similar to* α. *Moreover, for an arbitrary relation system* α *of rank* 1*:*

$$x^{-1}\big(p\alpha - (p\alpha)^{(1)}\big) = p\big(x^{-1}(\alpha - \alpha^{(1)})\big) \sim \alpha. \tag{2}$$

Proof. By Proposition 2.1 there is a basic generating system $B = \{a_{i-1}, x^v_i \mid i \in \mathbb{N}, v \in I_i\}$ of G such that $pa_i = a_{i-1} + \sum_{j\in\mathbb{N}}\sum_{u\in I_j}\alpha^u_{i-1,j}x^u_j$ for all $i \in \mathbb{N}$. We may assume that $\alpha^u_{i-1,j} = 0$ for all $i \in \mathbb{N}$ and all pairs $(j,u) \in \mathbb{N}\times I_j$ if $\alpha^u_{i-1,j} \in p^jR$, since in this case $\alpha^u_{i-1,j}x^u_j = 0$.

If $j \in \mathbb{N}, u \in I_j$ then define $\gamma^u_{i-1,j} \in R$ recursively as follows:

$$\begin{aligned}
\gamma^u_{0,j} &= 0,\\
\gamma^u_{i,j} &= 0, \text{ if } \alpha^u_{i-1,j} - \gamma^u_{i-1,j} \in \{0\}\cup(R\setminus pR),\\
\gamma^u_{i,j} &= -p^{-1}(\alpha^u_{i-1,j} - \gamma^u_{i-1,j}), \text{ if } \alpha^u_{i-1,j} - \gamma^u_{i-1,j} \in pR\setminus\{0\}.
\end{aligned}$$

Now we show by induction on i that $(\gamma^u_{i-1,j})$ is a relation array. To start with, observe that for all pairs $(j,u) \in \mathbb{N}\times I_j$ we have $\gamma^u_{0,j} = 0$. Assume that $i > 1$ and that $\gamma^u_{k,j} = 0$

for almost all pairs $(j,u) \in \mathbb{N} \times I_j$, where $k < i$. Since $\alpha^u_{i-1,j}$ as well as $\gamma^u_{i-1,j}$ are zero for almost all pairs $(j,u) \in \mathbb{N} \times I_j$ using the assumption and the fact that $(\alpha^u_{i-1,j})$ is a relation array. Thus the same holds true for $\alpha^u_{i-1,j} - \gamma^u_{i-1,j}$ and hence for $\gamma^u_{i,j}$. Now let $(\beta^u_{i,j})$ be the relation array defined by

$$\beta^u_{i-1,j} = \alpha^u_{i-1,j} - \gamma^u_{i-1,j} + p\gamma^u_{i,j}.$$

Since $\gamma^u_{i-1,j}$ is the carryover mod p, $\beta^u_{i,j} \in \{0\} \cup R \setminus pR$, and so the corresponding relation system β is restricted. But $\alpha - \beta = \gamma - px^{-1}(\gamma - \gamma^{(1)})$ and by [5, Lemma 5] we get $\alpha - \beta$ small. Again by [5, Lemma 6] the relation system β is also realized in G and by [4, Theorem 6] we have that β is small if and only if α is small.

Let α be an arbitrary relation system of rank 1. Applying the restriction procedure to $p\big(x^{-1}(\alpha - \alpha^{(1)})\big)$ and to $\alpha - \alpha^{(1)}$ results in the same relation array, hence $x^{-1}\big(p\alpha - (p\alpha)^{(1)}\big) = p\big(x^{-1}(\alpha - \alpha^{(1)})\big) \sim \alpha - \alpha^{(1)} \sim \alpha$. □

Theorem 3.2. *A restricted relation system α of rank 1 is small if and only if*

$$\lim_{j\to\infty} \big(\mathfrak{o}(\alpha_j) - j\big) = \infty. \tag{3}$$

Proof. Assume (3) and let $i \in \mathbb{N}_0$ be fixed, then by (3), $\mathfrak{o}(\alpha_j) - j \geq i$, i. e. $\alpha^u_{s,j} \in p^j R$ for all sufficiently large j and all $s < i + j$. Hence $\sum_{s=i}^{i+j-1} \alpha^u_{s,j} p^{s-i} \in p^j R$ for all sufficiently large j, and hence $\left(\sum_{s=i}^{i+j-1} \alpha^u_{s,j} p^{s-i}\right)^u_{i,j}$ is a relation array; thus α is small. Note that so far the hypothesis that α is restricted, was not used.

Conversely, let α be small and restricted. Let $i \in \mathbb{N}_0$ be fixed. Then for almost all pairs $(j,u) \in \mathbb{N} \times I_j$ we have that $\sum_{s=i}^{i+j-1} \alpha^u_{s,j} p^{s-i} \in p^j R$. Hence there exists a minimal $j_i \in \mathbb{N}$ such that $\sum_{s=i}^{i+j-1} \alpha^u_{s,j} p^{s-i} \in p^j R$ for all pairs (j,u), $j \geq j_i, u \in I_j$. Let $j \geq j_i$. Assume there is a minimal $i_0 \in \{i, \ldots, i+j-1\}$ such that $\alpha^{u_0}_{i_0,j} \neq 0$ for some $u_0 \in I_j$. Then

$$\alpha^{u_0}_{i_0,j} p^{i_0 - i} + \sum_{s=i_0+1}^{i+j-1} \alpha^{u_0}_{s,j} p^{s-i} = \sum_{s=i}^{i+j-1} \alpha^{u_0}_{s,j} p^{s-i} \in p^j R.$$

So by $i_0 - i < j$ this contradicts $\alpha^{u_0}_{i_0,j} \in R \setminus pR$ for our restricted α. This shows $\alpha^u_{s,j} = 0$ for all pairs (j,u), $j \geq j_i, u \in I_j$, where $s = i, \ldots, i+j-1$.

In particular, if $i = 0$ there is a $j_0 \in \mathbb{N}_0$ such that for all pairs (j,u), $j \geq j_0, u \in I_j$ we have $\alpha^u_{s,j} = 0$, for all $s = 0, \ldots, j-1$. Let $\hat{\jmath}_i = \max\{j_0, \ldots, j_i\}$ for $i \in \mathbb{N}$. In this way we obtain an increasing sequence $(\hat{\jmath}_i)_{i\in\mathbb{N}}$ with $\alpha^u_{s,j} = 0$ for all pairs (j,u), $j \geq \hat{\jmath}_i, u \in I_j$, and all $s = 0, \ldots, i+\hat{\jmath}_i - 1$. Consequently $\mathfrak{o}(\alpha_j) - j \geq i$ for all $j \geq \hat{\jmath}_i$ and $\lim_{j\to\infty}\big(\mathfrak{o}(\alpha_j) - j\big) = \infty$. □

The next example shows that the hypothesis that α is a restricted relation system is necessary to prove the converse.

Example 3.3. Let $\lambda = (1, 1, \ldots)$ be the torsion type, let $\alpha_{i,i+1} = p^i$ for all $i \in \mathbb{N}$ and $\alpha_{i,j} = 0$ otherwise. Then $\mathfrak{o}(\alpha_j) = j - 1$ for all $j \in \mathbb{N}$, hence $\lim_{j\to\infty}\big(\mathfrak{o}(\alpha_j) - j\big) = -1$. But the relation system α corresponding to the array $(\alpha_{i,j})$ has by Lemma 3.1 a restricted expansion $(\beta_{i,j})$, where $\beta_{2i,i+1} = 1$ and $\beta_{i,j} = 0$ otherwise. Thus α is small. This is also follows from the fact that $\left(\sum_{i=0}^{j-1} \alpha_{s+i,j} p^i\right)_{s,j} = \left(p^{j-1} p^{j-1-s}\right)_{s,j}$ is row finite. Alternatively we can use $p\alpha = 0$, to obtain, by (2), that α is small. This example, and the three proofs that α is small, indicate, that for a relation array the shape and the divisibility of the entries by powers of p, imply a certain type of relation system. This leads us into the next definition.

A relation system α is called *of finite-type*, if it is similar to a restricted relation system α such that $\liminf_{j\to\infty}\left(\mathfrak{o}(\alpha_j) - j\right) = -\infty$; and it is called *of ω-type*, if it is similar to a restricted relation system α such that $\liminf_{j\to\infty}\left(\mathfrak{o}(\alpha_j) - j\right) \in \mathbb{Z}$. The choice of notation here is motivated by the corresponding indicators.

Two relation systems of format $(\lambda, 1)$ are called *shift equivalent*, if there are natural numbers $k, l \in \mathbb{N}_0$ such that $x^{-k}(\alpha - \alpha^{(k)}) = x^{-l}(\beta - \beta^{(l)})$. This is an equivalence relation.

Lemma 3.4. *Let G be a module in the class $\mathcal{H}^1$ with relation system α. If β is a relation system which is shift equivalent to α, then β is also realized in G.*

Relation systems which are shift equivalent are either both small, or both of ω-type or both of finite-type, respectively. In particular,

$$\liminf_{j\to\infty}\left(\mathfrak{o}\left(\alpha_j\right) - j\right) = 1 + \liminf_{j\to\infty}\left(\mathfrak{o}\left(x^{-1}(\alpha_j - \alpha_j^{(1)})\right) - j\right).$$

Proof. Let $A = \{a_{i-1}, x_i^v \mid i \in \mathbb{N}, v \in I_i\}$ be a basic generating system of G with relation system α. Let $\hat{a}_i = a_{i+m}$ for all $i \in \mathbb{N}_0$. Then $\hat{A} = \{\hat{a}_{i-1}, x_i^v \mid i \in \mathbb{N}, v \in I_i\}$ is a basic generating system of G, since $p\hat{a}_i = \hat{a}_{i-1} + \sum_{j\in\mathbb{N}}\sum_{u\in I_j}\alpha_{i-1+m,j}^u x_j^u$ for all $i \in \mathbb{N}$, and $x^{-m}\left(\alpha - \alpha^{(m)}\right)$ is the corresponding relation system of the basic generating system $\hat{A}$.

Let H be a module in the class $\mathcal{H}^1$ with relation system γ of the same format as α. If there are $m, n \in \mathbb{N}_0$ such that $x^{-m}(\alpha - \alpha^{(m)}) = x^{-n}(\gamma - \gamma^{(n)})$, then G and H have the same relation systems, hence they are isomorphic. This shows that in a module G the relation system α is realized if and only if $x^{-m}\left(\alpha - \alpha^{(m)}\right)$ is realized for all natural numbers m.

The formula follows from Lemma 3.1 and the definitions of ω-type and of finite-type. □

Lemma 3.5. *$p^\omega G \neq 0$ for a module G in the class $\mathcal{H}^1$ with a relation system of ω-type.*

Proof. Let α be a relation system of G of ω-type corresponding to the basic generating system $A = \{a_{i-1}, x_i^v \mid i \in \mathbb{N}, v \in I_i\}$. Then, by Lemmas 3.1 and 3.4 we may assume α to be restricted and

$$\liminf_{j\to\infty}\left(\mathfrak{o}(\alpha_j) - j\right) = 0 \text{ and } \mathfrak{o}(\alpha_j) - j \geq 0 \text{ for all } j \in \mathbb{N}.$$

Thus we have $a_0 = p^k a_k - \sum_{j\in\mathbb{N}}\sum_{u\in I_j}\sum_{s=0}^{k-1}\alpha_{s,j}^u p^s x_j^u = p^k a_k$, for all $k \in \mathbb{N}$, since $p^s x_j^u = 0$ if $s \geq j$. Moreover, $\alpha_{s,j}^u \in p^j R$, for all $s < j$, by $\mathfrak{o}(\alpha_j) \geq j$. Hence $\mathrm{h}^*(a_0) \geq \omega$.

To show equality assume $pf = a_0$, where $f = a_1 + t \in p^\omega G$ and $t \in \bigoplus_{j\leq l}\bigoplus_{u\in I_j} Rx_j^u$ for some $l \in \mathbb{N}$. The latter follows from $pa_1 = a_0$ modulo torsion. By the choice of the basic generating system and the corresponding relation system α, there are infinitely many j with $\mathfrak{o}(\alpha_j) = j$. Let $k \in \mathbb{N}$ with $k > l + 1$ and $\mathfrak{o}(\alpha_{k-1}) = k - 1$. Then

$$z = \sum_{u\in I_{k-1}}\sum_{s=1}^{k-1}\alpha_{s,k-1}^u p^{s-1}x_{k-1}^u = \sum_{u\in I_{k-1}}\alpha_{k-1,k-1}^u p^{k-2}x_{k-1}^u \neq 0, \tag{4}$$

and since α is restricted there is some $u_0 \in I_{k-1}$ with $\alpha_{k-1,k-1}^{u_0} \neq 0$. Now note that $\mathrm{h}^*(z) = k - 2$. Since $l \leq k - 2$, we obtain modulo $\bigoplus_{j\leq k-2}\bigoplus_{u\in I_j} Rx_j^u$

$$\begin{aligned} f &\equiv a_1 = p^{k-1}a_k - \sum_{j\in\mathbb{N}}\sum_{u\in I_j}\sum_{s=1}^{k-1}\alpha_{s,j}^u p^{s-1}x_j^u = p^{k-1}a_k - \sum_{j=1}^{k-1}\sum_{u\in I_j}\sum_{s=j}^{k-1}\alpha_{s,j}^u p^{s-1}x_j^u \\ &\equiv p^{k-1}a_k - \sum_{u\in I_{k-1}}\alpha_{k-1,k-1}^u p^{k-2}x_{k-1}^u = p^{k-1}a_k - z. \end{aligned}$$

Thus $f - p^{k-1}a_k \equiv -z$ modulo $\bigoplus_{j\le k-2}\bigoplus_{u\in I_j} Rx_j^u$ and $\mathrm{h}^*\big(f - p^{k-1}a_k\big) = \mathrm{h}^*(z) \le k-2$, by (4). This implies $\mathrm{h}^*(f) \le k-2 < \omega$, since $\mathrm{h}^*(p^{k-1}a_k) \ge k-1$. Hence $f \notin p^\omega G$ and $a_0 \notin p^{\omega+1}G$. □

The following lemma shows that the property of a particular type of relation system, small, ω-type or finite-type, is in fact an invariant of the module. Moreover, if the relation system is of finite-type then the indicator of a_0 is of finite-type and can be directly read off the relations.

Lemma 3.6. *The relation systems of a module G in the class $\mathcal{H}^1$ are either all small, all of ω-type or all of finite-type. Moreover, if $\{b_{i-1}, x_i^v \mid i\in\mathbb{N}, v\in I_i\}$ is a basic generating system with relation system β of finite type, then for each $l\in\mathbb{N}_0$ and all sufficiently large k there is $\mathrm{h}^*(p^l b_0) = \mathrm{h}^*\big(\sum_{j\in\mathbb{N}}\sum_{u\in I_j}\sum_{s=0}^{k-1}\beta_{s,j}^u p^{s+l}x_j^u\big)$.*

Proof. By [4, Theorem 6] and Lemma 3.1 it suffices to get a contradiction upon assuming that there are two restricted relation systems α and β of G, where α is of ω-type and β is of finite-type. By Lemma 3.5, $p^\omega G \ne 0$.

Let $B = \{b_{i-1}, x_i^v \mid i\in\mathbb{N}, v\in I_i\}$ be a basic generating system corresponding to β, and define $\mathfrak{o}_k = \mathfrak{o}(\beta_k)$. Since β is of finite-type we have $\liminf_{j\to\infty}\big(\mathfrak{o}(\beta_j) - j\big) = -\infty$. Let $l\in\mathbb{N}_0$ be fixed. Then there are infinitely many $k\in\mathbb{N}$ such that $\mathfrak{o}_k + l < k$. For such a k there is a $u_l\in I_k$ with $\beta_{0,k}^{u_l} = \beta_{1,k}^{u_l} = \cdots = \beta_{\mathfrak{o}_k-1,k}^{u_l} = 0 \ne \beta_{\mathfrak{o}_k,k}^{u_l}$, and

$$\sum_{s=0}^{k-1}\beta_{s,k}^{u_l}p^{s+l} = \sum_{s=\mathfrak{o}_k}^{k-1}\beta_{s,k}^{u_l}p^{s+l} \notin p^k R \tag{5}$$

since β is restricted. By Proposition 2.1 we have $p^l b_0 = p^{k+l}b_k - s_k^l$ where

$$s_k^l = \sum_{j\in\mathbb{N}}\sum_{u\in I_j}\sum_{s=0}^{k-1}\beta_{s,j}^u p^{s+l}x_j^u \ne 0,$$

and $\mathrm{h}^*(s_k^l) < k$ by (5). Hence $\mathrm{h}^*(p^{k+l}b_k) \ge k+l \ge k > \mathrm{h}^*(s_k^l)$ and $\mathrm{h}^*(p^l b_0) = \mathrm{h}^*(s_k^l) \in \mathbb{N}_0$. This shows the second part of the lemma. Thus the indicator of b_0 has natural numbers as entries, contradicting the fact that $p^\omega G \ne 0$. □

4. Indicators

An indicator is called *of finite-type* if all the entries are natural numbers and there are infinitely many gaps; it is called *of ω-type* if there is an entry $\omega + k$, $k\in\mathbb{N}_0$ with no gaps beyond this entry; and it is called *of ∞-type* if there is an ∞ in the indicator. Equivalent indicators are either both of finite-type, both of ω-type or both of ∞-type, respectively. An equivalence class of indicators is called *of finite-type, of ω-type* or *of ∞-type*, respectively, if it represents, indicators of finite-type, of ω-type or of ∞-type.

Let $p^{\infty}G$ denote the maximal divisible submodule of a module G and $p^{\omega}G = \bigcap_{n\in\mathbb{N}} p^n G$, cf. [1]. Note that elements in $p^{\omega}G$ are torsion-free in a module in the class $\mathcal{H}$, since the torsion submodule is a direct sum of cyclics. The following lemma is obvious.

Lemma 4.1. *A module G in the class $\mathcal{H}^1$ splits if and only if $p^{\infty}G \neq 0$.*

The class $\mathcal{H}^1$ contains three subclasses, namely, the class of all splitting modules, the class of all reduced modules with $p^{\omega}G \neq 0$ and the class of modules with $p^{\omega}G = 0$. It turns out that these subclasses are essential in our context.

Theorem 4.2. *For a module G in the class $\mathcal{H}^1$ with relation system α the following are equivalent:*

(1) *G splits.*
(2) *α is small.*
(3) *The indicator of G is of ∞-type.*
(4) *There is an element $g \in G \setminus \{0\}$ with $\mathrm{h}^*(g) = \infty$.*

Proof. (1) and (2) are equivalent by [4, Theorem 6]. The rest is obvious. □

Theorem 4.3. *For a reduced module G in the class $\mathcal{H}^1$ with relation system α the following are equivalent:*

(1) *$p^{\omega}G = 0$.*
(2) *α is of finite-type.*
(3) *The indicator of G is of finite-type.*
(4) *All non-zero elements have finite height.*

Proof. (1), (3) and (4) are obviously equivalent and the equivalence of (2) and (3) is implied by Lemma 3.6 and the fact that the indicators of all non-zero elements are equivalent. □

Theorem 4.4. *For a reduced module G in the class $\mathcal{H}^1$ with relation system α the following are equivalent:*

(1) *$p^{\omega}G \neq 0$.*
(2) *α is of ω-type.*
(3) *The indicator of G is of ω-type.*
(4) *There is an element $g \in G \setminus \{0\}$ with $\mathrm{h}^*(g) = \omega$.*
(5) *$p^{\omega}G \neq 0$ is cyclic.*

Proof. By [3] or [1, Section 103] we know that the indicators of modules in the class $\mathcal{H}^1$ are either of ∞-type, or of ω-type or of finite-type. Thus (1), (3) and (4) are equivalent.
(1) and (5) are equivalent since $p^{\omega}G \neq 0$ is cyclic and not divisible.
(2) implies (4) by Lemma 3.5, and (4) implies (2) by Lemma 3.6. The additional statement follows from (5). □

Example 4.5. Let $\mathbb{H} = (\beta_0, \beta_1, \ldots)$ be an indicator of finite-type with gaps precisely at $0 < k_0 < k_1 < k_2 < \ldots$, i. e. $\delta_i = \beta_{k_i} - \beta_{k_i - 1} - 1 \geq 1$ for $i \in \mathbb{N}_0$. Let $\Delta_0 = \beta_0$, $\Delta_l = \beta_0 + \delta_1 + \cdots \delta_l$ for all $l \in \mathbb{N}$. Let G be a module in the class $\mathcal{H}^1$ with basic generating system $\{x_i, a_{i-1} \mid i \in \mathbb{N}\}$ and corresponding relation array $\left(\alpha_{i,j}\right)_{i\geq 0, j\geq 1}$. Then $\mathbb{H}(a_0) = \mathbb{H}$ since $\alpha_{i,j} = 0$, except for

$$\alpha_{\Delta_l, \Delta_l + k_l} = \alpha_{\Delta_l, \Delta_l + k_l + 1} = \cdots = \alpha_{\Delta_l, \Delta_l + k_{l+1} - 1} = 1, \quad l \in \mathbb{N}.$$

An example of relations of a module with elements of height ω can be found in [4].

5. Structure of Modules in the Class $\mathcal{H}$

In this section we fix the torsion type λ. The set A of all relation systems of format $(\lambda, 1)$ is an R-module relative to the usual operations. Let A_{small} denote the subset of all small relation systems and let A_ω denote the submodule of A generated by the relation systems of ω-type.

Theorem 5.1. *The submodules A_{small} and A_ω are pure in A with $\mathbf{t}\,A \subset A_{small} \subset A_\omega$. Moreover, the quotient module $\mathfrak{A} = A/A_{small}$ is a vector space over the field K and $\mathfrak{A}_\omega = A_\omega/A_{small}$ is a subspace of $\mathfrak{A}$.*

Proof. We begin by noting that A contains torsion elements, cf. Example 3.3, where a relation array of order p is given. Lemma 3.1 (2) and Lemma 3.4 show most of the statements. Observe that, $\mathbf{t}\,A \subset A_{small}$, since $p\alpha = 0$ implies that $\alpha \sim 0$ by (2), i. e. α is small. Moreover if $p\alpha$ is small, then by Lemma 3.4 we obtain that $x^{-1}\big(p\alpha - (p\alpha)^{(1)}\big)$ is small, and by (2) α is small. Hence the submodule A_{small} is pure. A similar argument shows that A_ω is pure. Hence $\mathfrak{A} = A/A_{small}$ is torsion-free. By Lemma 3.1 (2) each element α is divisible by p modulo A_{small}, hence $\mathfrak{A}$ is a vector space over K.

The sum $\alpha + \beta$ of a relation system α of ω-type and an arbitrary small relation system β is again of ω-type. Thus $\beta = (\alpha + \beta) - \alpha \in A_\omega$ shows that A_ω contains A_{small}. Hence $\mathfrak{A}_\omega = A_\omega/A_{small}$ is a subspace of $\mathfrak{A}$. $\square$

Let G be a module in the class $\mathcal{H}$ with basic generating system $\{a_{i-1}^k, x_i^v \mid i \in \mathbb{N}, v \in I_i, k \in I\}$ and corresponding relation system $\alpha = (\alpha_j \mid j \in \mathbb{N})$ of format (λ, d) with arbitrary d. Then for each $k \in I$ the system $\alpha^k = (\alpha_j^k \mid j \in \mathbb{N})$ of format $(\lambda, 1)$ is the relation system corresponding to the basic generating system $\{a_{i-1}^k, x_i^v \mid i \in \mathbb{N}, v \in I_i\}$ of the pure submodule G^k of G of torsion-free rank 1. The subsystems α^k are small, of ω-type or of finite-type for each $k \in I$, respectively, and independent of each other. Instead of reading the relation system α as a collection of α^k we can equivalently consider $G = \sum_{k \in I} G^k$ as a sum of pure submodules G^k of type $(\lambda, 1)$, which are independent modulo $\mathbf{t}\,G$. Note that $G = \sum_{k \in I} G^k$, since $G/\mathbf{t}\,G$ is a vector space. Then by $\bar{\alpha}^k = \alpha^k + A_{small} \in \mathfrak{A}$ the definition

$$\mathfrak{M}(\alpha) = \langle \bar{\alpha}^k \mid k \in I \rangle$$

of the subspace $\mathfrak{M}(\alpha)$ of $\mathfrak{A}$ generated by all $\bar{\alpha}^k$, defines a mapping either $\alpha \mapsto \mathfrak{M}(\alpha)$ or of a representation $G = \sum_{k \in I} G^k \mapsto \mathfrak{M}(\alpha)$. This mapping is induced by the linear mapping

$$\varphi : G/\mathbf{t}\,G \longrightarrow \mathfrak{A}, \quad \varphi(a_0^k + \mathbf{t}\,G) = \bar{\alpha}^k.$$

φ is well defined since $\{a_0^k + \mathbf{t}\,G \mid k \in I\}$ is a basis of $G/\mathbf{t}\,G$ and $\mathfrak{M}(\alpha) = \varphi\big(G/\mathbf{t}\,G\big)$.

Lemma 5.2. *Let G be a module in the class $\mathcal{H}$ with torsion type λ and arbitrary torsion-free rank d. The image $\mathfrak{M}(G) = \mathfrak{M}(\alpha)$ of $G/\mathbf{t}\,G$ under φ doesn't depend on the relation system α or the representation $G = \sum_{k \in I} G^k$, if the torsion generators are fixed.*

Proof. A new basis $\{b_0^k + \mathbf{t}\,G \mid k \in I\}$ of $G/\mathbf{t}\,G$ is given with an invertible $d \times d$-matrix $D = (d_{k,h})$ over K in the form $b_0^k + \mathbf{t}\,G = \sum_{h \in I} d_{k,h}(a_0^h + \mathbf{t}\,G)$. The corresponding relation system is $D\alpha$, cf. [5, Lemma 8 and Proposition 9]. Then

$$\varphi\Big(\sum_{h \in I} d_{k,h}\big(a_0^h + \mathbf{t}\,G\big)\Big) = \sum_{h \in I} d_{k,h}\varphi\big(a_0^h + \mathbf{t}\,G\big), \quad k \in I,$$

and $\mathfrak{M}(\alpha) = \mathfrak{M}(D\alpha) = \varphi\big(G/\mathbf{t}\,G\big)$, since D is invertible, i. e. $\mathfrak{M}(G) = \mathfrak{M}(\alpha)$ is independent of α. □

Note that $G \cong H$ doesn't imply $\mathfrak{M}(G) = \mathfrak{M}(H)$. But following the considerations here and in [5], the inequality of $\mathfrak{M}(G)$ and of $\mathfrak{M}(H)$ results from changes of the torsion generators.

Theorem 5.3. *Let G be a module in the class $\mathcal{H}$ with torsion type λ and arbitrary torsion-free rank d, then*

$$\ker\varphi = \big(p^\infty G \oplus \mathbf{t}\,G\big)\big/\mathbf{t}\,G.$$

Proof. If $a \in p^\infty G$ then the corresponding relation system α of format $(\lambda, 1)$ is 0, hence $\bar{\alpha} = 0$ and $\varphi\big(a+\mathbf{t}\,G\big) = 0$. Conversely, if $\varphi\big(a+\mathbf{t}\,G\big) = 0$ then the corresponding relation system α of format $(\lambda, 1)$ is small, hence a is a divisible element by [4, Theorem 6]. □

Theorem 5.4. *Let G be a module in the class $\mathcal{H}$ with torsion type λ and arbitrary torsion-free rank d, then the subspace of $G/\mathbf{t}\,G$ generated by*

$$\big(p^\omega G \oplus \mathbf{t}\,G\big)\big/\mathbf{t}\,G$$

is the inverse image of $\mathfrak{A}_\omega$ under φ.

Proof. If $a \in p^\omega G$ then the corresponding relation system α of format $(\lambda, 1)$ is of ω-type, hence $\bar{\alpha} \in \mathfrak{A}_\omega$ and $\varphi\big(a+\mathbf{t}\,G\big) \in \mathfrak{A}_\omega$. Conversely, if $\varphi\big(a+\mathbf{t}\,G\big) \in \mathfrak{A}_\omega$ then the corresponding relation system α of format $(\lambda, 1)$ is of ω-type or small, hence the submodule $\langle a_{i-1}, x_i^u\rangle$ contains an element which is either divisible or of height ω. □

An application of the dimension formula for a linear mapping immediately shows the following corollary.

Corollary 5.5. *Let G be a module in the class $\mathcal{H}$ with torsion type λ and arbitrary torsion-free rank d and relation system α.*

(1) *G is reduced if and only if φ is injective, i. e. if and only if $\{\alpha^k \mid k \in I\}$ is independent modulo A_{small}.*
(2) *The torsion-free rank of $G/p^\infty G$ and the dimension of $\mathfrak{M}(G)$ are equal.*
(3) *$p^\omega G = 0$ if and only if $\mathfrak{M}(G)\cap\mathfrak{A}_\omega = 0$, i. e. if and only if $\{\alpha^k \mid k \in I\}$ is independent modulo A_ω.*
(4) *The torsion-free rank of $G/p^\omega G$ and the dimension of $\mathfrak{M}(G)/\big(\mathfrak{M}(G)\cap\mathfrak{A}_\omega\big)$ are equal.*

Example 5.6. Let $p \neq 2$, let $G = \langle a_{i-1}^k, x_i \mid i \in \mathbb{N}, k \in \{1,2\}\rangle$ and let a corresponding relation array be defined by: $\alpha_{i,2i+1}^1 = \alpha_{i,2i+1}^2 = 1, i \in \mathbb{N}_0$, and $\alpha_{i,j}^1 = \alpha_{i,j}^2 = 0$ otherwise. Then $p^\infty G = \langle a_i^1 - a_i^2 \mid i \in \mathbb{N}\rangle$.

Let a second corresponding relation array be defined by: $\alpha_{i,2i+1}^1 = \alpha_{i,2i+1}^2 = 1, \alpha_{i,i+1}^2 = 1, i \in \mathbb{N}_0$, and $\alpha_{i,j}^1 = \alpha_{i,j}^2 = 0$ else. Then $p^\omega G = \langle a_0^1 - a_0^2\rangle$.

Let $T = \bigoplus_{i\in\mathbb{N}}\bigoplus_{v\in I_i} Rx_i^v$ be a torsion module which is a direct sum of cyclics and of isomorphism type λ. Let $\mathcal{H}_\lambda^1$ be the class of modules G having torsion submodule T and quotient $G/T \cong K$. We follow the notation in [1, Section 49]. Let $G \in \mathcal{H}_\lambda^1$ and define $g : K \longrightarrow G$ such that $g(u)$ is a representative of the coset of u, and in particular $g(p^{-i}) = a_i$. Then $\{a_{i-1}, x_i^v \mid i \in \mathbb{N}, v \in I_i\}$ is a basic generating system of G with corresponding relation system α of format $(\lambda, 1)$. We can associate g with α, i. e. $\alpha = \alpha(g)$. Moreover, the function g also determines a factor set $f = f(g)$ by

$$f(u,v) = g(u) + g(v) - g(u+v).$$

Let $g' : K \longrightarrow G$ be another function inducing the same factor set, $f(g) = f(g')$ with $g'(p^{-i}) = b_i$. Then $\{b_{i-1}, x_i^v \mid i \in \mathbb{N}, v \in I_i\}$ is another basic generating system with corresponding relation system β. By the definition of the functions g, g' we have $a_i - b_i \in T$. This implies by [5, Proposition 7] that α and β are similar, hence $\alpha + A_{small} = \beta + A_{small}$. Thus a mapping is defined

$$\mu : \mathrm{Fact}(K,T) \longrightarrow \mathfrak{A}, \quad \text{given by } \mu(f) = \alpha + A_{small}.$$

Next we show that this mapping μ induces an isomorphism between $\mathrm{Ext}(K,T)$ and $\mathfrak{A}$.

Theorem 5.7. *μ is an epimorphism with kernel* $\mathrm{Trans}(K,T)$ *and thus* $\mathrm{Ext}(K,T) \cong \mathfrak{A}$.

Proof. Let α be a relation system of format $(\lambda, 1)$. Then α is realized by a module $G \in \mathcal{H}_\lambda^1$ with corresponding basic generating system $\{a_{i-1}, x_i^v \mid i \in \mathbb{N}, v \in I_i\}$, cf. [4, Theorem 4]. Define $g : K \longrightarrow G$ by $g(p^{-i}) = a_i$ and extend the definition of g arbitrarily such that g maps elements to their cosets. We thus obtain that the image of the factor set $f(g)$ under μ is $\alpha + A_{small}$, i. e. $\mu(f) = \alpha + A_{small}$. Hence μ is epic.

The kernel of μ is the set $\mathrm{Trans}(K,T)$ of transformation sets, since the transformation sets are the factor sets of the splitting extension and by [4, Theorem 6] the corresponding relation systems are precisely all the small relation systems. Thus $\mu\big(\mathrm{Trans}(K,T)\big) = A_{small}$.

Next we show that μ is a homomorphism. For this we write down precisely how the factor set f can be determined by α. Let $u, v \in K \setminus pR$ and

$$f(u,v) = g(u) + g(v) - g(u+v) = up^{i(u)}a_{i(u)} + vp^{i(v)}a_{i(v)} - (u+v)p^{i(u+v)}a_{i(u+v)},$$

where $up^{i(u)} \in R \setminus pR$. We know that $f(u,v) \in T$, hence the sum representation of f on the right-hand side of the equation above must be of the form

$$r\left(p^l a_{l+h} - a_h\right) = r\left(\sum_{j\in\mathbb{N}}\sum_{u\in I_j}\sum_{s=h}^{l+h-1} \alpha_{s,j}^u p^{s-h} x_j^u\right),$$

where $r \in R$. In particular, r and the indices l, h depend only on u, v. Now let $f = f(g), f' = f'(g')$ be two factor sets with $\mu(f) = \alpha + A_{small}$ and $\mu(f') = \beta + A_{small}$. The sum of f and f' is given by

$$f(u,v) + f'(u,v) = \big(g(u) + g(v) - g(u+v)\big) + \big(g'(u) + g'(v) - g'(u+v)\big).$$

Writing this in terms of α and β and observing that the parameters r and the indices h, l depend only on u, v, we get

$$\begin{aligned} f(u,v) + f'(u,v) &= r\left(\sum_{j\in\mathbb{N}}\sum_{u\in I_j}\sum_{s=h}^{l+h-1} \alpha_{s,j}^u p^{s-h} x_j^u + \sum_{j\in\mathbb{N}}\sum_{u\in I_j}\sum_{s=h}^{l+h-1} \beta_{s,j}^u p^{s-h} x_j^u\right) \\ &= r\left(\sum_{j\in\mathbb{N}}\sum_{u\in I_j}\sum_{s=h}^{l+h-1} (\alpha_{s,j}^u + \beta_{s,j}^u) p^{s-h} x_j^u\right). \end{aligned}$$

This shows that the mapping μ is additive. The proof of $\mu(rf) = r\mu(f)$ for $r \in R$ is analogous. Hence μ is a homomorphism and induces an isomorphism between $\text{Ext}(K,T)$ and $\mathfrak{A}$. $\square$

REFERENCES

1. L. Fuchs, *Infinite Abelian Groups* I+II, Academic Press (1970, 1973).
2. N. Jacobson, *Basic Algebra* I+II, Freeman and Company (1974, 1980).
3. *C. Megibben*, On mixed groups of torsion-free rank one, Ill. J. Math. **11** (1967), 134–144.
4. *O. Mutzbauer and E. Toubassi*, A splitting criterion for a class of mixed modules, Rocky Mountain J. Math. **24** (1994), 1533 - 1543.
5. *O. Mutzbauer and E. Toubassi*, Classification of mixed modules, Acta Math. Hung. **72** (1996).

RALF JARISCH, MATHEMATISCHES INSTITUT, UNIVERSITÄT WÜRZBURG, AM HUBLAND, 97074 WÜRZBURG, GERMANY

OTTO MUTZBAUER, MATHEMATISCHES INSTITUT, UNIVERSITÄT WÜRZBURG, AM HUBLAND, 97074 WÜRZBURG, GERMANY
E-mail address: mutzbauer@mathematik.uni-wuerzburg.de

ELIAS TOUBASSI, DEPARTMENT OF MATHEMATICS, UNIVERSITY OF ARIZONA, TUCSON, ARIZONA 85721, U. S. A.
E-mail address: elias@math.arizona.edu

A New Proof for a Theorem of Müller and Singh

MENAKA LELWALA Department of Mathematics, University of Sri Jayewardenepura, Gangowila, Nugegoda, Sri Lanka

BRUNO J. MÜLLER Department of Mathematics, McMaster University, Hamilton, Ontario, L8S 4K1, Canada

We give a new proof for the following result ([4], Theorem 2) :

THEOREM 1 Let M be a uniform P-tame module over a serial ring R , and let T be the intersection of the clique of P . Then $\mathrm{ann}_M(\bigcap_{n\in\mathbb{N}} T^n)$ is uniserial.

Here M is called P-tame if it possesses an associated prime ideal P , if P is Goldie, and if $\mathrm{ann}_M(P)$ is non-singular as R/P-module.

The original proof in [4] , as well as those of various special cases in [7], [8] and [3] , are technical and rather complicated. The present argument uses the fact (which ought to be of independent interest) that serial rings can be localized at Goldie semiprime ideals ; such a localization reduces here to noetherian serial rings where the assertion is well known.

Specifically, Chatters [1] showed that, for a serial ring R and a Goldie semiprime ideal S which does not contain any nonzero idempotent, the set $\mathcal{C}(S)$ of elements regular modulo S is left- and right-Ore, and S is divisible (ie. S = cS = Sc for all $c\in\mathcal{C}(S)$).

In general, $\mathcal{C}(S)$ is not reversible. Utilizing the Ore condition, we define ideals $K=\{r\in R : c_1rc_2=0\}$, $'K=\{r\in R : c_1r=0\}$ and $K'=\{r\in R : rc_2=0\}$, all for suitable $c_i\in\mathcal{C}(S)$ depending on r . Reversibility amounts to K = 'K = K' , and this can be established in two cases which suffice for our application :

LEMMA 2 In the situation described above, $\mathcal{C}(S)$ is reversible if R is Goldie prime, or if S is nilpotent.

PROOF The first case is trivial : if $K'\neq 0$, then it contains a regular element a , contrary to $ac_2=0$. We conclude K' = 0 ; similarly 'K = 0 , and therefore $K = 0$.

In the second case, if $K' \subsetneqq K$, there exists m such that KS^m is not contained in K' but KS^{m+1} is . Then $X = (KS^m+K')/K'$ is a nonzero divisible R/S-bimodule which is torsion on the left and torsionfree on the right. [To see $SX = 0$, consider $k \in K$; then $c_1kc_2 = 0$ hence $0 = Sc_1kc_2 = Skc_2$ hence $Sk \subseteq K'$. The divisibility follows once K is seen to be left-divisible : $k \in \bar{K} \subseteq S = cS$ yields $k = cs$, hence $0 = c_1kc_2 = c_1csc_2$ hence $s \in K$. Left-torsion is established as follows : given $\bar{x} = \sum k_i\bar{s}_i \in KS^m$, we have $c_1k_ic_2 = 0$ (using the common multiple property of the Ore set), and $s_i = c_2s_i'$ (using the divisibility of S), hence $c_1\bar{x} = \sum c_1k_ic_2s_i' = 0$. Finally the right-torsionfreeness is trivial: $\bar{x}c = 0$ yields $xcc_2 = 0$ hence $\bar{x} = 0$.]

In consequence X becomes a right-module over the semisimple artinian quotient ring of R/S . The seriality of R implies that X is right-finite dimensional hence finitely generated. Therefore (again using the common multiple property) a single $c \in \mathcal{C}(S)$ annihilates X from the left : $0 = cX = X \neq 0$; a contradiction.

PROOF OF THEOREM 1 We may assume $\bigcap_{n\in\mathbb{N}} T^n = 0$. The tameness yields, first of all, that P and hence T are Goldie (by [3], 3.8). Then ([4], Theorem 6) implies that either R is Goldie prime, or T is nilpotent (according to whether the clique is "circular" or "linear"). In both cases, Lemma 2 establishes reversibility, and so the quotient ring R_T exists. Again by tameness, M is torsionfree, and therefore embeds into the quotient module M_T which is a uniform R_T-module.

The Jacobson readical of R_T is $JR_T = TR_T = T/K$ (cf. [2], p.252). In the Goldie prime case where $K = 0$, we have $\bigcap(JR_T)^n = \bigcap T^n = 0$. In the nilpotent case where $T^m = 0$ for some m , we get

$$(JR_T)^m = (T+K)^m/K = T^m+K/K = 0 .$$

Thus by ([5], 2.1), R_T is noetherian (in fact artinian in the second case). By ([6], 5.1) , M_T is a uniserial R_T-module. The proof is completed by the following lemma, the proof of which is verbatim the same as the one given in ([2], Lemma 3) for the special case $M = R$) :

LEMMA 3 Let $JS \subseteq R \subseteq S$ be two rings (where JS is the Jacobson radical of S). Assume that R/JS is serial, and is an order in the semisimple artinian ring S/JS . Then any uniserial S-module is uniserial as R-module.

REFERENCES

[1] A.W.Chatters, Serial rings with Krull dimension, Glasgow Math.J. 32 (1990), 71-78.

[2] B.J.Müller, The structure of serial rings, Methods in Module Theory (ed. G.Abrams et al), Marcel Dekker (1992), 249-270.

[3] B.J.Müller and Surjeet Singh, Uniform modules over serial rings, J. Algebra 144 (1991), 94-109.

[4] B.J.Müller and Surjeet Singh, Uniform modules over serial rings II, Springer Lecture Notes in Math. 1448 (1990), 25-32.

[5] Surjeet Singh, Serial right noetherian rings, Canad.J.Math. 36 (1984), 22-37.

[6] R.B.Warfield, Serial rings and finitely presented modules, J.Algebra 37 (1975), 187-222.

[7] M.H.Wright, Certain uniform modules over serial rings are uniserial, Comm.Algebra 17 (1989), 441-469.

[8] M.H.Wright, Uniform modules over serial rings with Krull dimension, Comm.Algebra 18 (1990), 2541-2557.

On Maximal Subgroups of the Semigroup of Nonnegative Matrices

AR. MEENAKSHI, Department of Mathematics,
Annamalai University, Annamalai Nagar - 608 002, INDIA

Abstract

A characterization of the maximal subgroups of non-negative matrices is obtained. It is shown that the Moore-Pentose inverse $A^{\dagger}$ of a non-negative singular matrix A of order n is covariant under the group $G_N(A) = \{T: T$ is monomial and commutes with $AA^{\dagger}$ and $A^{\dagger}A\}$, a subset of the covariance set $\vartheta(A)$, the set of all invertible matrices T such that $(TAT^{-1})^{\dagger} = TA^{\dagger}T^{-1}$. It is proved that $G_N(A)$ is non-abelian if and only if n > 2.

1. INTRODUCTION

For $A \in N_n$ the set of non-negative matrices of order n, let A^T be the transpose, R(A) be the range space and $A^{\dagger}$ be the Moore-Penrose inverse[7]. $A^{\dagger}$ is the unique solution of the following four Penrose equations:

$$A X A = A; \quad X A X = X; \quad (AX)^T = AX; \quad (XA)^T = XA \qquad (1.1)$$

In [2], it has been established that for $A \in N_n$, $A^{\dagger} \in N_n \Leftrightarrow A$ and A^T are row-monotone. (A real matrix A is row-monotone if $Ax \geq 0$, $x \in R(A^T) \Rightarrow x \geq 0$, that is all the components of x are non-negative). Row-monotone matrices have found many applications in numerical analysis and linear programming. The maximal subgroup H(E) of the multiplicative semigroup of N_n containing a symmetric idempotent E is characterized and the structure of each element of H(E) is obtained.

For a given A, the covariant set is the set of all invertiable matrices T such that $(TAT^{-1})^{\dagger} = TA^{\dagger}T^{-1}$. Covariant set of complex matrices of rank one (or) two was studied by H. Schwerdtfeger [9] and for more general case without any restriction on the ranks of A was studied by D. W. Robinson [8]. In my earlier work [5], Covariant sets for unitarily similar and EP matrices with common range space were determined. In this paper, non-negative covariant set $\vartheta_N(A)$ for $A \in N_n$ is characterized. It is proved that, $\vartheta_N(A)$ is the same for each element of H(E). Further it is shown that $A^{\dagger}$ is covariant under the group $G_N(A) = \{T : T$ is monomial and commutes with $AA^{\dagger}$ and $A^{\dagger}A\}$ and $G_N(A)$ is non-abelian $\Leftrightarrow n>2$. $A \in N_n$ is said to be monomial if A is invertible and $A^{-1} \in N_n$.

2. ON MAXIMAL SUBGROUPS OF N_n

First, certain results on row-monotone matrices are proved which are useful in characterizing maximal subgroups of the multiplicative semigroup N_n containing a symmetric idempotent.

THEOREM 2.1 If A and B are row-monotone such that $R(B) \subseteq R(A^T)$ then AB is row-monotone.

Proof: Let $(AB)x \geq 0$ with $x \in R(AB)^T$. Then $A(Bx) \geq 0$ and $x \in R(B^TA^T) \subseteq R(B^T)$. Put $z = Bx$ then $z \in R(B)$. Since $R(B) \subseteq R(A^T)$, $Z \in R(A^T)$. Now A is row-monotone,
$Az \geq 0$, $z \in R(A^T) \Rightarrow z \geq 0$. By row monotonicity of B,
$Bx = z \geq 0$, $x \in R(B^T) \Rightarrow x \geq 0$. Thus $(AB)x \geq 0$,
$x \in R(AB)^T \Rightarrow x \geq 0 \Rightarrow AB$ is row-monotone.
Hence the theorem.

Corollary 2.1 For $A \in N_n$ and EP, the following are equivalent:

(i) A is row-monotone

(ii) A^T is row-monotone.

(iii) $A^\dagger \geq 0$.

Proof: (i) $\Rightarrow$ (iii) Since A is row-monotone, by Theorem 3 in [2], there exists $Y \geq 0$ such that $YA = A^\dagger A$. Since $A \geq 0$ and EP, $AA^\dagger = A^\dagger A$. $A^\dagger = A^\dagger AA^\dagger = YAA^\dagger = Y(A^\dagger A) = YYA \geq 0$.

Thus (iii) holds.

(ii) $\Rightarrow$ (iii) Since A^T is row monotone, there exists $X \geq 0$ such that $AX = AA^\dagger$ and $A^\dagger = A^\dagger AA^\dagger = A^\dagger AX = AA^\dagger X = AXX \geq 0$.

Thus (iii) holds.

(iii) $\Rightarrow$ (i) and (ii) follows from Theorem 4 in [2], that is, for $A \geq 0$ with $A^\dagger \geq 0 \Rightarrow A$ is row-monotone and A^T is row-monotone. The proof is complete.

Corollary 2.2 : If A and B are EP_r row-monotone then AB is EP_r row monotone $\Leftrightarrow R(A) = R(B)$.

Proof : Since A and B are EP_r. AB is $EP_r \Leftrightarrow R(A) = R(B)$ isthe content of Theorem 3 in [1]. Since A and B are row-monotone, with $R(B) = R(A) = R(A^T)$, by

Theorem 2.1, AB is row-monotone. Hence the corollary. Now we shall prove the Main result in the following.

THEOREM 2.2 Let E be a symmetric idempotent ($E = E^2 = E^T$) real matrix. Then, $H(E) = \{A : A \in N_n$, A is EP row-monotone and $R(A) = R(E)\}$ is a maximal subgroup of N_n containing E.

Proof: Let $A \in H(E)$, then $A \geq 0$, A is EP row-monotone and $R(A) = R(A^T) = R(E)$. Hence $AA^\dagger = A^\dagger A = E$. By Corollary 2.1, $A^\dagger$ is row-monotone and by Theorem 4 in [2], $A^\dagger \geq 0$. Since $A^\dagger \geq 0$ and $(A^\dagger)^\dagger = A \geq 0$. $A^\dagger$ is also row-monotone. Clearly $A^\dagger$ is EP. Therefore $A^\dagger \in H(E)$. By Theorem 2.1, $E = AA^\dagger \geq 0$ is also EP row-monotone, hence $E \in H(E)$. Since $AA^\dagger = A^\dagger A = E$ and $AE = A(A^\dagger A) = A = (AA^\dagger)A = EA$, for each $A \in H(E)$, E is the identity element and $A^\dagger$ is the inverse of A. For $A, B \in H(E)$, $AB \geq 0$. Since $R(A) = R(E) = R(B)$, by corollary 2.2, AB is EP row-monotone. Further $(AB)^\dagger = B^\dagger A^\dagger \geq 0$ and $(AB)(AB)^\dagger = A(BB^\dagger)A^\dagger = AEA^\dagger = AA^\dagger = E$. Therefore $R(AB) = R(E)$ and $AB \in H(E)$. Thus H(E) is a subgroup of N_n having E as dentity. The maximality of H(E) can be proved along the same lines as in the proof of Theorem 1 in [5].

Corollary 2.3 Let $A \in N_n$ be of rank r. Then $A \in H(E)$ symmetric idempotent E $\Leftrightarrow$ there exists a permutation P and a monomial r x r matrix D ($D \geq 0$, $D^{-1} \geq 0$)

such that $PAP^T = \begin{bmatrix} D & 0 \\ 0 & 0 \end{bmatrix}$

Proof: Let $A \in H(E)$. Then $A \geq 0$, $A^\dagger \geq 0$ and A is EP. $AA^\dagger = A^\dagger A = E$ is symmetric idempotent and non-negative. Hence by Flor [3], there exists a permutation matrix P such that $PEP^T = \begin{bmatrix} S & 0 \\ 0 & 0 \end{bmatrix}$ where $S = \text{diag}\{x_1x_1^T, x_2x_2^T, \ldots, x_rx_r^T\}$, x_i's are positive vectors with $x_i^Tx_i = 1$ for each $i = 1$ to r. Now setting $X = PAP^T$, $X^\dagger = PA^\dagger P^T$, using Theorem 1 in [4] we get, $PAP^T = X = \begin{bmatrix} D & 0 \\ 0 & 0 \end{bmatrix}$. Since P is a permutation matrix and $A \geq 0$, $X \geq 0$. Hence $D^{-1} \geq \emptyset$). Again by Theorem 1 in [4], $X^\dagger = P\,A^\dagger P^T = \begin{bmatrix} D^{-1} & 0 \\ 0 & 0 \end{bmatrix} \geq 0 \Rightarrow D^{-1} \geq 0$. Thus D is monomial. Conversely, suppose D is monomial, $r \times r$ matrix, P is a permutation matrix such that $PAP^T = \begin{bmatrix} D & 0 \\ 0 & 0 \end{bmatrix}$, both A and $A^\dagger \in N_n$. By Theorem 4 in [2], A is row-monotone and by a Theorem of Baskett and Katz [1], A is EP. Thus $A \in H(E)$ where $AA^\dagger = A^\dagger A = E$.

REMARK 2.1. The group H(E) is isomorphic with the group of monomial matrices of order r where rk(E) = r.

REMARK 2.2. In particular, if E = In, the above group reduces to the group of monomial matrices of order n.

3. NON-NEGATIVE COVARIANT SETS

For a given matrix $A \in N_n$, the set of invertible complex matrices T such that $(TAT^{-1})^\dagger = TA^\dagger T^{-1}$ is the covariant set of A denoted as $\vartheta(A)$. In our earlier work [6], the following characterization of covariant sets were determined.

Result 3.1 : If A and B are EP matrices with R(A) = R(B) then $\vartheta(A) = \vartheta(B)$.

Result 3.2 : For a given complex matrix A and P unitary, $\vartheta(PAP^{-1}) = P\,\vartheta(A)\,P^{-1}$

Definition 3.1 :

For $A \in N_n$, the non-negative covariant set $\vartheta_N(A)$ is the set of all invertible non-negative matrices T such that $(TAT^{-1})^{\dagger} = TA^{\dagger}T^{-1}$

$$\vartheta_N(A) = \{T : T \in N_n,\ (TAT^{-1})^{\dagger} = TA^{\dagger}T^{-1}\}$$

Theorem 3.1 : If A and B are EP matrices with $R(A) = R(B)$, then $\vartheta_N(A) = \vartheta_N(B)$.

Proof : $T \in \vartheta_n(A) \Leftrightarrow T \in N_N$ and $T \in \vartheta(A)$

$\Leftrightarrow T \in N_n$ and $T \in \vartheta(B)$ [By Result 3.1]

$\Leftrightarrow T \in \vartheta_N(B)$

Hence the theorem

Corollary 3.1. Let $A \in N_n$ be EP. Then there exists a symmetric idempotent $E \in N_n$ such that $\vartheta_N(A) = \vartheta_N(AE) = \vartheta_N(EA) = \vartheta_N(E)$.

Proof : Since A is EP, be Theorem 2.3 of [5], there exists a hermitian idempotent E such that $AE = EA = A$ and $R(A) = R(E)$. Hence by Theorem 3.1, it follows that,

$$\vartheta_N(A) = \vartheta_N(AE) = \vartheta_N(EA) = \vartheta_N(AA^{\dagger}) = \vartheta_N(E).$$

By applying Theorem 3.1 and corollary 3.1, we can determine the non-negative covariant set of each element of H(E) in the following way:

Corollary 3.2 Let E $(E = E^2 = E^T)$ be a symmetric idempotent in N_n. Then $\vartheta_N(E)$ is the non-negative covariant set of each element in the maximal subgroup $H(E) = \{A : A \in N_n$, A is EP row-monotone, $R(A) = R(E)\}$

THEOREM 3.2 Let $A \in N_n$ and $P \in N_n$ be orthogonal. Then $\vartheta_N(PAP^{-1}) = P\,\vartheta_N(A)P^{-1}$

Proof : $X \in P\,\vartheta_N(A)P^{-1} \Rightarrow X = PTP^{-1}$ for some $T \in \vartheta_N(A)$

$\Rightarrow X = PTP^{-1}$ $T \in N_n$ and $T \in \vartheta(A)$

$$\Rightarrow X = PTP^{-1}, \quad X \in \vartheta\,(PAP^{-1})$$

[By Result 3.2]

$$\Rightarrow X = N_{n}, X \in \vartheta\,(PAP^{-1})$$

$$\Rightarrow X \in \vartheta_{N}(PAP^{-1})$$

Thus $P\,\vartheta_{N}(A)P^{-1} \subseteq \vartheta_{N}(PAP^{-1})$. Conversely, let $X \in \vartheta_{N}(PAP^{-1})$ then by the above part,

$$P^{-1}\vartheta_{N}(PAP^{-1})P \subseteq \vartheta_{N}(A)$$

$$\vartheta_{N}(PAP^{-1}) \subseteq P\vartheta_{N}(A)P^{-1}$$

Hence $$\vartheta_{N}(PAP^{-1}) = P\,\vartheta_{N}(A)P^{-1}$$

THEOREM 3.3 For $A \in N_n$, let $G_N(A) = \{T : T$ is monomial and T commutes with $AA^{\dagger}$ and $A^{\dagger}A\}$. Then the following hold:

(i) $G_N(A) \subseteq \vartheta_N(E)$ where $E = AA^{\dagger}$ is symmetric idempotent

(ii) $G_N(A)$ forms a group

(iii) Each element of $G_N(A)$ is orthogonally similar to a diagonal block matrix $\begin{bmatrix} S_1 & 0 \\ 0 & S_2 \end{bmatrix}$ where S_1 is $r \times r$, S_2 is $(n-r) \times (n-r)$ monomial matrices.

Proof : $T \in G_N(A) \Rightarrow T \in N_n$ and $T \in \vartheta(A)$

$\Rightarrow T \in N_n$ and $T \in \vartheta(E)$ (by cor. 3.1)

$\Rightarrow T \in \vartheta_N(E)$

Since T is arbitary, $G_N(A) \subseteq \vartheta_N(E)$. Thus (i) holds. Since T commutes with $AA^{\dagger}$ and $A^{\dagger}A$, T^{-1} also commutes with $AA^{\dagger}$ and $A^{\dagger}A$. Further T is monomial $\Rightarrow T^{-1} \in N_n$ and $T = (T^{-1})^{-1} \in N_n$. Hence T^{-1} is monomial and commutes with $AA^{\dagger}$ and $A^{\dagger}A$. Therefore $T^{-1} \in G_N(A)$. One can verify that T_1T_2

$\in G_N(A)$ whenever $T_1 \in G_N(A)$ and $T_2 \in G_N(A)$. Clearly $I_n \in G_N(A)$ Hence $G_N(A)$ is a group. Thus (ii) holds. $T \in G_N(A) \Rightarrow TE = ET$ where $E = AA^{\dagger}$ is a symmetric idempotent.

Hence there exists an orthogonal matrix P such that

$E = P\begin{bmatrix} Ir & 0 \\ 0 & 0 \end{bmatrix}P^{-1}$. Since T is monomial and $TE = ET \Rightarrow P^{-1}TP =$

$\begin{bmatrix} S_1 & 0 \\ 0 & S_2 \end{bmatrix}$ where S_1 is $r \times r$ and S_2 is $(n-r) \times (n-r)$

monomial matrices. Thus (iii) holds.

Remark 3.1 We note that $A^{\dagger}$ is non-negative covariant under the group $G_N(A)$. In particular if A invertible, then $E = In$, $\vartheta_N(A)$ reduces to the group of monomial materices of order n.

Corollary 3.3 For $A \in N_n$, $G_N(A)$ is non-abelian if and only if $n > 2$.

Proof:

If $G_n(A)$ is non-abelian, then $n > 2$. For, if $n = 2$ then the only matrices that commutes with $AA^{\dagger}(A^{\dagger}A)$ are polynomials in $AA^{\dagger}(A^{\dagger}A)$. Hence $G_N(A)$ is abelian which is a contradiction. Therefore $n>2$. Conversely, if $n > 2$, then by Theorem 3.3 (iii), any element $T \in G_N(A)$ is of the form

$P\begin{bmatrix} S_1 & 0 \\ 0 & S_2 \end{bmatrix}P^{-1}$, where $P \in N_n$ is orthogonal, given by

$$E = AA^{\dagger} = P\begin{bmatrix} Ir & 0 \\ 0 & 0 \end{bmatrix}P^{-1}$$

$$T \in G_N(A) \Rightarrow \begin{bmatrix} S_1 & 0 \\ 0 & S_2 \end{bmatrix} = P^{-1}TP \in P^{-1}G_N(A)P$$

$$\in G_N(P^{-1}AP)$$
$$\subseteq \vartheta_N(P^{-1}AP)$$

It is clear that $\vartheta_N(P^{-1}AP)$ is non-abelian for n>2. Thus $G_N(A)$ is non-abelian for n>2.

Hence the corollary.

REFERENCES

1. T.S. Baskett and I.J. Katz, Theorems on Products of EP matrices, Lin. Alg.Appl 2: 87-103 (1969).

2. A. Berman and R.J. Plemmons, Monotonicity and the generalized inverse, SIAM.J.Appl. Math 22: 155-161 (1972).

3. P. Flor, On groups of non-negative matrices, compo. Math 21: 376-382 (1969).

4. S.K. Jain, V.K. Goel and E.K. Kwak, Non-negative matrices having same non-negative Moore-Penrose and group inverses, Lin and Mult. Alg 7: 59-72 (1979).

5. AR. Meenakshi, On EP_r matrices with entries from an arbitrary field, Lin, and Mult. Alg. 9: 159-164 (1980).

6. AR. Meenakshi and V. Chinnadurai, Some remarks on the covariance of the Moore-Penrose inverse, Houston J.Math, 16: 167-174 (1992)

7. R. Penrose, A generalized inverse for matrices, Proc camb Philos Soc 51: 406-413 (1955).

8. D.W. Robinson, On the covariance of the Moore-Penrose inverse, Lin. Alg. Appl 61: 91-99 (1984).

9. H. Schwerdtfeger, On the covariance of the Moore-Penrose inverse, Lin Alg Appl 52/53: 629-644 (1983).

Local Completions of Torsion-free p-adic Modules

OTTO MUTZBAUER AND HELMUT SCHÄFER
Universitat Würzburg, Würzburg, Germany

ABSTRACT. In 1932 Kurosh, Mal'cev and Derry discovered the Local-Global principle. Applied to torsion-free abelian groups it has the form

$$A = \bigcap_p (\widehat{\mathbb{Z}}_p A \cap \mathbb{Q}A),$$
$$\mathrm{End}_{\mathbb{Z}} A = \bigcap_p (\mathrm{End}_{\widehat{\mathbb{Z}}_p} \widehat{\mathbb{Z}}_p A \cap \mathrm{End}_{\mathbb{Q}} \mathbb{Q}A),$$

where $\widehat{\mathbb{Z}}_p$ is the ring of p-adic integers and $\widehat{\mathbb{Z}}_p A$ is the p-adic localization of A. This principle can be used to construct torsion-free abelian groups. We introduce a topology on p-adic modules by defining what we call the *p-adic maximum norm* on a suitable vector space containing the module. This enables us to prove a generalization of the Local-Global principle by replacing the p-adic localizations by p-adic completions. As for the algebraic structure of the completion, we show that the completion of a reduced module can contain divisible elements. Describing modules in terms of generating systems and relations we can use the structure of these relations to decide whether the completion of the module is reduced.

1. p-ADIC TOPOLOGIES

Throughout the paper let R be a discrete valuation ring and denote its field of fractions by K. The ring R determines an exponential valuation $h : K \setminus 0 \to \mathbb{Z}$ on the quotient field K of R defined by $h(a) = x$ if $Ra = p^x R$, $a \neq 0$, and $h(0) = \infty$, where p is the prime element in R. Let γ be a real number, $0 < \gamma < 1$. Then $|a| = \gamma^{h(a)}$ defines a non-archimedean absolute value on K called the *p-adic absolute value* on K (see [3, Volume II, Chapter 9]). Let $\hat{K}$ be the completion of K relative to this absolute value. We call the elements $\beta \in \hat{K}$ *p-adic numbers*. As in [2, Section 3] we get expansions $\beta = \sum_{i=x}^{\infty} \beta_i p^i$ with $\beta_i \in R$ and $x \in \mathbb{Z}$. The ring

$$\hat{R} = \{\beta \in \hat{K} \mid |\beta| \leq 1\} = \{\beta \in \hat{K} \mid \beta = \sum_{i=0}^{\infty} \beta_i p^i,\ \beta_i \in R\}$$

is called the *ring of p-adic integers over R*. A p-adic integer β is a unit in $\hat{R}$ if and only if $|\beta| = 1$ or equivalently if β_0 is a unit in R. Note that $\hat{R}$ is a discrete valuation ring as well. The usual p-adic numbers are a special case of the above.

On a K-vector space V we define the *p- adic maximum norm* as

$$\| v \| = \max\{|\alpha_j| \mid j \in J\},$$

if $\{v_j \mid j \in J\}$ is a fixed basis of V, and $v = \sum_{j\in J} \alpha_j v_j$. We will say that the basis $\{v_j\}_{j\in J}$ *induces* this p-adic maximum norm and the topology on V.

The induced topology on V has as a base the set of open balls $B_x(a) = \{v \in V \mid \| a - v \| < x\}$, where $a \in V$, x a positive real. Hence this topology is linear and can also be given by means of a fundamental system of neighborhoods of 0 which consists of the sets $B_x(0)$, $x > 0$. Note that the topology on V has the 0-neighborhood basis $\{p^x W \mid x \in \mathbb{N}\}$, where $W = \oplus_{j\in J} Rv_j$.

At this point let us clarify the connection between the usual p-adic topology on R-modules, cf. [4], and our concept here. In the usual p-adic topology on an R-module L the submodules $p^x L$ form a neighborhood basis of 0. The corresponding topology is equal to our topology if and only if L and W are quasi-equal, cf. Proposition 1.1. With other words: Our topology is in general **not** the usual p-adic topology.

A topology on a group G compatible with the group structure of G defines a uniform structure on G, cf. [1, Chapter II], and if, additionally, G is commutative, then the completion of G exists (see [1, Chapter III]). If G is a commutative group and $\mathcal{T}_1$, $\mathcal{T}_2$ are topologies on G compatible with the group structure of G, then the corresponding completions coincide if and only if the topologies are the same. Hence it is enough to consider the topology induced by a basis if we want to determine the dependence of the completion of the chosen basis.

Let $\{v_i \mid i \in I\}$ and $\{v'_j \mid j \in J\}$ be two bases of a K-vector space V and put $W = \oplus_{i\in I} Rv_i$, respectively $W' = \oplus_{j\in J} Rv'_j$. We call the two bases *topology-equivalent* if both $\operatorname{ann}\left(W/W \cap W'\right)$ and $\operatorname{ann}\left(W'/W \cap W'\right)$ are non-trivial. Note that such an annihilator is an ideal of the form $p^k R$, $k \in \mathbb{N}$. The modules W and W' are called *quasi-equal*.

Proposition 1.1. *Two bases of a K-vector space V are topology-equivalent if and only if the topologies induced by the corresponding p-adic norms are the same.*

Proof. Let $\{v_i \mid i \in I\}$ and $\{v'_j \mid j \in J\}$ be two bases of V and denote by V_1, respectively V_2, the set V with the topology induced by the corresponding p-adic norm. Set $W = \oplus_{i\in I} Rv_i$ and $W' = \oplus_{j\in J} Rv'_j$.

To begin with, let the two bases be topology-equivalent. By symmetry it is enough to show that for each set of a 0-neighborhood basis of V_2 there is a 0-neighborhood of V_1 contained in this set. Take any neighborhood $p^{x'} W'$, $x' \in \mathbb{N}$, of 0 in V_2. By hypothesis there is positive integer x such that $p^x W \subset W \cap W' \subset W'$. Hence $p^{x+x'} W$ is contained in $p^{x'} W'$.

Conversely, suppose that the topological spaces V_1 and V_2 are identical. Then W and W' are neighborhoods of 0 in any of the two spaces and so is $W \cap W'$. Since $\{p^x W \mid x \in \mathbb{N}\}$ is a fundamental system of neighborhoods there is a positive integer x such that $p^x W \subset W \cap W'$. Hence $\operatorname{ann}\left(W/W \cap W'\right)$ is not trivial. The same argument shows that $\operatorname{ann}\left(W'/W \cap W'\right)$ is non-trivial. □

By Proposition 1.1 the topology induced by some p-adic maximum norm depends on the quasi-equality class of the free submodule $W = \oplus_{i\in I} Rv_i$ generated by the relevant basis.

Example. Let p be a prime number, let $\mathbb{Z}_p$ be the localization of the integers at p and let $V = \oplus_{i=0}^{\infty} \mathbb{Q} v_i$ be a $\mathbb{Q}$-vector space with basis $\{v_i \mid i \in \mathbb{N}\}$. Put $w_i = p^i v_i$, $i \in \mathbb{N}$. Then $V = \oplus_{i=0}^{\infty} \mathbb{Q} w_i$ and $\{w_i \mid i \in \mathbb{N}\}$ is also a basis of V. We have

$$\oplus_i \mathbb{Z}_p v_i \big/ \oplus_i \mathbb{Z}_p w_i \cong \bigoplus_i \mathbb{Z}_p v_i \big/ \mathbb{Z}_p w_i$$

and these two bases are not topology-equivalent.

2. A Generalization of Derry's Formula

We want to describe the structure of a completion. Let us start by explaining the notation. Let $V = \oplus_{j\in J} K v_j$ be a vector space over a field K with basis $\{v_j \mid j \in J\}$. We write the elements of the direct product $V^{\times} = \prod_{j\in J} K v_j$ in the form $(k_j \mid j \in J)$ with components $k_j \in K$. In an analogous way, we get the direct product $\hat{V}^{\times} = \prod_{j\in J} \hat{K} v_j$ consisting of the tuples $(k_j \mid j \in J)$ with $k_j \in \hat{K}$. The subspace of $\hat{V}^{\times}$ that consists of the tuples with only a countable number of non-zero components will be denoted by $\hat{V}_{\mathbb{N}}^{\times}$.

We consider the vector space $V = \oplus_{j\in J} K v_j$ as the subspace of $V^{\times}$ that consists of the tuples with only a finite number of non-zero components. Hence V also is a subspace of $\hat{V}^{\times}$.

Theorem 2.1. *Let V be a vector space over K and let $\{v_j \mid j \in J\}$ be a basis of V inducing the p-adic maximum norm on V.*

Then the completion $\hat{V}$ of V consists of all the tuples of $\hat{V}_{\mathbb{N}}^{\times}$ satisfying the property that for each $n \in \mathbb{N}$ almost all components of each tuple are contained in $p^n \hat{R}$.

Proof. V is a Hausdorff topological space with the countable 0-neighborhood basis $\{p^n W \mid n \in \mathbb{N}\}$, where $W = \oplus_{j\in J} R v_j$. Thus we need only consider Cauchy sequences for the completion $\hat{V}$ and not Cauchy filters as is the case in the general situation.

If $(k^{(n)})_{n\in\mathbb{N}} = \left((k_j^{(n)} \mid j \in J)\right)_{n\in\mathbb{N}}$ is a Cauchy sequence in V, then all components $(k_j^{(n)})_{n\in\mathbb{N}}$ are Cauchy sequences in K for each $j \in J$ with limit $k_j \in \hat{K}$, say. Hence $(k_j \mid j \in J)$ is the limit of $(k^{(n)})_{n\in\mathbb{N}}$. This limit has only a countable number of non-zero components, since each $k^{(n)}$ has only a finite number of non-zero components, i. e. the limit is in $\hat{V}_{\mathbb{N}}^{\times}$ and if n is fixed we have by straightforward arguments that k_j is contained in $p^n \hat{R}$ for all but finitely many $j \in J$.

Conversely, suppose that $(k_j \mid j \in J)$ is an element of $\hat{V}_{\mathbb{N}}^{\times}$ such that for each $n \in \mathbb{N}$ the components k_j are in $p^n \hat{R}$ for all but finitely many j. Since only countably many entries k_j are $\neq 0$, it is enough to consider the subcollection $(k_i \mid i \in \mathbb{N})$, where $k_i \neq 0$. By hypothesis almost all $k_i \in p^n \hat{R}$ for fixed n. So we may assume that the k_i are limits of Cauchy sequences $(k_i^{(l)})_{l\in\mathbb{N}}$, where $k_i^{(l)} \in K$ and $k_i^{(l')} - k_i^{(l)} \in p^l R$ for all i and all l, l' with $l' > l$. Hence the sequence

$$\left(k_1^{(i)}, k_2^{(i)}, \ldots, k_i^{(i)}, 0, \ldots\right)_{i\in\mathbb{N}}$$

forms a Cauchy sequence with limit $(k_i \mid i \in \mathbb{N})$, as desired. □

Example. Let $V = K^{\mathbb{N}}$. Then the tuple $(1, 1, 1, \ldots)$, whose components are all 1, is not in $\hat{V}$, but for instance, $(1, p, p^2, p^3, \ldots)$ is an element of $\hat{V}$.

Note that $\hat{R}$ and $\hat{K}$ with the usual p-adic topology are the respective completions.

Theorem 2.2. *Let L be a torsion-free R-module and let $\{v_j \mid j \in J\}$ be a maximal independent system of L. This system is a basis of the divisible hull $V = KL$ of L and induces the p-adic maximum norm on V. Denote the completion of V by $\hat{V}$. Then*

$$\hat{K}L \cap \overline{L} = \hat{R}L,$$

where $\overline{L}$ denotes the closure of L in $\hat{V}$.

Proof. As described before, we assume by suitable identification

$$L \subset KL \subset \hat{K}L \subset \hat{V} \subset \hat{V}^{\times}$$

and we consider all elements as certain tuples of $\hat{V}^{\times}$.

Obviously $\hat{R}L \subset \hat{K}L \cap \overline{L}$. Conversely, let $x \in \hat{K}L \cap \overline{L}$, say $x = (k_j \mid j \in J)$ with $k_j \in \hat{K}$. Let $(k^{(n)})_{n\in\mathbb{N}} = \left((k_j^{(n)} \mid j \in J)\right)_{n\in\mathbb{N}}$, $k_j^{(n)} \in K$ be a Cauchy sequence in L that converges to x. Then $(k_j^{(n)})_{n\in\mathbb{N}}$ is a Cauchy sequence in K with limit k_j, since the components are also Cauchy sequences. $\hat{K}L$ consists of the tuples with only a finite number of non-zero components, hence $k_j \neq 0$ for finitely many $j \in J$. As $(k^{(n)})_{n\in\mathbb{N}}$ is a Cauchy sequence there is an integer M such that

$$k^{(m)} - k^{(M)} \in W \qquad \text{for all } m > M$$

for the 0-neighborhood $W = \oplus_{j\in J} Rv_j$. Hence

$$k_j^{(m)} - k_j^{(M)} \in R \qquad \text{for all } m > M,\ j \in J,$$

so that

$$k_j - k_j^{(M)} \in \hat{R} \qquad \text{for all } m > M,\ j \in J.$$

Put $\alpha_j = k_j - k_j^{(M)}$ for all $j \in J$ and $\alpha = (\alpha_j \mid j \in J)$. Then $\hat{R}L$ consists of the tuples with only a finite number of non-zero components and entries in $\hat{R}$. As both $x \in \hat{K}L$ and $k^{(M)} \in L$ have only a finite number of non-zero components, the element $\alpha = x - k^{(M)}$ also has a finite number of non-zero components. Since all the components $k_j - k_j^{(M)}$, $j \in J$, are in $\hat{R}$, $\alpha \in \hat{R}L$. Hence $x = \alpha + k^{(M)} \in \hat{R}L$. □

Using the same arguments as in the proof of Theorem 2.2 we obtain Derry's formula.

Corollary 2.3. ([2, 93.2]) *Let L be a torsion-free R-module. Then*

$$KL \cap \hat{R}L = L.$$

Corollary 2.4. *Let L be a torsion-free R-module and let $\{v_j \mid j \in J\}$ be a maximal independent system of L. This system is a basis of the divisible hull $V = KL$ of L and induces the p-adic maximum norm on V. Denote the completion of V by $\hat{V}$. Then*

$$KL \cap \overline{L} = L.$$

Proof. Combining Theorem 2.2 and Corollary 2.3 we get the equality:

$$L \subset KL \cap L \subset KL \cap \hat{K}L \cap \overline{L} = KL \cap \hat{R}L = L. \ \square$$

In Theorem 2.2 we stated and proved that Derry's formula of intersection, $KL \cap \hat{R}L = L$, can be lifted to $\hat{K}L \cap \overline{L} = \hat{R}L$. The situation and the results are depicted in the following diagram of inclusions where the arrows point to the intersection of the sets on the same level as the symbol "$\cap$".

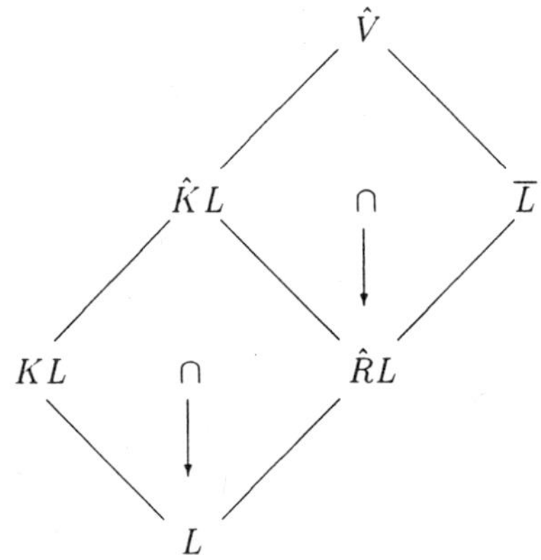

The results can be used to state a Local-Global principle for completions. In Derry's situation the Local-Global principle can be written as

$$A = \bigcap_p (\widehat{\mathbb{Z}}_p A \cap \mathbb{Q} A)$$

where A is torsion-free abelian group, $\widehat{\mathbb{Z}}_p$ is the ring of p-adic integers and the intersection is taken over all primes $p \in \mathbb{N}$. For completions this results in

$$\bigcap_p (\overline{\widehat{\mathbb{Z}}_p A} \cap \mathbb{Q} A) = \bigcap_p (\widehat{\mathbb{Z}}_p A \cap \mathbb{Q} A) = A.$$

For a topological module M over an integral domain R we denote by $\mathrm{CEnd}_R M$ the set of R-linear *continuous* endomorphisms of M.

Theorem 2.5. *Let L be a torsion-free R-module and let $\{v_j \mid j \in J\}$ be a maximal independent system of L. This system is a basis of the divisible hull $V = KL$ of L and induces the p-adic maximum norm on V. Then*

$$\mathrm{CEnd}_{\hat{R}} \overline{L} \cap \mathrm{End}_K KL = \mathrm{CEnd}_{\hat{R}} \hat{R}L \cap \mathrm{End}_K KL.$$

Proof. Let φ be an endomorphism in $\mathrm{End}_K KL$ which can be extended to a $\hat{K}$-linear endomorphism $\bar{\varphi}$ of $\hat{K}\overline{L}$ such that the restriction of $\bar{\varphi}$ to $\overline{L}$ is in $\mathrm{CEnd}_{\hat{R}}\overline{L}$. Note that this restriction is uniquely determined by φ since φ is necessarily a continuous endomorphism on a dense subspace of $\overline{L}$. Then

$$\bar{\varphi}(L) \subset \bar{\varphi}(\overline{L}) \subset \overline{L} \qquad \text{and} \qquad \bar{\varphi}(L) \subset \bar{\varphi}(KL) \subset \varphi(KL) \subset KL.$$

As $KL \cap \overline{L} = L$ we get $\bar{\varphi}(L) \subset L$ so that $\bar{\varphi}(\hat{R}L) \subset \hat{R}L$. Hence the restriction of $\bar{\varphi}$ to $\hat{R}L$, which is obviously continuous, is an element of $\mathrm{CEnd}_{\hat{R}}\hat{R}L$.

Conversely, let φ be an endomorphism in $\mathrm{End}_K KL$ which can be extended to a $\hat{K}$-linear endomorphism $\bar{\varphi}$ of $\hat{K}L$ such that the restriction of $\bar{\varphi}$ to $\hat{R}L$ is a continuous endomorphism in $\mathrm{CEnd}_{\hat{R}}\hat{R}L$. Then this restriction can be uniquely extended to an endomorphism in $\mathrm{CEnd}_{\hat{R}}\overline{L}$ by [1, Proposition III.3.5]. □

As a special case of the previous theorem we see that the Local-Global principle for endomorphisms in Derry's situation, namely

$$\mathrm{End}_{\mathbb{Z}} A = \bigcap_p \left(\mathrm{End}_{\widehat{\mathbb{Z}}_p} \widehat{\mathbb{Z}}_p A \cap \mathrm{End}_{\mathbb{Q}} \mathbb{Q} A \right),$$

can be lifted to completions in the following form:

$$\mathrm{CEnd}_{\mathbb{Z}} A = \bigcap_p \left(\mathrm{CEnd}_{\widehat{\mathbb{Z}}_p} \overline{\widehat{\mathbb{Z}}_p A} \cap \mathrm{End}_{\mathbb{Q}} \mathbb{Q} A \right).$$

3. Relation Systems

All modules are understood to be R-modules. Let us quickly recall the main definitions and conventions. We adopt the description of mixed modules G in the class $\mathcal{H}$ by generators and relations introduced in [5] where this class of mixed modules $\mathcal{H}$ has the property that the torsion submodule is a direct sum of cyclics and the quotient modulo the torsion is divisible of arbitrary rank. In our setting the torsion submodule has to be considered *in lieu of* the basic submodule F of rank r. Let L be a torsion-free R-module with a basic submodule F. Then $F = \bigoplus_{l \in P} Rb_l$, for suitable $b_l \in L$ and $r = |P|$. Thus the quotient L/F is a vector space over K of dimension d. Such a pair (L, F) of a torsion-free module L with a basic submodule F is said to be of *format* (r, d). Let I be a set of cardinality d, then the subset

$$B = \{a_{i-1}^k, b_l \mid i \in \mathbb{N}, k \in I, l \in P\} \subset L$$

is called a *basic generating system of* L, if

(1) $\{b_l \mid l \in P\}$ is a basis of F,
(2) $L/F = \bigoplus_{k \in I} K\bar{a}_0^k$,

where $\bar{a}_{i-1}^k = a_{i-1}^k + F$ and $p\bar{a}_i^k = \bar{a}_{i-1}^k$ for all $i \in \mathbb{N}, k \in I$. The definition implies that $\{a_0^k, b_l \mid k \in I, l \in P\}$ is a maximal independent set in L. It is easy to see that

$$L = \langle a_{i-1}^k, b_l \mid i \in \mathbb{N}, k \in I, l \in P \rangle.$$

A basic generating system intrinsically defines a series of equations with coefficients in R that describe the relations among the generators. Modeling upon these relations we introduce the concept of an abstract array $(\eta_{i-1,l}^k)$ with $i \in \mathbb{N}, k \in I, l \in P$ having entries in R and which is *row finite in* l, i. e. for a fixed pair (k, i), $\alpha_{i,l}^k = 0$, for almost all l. Such arrays $(\eta_{i-1,l}^k)$ are called *relation arrays.* The relation array $(\eta_{i-1,l}^k)$ is said to be of *format* (r, d). Note that all relation arrays of L relative to the basic submodule F, such that the pair (L, F) has the format (r, d), inherit that format.

The following result gives a description of such modules L by generators and relations. The proof is almost verbatim that of [5, Proposition 3], with F replacing the torsion submodule.

Proposition 3.1. ([5, Proposition 3]) *Every torsion-free module L with basic submodule $F = \bigoplus_{l \in P} Rb_l$ has a basic generating system $B = \{a_{i-1}^k, b_l \mid i \in \mathbb{N}, k \in I, l \in P\}$ and a relation array $(\eta_{i-1,l}^k)$ relative to B given by*

$$pa_i^k = a_{i-1}^k + \sum_{l \in P} \eta_{i-1,l}^k b_l, \qquad i \in \mathbb{N}, k \in I. \tag{1}$$

The entries $\eta_{i-1,l}^k$ are uniquely determined by the choice of a_i^k. For a fixed k there are only countably many entries $\eta_{i,l}^k \neq 0$.

By [5, Theorem 4] all relation arrays are realized, i. e. there is a torsion-free module L with basic submodule F having a basic generating system with a prescribed relation array.

There is a compact way of discussing relations namely, by the so called relation systems. A collection $\alpha = (\alpha_l \mid l \in P)$ of p-adic integers $\alpha_l \in \hat{R}$ is called *summable* if for each positive real number δ there are only finitely many $l \in P$ such that $|\alpha_l| > \delta$; or in terms of the exponential valuation h, if for each integer m there are only finitely many $l \in P$ such that $h(\alpha_l) < m$. It is easy to see that a collection α is summable if and only if $\sum_{l \in P} \alpha_l \in \hat{R}$. For a p-adic integer $\alpha = \sum_{i \geq 0} \alpha_i p^i$ let $\alpha^{(s)} = \sum_{i=0}^{s-1} \alpha_i p^i$ denote the integer given by the s-th partial sum. We extend the formation of of partial sums to collections of p-adic integers $\alpha^{(s)} = (\alpha_l^{(s)} \mid l \in P)$.

The row finiteness of relation arrays can be characterized, cf. [6, Lemma 3].

Lemma 3.2. ([6, Lemma 3]) *Let $(\eta_{i-1,l}^k)$ be a relation array. Then the matrix $\eta = (\eta_l^k)_{k \in I, l \in P}$, where $\eta_l^k = \sum_{i \in \mathbb{N}_0} \eta_{i,l}^k p^i$, has rows $\eta^k = (\eta_l^k \mid l \in P)$, which are all summable.*

Conversely, an array $(\eta_{i-1,l}^k)$ such that the corresponding matrix $\eta = (\eta_l^k)$ has only summable rows η^k, $k \in I$, is a relation array.

Proof. Let $(\eta_{i-1,l}^k)$ be a relation array. The exponential valuation of η_l^k is $\leq m$, $m \in \mathbb{N}$, if and only if there exists an $i \leq m$ with $\eta_{i,l}^k \neq 0$. This is not the case since there are only finitely many $\eta_{i,l}^k \neq 0$ for a fixed pair (k, i). For the converse it is easy to verify that a corresponding array must be a relation array. □

Such a matrix $\eta = (\eta_l^k)$ with summable rows is called a *relation system of format* (r, d), corresponding to the relation array $(\eta_{i-1,l}^k)$. If $(\eta_{i-1,l}^k)$ is a relation array of a torsion-free module L for a basic generating system B then the relation system $\eta = (\eta_l^k)$ corresponding to $(\eta_{i-1,l}^k)$ is called a *relation system of L relative to B* and *to* $(\eta_{i-1,l}^k)$. If a relation system corresponds to an array $(\eta_{i-1,l}^k)$ then the array is called an *expansion of the relation system.* Expansions of relation systems are relation arrays.

Now we consider only torsion-free modules with a basic submodule F such that $L/F \cong K$. The corresponding relation systems are of format $(r, 1)$. Therefore we omit the index k and the relation arrays have the form $(\eta_{i-1,l})$. The corresponding relation systems are the summable collections $\eta = (\eta_l \mid l \in P)$. Let $l_1, l_2, \ldots$ be an indexing of the countable subset $P_0 = \{l \in P \mid \eta_{i,l} \neq 0$ for some $i \in \mathbb{N}_0\}$ of P. Note that $\eta_l \neq 0$ only for $l \in P_0$, i. e. only for countably many $l \in P$.

The relation array $(\eta_{i-1,l})$ is said to be *small* if the array $\left(\sum_{i=0}^{j-1} \eta_{s+i,l_j} p^i\right)_{s,l_j}$ is a relation array. Smallness is independent of the indexing of P_0. The last array is not necessarily a relation array since the row finiteness in l is not guaranteed. The relation system η is said to be *small* if it has one small expansion. It follows that all expansions are small, cf. [5]. Two relation systems of the same format $(r, 1)$ are called *similar*, $\eta \sim \rho$, if their difference is small. This is an equivalence relation.

The following theorem can be proved using the same arguments as in [5, Theorem 6], where in the respective proofs the basic submodule F has to be considered *in lieu of* the torsion part of the module.

Theorem 3.3. ([5, Theorem 6]) *Let L be a torsion-free module with basic submodule F and $L/F \cong K$. Then the following are equivalent:*

(1) *F is a direct summand of L.*
(2) *L has only small relation arrays relative to F.*
(3) *L has one small relation array relative to F.*

4. The Extension Group Ext(K,F)

The set A of all relation systems of format $(r,1)$ is a torsion-free R-module relative to the usual operations. Let A_{small} denote the subset of all small relation systems.

Theorem 4.1. *A_{small} is a pure submodule of A and $\mathfrak{A} = A/A_{small}$ is a vector space over the field K.*

Proof. Obviously A_{small} is pure in A and thus $\mathfrak{A} = A/A_{small}$ is torsion-free. Since $p\big(p^{-1}(\eta - \eta^{(1)})\big) = \eta - \eta^{(1)} \sim \eta$ for each relation system $\eta \in A$, the quotient $\mathfrak{A}$ is divisible, hence a vector space over K. □

Let $F = \bigoplus_{l\in P} Rb_l$ be a free module of rank r. Let L be a torsion-free module with basic submodule F and $L/F \cong K$. We follow the notation in [2, Section 49]. Let $g : K \longrightarrow L$ be an arbitrary function such that $g(u)$ is a representative of the coset u, i. e. $g(p^{-i}) = a_i$. Then $\{a_{i-1}, b_l \mid i \in \mathbb{N}, l \in P\}$ is a basic generating system of L with corresponding relation system η of format $(r,1)$, i. e. $\eta = \eta(g)$. Moreover, the function g also determines a factor set $f = f(g)$ by

$$f(u,v) = g(u) + g(v) - g(u+v).$$

If now $g' : K \longrightarrow L$ is another function inducing the same factor set, $f(g) = f(g')$, then if $g'(p^{-i}) = c_i$, so $\{c_{i-1}, b_l \mid i \in \mathbb{N}, l \in P\}$ is another basic generating system with corresponding relation system ρ. Now by the common property of the functions g, g' to map elements of K into their cosets we have $a_i - c_i \in F$. This implies that $\eta \sim \rho$ by the following. Let $(\eta_{i-1,l})$ and $(\rho_{i-1,l})$ be expansions of η and ρ, respectively. Let $c_i = a_i + f_i$ for all $i \in \mathbb{N}$, where $f_i = \sum_{l\in P} \gamma_{i,l} b_l \in F$. Then $(\gamma_{i-1,l})$ is a relation array which is an expansion of the relation system γ. Moreover,

$$\begin{aligned} c_{i-1} + \sum_{l\in P} \rho_{i-1,l} b_l &= pc_i = p(a_i + f_i) \\ &= a_{i-1} + \sum_{l\in P} \eta_{i-1,l} b_l + pf_i \\ &= c_{i-1} + \sum_{l\in P} \eta_{i-1,l} b_l - f_{i-1} + pf_i, \end{aligned}$$

hence $\rho_{i-1,l} = \eta_{i-1,l} + p\gamma_{i,l} - \gamma_{i-1,l}$. Thus $\eta - \rho = \gamma^{(1)}$ is small. Hence $\eta + A_{small} = \rho + A_{small}$. Thus a mapping is defined:

$$\mu : \mathrm{Fact}(K,F) \longrightarrow \mathfrak{A}, \quad \mu(f) = \eta + A_{small}.$$

Next we show that this mapping μ induces an isomorphism between $\mathrm{Ext}(K,F)$ and $\mathfrak{A}$.

Theorem 4.2. *μ is an epimorphism with kernel* $\mathrm{Trans}(K,F)$, *and* $\mathrm{Ext}(K,F) \cong \mathfrak{A}$.

Proof. Let $\eta \in A$ be a relation system. Then η is realized by a torsion-free module L with basic submodule F and $L/F \cong K$. Let the corresponding basic generating system be $\{a_{i-1}, b_l \mid i \in \mathbb{N}, l \in P\}$, cf. [5, Theorem 4]. Defining a function $g : K \longrightarrow L$ by $g(p^{-i}) = a_i$ and completing the definition of g arbitrarily such that g maps elements to their cosets, respectively, we obtain that the image of the factor set $f(g)$ under μ is $\eta + A_{small}$, i. e. $\mu(f) = \eta + A_{small}$. Hence μ is epic.

The kernel of μ is the set $\mathrm{Trans}(K, F)$ of transformation sets, since the transformation sets are the factor sets of the splitting extension and by Theorem 3.3 the corresponding relation systems are precisely all small relation systems, i. e. $\mu\big(\mathrm{Trans}(K, F)\big) = A_{small}$.

Next we show that μ is a homomorphism. For this we write down how the factor set f can be determined by η, namely, and let for simplicity $u, v \in K \setminus pR$,

$$f(u,v) = g(u) + g(v) - g(u+v) = up^{i(u)}a_{i(u)} + vp^{i(v)}a_{i(v)} - (u+v)p^{i(u+v)}a_{i(u+v)},$$

where $up^{i(u)} \in R \setminus pR$. We know that $f(u,v) \in F$, hence the sum representation of f on the right-hand side of the equation above must be of the form

$$z\left(p^j a_{j+h} - a_h\right) = z\left(\sum_{l\in P}\sum_{s=h}^{j+h-1} \eta_{s,l}p^{s-h}b_l\right),$$

where $z \in R$. In particular, z and the indices j, h depend only on u, v. For example let us consider the case $i(u) > i(v) \geq 0$. Then $i(u+v) = i(u)$ and

$$f(u,v) = -vp^{i(v)}\left(p^{i(u)-i(v)}a_{i(u)} - a_{i(v)}\right).$$

Hence $z = -vp^{i(v)}$, $j = i(u) - i(v)$ and $h = i(v)$. Let now $f = f(g), f' = f'(g')$ be two factor sets and let $\mu(f) = \eta + A_{small}$ and $\mu(f') = \rho + A_{small}$. The sum of f and f' is given by

$$f(u,v) + f'(u,v) = \big(g(u) + g(v) - g(u+v)\big) + \big(g'(u) + g'(v) - g'(u+v)\big).$$

Writing this in terms of η and ρ and observing that the included parameters z and the indices j, h depend only on u, v, i. e. they are equal, we get:

$$f(u,v) + f'(u,v) = z\left(\sum_{l\in P}\sum_{s=h}^{j+h-1} \eta_{s,l}p^{s-h}b_l + \sum_{l\in P}\sum_{s=h}^{j+h-1} \rho_{s,l}p^{s-h}b_l\right) \tag{2}$$

$$= z\left(\sum_{l\in P}\sum_{s=h}^{j+h-1} (\eta^u_{s,l} + \rho_{s,l})p^{s-h}b_l\right). \tag{3}$$

This shows that the mapping μ is additive. The proof of $\mu(zf) = z\mu(f)$ for $z \in R$ is analogue. Hence μ is a homomorphism and induces an isomorphism between $\mathrm{Ext}(K, F)$ and $\mathfrak{A}$. □

5. Structures of Local Completions

In the following we shall be concerned with summable families. We begin by recalling the definition as it will be important for our studies.

Let G be a Hausdorff commutative group, I any index set and $(x_i)_{i\in I}$ a family of points of G. With each finite subset J of I we associate the element $s_J = \sum_{i\in J} x_i$ of G, which we call the *finite partial sum* of the family $(x_i)_{i\in I}$ corresponding to the set J. If $\mathcal{F}(I)$ denotes the set of finite subsets of I, we thus have a mapping $J \mapsto s_J$ of $\mathcal{F}(I)$ into G. Observe that $\mathcal{F}(I)$ is a directed set with respect to the relation $\subset$. The family $(x_i)_{i\in I}$ is said to be *summable* if the mapping $J \mapsto s_J$ has a limit with respect to the section filter of the set $\mathcal{F}(I)$. This limit is then said to be the sum of the family $(x_i)_{i\in I}$ and is denoted by $\sum_{i\in I} x_i$. Note that the definition of a summable family does not involve any ordering of the index set I and we may therefore say that the notion of a sum so defined is "commutative".

Let M be a complete topological module with submodules M_1, M_2 such that $M = M_1 \oplus M_2$. Then M_i ($i = 1,2$) is a uniform space with the induced uniform structure, respectively topology. If we speak of the topological space M_i we mean M_i with its induced topology. The product topology on $M_1 \times M_2$ is again the topology of M. As M is complete, M_i is a complete space (by Proposition [1, Proposition II.3.9]), so that

$$M = M_1 \oplus M_2 = \overline{M_1} \oplus \overline{M_2},$$

i. e. M_1 and M_2 are closed.

Using similar arguments as in [2, Sections 24 and 32] one can show the existence of a divisible hull and that of a basic submodule N of a p-adic module M where N is a direct sum of cyclics pure submodule with M/N divisible.

Standing hypothesis:

Let L be a torsion-free submodule of V with basic submodule $F = \sum_{l \in P} Rb_l$ of rank $r = |P|$ and $L/F \cong K^d$. Then there is basic generating system $\{a_{i-1}^k, b_l \mid i \in \mathbb{N}, k \in I, l \in P\}$ of L relative to F and a corresponding relation array $(\eta_{i-1,l}^k)$ or relation system η of format (r,d). From here on out our setting will be

$$F \subset L \subset V \subset \hat{V},$$

where V has a p-adic maximum norm and $\hat{V}$ is its corresponding completion. Furthermore $\overline{F}$ and $\overline{L}$ denote the closures of F and L in $\hat{V}$, respectively. We want to mention explicitly that the basis $B = \{b_l \mid l \in P\}$ of the basic submodule F and the basis of V used to define the p-adic maximum norm are a priori independent of each other.

Let $F \subset L \subset V \subset \hat{V}$. Then for each $k \in I$ the subsystem $\eta^k = (\eta_l^k \mid l \in P)$ of format $(r,1)$ is the relation system corresponding to the basic generating system $\{a_{i-1}^k, b_l \mid i \in \mathbb{N}, l \in P\}$ of a pure submodule L^k of L with $L^k/F = K(a_0^k + F) \cong K$. Note that these subsystems η^k are independent of each other.

We obtain $L = \sum_{k \in I} L^k$ since $L/F = \bigoplus_{k \in I} L^k/F$. The relation system η^k of L^k relative to F is of format $(r,1)$. So $\bar{\eta}^k = \eta^k + A_{small} \in \mathfrak{A}$ and we may define the subspace

$$\mathfrak{M}(\eta) = \langle \bar{\eta}^k \mid k \in I \rangle$$

of $\mathfrak{A}$ generated by all $\bar{\eta}^k$. This subspace $\mathfrak{M}(\eta)$ depends a priori on η or equivalently on the representation $L = \sum_{k \in I} L^k$. Define a linear mapping

$$\varphi : L/F \longrightarrow \mathfrak{A}, \quad \varphi(a_0^k + F) = \bar{\eta}^k.$$

φ is well defined since $\{a_0^k + F \mid k \in I\}$ is a basis of L/F. In particular, $\varphi\big(L^k/F\big) = K\bar{\eta}^k \subset \mathfrak{M}(\eta)$, thus $\varphi\big(L/F\big) = \mathfrak{M}(\eta) = \mathfrak{M}(L)$.

Lemma 5.1. *Let L be a torsion-free module with basic submodule F and $L/F \cong K^d$. The image $\mathfrak{M}(L) = \mathfrak{M}(\eta)$ of L/F under φ doesn't depend on the relation system η or the representation $L = \sum_{k \in I} L^k$, if the basis of F is fixed.*

Proof. A new basis $\{b_0^k + F \mid k \in I\}$ of L/F is given with a row finite invertible $d \times d$-matrix $D = (d_{k,h})$ over K in the form $b_0^k + F = \sum_{h\in I} d_{k,h}(a_0^h + F)$. The corresponding relation system is $D\eta$, cf. [6, Lemma 8 and Proposition 9]. Then

$$\varphi\Big(\sum_{h\in I} d_{k,h}\big(a_0^h + F\big)\Big) = \sum_{h\in I} d_{k,h}\varphi\big(a_0^h + F\big), \quad k \in I,$$

and $\mathfrak{M}(\eta) = \mathfrak{M}(D\eta) = \varphi\big(L/F\big)$, since D is invertible, i. e. $\mathfrak{M}(L) = \mathfrak{M}(\eta)$ is independent of η. □

Theorem 5.2. *Let L be a torsion-free module with basic submodule F of rank r and $L/F \cong K^d$, then*

$$\ker\varphi = \big(D \oplus F\big)/F,$$

where D is the maximal divisible submodule of L.

In particular, L is reduced if and only if φ is injective, i. e. if and only if $\{\eta^k \mid k \in I\}$ is independent modulo A_{small}. The rank of the reduced part of L and the dimension of $\mathfrak{M}(L)$ are equal.

Proof. If $a \in D$ then the corresponding relation system η of format $(r,1)$ is 0, hence $\bar{\eta} = 0$ and $\varphi\big(a + F\big) = 0$. Conversely, if $\varphi\big(a + F\big) = 0$ then the corresponding relation system η of format $(r,1)$ is small, hence a is a divisible element by Theorem 3.3. An application of the dimension formula for a linear mapping immediately shows the rest. □

Example. Let $p \neq 2$, let $L = \langle a_{i-1}^k, x_i \mid i \in \mathbb{N}, k \in \{1,2\}\rangle$ be a torsion-free module and let a corresponding relation array be defined by: $\eta_{i,2i+1}^1 = \eta_{i,2i+1}^2 = 1, i \in \mathbb{N}_0$, and $\eta_{i,j}^1 = \eta_{i,j}^2 = 0$ otherwise. Then $D = \langle a_i^1 - a_i^2 \mid i \in \mathbb{N}\rangle$ is the divisible part of L.

Let V be a vector space over K. Let $\hat{V}$ be the completion of V relative to some p-adic maximum norm induced by this basis. Let $\{a_0, b_l \mid l \in P\}$ be a basis of V. Let F be the free submodule of V generated by the basis $B = \{b_l \mid l \in P\}$. Let $(\eta_{i-1,l})$ be a relation array of format $(r,1)$. If the family $(\eta_{i,l}p^i b_l)_{(i,l)\in\mathbb{N}_0\times P}$ is summable η is called *B-summable.* Summability is equivalent to $\sum_{l\in P}\sum_{i\in\mathbb{N}_0}\eta_{i,l}p^i b_l \in \hat{V}$. The corresponding relation systems are also called *B-summable.* The B-summable relation systems form a submodule A_{sum} of the module A of all relation systems of format $(r,1)$.

We will now give conditions ensuring that $(\eta_{i,l})$ is B-summable for some basis $B = \{b_l \mid l \in P\}$. Trivially, this is the case if there are only a finite number of non-zero elements. We now give a definition in order to deal with an infinite number of non-zero elements.

Let P be any index set. We call the array $(\eta_{i,l})_{i\in\mathbb{N}_0,l\in P}$ with entries $\eta_{i,l} \in R$ a *polynomially growing relation array* if the number of non-zero elements in each row is bounded by a polynomial, more precisely, if there are non-negative integers k and c such that for each $i \in \mathbb{N}_0$

$$|\{\eta_{i,l} \neq 0 \mid l \in P\}| \leq i^k + c.$$

This definition yields another criterion for summability.

Proposition 5.3. *Let $\hat{V}$ be a complete vector space over the field $\hat{K}$ with some p-adic maximum norm. If $(\eta_{i,l})_{i\in\mathbb{N}_0,l\in P}$ is a polynomially growing relation array and if there are elements $b_l \in \hat{V}$, $l \in P$ and a constant C such that $\| b_l \| < C$ for all $l \in P$, then $(\eta_{i,l}p^i b_l)_{(i,l)\in\mathbb{N}_0\times P}$ is summable.*

Proof. The set $(\eta_{i,l}p^i b_l)_{(i,l)\in\mathbb{N}_0\times P}$ is countable so that we can enumerate it. It is enough to show that the series defined by a certain enumeration is absolutely convergent, since this implies summability.

Since $\eta_{i,l}\in\hat{R}$, we have $|\eta_{i,l}|\le 1$ for all $i\in\mathbb{N}_0$, $l\in P$. For all $i\in\mathbb{N}_0$, the number of non-zero elements in $(\eta_{i,l})_{l\in P}$ is bounded by the polynomial i^k+c, say. Thus we have

$$\sum_{l\in P}\|\ \eta_{i,l}b_l\ \|\le C\sum_{l\in P}|\eta_{i,l}|\le C(i^k+c)$$

for all $i\in\mathbb{N}_0$, since the sum is finite and has at most i^k+c elements. Hence

$$\sum_{i=0}^{\infty}\sum_{l\in P}\|\ \eta_{i,l}p^i b_l\ \|=\sum_{i=0}^{\infty}|p^i|\sum_{l\in P}\|\ \eta_{i,l}b_l\ \|=\sum_{i=0}^{\infty}\gamma^i C(i^k+c)<\infty.$$

This proves the absolute convergence of the series $\sum_{i=0}^{\infty}\sum_{l\in P}\eta_{i,l}p^i b_l$ and the proof is complete. □

Lemma 5.4. *A_{sum} is a pure submodule of A. If there is a constant C such that $\|\ b_l\ \|<C$ for all $l\in P$ then A_{small} is contained in A_{sum}, and in particular, $\mathfrak{A}_{sum}=A_{sum}/A_{small}$ is a subspace of the extension group $\mathfrak{A}=\operatorname{Ext}(K,F)$, where $F=\bigoplus_{l\in P}Rb_l$.*

Proof. If $p(\eta_{i,l})$ is B-summable, $B=\{b_l\mid l\in P\}$, then also $(\eta_{i,l})$ is B-summable, hence A_{sum} is a pure submodule. By Proposition 5.3 small relation systems are in particular B-summable if $\|\ b_l\ \|$ is globally bounded. Furthermore $\mathfrak{A}_{sum}$ is a subspace since $\mathfrak{A}=\operatorname{Ext}(K,F)$ is a vector space by Theorem 4.1. □

Lemma 5.5. *Let $F\subset L\subset V\subset\hat{V}$ and $L/F\cong K$. Let $B=\{b_l\mid l\in P\}$ be a basis of F. If η is B-summable, then*

$$\overline{L}=\hat{K}y\oplus\overline{F},\tag{4}$$

where $y=a_0+\sum_{l\in P}\sum_{i\in\mathbb{N}_0}\eta_{i,l}p^i b_l\in\hat{V}$.

Proof. Since $(\eta_{i,l}p^i b_l)_{(i,l)\in\mathbb{N}_0\times P}$ is summable, $y\in\overline{L}$. Certainly $\hat{K}y\cap\overline{F}=0$ since $a_0\notin\overline{F}$. We first show the inclusion $\hat{K}y\oplus\overline{F}\subset\overline{L}$. As a closed submodule of a complete space, $\overline{L}$ is complete. Hence $\overline{F}$ is a subset of $\overline{L}$ since F is contained in L. It follows from the definition of the elements a_i that

$$p^h a_h=a_0+\sum_{l\in P}\Big(\sum_{i=0}^{h-1}\eta_{i,l}p^i\Big)b_l\tag{5}$$

for all $h\in\mathbb{N}$. This shows that y can be written as a sum of two elements in $\overline{L}$, both divisible in $\overline{L}$ by p^h:

$$y=\left(a_0+\sum_{l\in P}\Big(\sum_{i=0}^{h-1}\eta_{i,l}p^i\Big)b_l\right)+p^h\left(\sum_{l\in P}\Big(\sum_{i=h}^{\infty}\eta_{i,l}p^{i-h}\Big)b_l\right).\tag{6}$$

In view of (5) the element y is divisible in $\overline{L}$ by p^h for all $h\in\mathbb{N}$, i. e. y is a divisible element in $\overline{L}$. Consequently $\hat{K}y\oplus\overline{F}$ is a submodule of $\overline{L}$.

As for the reverse inclusion, note that each b_l, $l\in P$, is contained in the right-hand side of (4). Combining (5) and (6) yields

$$a_h=p^{-h}\left(a_0+\sum_{l\in P}\Big(\sum_{i=0}^{h-1}\eta_{i,l}p^i\Big)b_l\right)=p^{-h}y-\left(\sum_{l\in P}\Big(\sum_{i=h}^{\infty}\eta_{i,l}p^{i-h}\Big)b_l\right).$$

This shows that a_h is an element of $\hat{K}y \oplus \overline{F}$ for all $h \in \mathbb{N}$. Hence $\overline{L}$ is contained in the right-hand side of (4) and equality holds. □

Example. This example shows that by forming the closure of a module, we can get new divisible elements even if the module we start with is reduced.

There exists a torsion-free module L with basic submodule F such that L has a generating system $\{a_i, b_i \mid i \in \mathbb{N}\}$ relative to F with relations

$$pa_{i+1} = a_i + b_i \qquad \text{for all } i \in \mathbb{N}.$$

In other words the relation array $\eta = (\eta_{i,l})$ is of the form $\eta_{i,l} = 1$ if $l = i$ and $\eta_{i,l} = 0$ otherwise. The relation array η is obviously not small, so that L is reduced by Theorem 3.3. The system $\{a_0, b_i \mid i \in \mathbb{N}\}$ is a basis of the divisible hull $V = KL$ of L and induces the p-adic maximum norm on V. Then

$$y = a_0 + \sum_{i=0}^{\infty} p^i b_i \in \hat{V}$$

is a divisible element of $\overline{L}$ and $\overline{L}$ contains $\hat{K}y$, where the closure of L is formed in the completion $\hat{V}$ of V.

Note that as we have taken $(a_0, b_0, b_1, b_2, b_3, \dots)$ as a basis of the divisible hull $V = KL$, the element y corresponds to the tuple $(1, 1, p, p^2, p^3, \dots, p^i, \dots)$ under the convention of writing elements as tuples of $\hat{V}^{\times}$. Hence y is an element of the completion by Theorem 2.1.

Theorem 5.6. *Let $F \subset L \subset V \subset \hat{V}$ where L is a full torsion-free submodule of V with basic submodule F of rank r and $L/F \cong K^d$. Let $(\eta^k_{i-1,l})$ be a corresponding relation array. Let C be some constant with $\| b_l \| < C$ for all $l \in P$, where $B = \{b_l \mid l \in P\}$ is the fixed basis of F. Let $\{c_j \mid j \in J\}$ be a basis of the inverse image of $\mathfrak{A}_{sum}$ under φ, where $c_j = \sum_{k \in I} d_{j,k} a^k_0 + F$ and $d_{j,k} \in R$. Let*

$$y_j = \sum_{k \in I} d_{j,k} a^k_0 + \sum_{k \in I} \sum_{l \in P} \sum_{i \in \mathbb{N}_0} d_{j,k} \eta^k_{i,l} p^i b_l.$$

Then

$$\overline{L} \supset \Big(\bigoplus_{j \in J} \hat{K} y_j\Big) \oplus \overline{F}.$$

In particular, $\overline{L}$ is not reduced if $\mathfrak{M}(L) \cap \mathfrak{A}_{sum} \neq 0$.

Proof. Since the norm of the basis elements b_l is globally bounded we have the setting $\mathfrak{A}_{sum} \subset \mathfrak{A}$. Moreover, $L = \sum_{k \in I} L^k$, where each L^k is given by a relation array $\eta^k = (\eta_{i-1,l})$, k fixed, of format $(r, 1)$. Thus $\mathfrak{M}(L)$ is generated as a subspace of $\mathfrak{A}$ by $\{\eta^k \mid k \in I\}$. Certain finite linear combinations $\eta = \sum_{k \in I} d_k(\eta^k_{i,l})$, $d_k \in R$, of the η^k may be B-summable, $B = \{b_l \mid l \in P\}$. So the inverse image under φ of those $\bar{\eta} = \eta + A_{small} \in \mathfrak{A}_{sum}$ is a subspace of L/F. Let $\{c_j \mid j \in J\}$ be a basis of this subspace, i. e. $c_j = \sum_{k \in I} d_{j,k} a^k_0 + F$, since $\{a^k_0 + F \mid k \in I\}$ is a basis of L/F. For each j the relation array $\sum_{k \in I} d_{j,k}(\eta^k_{i,l})$ is B-summable. Hence by Lemma 5.5 the elements y_j are divisible in $\overline{L}$. Thus $\overline{L} \supset \left(\bigoplus_{j \in J} \hat{K} y_j\right) \oplus \overline{F}$. There are always divisible elements in $\overline{L}$ if the inverse image of $\mathfrak{A}_{sum}$ under φ intersects $\mathfrak{M}(L)$ non-trivially. This shows the additional statement. □

The inclusion in Theorem 5.6 is certainly proper if the set J is infinite since the closure of $\bigoplus_{j\in J}\hat{K}y_j$ is also contained in $\overline{L}$. But if L/F is of finite dimension, then the divisible elements of $\overline{L}$ are caused by the summability of the relation system corresponding to (L,F), and the converse can be shown, too.

Proposition 5.7. *Let L be a torsion-free module with basic submodule F such that L/F is of finite dimension d. Let $\{a_i^k, b_l \mid i\in \mathbb{N}_0, 1\le k\le d, l\in P\}$ be a basic generating system of L, relative to $F=\bigoplus_{l\in P} Rb_l$. The system $\{a_0^k, b_l \mid 1\le k\le d, l\in P\}$ is a basis of the divisible hull $V=\hat{K}L$ of L and induces the p-adic maximum norm on V. Let $\hat{V}$ be the corresponding completion of V and let $\overline{L}$ denote the closure of L in $\hat{V}$. Then $\overline{L}$ is not reduced if and only if $\mathfrak{M}(L)\cap\mathfrak{A}_{sum}\neq 0$.*

Moreover, let $\{c_j \mid 1\le j\le m\}$, $m\le d$, be a basis of the inverse image of $\mathfrak{A}_{sum}$ under φ, where $c_j=\sum_{k=1}^d d_{j,k}a_0^k+F$ and $d_{j,k}\in R$ and let

$$y_j=\sum_{k=1}^d d_{j,k}a_0^k+\sum_{k=1}^d\sum_{l\in P}\sum_{i\in\mathbb{N}_0} d_{j,k}\eta_{i,l}^k p^i b_l.$$

Then

$$\overline{L}=\Big(\bigoplus_{j=1}^m \hat{K}y_j\Big)\oplus\overline{F}.$$

Proof. The norm of the basis elements b_l is globally bounded, since they induce the p-adic maximum norm. Thus by Lemma 5.4 we have the setting $\mathfrak{A}_{sum}\subset\mathfrak{A}$. By Theorem 5.6 the closure $\overline{L}$ is not reduced if $\mathfrak{M}(L)\cap\mathfrak{A}_{sum}\neq 0$.

Conversely, if $\overline{L}$ is not reduced it contains a divisible element

$$y=\sum_{k=1}^d \eta_0^k a_0^k+\sum_{l\in P}\beta_l b_l,\qquad \eta_0^k\in\hat{K},\beta_l\in\hat{R}.$$

We may assume without loss of generality that $\eta_0^k\in\hat{R}$, by multiplying y with a suitable power of p and consider this product instead of y. Moreover we may assume that $\eta_0^1\neq 0$. For each $n\in\mathbb{N}_0$ almost all coefficients β_l are in $p^n\hat{R}$, so that there are disjoint finite subsets S_n of P such that $l\in S_n$ if $\beta_l\in p^n\hat{R}\setminus p^{n+1}\hat{R}$. Thus $\beta_l=0$ for all $l\in P\setminus\bigcup_{n\in\mathbb{N}_0}S_n$. Let

$$y_j=\sum_{k=1}^d \eta_0^k a_0^k+\sum_{l\in\bigcup_{\nu=0}^{j-1}S_\nu}\beta_l b_l,$$

i. e. $y_0=\sum_{k=1}^d \eta_0^k a_0^k$, $y_1=\sum_{k=1}^d \eta_0^k a_0^k+\sum_{l\in S_0}\beta_l b_l$, $y_2=\sum_{k=1}^d \eta_0^k a_0^k+\sum_{l\in S_0\cup S_1}\beta_l b_l\in L$, and so forth. Hence $(y_j)_{j\in\mathbb{N}_0}$ converges to y. Furthermore $y_j=y-\sum_{l\in\bigcup_{\nu\ge j}S_\nu}\beta_l b_l\in p^jL$ for all $j\in\mathbb{N}_0$, so that the elements

$$z_i=p^{-i}y_i\in L,\qquad i\in\mathbb{N}_0,$$

are well-defined. By setting $z_i^1=z_i$, $i\in\mathbb{N}_0$ and $z_i^k=a_i^k$ for $2\le k\le d$, $i\in\mathbb{N}_0$, we get a new basis of L/F, namely $\{z_0^k+F \mid 1\le k\le d\}$, and a new basic generating system

$\{z_i^k, b_l \mid i \in \mathbb{N}_0, 1 \le k \le d, l \in P\}$ of L, relative to F. Moreover,

$$\begin{aligned} pz_{i+1} &= p^{-i} y_{i+1} \\ &= p^{-i} y_i + p^{-i} \sum_{l \in S_i} \beta_l b_l \\ &= z_i + \sum_{l \in S_i} p^{-i} \beta_l b_l. \end{aligned}$$

Thus the basic generating system $\{z_i^k, b_l \mid i \in \mathbb{N}_0, 1 \le k \le d, l \in P\}$ of L has a relation array $\eta = (\eta_{i,l}^k)$ of the form

$$\eta_{i,l}^1 = \begin{cases} p^{-i}\beta_l & \text{for } l \in S_i \\ 0 & \text{otherwise} \end{cases}.$$

The other entries $\eta_{i,l}^k$, $k > 1$, remain unchanged. Hence

$$\sum_{i \in \mathbb{N}_0, l \in P} \eta_{i,l}^1 p^i b_l = \sum_{i \in \mathbb{N}_0} \sum_{l \in S_i} p^{-i} \beta_l p^i b_l = \sum_{l \in P} \beta_l b_l = y - \sum_{k=1}^{d} \eta_0^k a_0^k \in \overline{L},$$

i. e. $(\eta_{i,l}^1)$ is B-summable, $B = \{b_l \mid l \in P\}$. This shows that $\bar{\eta}^1 \in \mathfrak{M}(L) \cap \mathfrak{A}_{sum} \neq 0$.

The explicit form of the divisible elements $y_j \in \overline{L}$ follows from Theorem 5.6. The inclusion $\overline{L} \subset \left(\bigoplus_{j=1}^m \hat{K} y_j\right) \oplus \overline{F}$ is a consequence of the fact that all divisible elements in $\overline{L}$ come from non-zero elements of $\mathfrak{M}(L) \cap \mathfrak{A}_{sum}$. □

We thank the referee for many substantial suggestions.

REFERENCES

1. N. Bourbaki, *General Topology* I+II, Addison Wesley (1966).
2. L. Fuchs, *Infinite Abelian Groups* I+II, Academic Press (1970, 1973).
3. N. Jacobson, *Basic Algebra* I+II *Second Edition*, Freeman and Company (1985, 1989).
4. I. Kaplansky, *Infinite Abelian Groups* Ann Arbor, Michigan (1969).
5. O. Mutzbauer and E. Toubassi, *A splitting criterion for a class of mixed modules*, Rocky Mountain J. Math. **24** (1994), 1533 - 1543.
6. O. Mutzbauer and E. Toubassi, *Classification of mixed modules*, Acta Math. Hung. **92** (1996).

OTTO MUTZBAUER, MATHEMATISCHES INSTITUT, UNIVERSITÄT WÜRZBURG, AM HUBLAND, 97074 WÜRZBURG, GERMANY

E-mail address: mutzbauer@mathematik.uni-wuerzburg.de

HELMUT SCHÄFER, MATHEMATISCHES INSTITUT, UNIVERSITÄT WÜRZBURG, AM HUBLAND, 97074 WÜRZBURG, GERMANY

Balanced Subgroups of Completely Decomposable Groups

L.G. NONGXA, University of the Western Cape, Bellville 7535, South Africa

K.M. RANGASWAMY, University of Colorado, Colorado Springs, CO 80933

C. VINSONHALER, University of Connecticut, Storrs, CT 06269

0 INTRODUCTION: A torsion-free abelian group is called **completely decomposable** if it is isomorphic to a direct sum of subgroups of the additive rationals Q. The purpose of this note is to study balanced subgroups of completely decomposable groups. Specifically, we ask when such subgroups are direct summands.

For finite rank, the question has been answered, by the case n = 1 of Theorem 4.4 in [NV]. Recall that if G is a completely decomposable group, then the **critical typeset of G**, denoted by $T_{cr}(G)$, is the set of types of rank-one summands of G. Types are isomorphism classes of subgroups of Q. They form a distributive lattice under sum (join) and intersection (meet or inf).

THEOREM 0: Let G be a finite rank completely decomposable group.` Then every balanced subgroup of G is a summand if and only if for any non-empty subset S of $T_{cr}(G)$ with $\tau_S = \inf S$:

(i) if $\tau_S \in T_{cr}(G)$, then $\inf\{\tau \in S : \tau > \tau_S\} > \tau_S$

(ii) if $\tau_S \notin T_{cr}(G)$, then S is a disjoint union $S_1 \cup S_2$, with $\inf S_i > \tau_S$ for i=1,2.

We will say that a completely decomposable group G (or $T = T_{cr}(G)$) of arbitrary rank **satisfies THM0** if for each finite subset S of T, both (i) and (ii) of Theorem 0 hold. **Unless otherwise noted, G will always denote a completely decomposable group satisfying THM0.** In the infinite rank case, the THM0 condition is still necessary, but no longer sufficient for each balanced subgroup to be a summand.

In Section 1 we focus on balanced subgroups of G. We show that if the typeset of G satisfies the maximum condition, then all balanced subgroups of G are strongly pure and completely decomposable (Theorem 1.6). We also establish that an arbitrary torsion-free separable abelian group satisfies THM0 if and only if all balanced subgroups are *-pure in the sense of Hill and Megibben [HM]. In Section 2 we study the balanced quotients of G. If the maximal homogeneous components of G are of finite rank, then all balanced subgroups H of G such that the typeset of G/H satisfies the maximum condition are summands (Theorem 2.4). In Section 3, we give some results and examples in the case where the typeset of G contains an infinite ascending chain. These suggest that a theorem characterising completely decomposable groups all of whose balanced subgroups are summands might be untidy. A theorem on critical typesets may be more accessible - we conclude with a conjecture in this direction.

Recall that a pure subgroup H of a torsion-free abelian group G is **balanced** in G if, for any element $g \in G\backslash H$ there exists $h \in H$ such that $tp_{G/H}(g + H) = tp_G(g+h)$. By Baer's Lemma [F, Lemma 86.5], this is equivalent to the observation that the coset contains an element of the same height-sequence (as the given coset). It is well-known that an exact sequence of torsion-free groups is balanced if and only if (direct sums of) rank-one groups are projective with respect to the sequence. Denote by T(G) the typeset of a torsion-free abelian group G, that is, the set of types of the nonzero elements of G. For any type τ, $G(\tau)$ is the set of all elements of G of type at least τ - a fully invariant subgroup of G; $G^*(\tau)$ denotes the subgroup of G generated by those elements of type greater than τ; and $G^\#(\tau)$ is the pure subgroup generated by $G^*(\tau)$. We can realize $T_{cr}(G)$ as $\{\tau \in T(G) : G^\#(\tau) \neq G(\tau)\}$. Denote the height-sequence of $g \in G$ by $|g|^G$ or simply $|g|$, and the p-height of g by p-ht(g) (if the containing group is obvious). A decomposition of G into homogeneous components, called a **standard decomposition,** is written $G = \oplus\{G_\tau : \tau \in T_{cr}(G)\}$, where $G_\tau \cong G(\tau)/G^*(\tau)$; and $\pi_\tau : G \to G_\tau$ is the corresponding projection.

1. THE BALANCED SUBGROUP H. We begin with the following observation.

LEMMA 1.1: For any $\tau \in T_{cr}(G)$, $\tau \notin T(G^*(\tau))$.

Proof: The statement of the lemma is in fact equivalent to (THM0)(i), a routine verification. ♦

Recall [N] that a subgroup H of G is **strongly pure** if, for every $0 \neq h \in H$, there exists $\theta \in \mathrm{Hom}_Z(G,H)$ such that $\theta(h) = h$. It is easy to check that if H is strongly pure in G, then finite rank summands of H are summands of G.

LEMMA 1.2: Homogeneous balanced subgroups H of G are completely decomposable and strongly pure. Finite rank summands of H are summands of G.

Proof: Let H be a homogeneous balanced subgroup of G of type σ. We claim that $\sigma \in T_{cr}(G)$. If not, let $0 \neq h = \Sigma\{\pi_\tau(h) : \tau \in S\}$, for a non-empty subset S of $T_{cr}(G)$. From THM0(ii), since $tp_G(h) = \sigma = \inf S \notin T_{cr}(G)$, we can write $h = g_1+g_2$ where $g_i = \Sigma\{\pi_\tau(h) : \tau \in$

$S_i\}$ and $\tau_i = tp_G(g_i) = \inf S_i > \sigma$, $i = 1,2$. The coset $g_1 + H = -g_2 + H$ has type $\geq \tau_1$ and by balancedness contains an element g_1+h_1 of the same type. Then $tp(h_1) \geq \tau_1$. Since $H(\tau_1) = 0$, $h_1 = 0$ and this implies that $|g_1|^G \geq |g_1 + h'|^G$ for **every** $h' \in H$. If we put $h' = -h$ we obtain $|g_1|^G \geq |g_2|^G$ which will imply that $\sigma = \tau_1\wedge\tau_2 = \tau_2$, contradicting the assumption on the τ_i's. Thus $\sigma \in T_{cr}(G)$ and $\pi_\sigma(h) \neq 0$. By THM0(i), $h = \pi_\sigma(h) + g'$ where $tp_G(g') > \sigma$. By considering the coset $g' + H$ and following an argument similar to that above, we can conclude that $|g'|^G \geq |\pi_\sigma(h)|$. Thus $|h|^G = |\pi_\sigma(h)|^G$ and there is an isomorphism from $\langle\pi_\sigma(h)\rangle_*$ (a summand of G) to $\langle h\rangle_* \subseteq H$ which maps $\pi_\sigma(h)$ onto h. Any projection from G onto $\langle\pi_\sigma(h)\rangle_*$ followed by this isomorphism and the inclusion into H gives $\theta \in Hom_Z(G,H)$ which fixes h. Thus, H is strongly pure. By [N, Theorem 1], H is completely decomposable. The remark following the definition of strongly pure completes the proof. ♦

Let H be a balanced subgroup of G and let $T = \{\tau \in T(H): H(\tau)\cap G^*(\tau) \neq H(\tau)\}$. If $\tau \in T$, then $0 \neq H(\tau)/(H(\tau)\cap G^*(\tau)) \cong (G^*(\tau)+H(\tau))/G^*(\tau) \subseteq G(\tau)/G^*(\tau)$ implies that $T \subseteq T_{cr}(G)$.

LEMMA 1.3: For every type τ and every balanced subgroup H of G, $H(\tau)\cap G^*(\tau) = H^*(\tau)$.

Proof: Trivially, $H(\tau)\cap G^*(\tau) \supseteq H^*(\tau)$. First assume $\tau \in T_{cr}(G)$ and let $0 \neq h \in H(\tau)\cap G^*(\tau)$. By Lemma 1.1, $\tau \notin T(G^*(\tau))$ and this implies that $tp_H(h) > \tau$, i.e. $h \in H^*(\tau)$. Second suppose that $\tau \notin T_{cr}(G)$; then $H(\tau) \subseteq G(\tau) = G^*(\tau)$. If $h \in H(\tau) \subseteq G^*(\tau)$ with $tp_G(h) = \tau$, then use THM0(ii) to write $h = g_1+g_2$ where $tp_G(g_i) = \tau_i > tp_H(h) = \tau$ and $\tau = \tau_1\wedge\tau_2$. Since $H(\tau)$ is balanced in $G(\tau)$, there exists $h_0 \in H(\tau)$ such that $tp_G(g_1+h_0) = tp_{G(\tau)/H(\tau)}(g_1 + H(\tau))$. But $g_1 + H(\tau) = -g_2 + H(\tau)$ which implies that $tp_{G(\tau)/H(\tau)}(g_1 + H(\tau)) = tp_{G(\tau)/H(\tau)}(g_2 + H(\tau)) \geq \tau_1\vee\tau_2 > \tau_i$, $i = 1,2$. Thus $tp_G(g_1+h_0) \geq \tau_1\vee\tau_2$ which implies that $tp_G(h_0) \geq \tau_1 > \tau$, i.e. $h_0 \in H^*(\tau)$. Similarly, $tp_G(g_1+h_0) = tp(-g_2+h+h_0) \geq \tau_1\vee\tau_2 > \tau$ implies that $h+h_0 \in H^*(\tau)$. Hence $h \in H^*(\tau)$ and the proof is complete. ♦

Bowman and Rangaswamy [BR] defined a balanced subgroup H of an arbitrary torsion-free group G to be ***-balanced** if $H\cap G^\#(\tau) = H^\#(\tau)$ for every type τ. The previous lemma asserts that balanced subgroups of our group G are *-balanced.

LEMMA 1.4: For every balanced subgroup H of G

(a) $T_{cr}(H) = \{\tau \in T(H) : H(\tau)\cap G^*(\tau) \neq H(\tau)\} \subseteq T_{cr}(G)$;

(b) for every $\tau \in T_{cr}(H)$, $H(\tau) = H_\tau \oplus H^*(\tau)$, where H_τ is a homogeneous completely decomposable balanced (hence strongly pure) subgroup of G of type τ.

Proof: (a) Equality of the two sets follows from Lemma 1.3, since $H^{\#}(\tau) = H^{*}(\tau) = H(\tau) \cap G^{*}(\tau)$ for all types τ.

(b) The group $H(\tau)/H^{*}(\tau)$ consists of elements of type at least τ and can be embedded in $G(\tau)/G^{*}(\tau)$, a homogeneous completely decomposable group of type τ. Hence $H(\tau)/H^{*}(\tau)$ is homogeneous of type τ and therefore completely decomposable by [F, Theorem 86.6]. Furthermore, $H^{*}(\tau)$ is balanced in $H(\tau)$ by Baer's Lemma. Thus there is a splitting $H(\tau) = H_{\tau} \oplus H^{*}(\tau)$ with $H_{\tau} \cong H(\tau)/H^{*}(\tau)$ homogeneous completely decomposable. The summand H_{τ} is balanced in $H(\tau)$, which is balanced in $G(\tau)$ (a summand of G). Thus H_{τ} is balanced in G and Lemma 2 applies to complete the proof. ♦

LEMMA 1.5: For every balanced subgroup H of G, if $H(\tau) = H_{\tau} \oplus H^{*}(\tau)$ for each τ in $T_{cr}(H)$, then $\Sigma\{H_{\tau} : \tau \in T_{cr}(H)\} = \oplus\{H_{\tau} : \tau \in T_{cr}(H)\}$, is a strongly pure completely decomposable subgroup of G.

Proof: We first show that $\Sigma\{H_{\tau} : \tau \in T_{cr}(H)\}$ is a direct sum. Let $0 = \Sigma\{h_{\tau} : \tau \in I\}$, I a finite subset of $T_{cr}(H)$ and $h_{\tau} \in H_{\tau}$, $\tau \in I$. By strong purity of H_{τ} (Lemma 1.4), there exists $\theta_{\tau} : G \to H_{\tau}$ such that $\theta_{\tau}(h_{\tau}) = h_{\tau}$. If σ is any minimal member of I, then $0 = \theta_{\sigma}(\Sigma\{h_{\tau} : \tau \in I\} = h_{\sigma}$, which implies that the sum is direct.

To show strong purity of the sum, let $0 \neq h \in \oplus\{H_\tau : \tau \in T_{cr}(H)\}$ and put $h = \Sigma\{h_\tau : \tau \in S\} \in \oplus\{\langle h_\tau\rangle_* : \tau \in S\}$, where $\varnothing \neq S \subseteq T_{cr}(H) \subseteq T_{cr}(G)$. We prove by induction on $|S|$ that $\oplus\{\langle h_\tau\rangle_* : \tau \in S\}$ is a summand of G; the claim holds if $|S| = 1$ by Lemma 1.2. Suppose the result holds for all $S' \subseteq T_{cr}(H)$ with $|S'| < |S|$ and let σ be a minimal element of S. Then $S' = S\backslash\{\sigma\}$ is smaller than S. Moreover, $\langle h_\sigma\rangle_*$ is a summand of G and the complementary summand K must contain each h_τ , $\tau \in S'$. By induction hypothesis, $\oplus\{\langle h_\tau\rangle_* : \tau \in S'\}$ is a summand of K. Thus $\Sigma\{\langle h_\tau\rangle_* : \tau \in S\} = \langle h_\sigma\rangle_* \oplus (\oplus\{\langle h_\tau\rangle_* : \tau \in S'\})$ is a summand of G. Any projection from G onto $\oplus\{\langle h_\tau\rangle_* : \tau \in S\}$ followed by inclusion into $\oplus\{H_\tau : \tau \in T_{cr}(H)\}$ is a homomorphism that fixes h and we are done. ♦

THEOREM 1.6 If H is a balanced subgroup of a completely decomposable group G satisfying THM0 and T(H) satisfies the maximum condition, then H is completely decomposable and strongly pure in G.

Proof: We shall show that $H = \oplus\{H_\tau : \tau \in T_{cr}(H)\}$ for any choice of socle decompositions $H(\tau) = H_\tau \oplus H^*(\tau)$. Let $0 \neq h \in H$ with $tp_H(h) = \sigma$. Use the socle decomposition to write $h = h(\sigma) + \sum h_i$ with $h(\sigma) \in H_\sigma$, $h_i \in H(\sigma(i))$ and $\sigma<\sigma(i)$ for each of the finitely many indices i. In similar fashion, write $h_i = h(\sigma(i)) + \sum h_{ij}$ with $h(\sigma(i)) \in$

$H_{\sigma(i)}$, $h_{ij} \in H(\sigma(ij))$ and $\sigma(i)<\sigma(ij)$ for the finitely many index pairs ij. Then write $h_{ij} = h(\sigma(ij)) + \Sigma h_{ijk}$ and so on. Each chain $\sigma < \sigma(i) < \sigma(ij) < \sigma(ijk) < \ldots$ must terminate by the maximum condition, so that $h = h(\sigma) + \Sigma h(\sigma(i)) + \Sigma h(\sigma(ij)) + \ldots \in \Sigma H_\tau$ as desired. ♦

We note that one direction of Theorem 0 is an easy consequence of Theorem 1.6.

COROLLARY 1.7: Finite rank balanced subgroups of a completely decomposable group satisfying THM0 are summands.

Proof: Finite rank strongly pure subgroups are summands. ♦

We have not been able to determine whether or not Theorem 1.6 still holds if we remove the maximum condition on the typeset of H. Are balanced subgroups of G strongly pure and/or completely decomposable? It is easy to see that if balanced subgroups of G are separable, then they will be strongly pure in G.

For the final results of this section, we need some definitions from [HM]. If $\mathbf{s} = (s_p)$ is a characteristic sequence (a sequence of non-negative integers or ∞'s) and G is an arbitrary torsion-free abelian group, then $G(\mathbf{s})$ is the set of all elements in G whose height sequence is at least $\mathbf{s}$ and $G(\mathbf{s}^*)$ is the subgroup of $G(\mathbf{s})$

generated by elements g such that $tp_G(g) > tp(\mathbf{s})$. For any prime p, $G(\mathbf{s}^*,p) = G(\mathbf{s}^*) + pG(\mathbf{s})$.

A pure subgroup H of G is called ***-pure** in G if $H \cap G(\mathbf{s}^*) = H(\mathbf{s}^*)$ and $H \cap G(\mathbf{s}^*,p) = H(\mathbf{s}^*,p)$ for all height sequences **s** and primes p.

LEMMA 1.8: If G is homogeneous and $G(\mathbf{s}^*) \neq 0$, then $G(\mathbf{s}^*,p) = G(\mathbf{s}^*)$ for every prime p.

Proof: Suppose that $0 \neq x \in G(\mathbf{s}^*)$, and $tp_G(x) > tp(\mathbf{s})$. If $y \in pG(\mathbf{s}) \subseteq G(\mathbf{s})$, then $|y|^G \geq \mathbf{s}$ and $tp_G(x) = tp_G(y)$ since G is homogeneous. Thus $y \in G(\mathbf{s}^*)$ and this implies that $pG(\mathbf{s}) \subseteq G(\mathbf{s}^*)$, all that is needed for the proof. ♦

The critical typeset of a separable torsion-free abelian group can also be defined as the set of types of rank-one summands. The final result of Section 1 is placed in this more general setting.

THEOREM 1.9: Let G be a torsion-free separable abelian group. Then every balanced subgroup of G is *-pure if and only if $T_{cr}(G)$ satisfies condition THM0 above.

Proof: Assume every balanced subgroup of G is *-pure. Because G is separable, every finite rank subgroup is contained in a finite rank completely decomposable summand of G. It follows that every finite rank balanced subgroup of G is a summand, since finite rank $*$-pure subgroups are summands by [DR, Proposition 2]. This implies that $T_{cr}(G)$ satisfies THM0.

Conversely, suppose that $T_{cr}(G)$ satisfies THM0 and let $h \in H \cap G(\mathbf{s}^*)$, where $\mathbf{s}$ is a height sequence belonging to a type σ. It suffices to consider the case where $tp_G(h) = \sigma$. By definition we can write $h = \Sigma g_i$ where $tp(g_i) > \sigma$. Let $G' = \oplus G_\tau$ be a standard decomposition of a completely decomposable summand of G that contains each g_i. Then since $tp(g_i) > \sigma$, $\pi_\tau(g_i) \neq 0$ implies $\tau > \sigma$. By THM0(i), $tp_G(h) \notin T_{cr}(G)$. It follows from THM0(ii) that $h = x_1+x_2$ with $x_i \in G'$ and $tp_G(x_i) > tp_G(h) = \sigma$. Since H is balanced in G, there exist $h_i \in H$, such that $|x_i+h_i|^G = |x_i+H|^{G/H} \geq |x_1|^G \vee |x_2|^G > |x_i|^G$, $i = 1,2$. Then $|h_i|^G = |x_i|^G$ so that $|h_i|^G$ is not equivalent to $\mathbf{s}$. This implies that $h_i \in H(\mathbf{s}^*)$ for $i=1,2$. Now $|h+h_1+h_2|^G = |(x_1+h_1)+(x_2+h_2)|^G \geq |x_1|^G \vee |x_2|^G > \mathbf{s}$. Since $|x_i|^G$ is not equivalent to $\mathbf{s}$, this implies that $|h+h_1+h_2|^G$ is not equivalent to $\mathbf{s}$ which shows that $h+h_1+h_2 \in H(\mathbf{s}^*)$. Thus $h \in H(\mathbf{s}^*)$ as desired.

We now show that $H \cap G(\mathbf{s}^*,p) = H(\mathbf{s}^*,p)$. As before let $h \in H \cap G(\mathbf{s}^*,p)$ and write $h = x + py$, with $x \in G(\mathbf{s}^*)$ and $y \in G(\mathbf{s})$. Since H is balanced in G there exists $h_0 \in H$ such that $|x+h_0|^G = |x+H|^{G/H} = |-py+h+h_0|^G$. Using Lemma 1.1, $tp_G(x+h_0) = tp_G(x+H) \geq tp_G(x) > \sigma$, so that $h_0 \in H \cap G(\mathbf{s}^*) = H(\mathbf{s}^*)$. Moreover, $h+h_0 \in H \cap pG(\mathbf{s}) = pH(\mathbf{s})$ by purity. Thus, $h = -h_0+(h+h_0) \in H(\mathbf{s}^*)+pH(\mathbf{s})$ as desired. ♦

2 THE BALANCED QUOTIENT K:

Recall that, for any type $\tau \neq \text{type}(Q)$, there exists a countable homogeneous indecomposable group of type τ [F, Section 88]. Hence, if there is a type $\tau \neq \text{type}(Q)$ such that $\text{rank}(G(\tau)/G(\tau^*))$ is countably infinite, then G contains a balanced subgroup that is not a summand. Thus, for the rest of this note we assume our completely decomposable group G satisfies THM0 **and** $\text{rank}(G(\tau)/G(\tau^*))$ is finite for every type τ. We also fix a balanced exact sequence,

$$0 \to H \to G \to K \to 0.$$

By [NV, Lemma 2.2], $T_{cr}(K) = \{\tau \in T(K) : K^{\#}(\tau) \neq K(\tau)\} \subseteq T_{cr}(G)$.

LEMMA 2.1: For every type τ, $K(\tau^*) = K^{\#}(\tau)$.

Proof: Denote by β the map $G \to K$. We consider two possibilities.

(i) $\tau \in T_{cr}(G)$. If $x \in K^{\#}(\tau)$, then $nx \in K^*(\tau)$ for some non-zero integer n and there exists $g \in G^*(\tau)$ such that $\beta(g) = nx$. From Lemma 1.1, $\tau < tp_G(g) \leq tp_K(\beta(g)) = tp_K(nx) = tp_K(x)$ and this implies that $x \in K^*(\tau)$.

(ii) $\tau \notin T_{cr}(G)$. Choose $x \in K^{\#}(\tau)$ to be of type exactly τ (otherwise $x \in K^*(\tau)$). Then $x = \beta(g)$ for some $g \in G(\tau) = G^*(\tau)$ which implies that $g = \Sigma\{g_i : i \in I\}$ where $\tau < tp_G(g_i)$, $i \in I$. Hence $x = \Sigma\{\beta(g_i) : i \in I\} \in K^*(\tau)$. ♦

If T is a set of types and $\tau, \sigma \in T$, then **σ is a cover for τ in T** if $\tau < \sigma$ and $\{\rho \in T : \tau \leq \rho \leq \sigma\} = \{\tau, \sigma\}$.

LEMMA 2.2: Let $\tau \in T_{cr}(K)$. Then

(a) $K(\tau) = K_\tau \oplus K^*(\tau)$ where K_τ is a finite rank homogeneous completely decomposable group of type τ;

(b) if σ is a cover for τ in $T(K)$, then $K^*(\tau) = K(\sigma)$.

Proof: (a) This follows from the exactness of the sequence

$$0 \to H(\tau)/H^*(\tau) \to G(\tau)/G^*(\tau) \to K(\tau)/K^*(\tau) \to 0.$$

The sequence splits since $G(\tau)/G^*(\tau)$ is a finite rank homogeneous completely decomposable group. Hence so is $K(\tau)/K^*(\tau)$ and, by Baer's Lemma $K^*(\tau)$ is a summand of $K(\tau)$.

(b) Plainly, $K(\sigma) \subseteq K^*(\tau)$. Suppose that $x \in K$ has type $\rho > \tau$. If $\sigma < \rho$ then we're done. Otherwise, since σ is a cover for τ, $\rho \wedge \sigma = \tau$. Moreover, $\rho = \inf T_1$ and $\sigma = \inf T_2$ for some finite subsets T_1, T_2 of $T_{cr}(G)$. Then $S = T_1 \cup T_2$ violates THM0(i), implying that $\sigma < \rho$ as desired. ♦

LEMMA 2.3: (a) For each $\tau \in T_{cr}(K)$, the decomposition $K(\tau) = K_\tau \oplus K^*(\tau)$ can be chosen so that $K_\tau \subseteq \text{image}(G_\tau)$. If this is done then $\Sigma\{K_\tau$:

$\tau \in T_{cr}(K)\}$ is the completely decomposable direct sum $\oplus\{K_\tau : \tau \in T_{cr}(K)\}$.

(b) If $\tau \in T(K)$ has finite depth, then $K(\tau) = \oplus\{K_\sigma : \sigma\geq\tau\}$ is completely decomposable, where $K(\sigma) = K_\sigma \oplus K^*(\sigma)$ for every $\sigma \geq \tau$.

Proof: (a) By balancedness, $K(\tau) = \text{image}(G(\tau)) = \text{image}(G_\tau) + K^*(\tau)$. Moreover, if $I = \text{image}(G(\tau))$, the factor map $I \to I/I\cap K^*(\tau)$ splits by Baer's Lemma. Thus, we can choose $K_\tau \subseteq \text{image}(G_\tau)$. To show $\Sigma\{K_\tau : \tau \in T_{cr}(K)\}$ is direct, suppose $\Sigma\{x_\tau : \tau\in S\} = 0$ for $x_\tau \in K_\tau$ and S a finite subset of $T_{cr}(K) \subseteq T_{cr}(G)$. We use induction on $|S|$ to show each $x_\tau = 0$, the case $|S| = 1$ being clear. Assume each x_τ is nonzero and choose preimages $g_\tau \in G_\tau$ for x_τ. Then $h = \Sigma\{g_\tau : \tau\in S\} \in H$. If $\sigma = \inf S$ is an element of $T_{cr}(G)$, then $\sigma \in S$ by THM0(i). By Lemma 2.2(a), $x_\sigma = 0$. Thus we may assume that σ is not an element of $T_{cr}(G)$. By THM0(ii), there is a partition $S = T_1 \cup T_2$ with $\inf T_i > \sigma$, $i = 1,2$. Write $h = g_1 + g_2$ with $g_i = \Sigma\{g_\tau : \tau\in T_i\}$ and put $x_i = \Sigma\{x_\tau : \tau\in T_i\}$, the image of g_i in K. Since $x_1 + x_2 = 0$, $tp(x_1) = tp(x_2)$. But our induction hypothesis on $|S|$ implies that $tp(x_i) = \inf T_i$. Thus, $\inf T_1 = \inf T_2 = \inf S$, a contradiction. It follows that each x_τ must be zero, whence that $\Sigma\{K_\tau : \tau \in T_{cr}(K)\} = \oplus\{K_\tau : \tau \in T_{cr}(K)\}$ is completely decomposable.

(b) We use induction on $\text{depth}(\tau) = d_\tau$. The result holds if $d_\tau = 1$ (i.e. τ is maximal in $T(K)$, $K(\tau) = K_\tau$). Assume that the result holds for all types $\sigma > \tau$ ($d_\sigma < d_\tau$) and note that $K(\tau) = K_\tau \oplus K(\tau^*) = K_\tau \oplus (\Sigma\{K(\sigma) : \sigma > \tau\}$ and $K(\sigma) = \oplus\{K_\rho : \rho \geq \sigma\}$ by induction hypothesis. We claim that $\Sigma\{K_\rho : \rho > \tau\} = \oplus\{K_\rho : \rho > \tau\}$. Suppose that $0 = \Sigma\{x_\rho : \rho \in I\} \in \Sigma\{K_\rho : \rho > \tau\}$, $x_\rho \in K_\rho$, I a finite subset of $\{\rho \in T_{cr}(K) : \rho > \tau\}$. Put $\tau_I = \inf I \geq \tau$. If $\tau < \tau_I$, then $0 = \Sigma\{x_\rho : \rho \in I\} \in K(\tau_I) = \oplus\{K_\rho : \rho \geq \tau_I\}$ and this implies that $x_\rho = 0$ for all $\rho \in I$.

Otherwise $\tau = \tau_I$, and $\tau \notin T_{cr}(G)$ by THM0(i). Then $I = I_1 \cup I_2$ with I_1 and I_2 satisfying THM0(ii) and we let $\sigma_i = \inf I_i$, $i = 1,2$. Let $g_i = \Sigma\{x_\rho : \rho \in I_i\} \in K(\sigma_i) = \oplus\{K_\rho : \rho \geq \sigma_i\}$, $i = 1,2$. Then, $g_1 = -g_2 \in \oplus\{K_\rho : \rho \geq \sigma_1\} \cap \oplus\{K_\rho : \rho \geq \sigma_2\} \subseteq K(\sigma_1 \vee \sigma_2) = \oplus\{K_\rho : \rho \geq \sigma_1 \vee \sigma_2\}$. Since $I_1 \cap I_2 = \emptyset$, we must have $g_1 = g_2 = 0$ by uniqueness of linear combination of an element in $K(\sigma_1 \vee \sigma_2)$. Then $0 = g_i = \Sigma\{x_\tau ; \tau \in I_i\} \in \oplus\{K_\rho : \rho \geq \sigma_i\}$ implies that $x_\tau = 0$ for all $\tau \in I_i$, $i = 1,2$. Hence $\Sigma\{K_\rho : \rho > \tau\} = \oplus\{K_\rho : \rho > \tau\}$ and $K(\tau) = K_\tau \oplus \oplus\{K_\rho : \rho > \tau\}$. ♦

In view of Lemma 2.3(a), if $K = \Sigma K_\tau$ with $K_\tau \subseteq \text{image}(G_\tau)$, then the balanced sequence $0 \to H \to G \to K \to 0$ splits, because K is completely decomposable and so balanced projective. To date we have not been able to use this fact. However, assuming the maximum condition on K, we can use 2.3(b) to show splitting occurs.

THEOREM 2.4: Suppose that H is a balanced subgroup of G and

$T(G/H)$ satisfies the maximum condition. Then H is a summand of G.

Proof: Put $K = G/H$. If τ is a maximal element of $T(K)$, then $K(\tau) = K_\tau$. Using induction on the depth of τ in $T(K)$, via Lemmas 2.3(b) and 2.2 we have $K = \Sigma\{K_\tau : \tau \in T_{cr}(K)\}$. We next show the sum is direct. Suppose $\Sigma\{x_\tau : \tau\in S\} = 0$ for $x_\tau \in K_\tau$ and S a finite subset of $T_{cr}(K) \subseteq T_{cr}(G)$. If $\sigma = \inf S$, then by hypothesis, σ has finite depth in $T(K)$. By Lemma 2.3(b), $K(\sigma) = \oplus\{K_\tau : \tau \geq \sigma \in T_{cr}(K)\}$. Thus, each x_τ is zero and $K = \oplus\{K_\tau : \tau \in T_{cr}(K)\}$ is completely decomposable. It follows that K is balanced projective and the balanced epimorphism $G \to K$ splits, so that H is a summand of G. ♦

COROLLARY 2.5: Let G be a completely decomposable group whose typeset satisfies the maximum condition. Then every balanced subgroup of G is a summand if and only if G satisfies THM0 and $\text{rank}(G(\tau)/G(\tau^*))$ is finite for every $\tau \in T_{cr}(G)$. ♦

3 INFINITE ASCENDING CHAINS

The presence of ascending chains seems to complicate the situation.

LEMMA 3.1: Let G be a completely decomposable group and $C = \{\tau_n : n<\omega\}$, $\tau_n < \tau_{n+1}$ an ascending chain in $T_{cr}(G)$ and suppose there is a prime p such that $\text{type}(\mathbf{Z}_p) \geq \tau_n$ for every n. If $\text{rank}(G(\tau_n)/G^*(\tau_n)) \geq 2$ for infinitely many n, then G contains a balanced subgroup that is not a summand.

Proof: Let $X = \oplus_{i<\omega}X_i \oplus X'$, where $\mathrm{rank}(X_i) = \mathrm{rank}(X') = 1$ and $\mathrm{tp}(\mathbf{Q}) > \mathrm{tp}(X') > \mathrm{tp}(X_{i+1}) > \mathrm{tp}(X_i)$, $i < \omega$. Let p be a prime satisfying $pX' \neq X'$ and choose $0 \neq x \in X'$, $0 \neq x_i \in X_i$, such that $|x|^X > |x_{i+1}|^X > |x_i|^X$, $i < \omega$ and the p-height of x is zero. If we define K to be the pure subgroup of X generated by $\{px+x_i : i < \omega\}$, then K is **not** completely decomposable (see [Bi], Lemma 18). The pure rank one subgroups of K of minimal support (fewest nonzero projections) in X are $\langle px+x_i\rangle_*$ and $\langle x_i-x_j\rangle_*$; and it is routine to check (see the argument in the proof of Theorem 1.2 in [AV]), that the summation map from $G' = (\oplus_{i<\omega}\langle px+x_i\rangle_*)\oplus(\oplus_{1\le i<\omega}\langle x_i-x_{i+1}\rangle_*)$ onto K is a balanced epimorphism. The resulting exact sequence, $0 \to H \to G' \to K \to 0$, with $G' = \oplus_{i<\omega}(\langle px+x_i\rangle_* \oplus \langle x_i-x_{i+1}\rangle_*)$, is balanced exact; and $\mathrm{rank}(G(\tau)/G^*(\tau)) = 2$ for every $\tau \in T_{cr}(G') = \{\mathrm{tp}(X), \mathrm{tp}(X_i) : i < \omega\}$. In particular, G′ is isomorphic to a summand of G. Moreover, H is not a summand of G′ since K is not completely decomposable. This completes the proof. ♦

The next example shows the necessity of the hypothesis $\mathrm{rank}(G(\tau_n)/G^*(\tau_n)) \geq 2$ in 3.1.

Example 3.2: Let $G = \oplus\{G_\tau : \tau \in T_{cr}(G)\}$ where $T_{cr}(G) = \{\tau_n : n<\omega\}$ is a chain and $\mathrm{rank}(G_\tau) = 1$ for almost all $\tau \in T_{cr}(G)$. Then every balanced subgroup of G is a summand.

Let H be a balanced subgroup of G and write $H(\tau_n) = H_n \oplus H^*(\tau_n)$ for each n, using Lemma 1.4. Write $G_\tau = G_n$ for $\tau = \tau_n$. If $H_n \neq 0$, it follows that $G(\tau_n) = G_n \oplus G^*(\tau_n) = H_n \oplus G^*(\tau_n)$, since $\text{rank}(G_n) = 1$. If H has finite rank, then H is a summand by Lemma 1.1. Thus, by passing to a summand of G if necessary, we may assume $\text{rank}(H_n) = 1$ for all n. We claim that $G_n \subseteq H$ for all n. Suppose, by way of contradiction, that $0 \neq g \in G_n \setminus H$. Then since H is balanced, there is $h \in H$ such that $tp_G(g+h) = tp_{G/H}(g+H) \geq \tau_n \geq \tau_r$ for $n \geq r$. Now, for every $r \geq 0$, $G(\tau_n) = \oplus_{0\leq i\leq r} G_{n+i} \oplus G^*(\tau_{n+r}) = \oplus_{0\leq i\leq r} H_{n+i} \oplus G^*(\tau_{n+r})$ and so we can write $g = -h_r + g_r$ where $0 \neq g_r \in G^*(\tau_{n+r})$ and $-h_r \in \oplus_{0\leq i\leq r} H_{n+i}$. Then $tp_G(g+h) \geq tp_G(g+h_r) = tp_G(g_r) > \tau_{n+r}$. Thus, $tp_G(g+h) \geq \tau$ for every $\tau \in T(G)$, a contradiction. Therefore, $G_n \subseteq H$ for every n and G = H. We conclude that H is a summand of G, as desired.

A final example points the way to a conjecture.

Example 3.3: Let $G = \oplus\{G_\tau : \tau \in T_{cr}(G)\}$ be a standard decomposition where each $\text{rank}(G(\tau)/G^*(\tau))$ is finite; $T_{cr}(G) = \{\tau_k : k<\omega\}$ is an ascending chain; and for every prime p, there exists a positive integer k(p) such that $pG(\tau_{k(p)}) = G(\tau_{k(p)})$. If H is a balanced subgroup of G, then H is a summand

We show K = G/H is separable, whence completely decomposable since countable separable groups are completely decomposable. By Lemma

2.2, for every $k \geq 1$, $K(\tau_k) = K_k \oplus K(\tau_{k+1})$ where K_k is a homogeneous finite rank completely decomposable group (possibly zero). This implies that $K = \oplus_{1\leq i\leq k} K_i \oplus K(\tau_{k+1})$ for every $k \geq 1$. Without loss of generality, let $0 \neq x \in K$ be of type τ_1 and put $x = x_1 + x_2'$ where $0 \neq x_1 \in K_1$ and $x_2' \in K(\tau_2)$. There exists an integer n such that $|nx|^K = |x_1|^K$ since $tp_K(x) = tp_K(x_1)$. Then,

(3.3.1) For every prime p dividing n, $p\text{-ht}(x) = p\text{-ht}(x_2')$, and if $(n,p) = 1$, then $p\text{-ht}(x) = p\text{-ht}(x_1)$.

By assumption on $T_{cr}(G)$, there exists an integer $k \geq 2$ such that $nK(\tau_k) = K(\tau_k)$, and we let $K(\tau_2) = \oplus_{2\leq i\leq k-1} K_i \oplus K(\tau_k)$, $x_2' = \Sigma_{2\leq i\leq k-1} x_i + y$, where $x_i \in K_i$ and $0 \neq y \in K(\tau_k)$ is of type τ_k, say and $y = x_k + x_{k+1}$, $0 \neq x_k \in K_k$ and $x_{k+1} \in K(\tau_{k+1})$. Again, there exists an integer m such that $|my|^K = |x_k|^K$ and $(m,n) = 1$ and we have

(3.3.2) For every prime p dividing m, $p\text{-ht}(y) = p\text{-ht}(x_{k+1})$ and if $(p,m) = 1$, then $p\text{-ht}(y) = p\text{-ht}(x_k)$.

Let $1 = rm+sn$ and observe that

$$x = x_1 + \Sigma_{2\leq i\leq k-1} x_i + (x_k + rmx_{k+1} + snx_{k+1}).$$

Let $x_1' = x_1 + snx_{k+1}$ and $x_k' = x_k + rmx_{k+1}$ and note that $|x_1'|^K = \inf\{|x_1|^K, |snx_{k+1}|^K\}$ and $|x_k'|^K = \inf\{|x_k|^K, |rmx_{k+1}|^K\}$ since $K = K_1 \oplus (\oplus_{2\leq i\leq k-1} K_i) \oplus K_k \oplus K(\tau_{k+1})$. By (3.3.1), if $(p,n) = 1$, then

$$p\text{-ht}(x) = p\text{-ht}(x_1) \leq p\text{-ht}(snx_{k+1});$$

and if p divides n, then

$$p\text{-ht}(x_1) = p\text{-ht}_Z(n) + p\text{-ht}(x)$$

$= p\text{-}ht_Z(n) + p\text{-}ht(x_2') \leq p\text{-}ht_Z(n) + p\text{-}ht(y)$

$= p\text{-}ht(ny) \leq p\text{-}ht(nx_{k+1}) \leq p\text{-}ht(snx_{k+1})$.

Consequently, $|x_1'|^K = |x_1|^K$ and we can replace the summand $\langle x_1\rangle_*$ of K_1 by $\langle x_1'\rangle_*$ and rewrite $K = K_1' \oplus K(\tau_2)$, where $x_1' \in K_1'$. Similarly, by (3.3.2), if $(p,m) = 1$, then

$p\text{-}ht(y) = p\text{-}ht(x_k) \leq p\text{-}ht(x_{k+1}) \leq p\text{-}ht(rmx_{k+1})$;

and if p divides m, then

$p\text{-}ht(x_k) = p\text{-}ht_Z(m) + p\text{-}ht(y)$

$= p\text{-}ht_Z(m) + p\text{-}ht(x_{k+1}) \leq p\text{-}ht(rmx_{k+1})$.

It follows that $|x_k'|^K = |x_k|^K$ and we can write $K(\tau_k) = K_k' \oplus K(\tau_{k+1})$ where $x_k' \in K_k'$. Finally, $x \in K_1' \oplus (\oplus_{2\leq i\leq k-1} K_i) \oplus K_k'$, a finite rank completely decomposable summand of K. Thus, K is separable, as desired.

Lemma 3.1 and Example 3.3 motivate our concluding conjecture, which we phrase in terms of all groups with a given critical typeset. The "only if" part has been established by our work to this point.

Conjecture: Let T be a set of types. Each completely decomposable group G, with critical typeset T and $\text{rank}(G(\tau)/G^*(\tau))$ finite for every $\tau \in T$, has all balanced subgroups as summands if and only if:

(a) T satisfies THM0;

(b) Each ascending chain $\{\tau_k: k<\omega\}$ in T has the property that for each prime p, $\tau_k(p) = \infty$ for some k. That is, the chain eventually becomes p-divisible.

REFERENCES

[AV] D. Arnold and C. Vinsonhaler, Pure subroups of finite rank completely decomposable groups, Lecture Notes in Mathematics **1006**, Springer-Verlag, New York, 1983, 97-143.

[Bi] L. Bican, Completely decomposable groups any pure subgroup of which is completely decomposable, Czech. Math. J.(1974), 176-191.

[BR] H. Bowman and K.M. Rangaswamy, Special balanced subgroups of torsion-free separable abelian groups, Lecture Notes in Math.**874**, Springer-Verlag, New York, 1981, 32-40.

[DR] M. Dugas and K.M. Rangaswamy, On torsion-free abelian k-groups, Proc. Amer. Math. Soc. **99**(1987), 403-408.

[F] L. Fuchs, Infinite Abelian Groups, vol.II, Academic Press, New York, 1973.

[HM] P. Hill and C. Megibben, Torsion-free groups, Trans. Amer.Math. Soc. 295(1986), 735-751.

[N] Loyiso G. Nongxa: Strongly pure subgroups of separable torsion-free abelian groups; Trans. Amer. Math. Soc. **290**(1985), 363-373.

[NV] Loyiso G. Nongxa and C. Vinsonhaler, Balanced Butler groups, J. Algebra (to appear March, 1996).

Summable Subsets of Direct Products of Infinite Cyclic Groups

John D. O'Neill, Department of Mathematics, University of Detroit Mercy
Detroit, Michigan, 48219.

INTRODUCTION

Let Z^κ be the direct product of κ copies of the integers Z where κ is an infinite cardinal number. Suppose that Z^κ equals $A \oplus B$ as an abelian group. Many years ago in [1] and [8] it was shown that, if $\kappa < \mu$, the least measurable cardinal number, then A (also B) is a direct product of copies of Z. In [9] we extended this result to the case where $\kappa = \mu$. Recently [10] we showed, for any infinite κ, that A or B is isomorphic to Z^κ. In our work the concept of "summable subsets of direct products" (see Definition 2.1 below) has proved quite useful. In Section 2 we present some major properties of summable sets; then in Section 3 we illustrate their use by applying them to the study of summands of direct products. In Section 1 we list a few special set theoretic facts which we will need later on.

All unexplained notation and terminology may be found in [4] and [5] for abelian groups and set theory respectively. We accept the axiom of choice and hence every set can be well-ordered. An ordinal number is the set of its predecessors; a cardinal number is an ordinal which is not equinumerous with any predecessor. "Almost all" will mean "all but a finite number of" and <X> is the group generated by X. The letter μ will always be the least measurable cardinal number.

1. Set Theory.

Let I be an infinite set and let $\mathcal{P}(I)$ be the collection of its subsets. Then $(\mathcal{P}(I), \cup, \cap, -)$ is a Boolean Algebra. If K is an ideal on I (i.e. of $\mathcal{P}(I)$) and $B = \mathcal{P}(I)/K$ then $(B, +, \cdot, -)$ is a Boolean Algebra where the operations on B are as defined in Section 17 of [5]. Some important technical words which we will be using frequently and whose definitions are in [5] are: atom, atomic, κ -complete, κ -saturated, $satK$, inaccessible, weakly compact.

(1.1) LEMMA. Let I be an infinite set and suppose that S is a infinite collection of subsets of I such that, for each $i \in I$, i is in s for at most a finite number of elements s in S. Then there is a partition P of I such that $|P| = |S|$ and each s in S is the union of members of P.

Proof. By adjoining $I - \bigcup S$ to S we may assume that I equals $\bigcup S$. Let P equal $\{\bigcap F - \bigcup(S - F) : F$ is a finite subset of $S\}$. If $i \in I$, then i is in exactly one element $p(i)$ in P. Now, if $s \in S$, then $s = \bigcup\{p(i) : i \in s\}$. Since S is infinite, P and S have the same cardinality.

(1.2) PROPOSITION. Suppose I is an infinite set and K is a λ -saturated ν -complete ideal on I containing all singletons with $\lambda \leq \zeta$, the cofinality of $|I|$. Then I contains a subset M with $|M| < |I|$ such that, if $J = I - M$, then $K \cap J$ is a λ -saturated ν -complete ideal on J containing all singletons and containing all subsets of J of cardinality less than that of J. If $\nu >$ cf$|I|$, then $J = K$.

Proof. Let W be a maximal collection of pairwise disjoint subsets of $I - K$ such that $|w| < |I|$ for each $w \in W$. Let $M = \bigcup W$. Since $|W| < \lambda \leq \zeta$ which is a regular cardinal number [Lemma 3.5 in 5]. $|M| < |I|$ and we let $J = I - M$. Clearly $K \cap J$ is a ν-complete ideal in $\mathcal{P}(J)$. Let N be a collection of subsets of J whose images are pairwise disjoint in $\mathcal{P}(J)/(K \cap J)$. Then the elements in $M + N$ have pairwise disjoint images in $\mathcal{P}(I)/K$ Therefore $|N| < |M + N| < \lambda$ so $K \cap J$ is λ -saturated in $\mathcal{P}(J)$. Now $|J| = |I|$ and, by the definition of W, the ideal $K \cap J$ on J contains all subsets of J of smaller cardinality. Also, if $\nu >$cf$|J|$, then $J \subseteq K$.

The next result is a generalization of Lemma 27.5 and of Example 35.5 in [5]; see also Proposition 16.4 in [6]

(1.3) PROPOSITION. Let I be an infinite set and let K be a λ -saturated ν -complete ideal on I containing all singletons where ν is a regular uncountable cardinal, where $\lambda \leq \nu$ and where

(a) $2^{<\lambda} < \nu$ $(2^{<\nu} = \sup_{\alpha<\lambda} 2^{\alpha})$, or

(b) ν is weakly compact.

Then $\mathcal{P}(I)/K$ is atomic.

Proof. Let J be any subset of I not in K. To show $\mathcal{P}(I)/K$ is atomic we must show the element $J+K$ in $\mathcal{P}(I)/K$ contains an atom or equivalently that $\mathcal{P}(J)/J \cap K$ contains an atom. Now $K \cap J$ is a λ -saturated ν -complete ideal on J containing all singletons. Thus it will suffice to show $\mathcal{P}(I)/K$ contains an atom.

(a) We assume that $\mathcal{P}(I)/K$ is atomless and derive a contradiction. We construct a tree T as in the proof of Lemma 27.5 in [5]. Let the 0th level of T be $\{I\}$. Each level of T consists of pairwise disjoint subsets of I not in K. Each $X \in T$ has 2 immediate successors Y and W not in K such that $Y \cup W = X$ and $Y \cap W = \emptyset$. If α is a limit ordinal, then the αth level consists of intersections $X = \bigcap_{\xi<\alpha} X_\xi$ such that each X_ξ is on the ξth level of T and X is not in K. We observe that every branch of T has length $< \lambda$: If $\{X_\xi : \xi < \lambda\}$ is a branch of T, then the set $\{Y_\xi : \xi < \lambda\}$ where $Y_\xi = X_\xi - X_{\xi+1}$ is a disjoint collection of sets not in K. Consequently T has height $\leq \lambda$. Similarly each level of T has cardinality $< \lambda$. It follows that T has $\leq 2^{<\lambda}$ branches. Let $\{b_\alpha : \alpha < \beta\}$, $\beta \leq 2^{<\lambda}$, be an enumeration of all branches of the form $b = \{X_\xi : \xi < \gamma\}$ such that $\bigcap_{\xi<\gamma} X_\xi$ is nonempty. For each $\alpha < \beta$ let $V_\alpha = \bigcap\{X : X \in b_\alpha\}$. The collection $\{V_\alpha : \alpha < \beta\}$ is a partition of I into β sets, each of which is in K since $\mathcal{P}(I)/K$ is atomless. Since $\beta < \nu$, the set $\bigcup_{\alpha<\beta} V_\alpha$, which equals I, is in K, a contradiction.

(b) The proof is the same as in (a) if T has height less than ν since ν is strongly inaccessible [Lemma 29.5 in 5]. If T has height ν, it would have a branch $b = \{X_\xi : \xi < \nu\}$ of length ν since ν is weakly compact and has the tree property [Lemma 29.6 in 5]. But then the set $\{Y_\xi : \xi < \nu\}$ where $Y_\xi = X_\xi - X_{\xi+1}$ would be a disjoint set of cardinality ν, violating the fact that K is λ -saturated for $\lambda \leq \nu$.

(1.4) PROPOSITION. Let I be an infinite set and let K be a λ -saturated ν -complete ideal on I containing all singletons where $\lambda \leq \nu \leq |I|$. If $\mathcal{P}(I)/K$ is atomic, then there exists a partition A of I with $|A| < \lambda$ and maximal ideals K_a in $\mathcal{P}(a)$ for each $a \in A$ such that $K = \prod_{a\in A} K_a$ and $\mathcal{P}(I) = \prod_{a\in A} < a > \oplus K$ as an additive group.

Proof. Let A be a subset of $\mathcal{P}(I)$ such that $\{a+K : a \in A\}$ is the set of atoms of $\mathcal{P}(I)/K$. Since $|A| < \lambda \leq \nu$, the elements in A may be chosen to be pairwise disjoint. Let $X = I - \bigcup A$ and, since $X \in K$, adjoin X to any one element a in A. Now A is a partition of I. Since $|A| < \nu$, the group $\prod_{a\in A} K_a$ equals K and the decomposition of $\mathcal{P}(I)$ follows.

2. Summable Subsets of Z^I.

Let I be an infinite set and let Z^I be the direct product of $|I|$ copies of the integers Z. For $x \in Z^I$ we write $x = (x_i)$, $i \in I$, where $x_i \in Z$; $supp(x) = \{i \in I : x_i \neq 0\}$. If s is a subset of I, then $e_s = (x_i)$ where $x_i = 1$ for $i \in s$ and 0 otherwise. We write e_i for $e_{\{i\}}$. Let π_i be the projection of Z^I to its ith component

(2.1) DEFINITION. A subset X of Z^I is *summable* if, for each $j \in I, \pi_j(x) = 0$ for almost all $x \in X$. If $X = \{x_j : j \in J\}$ is summable with $x_j = (x_{ji})$, $i \in I$, then $\sum_{j\in J} n_j x_{ji}$ $(n_j \in Z)$ is a finite sum for each i and we write $\sum_{j\in J} n_j x_j$ for the element $(\sum_{j\in J} n_j x_{ji}), i \in I$, in Z^I.

(2.2) DEFINITION. A summable subset $X = \{x_j : j \in J\}$ is *completely independent* if, whenever $\sum_J n_j x_j$, $(n_j \in Z)$ equals 0, then $n_j = 0$ for each j. In this case the group $S = < \sum_J n_j x_j : n_j \in Z >$ is a direct product of copies of Z and we write $S = \prod_{j\in J} < x_j >$.

(2.3) PROPOSITION. If X is an infinite summable subset of Z^I, then
(a) $|X| \leq |I|$, and
(b) X contains a completely independent subset Y where $|Y| = |X|$.

Proof. (a) Let $X_i = \{x \in X : \pi_i(x) \neq 0\}$ for each $i \in I$. Each X_i is finite. So$|X| = |\bigcup_{i\in I} X_i| \leq \aleph_0 \cdot |I| = |I|$. (b) Let $|X| = \gamma$. By induction we can choose from X a subset Y =$\{x_j \in X : j \in I$ and $j < \gamma\}$ such that, for each j, $\pi_j(x_j) \neq 0$ but $\pi_i(x_j) = 0$ for all $i < j$. To show this suppose $k < \gamma$ and we have made the choices for all $j < k$. Let $W = \{x \in X : \pi_i(x) \neq 0$ for some $i < k\}$. Then $|W| \leq \aleph_0 \cdot |k| < \gamma$. Hence $X - W$ is nonempty and we choose x_k from this set and k from supp(x_k). Now Y satisfies (b).

(2.4) PROPOSITION. Let X be an infinite summable subset of Z^I such that each $x \in X$ generates a pure subgroup of Z^I. Then X contains a subset Y such that $|Y| = |X|$ and $Z^I = \prod_{y\in Y} < y > \oplus D$ where $D \cong Z^\kappa$ for some κ.

Proof. Write $X = \{x_j : j \in J\}$ and for each j let F_j be a finite subset of I such that $Z^I = < x_j > \oplus G_j \oplus \prod_{i\notin F_j} < e_i >$ (this can be done since Z^I is separable [see ch. 87 of [4]). Let $\zeta = |J| = |X|$ and let $\mathcal{F} = \{F_j : j \in J\}$. Suppose ordinal $\beta < \zeta$ and for each $\alpha < \beta$ we have chosen $\alpha \in J$ such that, for all $\gamma < \alpha, F_\alpha \cap F_\gamma = \emptyset =$ supp$(x_\alpha) \cap F_\gamma$. Since X is summable and each F_γ is finite, the set V of all $x \in X$ with supp$(x) \cap F_\alpha$ nonempty for some $\alpha < \beta$ has cardinality less than ζ. Therefore $|X - V| = \zeta$ and we choose $\beta \in J$ so that $x_\beta \in X - V$. Now for all $\alpha < \beta$ we have $F_\beta \cap F_\alpha = \emptyset =$ supp$(x_\beta) \cap F_\alpha$. In this manner we can choose $\beta \in J$ and $F_\beta \subseteq I$ for all ordinals $\beta < \zeta$. Let $Y = \{x_\alpha : \alpha < \zeta\}$, $G = \oplus_{\alpha<\zeta} G_\alpha$ and $L = \bigcup_{\alpha<\zeta} F_\alpha$. Now $|Y| = |X|$, each G_α is free of finite rank and supp$(u) \in F_\alpha$ for any u in G_α. Furthermore supp$(x_\alpha) \cap (\bigcup_{\beta<\alpha} F_\beta) = \emptyset$ for all α. It follows that $Z^I = \prod_{y\in Y} < y > \oplus \prod_{<\zeta} G_\alpha \oplus \prod_{i\notin L} < e_i >$. Let D be the sum of the two right-hand products.

(2.5) PROPOSITION. Let $X = \{x_j : j \in J\}$ be a summable subset of $Z^I$and write $x_j = \sum_{n\in Z} n e_{s_{jn}}$ where $\{s_{jn} : n \in Z\}$ is a partition of I. Then $S = \{e_{s_{jn}} : j \in J$ and $n \in Z\}$ is a summable subset of Z^I and, if X is infinite, $|S| = |X|$.

Proof. If $k \in I$, then $k \in$ supp(x_j) for only a finite number of j's and, for each such j, only one s_{jn} contains k. Therefore S is summable and, if X is infinite, $|S| \leq \aleph_0 \cdot |X| = |X| \leq |S|$ so $|X| = |S|$.

(2.6) COROLLARY. Let X be an infinite summable subset of Z^I. There is a partition P of I such that:

(a) $X \subseteq \prod_{p \in P} < e_p >$ with $|P| = |X|$, and

(b) $Z^I = \prod_{p \in P} < e_p > \oplus Z^{I-L}$ where $|L| = |P|$.

Proof. (a) Let X and S be as in Proposition 2.5. By Lemma 1.1, there is a partition P of I such that $|P|{=}|S| = |X|$ and such that each s_{jn} in S equals p_{jn}, a union of members of P. If $x_j \in X$, then $x_j = \sum_{n \in Z} e_{s_{jn}} = \sum_{n \in Z} ne_{pn}$, an element in $\prod_{p \in P} < e_p >$. (b) Let L consist of one element from each p in P.

(2.7) PROPOSITION. Let $\phi : Z^I \to Z^J$ be a homomorphism and let $K = \{s \subseteq I : \phi(\prod_{i \in s} < e_i >) = 0\}$.

(a) K is a μ-complete ideal on I where μ is the least measurable cardinal number.

(b) If X is an infinite summable subset of Z^I, then $\phi(X)$ is summable in Z^J and, if $\phi(x) \neq 0$ for all $x \in X$, then $|\phi(X)| = |X|$.

(c) Let S be an infinite subset of $\mathcal{P}(I)$ such that $W = \{s + K : s \in S\}$ is a set of nonzero orthogonal elements in $\mathcal{P}(I)/K$. Then $V = \{\phi(e_s) : s \in S\}$ is summable in Z^J and $|V| = |W|$.

(d) K is λ-saturated for infinite λ if $|Y| < \lambda$ for every summable subset Y of $\phi(Z^I)$.

Proof. Let ϕ_j be the projection of $\phi(Z^I)$ to the jth component of Z^J and let $K_j = \{s \in I : \phi_j(\prod_{i \in s} < e_i >) = 0\}$ for each $j \in J$.

(a) By Łoś' Theorem [Theorem 94.4 in 4] each K_j is a μ-complete ideal on I (even if it is a principal ideal) and K equals $\bigcap_{j \in J} K_j$.

(b) Suppose W is an infinite subset of X such that, for fixed j, $\phi_j(w) \neq 0$ for all x in W. By (2.3b) we may assume that W is a completely independent set. From the restriction $\phi : \prod_{w \in W} < w > \to Z^J$ we see that $\phi_j(w) = 0$ for almost all w—a contradiction. So $\phi(X)$ is summable. Choose in X a maximal well-ordered subset $Y = \{x_j \in X : j \in J \text{ and } j < \gamma\}$ such that, for each j, $\phi_j(x_j) \neq 0$ but $\phi_i(x_j) = 0$ for all $i < j$. Since $\{x \in X : \phi_j(x) \neq 0\}$ is finite for each j, $|\phi(Y)|$ must equal $|X|$. Since$|\phi(Y)| \leq |\phi(X)| \leq |X|$, it follows that $|\phi(X)| = |X|$

(c) Let T be a countable subset of S. Now $s \cap t \in K$ for $s \neq t$ in T. Let $u = \bigcup\{s \cap t : s \neq t \text{ in } T\}$. Since K is μ-complete, $u \in K$. Now $\{s' = s - u : s \in T\}$ is a pairwise disjoint set. So $\phi(e_s) = \phi(e_{s'}) = 0$ for almost all s$\in T$. Therefore $V = \{\phi(e_s) : s \in S\}$ is summable in $Z^J$and $|V| = |S| = |W|$.

(d) This follows immediately from (c).

(2.8) COROLLARY[Ex. 7 of ch. 3 in 3]. If $Z^I \cong Z^J$, then $|I| = |J|$.

Proof. Suppose that $|I| > |J|$. Let ϕ in (2.7) be an isomorphism. Let $X = \{e_i : i \in I\}$. Then $\phi(X)$ is a summable subset of Z^J by (2.7b) and $|\phi(X)| = |X| = |I|$. But $|\phi(X)| \leq |J|$ by (2.3a)—a contradiction.

Our next result is of independent interest.

(2.9) PROPOSITION. If X is a summable subset of Z^I, then it generates a free subgroup which has a summable basis.

Proof. We may assume that X and I are infinite sets. We will repeatedly, without mention, use the fact that Z^I is a separable group [ch. 87 of 4].

(a) $< X >$ is a free group. First we want to find a proper descending chain: $I = I_o \supset I_1 \supset \cdots$ and finite rank subgroups H_n for all $n < \omega$ such that $Z^{I_n} = H_n \oplus Z^{I_{n+1}}$ and $X \cap (Z^{I_n} - Z^{I_{n+1}}) \subseteq H_n \oplus H_{n+1}$ for each n. To begin we choose I_1 cofinite in $I = I_o$ and set $H_0 = Z^{I-I_1}$. Now $S = X \cap (Z^{I_0} - Z^{I_1})$ is finite since X is summable. We choose $Z^{I_1} = H_1 \oplus Z^{I_2}$ so that $H_o \oplus H_1$ contains S and I_2 is a proper cofinite subset of I_1. We repeat this process for all positive n. Let $F_\alpha = \oplus_{n<\omega} H_n$ and $Z^{I_\alpha} = \bigcap_{n<\omega} Z^{I_n}$. If $x \in X$ and $x \notin Z^{I_\alpha}$, then $x \in (Z^{I_n} - Z^{I_{n+1}})$ for some $n < \omega$ and as a result is in $H_n \oplus H_{n+1}$ which is in F_α. Similarly we find subgroups F_β and Z^{I_β} of Z^{I_α} such that F_β is free, I_β is a proper subset of I_α and, if $x \in X \cap Z^{I_\alpha}$, then $x \in F_\beta$ or Z^{I_β}. We continue this process until, since the subsets of I converge to $\emptyset$, $F = F_\alpha \oplus F_\beta \oplus \cdots \oplus F_\kappa \oplus \cdots$ contains X. Since F is free, so is its subgroup $< X >$.

(b) $< X >$ has a summable basis. In (a) we showed that $X \subseteq F = \oplus_J F_j$ where each F_j was countable and where, if $x \in X$, then x is in just one F_j. Hence there is a partition $\{X_j : j \in J\}$ of X where $X_j \subseteq F_j$ for each j. Therefore $< X >$ equals $\oplus_J < X_j >$. Thus, to complete the proof it will suffice to show that $< X >$ has a summable basis where X is countable. Let $Y = \{y_n : n < \omega\}$ be a basis of $< X >$ and let each y_n have a fixed representation as a linear combination of members of X. For any subset H of Y we define $supp(H)$ to be $\{x \in X : x \in \text{supp}(y)$ for some y in the subgroup $< H >\}$. Let $H_0 = \{y_0\}$. For each $n > 0$ let H_n be a finite subset of Y containing y_n and H_{n-1} such that $\text{supp}(H_{n-1})$ is contained in $< H_n >$. Now $Y = \bigcup_{n\in\omega} H_n$. Let $B_0 = H_0$ and $B_1 = H_1 - H_0$. For $n > 1$ and $m \leq n-2$ delete from any y in $H_n - H_{n-1}$ any x-term if $x \in X \cap \text{supp}(H_{n-2})$ to obtain y' (note that such an x is in $< H_{n-1} >$); let $B_n = \{y' : y \in H_n - H_{n-1}\}$. Now $B = \bigcup_{n<\omega} B_n$ is a basis of $< X >$ and each B_n is finite. Let $i \in I$ and let $S = \{x \in X : \pi_i(x) \neq 0\}$, a finite set. By construction $S \cap \text{supp}(B_n) = \emptyset$ for almost all n so $\pi_i(B_n) = 0$ for almost all n. Therefore B is summable.

3. Direct summands of $\mathbf{Z^I}$.

In this section we apply the results of Section 2 to the study of decompositions of Z^I. Again, μ is the least measurable cardinal number. By "G is a product" we mean $G \cong Z^\alpha$ for some α. We begin with some known results.

(3.1) PROPOSITION. Let $Z^I = A \oplus B$ with $|I| = \kappa$.

(a) [1 and 8; see also Ex. 11 of ch. 13 in 4] If $\kappa < \mu$, then A and B are both products.

(b) [Th. 12 in 9] If $\kappa = \mu$, then A and B are both products.

(c) [see page 39 of 2] If $|A| < \mu$, then A is a product.

(d)[10] If κ is infinite, then $A \cong Z^\kappa$ or $B \cong Z^\kappa$.

The following result is known but, since we couldn't find a suitable reference, we will include a proof.

(3.2) PROPOSITION. Let A be a subgroup of Z^I of cardinality α. Then A is contained in a direct summand C of Z^I where $C \cong Z^\beta$ for some $\beta \leq 2^\alpha$.

Proof. Write $a = (a_i), i \in I$, for a in A. For i and j in I define $i \sim j$ if $a_i = a_j$ for all $a \in A$. Now $\sim$ is an equivalence relation on I so there is a partition P of I where i and j are in the same p in P if and only if $i \sim j$. Each $p \in P$ defines a function $p : A \to Z$ by $p(a) = a_i$ for any i in P. If $|P| = \beta$, then $\beta \leq \left|Z^{|A|}\right| = 2^\alpha$. Let $G = \prod_{p \in P} < e_p >$ and let L consist of one element in I from each $p \in P$. Then $Z^I = G \oplus Z^{I-L}$ and $G \cong Z^\beta$. If $a = \sum_{n \in Z} n e_{s_n}$ is an element in A, then each s_n is the union p_n of members of P and $\{p_n : n \in Z\}$ is a partition of I. Ergo $a = \sum_{n \in Z} n e_{p_n}$, an element in G.

Our next result relates summable sets to summands of direct products.

(3.3) PROPOSITION. Suppose $Z^J = A \oplus B$ with $|J| \geq \kappa \geq \omega$. Then A contains a summable subset of cardinality κ iff A contains a direct summand isomorphic to Z^κ.

Proof. (a)$\Leftarrow$. Let $\phi : A \to Z^J$ be the identity map. Suppose X $\subseteq C \subseteq A$ where $C \cong Z^\kappa$, $|X| = \kappa$ and X is summable relative to Z^κ. Then $\phi(X) = X$ is summable relative to Z^J by (2.7b). (b)$\Rightarrow$. By Proposition 2.4 we may write $A \oplus B = Z^I \oplus E$ where $Z^I = \prod_I < e_i >$, each e_i is in A and $|I| = \kappa$. Let $\pi : A \oplus B \to B$ and $\rho : Z^I \oplus E \to Z^I$ be the natural projections and let $f = \pi\rho$. In Propositions 3—5 in [10] we showed, for some subset $S \subseteq I$ with $|S| = \kappa$, that $f : Z^S \to Z^S$ is an isomorphism. Therefore $P = Z^S \oplus D = A \oplus B$ where, for projections $\bar{\rho} : Z^S \oplus D \to Z^S$ and $\pi : A \oplus B \to B$, the map $\bar{\rho}\pi : Z^S \to Z^S$ is an isomorphism. By Lemma 7 in [10] $A = \pi(Z^S) \oplus \ker\bar{\rho}$. To show $\pi(^S) \cong Z^\kappa$ it suffices to show the kernel of the map $\pi : Z^S \to \pi(Z^S)$ is $\{0\}$. But, if $\pi(x) = 0$ for $x \in Z^S$, then $\bar{\rho}\pi(x) = 0$ and $x = 0$ since $\bar{\rho}\pi$ is one-to-one.

The previous proof was the essential ingredient used to prove (3.1d) above.

(3.4) COROLLARY. Suppose that $Z^I = A \oplus B$ for infinite I and A is not isomorphic to Z^I. Then every summable subset of A has cardinality less than $|I|$.

Remark. It would be nice in (3.4) if we could conclude that the subgroup A there is isomorphic to Z^α for some $\alpha < |I|$. Using induction on $|I|$ we could then conclude that A is a product. At this time we cannot show this for arbitrary set I. Perhaps it cannot be shown in ZFC. However, by means of (3.7) below, we shall in (3.8) obtain the desired result for a number of cases not covered by Proposition 3.1. Future set-theoretic considerations prompt the next definition.

(3.5) DEFINITION. A direct summand A of Z^I is *λ-saturated* in Z^I if every summable subset of A has cardinality less than λ. *Sat$A$* is the least λ such that A is λ-saturated.

If A is a summand of Z^I and $|A| = \lambda$, then A is λ-saturated by (3.3). Also by (3.3), if sat$A = \lambda$, then $|A| \geq 2^{<\lambda} = \sup_{\alpha<\lambda}(2^\alpha)$. In this case $|A|$ may be very large relative to λ. For example suppose that $2^\alpha = 2^\beta$ for some $\beta > \alpha$. Then sat$(Z^\alpha) = \alpha^+$ and sat$(Z^\beta) = \beta^+$ although $|Z^\alpha| = |Z^\beta|$. In any case sat$A \leq \gamma^+$ where $\gamma = \min(|I|, 2^{|A|})$ by (2.3a) and by (3.2). We summarize these facts below.

(3.6) PROPOSITION. Let A be a direct summand of Z^I where I is infinite.

(a) If $|A| < 2^\lambda$, then A is λ-saturated.

(b) If sat$A = \lambda$, then $|A| \geq 2^{<\lambda} = \sup_{\alpha<\lambda}(2^\alpha)$.

(c) If $\gamma = \min(|I|, 2^{|A|})$, then A is γ^+-saturated.

(3.7) PROPOSITION. Let $Z^J = A \oplus B$ where $|J| = \kappa > \mu$ the least measurable cardinal and suppose that A is not isomorphic to Z^J. Let $\phi : A \oplus B \to A$ be the natural projection. Then A is κ-saturated and Z^J has a decomposition $Z^J = Z^L \oplus Z^I$ where $|L| < \kappa = |I|$ such that, if $K = \{s \in \mathcal{P}(I) : \phi(\prod_{i\in s} < e_i >) = 0\}$, then

(a) K is a λ-saturated ν-complete ideal on I containing all singletons where $\lambda \leq \kappa$, $\mu \leq \nu$, and $\nu \leq \kappa$ unless $I = K$.

(b) If $\lambda \leq$ cfκ (e.g. if κ is a regular cardinal), then K contains all subsets of I of cardinality less than κ and $\nu \leq$ cfκ unless $I = K$.

(c) if $\lambda \leq \nu$ and $\mathcal{P}(I)/K$ is atomic, then A is isomorphic to Z^α for some $\alpha < \kappa$.

Proof. By (3.4) A is κ-saturated. Write $Z^J = \prod_{j\in J} < e_j >$.

(a) Let $L' = \{j \in J : \phi(e_j) \neq 0\}$. Since A is κ-saturated, $|L'| < \kappa$ by (2.7b). Let $I' = J - L'$ and let $K' = \{s \in \mathcal{P}(I') : \phi(\prod_{i\in s} < e_i >) = 0\}$. Then $|I'| = \kappa$ and K' is a λ-saturated ν-complete ideal on I' containing all singletons where $\lambda \leq \kappa$ and $\nu \geq \mu$ by (2.7d) and (2.7a). Since K' contains all singletons, $\nu \leq \kappa$ unless $K' = I'$. If κ is singular and $\lambda >$ cfκ we satisfy (a) by simply dropping the primes from I', L' and K'. Otherwise we continue on to (b).

(b) If $\lambda \leq$ cfκ, let M be as in Proposition 1.2. Let $L = L' \cup M$, $I = I' - M$ and $K = K' \cap I$. Now $|L| < \kappa = |I|$ and K is a λ-saturated ν-complete ideal on I containing all singletons which satisfies (b) as well as (a) by (1.2).

(c) By Proposition 1.4 there is a partition P of I with $|P| < \lambda \leq \kappa$ such that $\mathcal{P}(I)$ equals$\prod_{p\in P} < p > \oplus K$ as an additive group. As a result $Z^I = \prod_{p\in P} < e_p > \oplus G$ where G equals $\{\prod_{i\in s} < e_s >: s \in K\}$. Since $G \subseteq B$ and $A \oplus B$ equals $Z^L \oplus Z^I$, we obtain $A \oplus (B/G) \cong Z^L \oplus Z^{|P|}$ and $|L \cup P| < \kappa$.

(3.8) PROPOSITION. Assume Proposition 3.7 in all its particulars. Then A is isomorphic to a direct summand of Z^α for some $\alpha < \kappa$ if:

(a) $I = K$,

(b) ν is a regular cardinal and $2^{<\lambda} < \nu$,

(c) ν is a weakly compact cardinal and $\lambda = \nu$,

(d) $2^{|A|} < \kappa$, or

(e) $\lambda \leq \mu$.

Proof. (a) If $I = K$, then $Z^I \subseteq B$. As a result $A \oplus (B/Z^I) \cong Z^L$ with $|L| < \kappa$. (b) and (c) By Proposition 1.3 $\mathcal{P}(I)/K$ is atomic. Proposition 3.7c completes the proof. (d) Proposition 3.2. (e) Since $\lambda \leq \mu \leq \nu$, K is a μ-saturated μ-complete ideal on I and, since μ is weakly compact, (b) completes the proof.

Example. Suppose $\alpha_0 \geq \mu$, $\alpha_{n+1} = 2^{\alpha_n}$ and $\kappa = \lim_{n\to\infty} \alpha_n$. The ideal K in (3.7) is λ-saturated for some $\lambda \leq \kappa$. Now sat$B$ is finite or regular for any Boolean algebra B [Corollary 3.11 in 7]. As a result sat$K < \kappa$ and K is λ-saturated for $\lambda < \alpha_n$ for some n. Then $2^\lambda \leq 2^{\alpha_n} = \alpha_{n+1}$. Therefore (3.9a) is satisfied for $\nu > \alpha_{n+1}$.

REFERENCES

1. S. Balcerzyk, A. Białynicki-Birula and J. Łoś, *On direct decompositions of complete direct sums of groups of rank 1.* Bulletin de L'Academie Polonaise des Sciences, série des sciences math., astr., et phys., 9(1961), 451-454.

2. K. Eda, *A generalized direct product of abelian groups*, J. of Algebra, 92(1985), 33-43.

3. P. Eklof and A. Mekler, Almost Free Modules, North-Holland, Amsterdam, 1990.

4. L. Fuchs, Infinite Abelian Groups, vol. I(1970), vol. II(1973), Academic Press, New York.

5. T. Jech, Set Theory, Academic Press, New York, 1978.

6. A. Kanamori, The Higher Infinity, Springer-Verlag, Berlin, 1994.

7. S. Koppelberg, General Theory of Boolean Algebras, Volume 1 of Handbook of Boolean Algebras, edited by J.D. Monk, North-Holland, 1989.

8. R.J. Nunke, *On direct products of infinite cyclic groups.* Proc. Amer. Math. Soc., 13(1962), 66-71.

9. J.D. O'Neill, *Direct summands of* Z^κ *for large* κ, Abelian Group Theory and Related Topics, Contemporary Mathematics, vol. 171, edited by R. Göbel, P. Hill and W. Liebert, Amer. Math. Soc., 1994, 313-323.

10. J.D. O'Neill, *A result on direct products of infinite cyclic groups,* Comm. in Algebra, to appear.

On Baer's Lemma

J. D. Reid
Wesleyan University, Middletown, CT

1 Introduction.

If A and G are abelian groups we recall that $\mathrm{S}_A(G)$ denotes the subgroup $\Sigma_{f \in \mathrm{Hom}(A,G)} f(A)$ of G. It is clear that a necessary condition that an exact sequence

$$0 \longrightarrow B \longrightarrow G \longrightarrow A \longrightarrow 0$$

split is that $\mathrm{S}_A(G) + B = G$. In a classical paper Baer proved that this condition is also sufficient in case A is torsion-free of rank 1. (Indeed A need only be completely decomposable and homogeneous. Cf. Fuchs [7], Ch. XIII for this and for other references.)

Baer's result was recast and generalized by Arnold and Lady [2] to the case in which A is torsion-free of finite rank. To state their result we recall that the torsion-free group A is *faithful* if $IA \neq A$ for all proper right ideals I of the endomorphism ring of A, and that P is *finitely A-projective* if P is a summand of a direct sum of a finite number, say n, of copies of A. Arnold and Lady prove by induction on n that every exact sequence

$$0 \longrightarrow B \longrightarrow G \longrightarrow P \longrightarrow 0, \ \mathrm{S}_A(G) + B = G$$

with P finitely A-projective splits if and only if A is faithful. Their proof uses a certain equivalence, valid in the case of finite rank A, between the category of A-projective groups and the category of projective right $\mathrm{End}(A)$-modules.

Finite rank torsion-free groups are examples of *self small* groups, i.e. groups A for which $\mathrm{Hom}(A, \oplus_{\nu \in N} A) = \oplus_{\nu \in N}\mathrm{Hom}(A, A)$ for arbitrary index sets N. Arnold and Murley [3] point out that the category equivalence mentioned above is valid for self small groups, and Albrecht [1] uses this to prove that, for self small A, sequences as above (i.e. with P finitely A-projective) always split if and only if A is faithful. Again, the proof is by induction. In addition, removing the finiteness condition, Albrecht proves that for self small A, all such sequences (i.e. P a summand of an arbitrary direct sum of copies of A) split if and only if A is *fully faithful*, i.e. $M \otimes_{\mathrm{End}(A)} A \neq 0$ for all right $\mathrm{End}(A)$-modules M, not just the finitely generated ones.

Faticoni [5], continuing the applications of the category equivalence, makes a detailed study of the various finiteness conditions involved in these splitting problems and generalizes by working with modules over arbitrary rings rather than abelian groups. We remark that it seems to have been here that the inductive argument implicit in the work of Arnold-Lady (P finitely A-projective, finite rank faithful A) and Albrecht (P finitely A-projective, A faithful) was made explicit. Faticoni also suggests the question of the existence of modules that are faithful but not fully faithful.

In this note we take a rather different approach to these questions. We show that these various results follow in a natural way from a very simple lemma (Proposition 2.2) which seems to be new and which is useful in other contexts as well as that here. We also work with modules over arbitrary rings. Of course in this generality self small S-modules are those for which $\mathrm{Hom}_S(A, \oplus_{\nu \in N} A) = \oplus_{\nu \in N} \mathrm{Hom}_S(A, A)$; analogously, $\mathbf{S}_A(G) = \Sigma_{f \in \mathrm{Hom}_S(A,G)} f(A)$ for S-modules G. A sample result is the following.

Theorem 1.1 *Let S be a ring and let A be a self-small right S-module with endomorphism ring $R = \mathrm{End}_S(A)$. Let P be a summand of a direct sum of κ-many copies of A and let F be a free right R-module of rank κ. Assume that, for all right ideals I of $\Lambda = \mathrm{End}_R(F)$, $I(F \otimes_R A) = F \otimes_R A$ only if $IF = F$. Then every exact sequence*

$$0 \longrightarrow B \longrightarrow G \longrightarrow P \longrightarrow 0$$

of S-modules with $\mathbf{S}_A(G) + B = G$ splits.

For the connection of the hypothesis here with the basic lemma (Proposition 2.2) and the relation of this hypothesis to the results quoted above, see Proposition 3.4; see Theorem 4.2 for a general statement.

We thank C. I. Vinsonhaler and W. J. Wickless for sharing our interest in these matters and for suggestions which have helped clarify the exposition here. We also thank the referee for several suggestions.

We adopt and hold fixed the following notation and point of view. S is a ring and A is a right S-module. Put $R = \mathrm{End}_S(A)$ so that A is an R-S-bimodule.

Let F be the free right R-module with basis $\{x_\nu | \nu \in N\}$ for some index set N. Let $\Lambda = \mathrm{End}_R(F)$ so F is a Λ-R-bimodule. Then $F \otimes_R A$ is an S-module by $(x \otimes a)s = x \otimes as$ and it is clear that

$$F \otimes_R A \cong \oplus_{\nu \in N} A.$$

2 On Free Modules.

The following is undoubtedly well known. As a vehicle for establishing notation we attach a proof but we hope that this serves the readers' [1] convenience as well.

Lemma 2.1 *Let F be the free R-module on $\{x_\nu | \nu \in N\}$. For any $\nu \in N$, the Λ-exact sequence*

$$0 \longrightarrow L_\nu \longrightarrow \Lambda \longrightarrow F \longrightarrow 0, \ \lambda \rightsquigarrow \lambda x_\nu$$

is split exact. Indeed if $\{\epsilon_\nu\}$ are the projections of F onto the $x_\nu R$, then $L_\nu = \Lambda(1 - \epsilon_\nu)$ and $F \cong \Lambda\epsilon_\nu$ as left Λ-module. (i.e. F is cyclic projective over Λ with each of the x_ν as generator.)

Proof: Since F is free over R it is clear that the map $\phi : \Lambda \longrightarrow F$ defined as above by $\phi(\lambda) = \lambda x_\nu$ is an epimorphism and that $\Lambda(1 - \epsilon_\nu) \subseteq L_\nu$. Moreover $\phi(\lambda\epsilon_\nu) = \lambda\epsilon_\nu x_\nu = \lambda x_\nu$ for $\lambda \in \Lambda$ so ϕ takes $\Lambda\epsilon_\nu$ onto F. The map $\lambda x_\nu \rightsquigarrow \lambda\epsilon_\nu$ of F into Λ is easily seen to be well defined and is then a splitting map for the sequence. The Lemma follows from these remarks. □

The folowing lemma is basic to our approach, and has other applications as well (see the *Remarks* below).

Proposition 2.2 *If I is a right ideal of Λ, then $IF = F$ if and only if I contains all the R-endomorphisms of F of finite type: i.e. all λ such that λF is contained in some finitely generated R-submodule of F.*

[1] Observe the note of optimism implicit in the grammar here.

Proof: In the notation of the Lemma, $F = IF = I\Lambda x_\nu = Ix_\nu$ so the map ϕ above takes I *onto* F. Thus $L_\nu + I = \Lambda$ so $\epsilon_\nu = \alpha_\nu - \lambda_\nu$ for some $\alpha_\nu \in I, \lambda_\nu \in L_\nu$. Then

$$\lambda_\nu = \lambda_\nu(1-\epsilon_\nu) = (\epsilon_\nu + \lambda_\nu)(1-\epsilon_\nu) = \alpha_\nu(1-\epsilon_\nu) \in I.$$

We conclude that $\epsilon_\nu = \alpha_\nu - \lambda_\nu \in I$ and this is true for all ν. Now if λF is contained in $\Sigma_{i=1}^m y_i R$ say, we may write

$$y_i = \Sigma_{\nu\in M} x_\nu r_{\nu i}$$

for all i and some finite set M of indices. Since $\Sigma_{\nu\in M}\epsilon_\nu$ is the identity on $\oplus_{\nu\in M} x_\nu R$ and $\lambda F \subseteq \oplus_{\nu\in M} x_\nu R$, we have $\lambda = (\Sigma_{\nu\in M}\epsilon_\nu)\lambda = \Sigma_{\nu\in M}\epsilon_\nu\lambda \in I$ as asserted. □

If F is given the discrete topology and the set of functions F^F the product topology then $\Lambda = \mathrm{End}_R(F)$ is a closed subset of F^F. We recall that the induced topology on Λ is called the *finite topology.* A base for the system of neighborhoods of $\lambda \in \Lambda$ is given by the collection of finite subsets X of F by $N_X(\lambda) = \{\mu \in \Lambda | \lambda(x) = \mu(x) \text{ for all } x \in X\}$. Thus for example a base for the neighborhoods of 0 is the set of all annihilators of finite subsets of F. As is well known, Λ is a topological ring relative to this structure.

Corollary 2.3 *If the right ideal I is closed in the finite topology given by Λ acting on F then $IF = F$ if and only if $I = \Lambda$.*

As an example, we observe

Proposition 2.4 *Let I be a finitely generated right ideal of Λ. Then I is closed in the finite topology given by Λ acting on F.*

Proof: Suppose that λ belongs to the closure of I in the finite topology. Then for each ν there is $\phi_\nu \in I$ such that $\lambda x_\nu = \phi_\nu x_\nu$. Since I is finitely generated, $I = \Sigma_{j=1}^m \lambda_j \Lambda$ for certain $\lambda_j \in \Lambda$. Write

$$\phi_\nu = \Sigma_{j=1}^m \lambda_j \phi_{j\nu},\ \phi_{j\nu} \in \Lambda.$$

Since F is free we may define for each j a map $\phi_j \in \Lambda$ by the condition that $\phi_j x_\nu = \phi_{j\nu} x_\nu$ for all ν. Then for each ν

$$\lambda x_\nu = \phi_\nu x_\nu = \Sigma_{j=1}^m \lambda_j \phi_{j\nu} x_\nu = \Sigma_{j=1}^m \lambda_j \phi_j x_\nu = (\Sigma_{j=1}^m \lambda_j \phi_j) x_\nu$$

so $\lambda = \Sigma_{j=1}^{m} \lambda_j \phi_j \in I$ as required. □

Remarks: If I is a finitely generated right ideal of Λ then $IF = F$ if and only if $I = \Lambda$ (cf. also [5], 7.1a); or if F has finite rank then the finite topology is discrete so for any right ideal I of Λ, $IF = F$ if and only if $I = \Lambda$. In case R is Noetherian, this follows as well from the fact that Λ is Noetherian for finite rank F. It is a classical result that right (and left) ideals in the ring of endomorphisms of free modules of finite rank over principal ideal domains are principal. For a generalization of this, which in fact motivated our approach here, see [8].

3 Self Small Modules.

We keep the notation introduced above and recall that the module A is said to be *self small* (compare the *dually slender* modules [4]) if for every $g \in \mathrm{Hom}_S(A, \oplus_{\nu \in N} A)$, there exists a finite set $M = M(g) \subseteq N$ such that $g(A) \subseteq \oplus_{\nu \in M} A$. In the following we use the well known fact that, having chosen the basis $\{x_\nu | \nu \in N\}$, each element of $F \otimes_R A$ has a *unique* expression of the form $\Sigma x_\nu \otimes a_\nu$ with the $a_\nu \in A$ and almost all $a_\nu = 0$.

Proposition 3.1 *For self small A, $\Lambda \cong \mathrm{End}_S(F \otimes_R A)$ under the action induced by $\lambda(x \otimes a) = \lambda x \otimes a$.*

Proof: This follows easily from the fact that $F \otimes_R A$ is A-solvable (in the sense of Albrecht) for self small A, but we include a simple direct proof for the reader's convenience.

It is clear that each $\lambda \in \Lambda$ induces a map $T_\lambda \in \mathrm{End}_S(F \otimes_R A)$ by $T_\lambda(x \otimes a) = \lambda x \otimes a$ and that this yields an embedding of Λ into $\mathrm{End}_S(F \otimes_R A)$. On the other hand let $\rho \in \mathrm{End}_S(F \otimes_R A)$ and suppose that A is self small. Then $a \rightsquigarrow \rho(x_\nu \otimes a)$ is an S-map of A into $F \otimes_R A = \oplus_\nu A$ so there exists a finite set $M = M(\rho, \nu)$ of indices such that

$$\rho(x_\nu \otimes a) = \Sigma_{\mu \in M} x_\mu \otimes a_{\mu\nu}, \; a_{\mu\nu} \in A.$$

Define $r_{\mu\nu}$ by $r_{\mu\nu} a = a_{\mu\nu}$. Then $r_{\mu\nu} \in R$, i.e. $r_{\mu\nu}$ is an S-map, because ρ is an S-map and we have

$$\rho(x_\nu \otimes a) = \Sigma_{\mu \in M} x_\mu \otimes r_{\mu\nu} a = \Sigma_{\mu \in M} x_\mu r_{\mu\nu} \otimes a.$$

Since for each ν the set $M = M(\rho, \nu)$ is finite and F is free, we may define $\lambda \in \Lambda$ by $\lambda x_\nu = \Sigma_{\mu \in M} x_\mu r_{\mu\nu}$. Then, for all ν, $\rho(x_\nu \otimes a) = \lambda x \otimes a$ so that $\rho = T_\lambda$ as required. □

Corollary 3.2 *For self small S-modules A, the A-projective S-modules are exactly the modules $P \otimes_R A$, P a projective R-module.*

A direct sum of copies of the S-module A is called an A-*free* module. Corollary 3.2 implies that for self small A, every A-projective S-module is A-free if and only if every projective R-module is free. This generalizes a result established by Arnold-Lady in the case of torsion-free abelian groups A of finite rank (cf. [2]).

Definition 3.3 The right S-module A is *faithful* if $IA \neq A$ for all proper right ideals I of $R = \text{End}_S(A)$. A is *fully faithful* if $M \otimes_R A \neq 0$ for all right R-modules M.

We know that the endomorphism ring Λ of F acts on $F \otimes_R A$ but to connect with the fundamental Proposition 2.2 we need to relate the action of a right ideal I of Λ on $F \otimes_R A$ with its action on F itself as in the following

Proposition 3.4 *Suppose that A is self small and faithful. Let F be free over R as above and let I be a right ideal of Λ. Assume that either*

- *(i) F is a finitely generated R-module; or*
- *(ii) A is fully faithful; or*
- *(iii) A is flat over its endomorphism ring R.*

Then for any right ideal I of Λ, $I(F \otimes_R A) = F \otimes_R A$ if and only if $IF = F$.

Proof: In case (i), let $\{x_1, \ldots, x_n\}$ be a basis of F over R with associated projection maps $\{\epsilon_1, \ldots, \epsilon_n\}$, and let I be a right ideal of Λ with $I(F \otimes_R A) = F \otimes_R A$. Put $k_I = |\{\nu | \epsilon_\nu \notin I\}|$. Evidently $IF = F$ if $k_I = 0$. Assume that $k_I = l > 0$ and that $I'F = F$ for all right ideals I' of Λ with $I'(F \otimes_R A) = F \otimes_R A$ and $k_{I'} < l$. We may assume the indices chosen so that $\epsilon_\nu \notin I, 1 \leq \nu \leq l$ and that $\epsilon_\nu \in I$ for $\nu > l$.

Write $I' = I + \epsilon_l \Lambda$ so $k_{I'} < l$ and clearly $I'(F \otimes_R A) = F \otimes_R A$ so by hypothesis, $I'F = F$. Again we have $\epsilon_l F = x_l R \cong R$ and we write I_0 for the right ideal of R that is the image of $IF \cap x_l R$ under this isomorphism. Then $\frac{I'F}{IF} = \frac{F}{IF} \cong \frac{R}{I_0}$ and the sequence

$$0 \longrightarrow IF \longrightarrow F \longrightarrow R/I_0 \longrightarrow 0$$

yields

$$IF \otimes_R A \longrightarrow F \otimes_R A \longrightarrow R/I_0 \otimes A \longrightarrow 0.$$

Since the map on the left has $I(F \otimes_R A) = F \otimes_R A$ as image we conclude that $R/I_0 \otimes A = 0$ contradicting the fact that A is faithful.

For case (ii), if I is a right ideal of Λ with $I(F \otimes_R A) = F \otimes_R A$, the exact sequence

$$0 \longrightarrow IF \longrightarrow F \longrightarrow F/IF \longrightarrow 0$$

yields

$$IF \otimes_R A \longrightarrow F \otimes_R A \longrightarrow F/IF \otimes A \longrightarrow 0.$$

and it follows as above that $F/IF \otimes A = 0$ contradicting the full faithfulness of A unless $IF = F$.

∎

The fact that the conclusion of the Proposition is valid under the hypothesis in (iii) follows immediately from part (ii) together with Albrecht's observation that a faithful R-flat module is necessarily fully faithful. However, since the precise conditions under which faithfulness implies full faithfulness appear to be unknown (e.g. [6]), it may be of interest to pinpoint the role that flatness plays in our approach. It is our hope that this might prove useful in seeking such conditions so we beg the reader's indulgence.

Assume then that $I(F \otimes_R A) = F \otimes_R A$. If $IF \neq F$ then for some ν, $x_\nu R \not\subseteq IF$ so the projection ϵ_ν of F onto $x_\nu R$ is not in I. Then I is properly contained in $I' = I + \epsilon_\nu \Lambda$ and $x_\nu R \subseteq I'F$ so IF is properly contained in $I'F$ as well. Let I_0 be the right ideal of R given by the image of the injection $x_\nu R \cap IF \hookrightarrow x_\nu R \cong R$. Then $I_0 \neq R$ and the exact sequence of right R-modules

$$0 \longrightarrow IF \longrightarrow I'F \longrightarrow I'F/IF \cong R/I_0 \longrightarrow 0$$

yields

$$\begin{array}{ccccccc} IF \otimes_R A & \longrightarrow & I'F \otimes_R A & \longrightarrow & R/I_0 \otimes A & \longrightarrow & 0 \\ \downarrow & & \downarrow & & & & \\ I(F \otimes_R A) & \longrightarrow & I'(F \otimes_R A) & \longrightarrow & 0. & & \end{array}$$

Since A is R-flat, the vertical arrows in the left square are isomorphisms. Further, the left square is commutative and $I(F \otimes_R A) = F \otimes_R A$ so the bottom row is exact. It follows that $R/I_0 \otimes_R A = 0$ and from the fact that A is faithful we conclude that $I_0 = R$, a contradiction.□

4 Baer's Lemma.

Recall again that if G is another S-module, then

$$\mathbf{S}_A(G) = \Sigma_{f \in \mathrm{Hom}_S(A,G)} f(A).$$

Since $F \otimes_R A \cong \oplus_\nu A$ as S-modules it is clear that $\mathbf{S}_A(G) = \Sigma h(F \otimes_R A)$, the sum taken over all $h \in \mathrm{Hom}_S(F \otimes_R A, G)$, as well.

Proposition 4.1 *Let*

$$0 \longrightarrow B \longrightarrow G \xrightarrow{\beta} F \otimes_R A \longrightarrow 0$$

be an exact sequence of S-modules and assume that $\mathbf{S}_A(G) + B = G$. Put $I = \beta Hom_S(F \otimes_R A, G)$. Then I is a right ideal of $End_S(F \otimes_R A)$ and $I(F \otimes_R A) = F \otimes_R A$.

Proof: This is clear from the definitions and the remarks above. □

Theorem 4.2 *Let A be a self small S-module such that $IF = F$ for all right ideals I of Λ that satisfy $I(F \otimes_R A) = F \otimes_R A$ (Cf. Proposition 3.4). Then any exact sequence of S-modules*

$$0 \longrightarrow B \longrightarrow G \xrightarrow{\beta} F \otimes_R A \longrightarrow 0, \ \mathbf{S}_A(G) + B = G, \tag{1}$$

is split exact.

Proof: Let I be the ideal of $\mathrm{End}_S(F \otimes_R A)$ as in Proposition 4.1. We may identify I with an ideal, which we also call I, of Λ (Proposition 3.1) satisfying $IF = F$ (Proposition 4.1). It suffices to show that I is closed in the finite topology, for then we will have $I = \Lambda$ (Corollary 2.3) and the identity map in Λ factors through β so the sequence splits.

Suppose that $\lambda \in \Lambda$ belongs to the closure of I in the finite topology. Then for each ν there is $\phi_\nu \in I$ such that $\lambda x_\nu = \phi_\nu x_\nu$. Since I consists of all the endomorphisms of the form βg for $g \in \mathrm{Hom}_S(F \otimes_R A, G)$, we have $\phi_\nu = \beta g_\nu$ for some $g_\nu \in \mathrm{Hom}_S(F \otimes_R A, G)$. By the universal property of the direct sum, there is an element $g \in \mathrm{Hom}_S(F \otimes_R A, G)$ such that $g(x_\nu \otimes a) = g_\nu(x_\nu \otimes a)$ for each ν. Then $\beta g(x_\nu \otimes a) = \beta g_\nu(x_\nu \otimes a) = \phi_\nu(x_\nu \otimes a) = \lambda(x_\nu \otimes a)$ so $\lambda = \beta g \in I$ as required, i.e. I is closed. This completes the proof. □

The modern versions of Baer's Lemma usually allow the quotient module in the sequence (1) to be A-projective, a summand of a direct sum of copies of A, rather than an actual direct sum as in Theorem 4.2. However the more general case follows from that treated here by well known arguments. We sketch one way to see this. We are assuming of course that A is self small over S.

As indicated in Corollary 3.2 the A-projective modules have the form $P_0 \otimes_R A$ where P_0 is a projective R-module. Then we have a free right R-module of the form $F = P_0 \oplus P_1$ with the projection map $\pi_0 : F \longrightarrow P_0$ inducing the projection of $F \otimes_R A$ onto $P_0 \otimes_R A$. Given an exact sequence

$$0 \longrightarrow B \longrightarrow G \xrightarrow{\beta} P_0 \otimes_R A \longrightarrow 0, \tag{2}$$

we consider the pullback

$$0 \longrightarrow B \longrightarrow X \longrightarrow F \otimes_R A \longrightarrow 0 \tag{3}$$

defined by π_0. Thus $X = \{(g,y) | g \in G, y \in F \otimes_R A, \beta(g) = \pi_0(y)\}$ and we have the map ξ of X onto G, given by $\xi(g,y) = g$, which is split by $\alpha : g \rightsquigarrow (g, i_0\beta(g))$. Here i_0 is the injection of $P_0 \otimes_R A$ into $F \otimes_R A$ and satisfies $\pi_0 i_0 = 1_{P_0 \otimes_R A}$ so that $\alpha(g)$ is indeed in X. Put $X_1 = \{(0,y) | y \in F \otimes_R A$ and $\pi_0(y) = 0\}$. Then $X_0 \cong G$, $X_1 \cong P_1 \otimes_R A$ and $X = X_0 \oplus X_1$. It follows easily from this that if $\mathbf{S}_A(G) + B = G$ then $\mathbf{S}_A(X) + B = X$. Thus if $\mathbf{S}_A(G) + B = G$ and A satisfies any of the three conditions of Proposition 3.4, for example, then by Theorem 4.2 the sequence (3) splits and it follows that the sequence (2) splits as well.

For example, this establishes Theorem 1.1 as stated in Section 1.

References

[1] Albrecht, U., Faithful abelian groups of infinite rank, Proc. AMS, 103, (1), 1988, 21-26.

[2] Arnold, D. M. and Lady, E. L., Endomorphism rings and direct sums of torsion-free abelian groups, Trans. AMS, 211, 1975, 225-237.

[3] Arnold, D. M. and Murley, C. E., Abelian groups A such that Hom(A, ·) preserves direct sums of copies of A, Pac. J. Math. 56, (1), 1975, 7-20.

[4] Eklof, Paul C., Goodearl, K. R., Trlifaj, Jan, Dually slender modules and steady rings, These Proceedings.

[5] Faticoni, Theodore G., The endlich Baer splitting property, Pacific J. Math. 157 (1993), 225-240.

[6] Faticoni, Theodore G., Torsion-free abelian groups torsion over their endomorphism rings, Bull. Austral, Math Soc., (50) 1994, 177-195.

[7] Fuchs, L., *Infinite Abelian Groups*, Pure and Appl. Math., vol. 36-II, Academic Press, New York, 1970.

[8] Reid, J. D., A note on endomorphism rings of free modules, preprint.

Two Problems of Ziegler and Uniform Modules Over Regular Rings

JAN TRLIFAJ
Charles University, Prague, Czech Republic

ABSTRACT. We deal with two well-known problems of the model theory of modules, due to Ziegler, in the particular case of modules over (von Neumann) regular rings. In this case, there is a complete elimination of quantifiers, so the essence of the problems is algebraic. In §1, we solve the first problem by proving that the complete theory of a (right R-) module M has continuous part zero iff the lattice of all pp-definable subgroups of M has width, for any module M over any regular ring R. In §3, we deal with the second problem of whether isolated points of the space of indecomposables, $\mathcal{I}(T)$, are isolated by T-minimal pairs, for each complete theory of modules T. We show that this holds for any regular ring R iff there is no simple non-artinian regular ring over which all indecomposable injective modules are isomorphic. The latter is an open question of the theory of regular rings. Despite this, we provide a positive solution of the second problem in several particular cases: when all primitive factors of R are artinian, or R is semiartinian, or card$(R) < 2^\omega$.

The study of first-order properties of modules over an associative ring with unit, R, is close to the study of their algebraic properties. Namely, by a result of Baur et al. [P1, Corollary 2.16], if T is any complete theory of modules, then any formula of the language of modules is equivalent modulo T to a boolean combination of pp-formulas. Recall that a formula $\varphi(v_0, \ldots, v_{n-1})$ in free variables $v_0, \ldots, v_{n-1}$ is called a *pp-formula* provided that $\varphi(v_0, \ldots, v_{n-1})$ has the form

$$\exists w_0, \ldots, w_{l-1} \wedge_{j<m} (\sum_{i<n} v_i r_{ij} + \sum_{k<l} w_k s_{kj} = 0),$$

where $r_{ij}, s_{kj} \in R$ for all $i < n$, $j < m$ and $k < l$.

Research supported by a Postdoctoral Fellowship of the Royal Society at University of Manchester, UK. The author thanks Mike Prest for many helpful comments on the subject of this paper.

Denote by T^* the *largest complete theory of modules* [P1, p.43]. If R is regular, there is complete elimination of quantifiers modulo T^*. In fact, any pp-formula is then equivalent to a conjunction of atomic formulas, of the form $\wedge_{j<n}(\sum_{i<n} v_i r_{ij} = 0)$, $r_{ij} \in R$, [P1, Proposition 16.17]. So in this case, all first-order properties are algebraic. Using this fact, we deal with two well-known problems of the model theory of modules, going back to the fundamental paper of Ziegler [Z].

The first problem ([Z, Problem on p.183]) relates properties of the lattice, $\mathcal{L}_M$, of all *pp-definable subgroups* of a module M (= the subgroups of M defined by pp-formulas in one free variable) to the properties of the complete theory, Th(M), of M. The problem is whether for any ring R and any module M, the lattice-theoretic condition "$\mathcal{L}_M$ has width" is equivalent to "Th(M) has continuous part zero" (see §0 and §1 for unexplained terminology). Ziegler proves that the first condition always implies the second, and that the second implies the first provided that R is countable and M arbitrary [Z, Theorem 7.1].

Let T be a complete theory of modules. The second problem concerns a characterization of isolated points of the topological *space of indecomposables*, $\mathcal{I}(T)$. Recall [P1, §4.7] that the points of $\mathcal{I}(T)$ form a representative set of indecomposable pure-injective summands of models of T. In particular, $\mathrm{Zg}_R = \mathcal{I}(T^*)$ is the right *Ziegler spectrum* of R, [P2, p.197], consisting of a representative set of all indecomposable pure-injective modules. W.l.o.g., we assume that $\mathcal{I}(T) \subseteq \mathrm{Zg}_R$. The topology on $\mathcal{I}(T)$ is defined as follows:

If $\varphi(v)$ is a pp-formula in one free variable v, we let $\varphi(N) = \{n \in N \mid N \models \varphi(n)\}$. If $\psi(v)$ is a pp-formula such that $\psi \leq_T \varphi$ (i.e. $\psi(N) \subseteq \varphi(N)$ for each module N which is a model of T), then the pair φ/ψ defines a subset $(\varphi/\psi)_T = \{N \in \mathcal{I}(T) \mid \psi(N) \subset \varphi(N)\}$ of $\mathcal{I}(T)$. The set $\{(\varphi/\psi)_T \mid \psi \leq_T \varphi \,\&\, \neg(\varphi \leq_T \psi)\}$ is a basis of the topology on $\mathcal{I}(T)$.

The fundamental role of the Ziegler spectrum comes from the fact [P1, Theorem 4.67] that closed subsets of Zg_R are in one-one correspondence with complete theories of modules closed under products.

Moreover, the pair φ/ψ is said to be *T-minimal* provided that $\neg(\varphi \leq_T \psi)$, and $\psi \leq_T \tau \leq_T \varphi$ implies $\tau \leq_T \psi$ or $\varphi \leq_T \tau$, for any pp-formula $\tau(v)$. If T^* is the largest complete theory of modules, then the T^*-minimal pairs are called *minimal*. If φ/ψ is a T-minimal pair, then $(\varphi/\psi)_T = \{N\}$ for some $N \in \mathcal{I}(T)$, [P1, p.189]. In other words, any T-minimal pair isolates a point of $\mathcal{I}(T)$. We say that the ring R satisfies the *isolation condition* (= Condition ($\wedge$) in [P1, p.211]) if the converse is also true, i.e. the following assertion holds:

" a module $N \in \mathcal{I}(T)$ is an isolated point of $\mathcal{I}(T)$ iff there is a T-minimal pair φ/ψ of pp-formulas such that $(\varphi/\psi)_T = \{N\}$, for any complete theory of modules T ".

If R satisfies the isolation condition, then there is a close relationship between topological properties of Zg_R and model-theoretic properties of *Mod-R*, [P1, §10.4]. Ziegler proved that the isolation condition holds for any countable ring [Z, Lemma 8.3]. The second problem is whether it holds for an arbitrary ring R.

We consider these problems in the algebraic case, i.e. when R is a regular ring. As mentioned above, it is exactly this case for which there is a complete elimination of quantifiers. So the two problems can be dealt with by algebraic methods.

Using this approach in Section 1, we obtain a full positive answer to the first problem of Ziegler in our setting. That is, we prove that $\mathcal{L}_M$ has width iff Th(M) has continuous part zero, for any regular ring and any $M \in$ Mod-R (Theorem 1.5). The main points of the proof are Lemmas 1.2 and 1.3, showing that semiartinian regular rings can be characterized both as the regular rings such that $\mathcal{L}_M$ has width for any $M \in$ Mod-R, and as the regular rings such that the continuous part of any pure-injective module is zero.

In Section 2, we study uniform modules over regular rings in order to obtain more information about Zg_R. We prove e.g. that $\mathrm{Zg}_R =$ simp-R provided that R is regular and all primitive factors of R are artinian (Theorem 2.2).

In Section 3, we show that the isolation condition can be expressed as a condition on Ziegler spectra of all factor-rings of R (Lemma 3.3). Then we prove that the existence of a regular ring not satisfying the isolation condition is equivalent to the existence of a simple non-artinian regular ring R such that $\mathrm{card}(\mathrm{Zg}_R) = 1$ (Theorem 3.5). However, the latter is an open problem in regular ring theory, with negative answers known in some cases. Using this, we obtain a positive answer to the second problem for several important classes of regular rings: those with all primitive factors artinian, the semiartinian ones, and the ones of cardinality $< 2^\omega$ (Corollary 3.6).

0 PRELIMINARIES

Our basic references are [P1] and [P2] on model theory of modules and [G1] on regular rings. For a ring R, we denote by $\mathcal{E}_R$ the set of all idempotents of R and by Mod-R the category of all (unitary right R-) modules. If $M \in$ Mod-R, then $E(M)$ denotes an injective hull of M. Moreover, simp-R denotes a representative set of all simple modules such that $\mathrm{Zg}_R \supseteq \{E(S) \mid S \in \text{simp-}R\}$. Similarly, classes of left R-modules, R-Mod, and R-simp, are defined. A ring R is a right *V-ring* if each element of simp-R is injective. A module M is *uniform* provided that $N \cap N' \neq 0$ for all non-zero submodules N, N' of M. If $M \in$ Mod-R, then gen(M) denotes the minimal cardinality of a generating set for M.

We start with several basic properties of regular and semiartinian rings which we shall often use in the sequel. Recall that a ring R is (von Neumann) *regular* if $\forall x \in R\, \exists y \in R : xyx = x$. If y can always be taken to be invertible, then R is called *unit regular.* Further, R is *abelian regular* provided that all idempotents of R are central. For example, any commutative regular ring is abelian regular. On the other hand, any abelian regular ring has all of its primitive factors artinian (see [G1, p.60]). A module M is *semiartinian* provided that $\mathrm{Soc}(M/N) \neq 0$ for all $N \subset M$ (or, equivalently, provided that the Loewy (= socle) sequence of M terminates at M). A ring R is right (left) *semiartinian* if it is semiartinian as a module (left R-module). It is well-known that if R is right (left) semiartinian, then each $M \in$ Mod-R ($M \in R$-Mod) is semiartinian.

LEMMA 0.1. *Let R be a regular ring.*

(i) *Let g and f be idempotents of R such that $f \in gR$ ($f \in Rg$). Then there exist orthogonal idempotents $e, e' \in R$ such that $e + e' = g$ and $fR = eR$ ($Rf = Re$). In particular, $eR \oplus e'R = gR$ ($Re \oplus Re' = Rg$), and $e = geg$.*

(ii) Let $g \in R$. Then $gR \in$ Mod-R is indecomposable iff $gR \in$ simp-R iff $Rg \in R$-simp iff $Rg \in R$-Mod is indecomposable.

(iii) The left and the right Loewy sequences of R coincide. In particular, R is right semiartinian iff it is left semiartinian.

(iv) The classes of all pure-injective and all injective modules coincide.

Proof. (i) and (ii) follow from [G1, Theorem 1.1 and Proposition 2.11].

(iii) This follows from (ii) by induction on the Loewy length of R.

(iv) By [P1, Corollary 16.14(b)]. □

For any ring R and any pure-injective module E, denote by E_C and E_D the *continuous* and the *discrete*, respectively, *part* of E (i.e. E_C and E_D are the unique summands of E such that E_C has no indecomposable summands, and E_D is a pure-injective hull of a direct sum of indecomposable summands of E, [P1, Corollary 4.A14]). Recall that given a module M, the complete theory $\text{Th}(M)$ is said to have *continuous part zero* provided that $E_C = 0$ for any pure-injective module E which is a model of $\text{Th}(M)$.

LEMMA 0.2. *Let R be a right semiartinian ring and E be an injective module. Then $E_C = 0$.*

Proof. By the premise, $\text{Soc}(E)$ is an essential submodule of E, whence $E = E(\text{Soc}(E))$. But $\text{Soc}(E) = \oplus\sum_{\alpha<\kappa} S_\alpha$ for a cardinal κ and a sequence $(S_\alpha \mid \alpha < \kappa)$ of simple modules. So $E = E(\oplus\sum_{\alpha<\kappa} E(S_\alpha))$, and $E(S_\alpha)$ is indecomposable for each $\alpha < \kappa$. So $E = E_D$, and $E_C = 0$ by [P1, Corollary 4.A5]. □

In general, Lemma 0.2 does not extend to pure-injective modules. For example, if R is a finite dimensional algebra of wild representation type, then there do exist continuous pure-injectives in Mod-R (see [P2, p.196]).

1 MODULES WITH WIDTH

Let R be a regular ring. Then the set of all finitely generated left ideals of R forms a sublattice, $\mathcal{L}_R$, of the lattice of all left ideals of R, [G1, Theorem 2.3]. For each $I \in \mathcal{L}_R$ there is an $e \in \mathcal{E}_R$ such that $I = Re$ (but e is not unique in general).

Let $M \in$ Mod-R. Let $\mathcal{L}_M$ be the lattice of all pp-definable subgroups of M. By [P1, Proposition 16.17], $\mathcal{L}_M = \{Me \mid e \in \mathcal{E}_R\} = \{MI \mid I \in \mathcal{L}_R\}$, and the lattice operations are given by $Me \vee Mf = Me + Mf = M(Re + Rf)$ and $Me \wedge Mf = Me \cap Mf = M(Re \cap Rf)$ for all $e, f \in \mathcal{E}_R$. Hence, $\varphi : \mathcal{L}_R \to \mathcal{L}_M$ defined by $\varphi(I) = MI$ is a lattice homomorphism. If $M = R$, then $\varphi = id$, so our notation for rings and for modules is consistent. Moreover, φ is an isomorphism iff M is a faithful module iff R embeds into a direct product of copies of M. Since R is regular, [P1, Theorem 16.A(iv) and Corollary 16.19] show that the latter condition implies $\text{Th}(M^{\aleph_0}) = \text{Th}(R^{\aleph_0})$, i.e. that $M^{\aleph_0}$ and $R^{\aleph_0}$ are elementarily equivalent.

If $\cong$ is a congruence on the lattice $\mathcal{L}_M$, denote by $\overline{Me}$ the congruence class containing Me (and $\overline{0} = \overline{M0}$). If $\cong \neq \mathcal{L}_M \times \mathcal{L}_M$ we say that $\cong$ is non-trivial.

If L is a modular lattice, we define a *width of an interval* $[a, b]$ in L as follows (cf. [P1, p.205]):

$\text{w}([a, b]) = 0$ iff $a = b$;

$\text{w}([a, b]) = \alpha$ iff ($\text{w}([a, b])$ is not less than α and, for all $c, d \in L$ with $a \leq c \leq b$ and $a \leq d \leq b$ we have either $\text{w}([c \wedge d, c]) < \alpha$ or $\text{w}([c \wedge d, d]) < \alpha$).

Clearly, $\mathrm{w}([a,b]) = 1$ iff $[a,b]$ is a chain in L. If $\mathrm{w}([a,b]) \neq \alpha$ for all ordinals α, we put $\mathrm{w}([a,b]) = \infty$. Finally, the *width of the lattice* L is defined by $\mathrm{w}(L) = \mathrm{w}([0,1])$. The width of the lattice $\mathcal{L}_M$ will simply be denoted by $\mathrm{w}(M)$.

LEMMA 1.1. *Let R be a regular ring and $M \in$ Mod-R. The following conditions are equivalent:*

(i) $\mathrm{w}(M) = \infty$;

(ii) there is a non-trivial congruence $\cong$ on $\mathcal{L}_M$ such that the partial order induced by lattice operations on $\mathcal{L}_M/\cong$ is dense;

(iii) there is a non-trivial congruence $\cong$ on $\mathcal{L}_M$ such that for each $e \in \mathcal{E}_R$ with $\overline{Me} \neq \overline{0}$ there exist orthogonal idempotents $f, f' \in \mathcal{E}_R$ such that $e = f + f'$ and $\overline{0} \subset \overline{Mf} \subset \overline{Me}$.

Proof. (i) $\Longrightarrow$ (ii): By [P1, Corollary 10.8].

(ii) $\Longrightarrow$ (iii): By the premise, for each $e \in \mathcal{E}_R$ with $\overline{Me} \neq \overline{0}$ there exists $g \in \mathcal{E}_R$ such that $\overline{0} \subset \overline{Mg} \subset \overline{Me}$. Let h be such that $Rh = Rg \cap Re$. Then $\overline{Mh} = \overline{Mg \cap Me} = \overline{Mg}$. The left-hand version of Lemma 0.1(i) implies existence of orthogonal idempotents $f, f' \in \mathcal{E}_R$ satisfying $e = f + f'$ and $Rf = Rh$. Hence, $\overline{0} \subset \overline{Mf} = \overline{Mg} \subset \overline{Me}$.

(iii) $\Longrightarrow$ (i): It suffices to prove that condition (iii) of [P1, Corollary 10.8] holds in $\mathcal{L}_M/\cong$. Let $g, h \in \mathcal{E}_R$ be such that $\overline{Mg} \subset \overline{Mh}$. By the left-hand version of Lemma 0.1(i), there exist orthogonal idempotents $e, e' \in \mathcal{E}_R$ such that $e + e' = h$ and $Rg = Re'$. Note that $Mh = Mg \oplus Me$, so $\overline{Me} \neq \overline{0}$. By the premise, there exist orthogonal idempotents $f, f' \in \mathcal{E}_R$ such that $e = f + f'$ and $\overline{0} \subset \overline{Mf} \subset \overline{Me}$. Put $k = e' + f$ and $k' = e' + f'$. Then $\overline{Mg} = \overline{Mk \cap Mk'} \subset \overline{Mk} \subset \overline{Mk + Mk'} = \overline{Mh}$. □

LEMMA 1.2. *Let R be a regular ring. Then the following conditions are equivalent:*

(i) R is semiartinian;

(ii) $E_C = 0$ for each (pure-) injective module E.

Proof. (i) $\Longrightarrow$ (ii): By Lemmas 0.1(iv) and 0.2.

(ii) $\Longrightarrow$ (i): Let $(S_\alpha \mid \alpha \leq \kappa)$ be the right Loewy sequence of R. By Lemma 0.1(iii) it suffices to show that $S_\kappa = R$. Assume $S_\kappa \neq R$ and put $E = E(R/S_\kappa)$. By the premise, there exist indecomposable injective modules E_β, $\beta < \lambda$ such that $E = E(\oplus \sum_{\beta<\lambda} E_\beta)$. Since $(R/S_\kappa) \cap E_0 \neq 0$, there is some $e \in \mathcal{E}_R$ such that $0 \neq e + S_\kappa \in E_0$. So $E((e+S_\kappa)R) = E_0$ and $(e+S_\kappa)R$ is an indecomposable, and hence simple, right R/S_κ-module, by Lemma 0.1(ii). So $0 \neq \mathrm{Soc}_{R/S_\kappa}(R/S_\kappa) = \mathrm{Soc}(R/S_\kappa)$, a contradiction with S_κ being the last term of the Loewy sequence. □

LEMMA 1.3. *Let R be a regular ring. Then the following conditions are equivalent:*

(i) R is semiartinian;

(ii) $\mathrm{w}(R) < \infty$;

(iii) $\mathrm{w}(M) < \infty$ for each $M \in$ Mod-R.

Proof. (i) $\Longrightarrow$ (ii): We shall prove that the condition (iii) of Lemma 1.1 is not satisfied for $M = R$. Let $\cong$ be a non-trivial congruence on $\mathcal{L}_R$. Take $g \in \mathcal{E}_R$ such that $\overline{Rg} \neq \overline{0}$. By the premise, Rg is a semiartinian left R-module, with Loewy sequence $(L_\alpha \mid \alpha \leq \kappa)$. Let $\beta \leq \kappa$ be the smallest ordinal such that there exists $h \in \mathcal{E}_R \cap L_\beta$ with $\overline{Rh} \neq \overline{0}$. Clearly, β is non-limit, and $\overline{Rx} = \overline{0}$ for all $x \in \mathcal{E}_R \cap L_{\beta-1}$. Since $L_\beta/L_{\beta-1}$ is semisimple, there exists $e \in \mathcal{E}_R \cap L_\beta$ such that $Re + L_{\beta-1}$ is a simple left R-module. Assume the condition (iii) of Lemma 1.1 is satisfied in our setting, i.e. there exist orthogonal idempotents $f, f' \in \mathcal{E}_R$ such that $e = f + f'$ and $\overline{0} \subset \overline{Rf} \subset \overline{Re}$. Since $f \notin L_{\beta-1}$, $Rf + L_{\beta-1}$ is a non-zero submodule of $Re + L_{\beta-1}$. This implies $Rf + L_{\beta-1} = Re + L_{\beta-1}$, and $\overline{Rf} = \overline{Re}$, a contradiction.

(ii) $\Longrightarrow$ (iii): This is immediate by Lemma 1.1, as $\mathcal{L}_M$ is a homomorphic image of $\mathcal{L}_R$.

(iii) $\Longrightarrow$ (i): Let $(S_\alpha \mid \alpha \leq \kappa)$ be the right Loewy sequence of R. Assume $S_\kappa \subset R$ and put $M = R/S_\kappa$. We shall prove that condition (iii) of Lemma 1.1 is satisfied for M and $\cong = id_{\mathcal{L}_M}$. Let $I = \mathrm{Ann}_R M (\supseteq S_\kappa)$. Take any $e \in \mathcal{E}_R$ such that $Me \neq 0$, i.e. $e \notin I$. Note that $S = R/I$ is a regular ring, $M \in$ Mod-S is faithful, and $\mathrm{Soc}_S(M) = \mathrm{Soc}(M) = 0$. In particular, $(e+I)S$ is not indecomposable. By [G, Proposition 2.18], there exist orthogonal idempotents $f, f' \in \mathcal{E}_R$ satisfying $e = f + f'$, $(e+I)S = (f+I)S \oplus (f'+I)S$, $f \notin I$ and $f' \notin I$. It follows that $0 \neq Mf \subset Me = Mf \oplus Mf'$. By Lemma 1.1(iii), $\mathrm{w}(M) = \infty$, a contradiction. □

REMARK 1.4. The width of a modular lattice L can be viewed as a degree of its complexity, or its "dimension". Note that there are other notions of "dimension": the minimal dimension, m-dim, the breadth, br, and the Krull dimension, Kdim (see [P1, pp.205 and 217]). By [P1, Corollary 10.8], $\mathrm{w}(L) < \infty$ iff $\mathrm{br}(L) < \infty$ iff $\mathrm{Kdim}(L) < \infty$. Also, by [P1, Proposition 10.5], $\text{m-dim}(L) < \infty$ implies $\mathrm{w}(L) < \infty$. If R is a regular ring and $L = \mathcal{L}_R$, then Lemma 1.3 shows that the opposite is true: if $\mathrm{w}(R) < \infty$ then R is semiartinian, and hence $\text{m-dim}(\mathcal{L}_R) < \infty$. It follows that $\mathcal{L}_R$ has any of these "dimensions" iff R is semiartinian.

THEOREM 1.5. *Let R be a regular ring and M be a module. Then the following conditions are equivalent:*

(i) $\mathrm{Th}(M)$ *has continuous part zero;*

(ii) $\mathrm{w}(M) < \infty$.

Proof. (i) $\Longrightarrow$ (ii): Put $I = \mathrm{Ann}_R M$ and $S = R/I$. Then $M \in$ Mod-S is faithful. The regularity of S implies that $\mathrm{Th}(M^{\aleph_0}) = \mathrm{Th}(S^{\aleph_0})$, and the models of $\mathrm{Th}(M^{\aleph_0})$ are exactly the faithful right S-modules. Since injective right S-modules are injective as (right R-) modules, (i) and [P1, Corollary 4.40] imply that $E_C = 0$ for each injective right S-module E. By Lemma 1.2, S is semiartinian. By Lemma 1.3, $\mathrm{w}(M_S) < \infty$. Since $\mathcal{L}_{M_S} = \mathcal{L}_M$, we infer that $\mathrm{w}(M) < \infty$.

(ii) $\Longrightarrow$ (i): Let E be a pure-injective module which is a model of (the complete theory) $\mathrm{Th}(M)$. Then $\mathrm{Th}(E) = \mathrm{Th}(M)$, whence $\mathcal{L}_E$ is isomorphic $\mathcal{L}_M$. So $\mathrm{w}(E) < \infty$ and [P1, Theorem 10.9] gives $E_C = 0$. □

2 UNIFORM MODULES

If R is a ring and E is an indecomposable injective module, then exactly one of the following two cases occurs:

(I) $E = E(S)$ for some $S \in$ simp-R,

(II) $E = E(U)$ where U is a cyclic uniform module with $\mathrm{Soc}(U) = 0$.

Of course, a classification for case (I) requires a classification of simp-R. The latter is a hard open problem when R is an arbitrary regular ring. Even the very particular case of full endomorphism rings of infinite dimensional linear spaces is understood only in part (see e.g. [OR], [R1] and [R2]).

For some regular rings, case (II) does not occur. Namely, if R is regular and E is in case (II) with $U = R/I$ for a right ideal I, then it is easy to see that I is essential in R, and I contains the last term (and hence all members) of the Loewy sequence of R. In particular, if R is regular and semiartinian, then $\mathrm{Zg}_R = \{E(S) \mid S \in \text{simp-}R\}$. The same occurs for abelian regular rings, and more generally, for regular rings with all primitive factors artinian:

LEMMA 2.1. *(i) Let R be a ring, I be a right ideal of R and $e \in R \setminus I$ be such that $e^2 = e$. Then $(eR + I) \cap ((1-e)R + I) = I$ iff $eI \subseteq I$.*

(ii) Let R be a regular ring and I be a two-sided ideal of R. Then $U = R/I$ is uniform iff I is a maximal right ideal of R.

(iii) If R is a full matrix ring over an abelian regular ring, then $\mathrm{Zg}_R = \text{simp-}R$.

Proof. (i) Assume $eI \subseteq I$. Let $x \in (eR+I)\cap((1-e)R+I)$, i.e. $x = er + i = (1-e)r' + i'$ for some $r, r' \in R$ and $i, i' \in I$. Then $ex = ei' \in I$ and $(1-e)x = (1-e)i \in I$, so $x \in I$. Conversely, if $ei \notin I$ for some $i \in I$, then $ei = (1-e)(-i)+i \in ((eR+I)\cap((1-e)R+I))\setminus I$.

(ii) By part (i), if U is uniform then $eR + I = R$ for any $e \in R \setminus I$ with $e^2 = e$. So I is a maximal right ideal of R.

(iii) First, assume that R is abelian regular. Then, by (ii), any cyclic uniform module, and hence any uniform module, is simple. So $\mathrm{Zg}_R = \text{simp-}R$. Clearly, the latter equality is preserved by Morita equivalence. □

THEOREM 2.2. *Let R be a regular ring such that all primitive factors of R are artinian. Then $\mathrm{Zg}_R = \text{simp-}R$.*

Proof. Assume there is a cyclic uniform module $U = R/I$ such that U is not simple. Let C_R be the set of all non-zero central idempotents, e, of R such that eR is isomorphic to a full matrix ring over an abelian regular ring. By [G1, Theorem 6.6], $C_R \neq \emptyset$.

If $e \in C_R \setminus I$, then $1 - e \in I$ by Lemma 2.1(i), so $U \simeq eR/(eR \cap I)$. Then U is a cyclic uniform eR-module which is not simple, in contradiction with Lemma 2.1(iii). This proves that $0 \subset J_0 \subset I \subset R$, where J_0 is the ideal of R generated by C_R.

Now, $\bar{R} = R/J_0$ is a regular ring with all primitive factors artinian and U is a cyclic uniform right $\bar{R}$-module which is not simple. If $\bar{R}$ is not a simple ring, we proceed similarly and obtain a proper ideal J_1 generated by $\{e \in R \mid e + J_0 \in C_{\bar{R}}\}$ such that $0 \subset J_0 \subset J_1 \subset I \subset R$. In this way, we get a cardinal κ and a strictly increasing continuous sequence $\{J_\alpha \mid \alpha \leq \kappa\}$ of proper ideals of R such that $R' = R/J_\kappa$ is simple and U is a cyclic uniform right R'-module which is not simple. By the premise, R' is artinian, a contradiction. □

In particular, if R is regular and has all primitive factors artinian, then R is a right V-ring (in fact, by [G1, Proposition 6.18], a regular ring has all primitive factors artinian iff all homogenous semisimple modules are injective). Hence, Theorem 2.2 says that for each right ideal I of R such that I is not maximal there exist two non-empty sets of maximal right ideals, $\mathcal{M}_1$ and $\mathcal{M}_2$, such that $I = \bigcap_{M \in \mathcal{M}_1 \cup \mathcal{M}_2} M$, but $I \neq \bigcap_{M \in \mathcal{M}_i} M$ for $i = 1, 2$. It is an open problem whether this fact (i.e. Theorem 2.2) remains true for any regular right V-ring.

Nevertheless, if some of the primitive factors of R are not artinian, the case (II) may occur:

EXAMPLE 2.3. By [G2, Theorem 3.8], for each ordinal $\alpha > 1$ there exists a prime unit regular ring R and a faithful cyclic module U such that the lattice of all submodules of U is anti-isomorphic to the interval of ordinals $[0, \alpha]$. In particular, if $\alpha \geq \omega$ is a limit ordinal, then U is a uniserial, hence uniform, module such that $\mathrm{Soc}(U) = 0$. This implies that $E = E(U)$ is in case (II).

If R is any regular ring and I is a right ideal of R, then I has many "trivial" decompositions: namely, any finitely generated submodule of I is generated by an idempotent, and hence splits. If these are the only decompositions, I is said to be almost indecomposable:

DEFINITION 2.4. Let R be a ring. A right ideal I is said to be *almost indecomposable* provided that either $\mathrm{gen}(I_0) < \aleph_0$ or $\mathrm{gen}(I_1) < \aleph_0$ whenever $I = I_0 \oplus I_1$.

If R is a regular self-injective ring and $U = R/I$ is a non-simple uniform module, then I must be a large (uncountably generated) almost indecomposable right ideal of R:

THEOREM 2.5. *Let R be a regular right self-injective ring. Let $U = R/I$ be a cyclic uniform module. Then*

(i) I is almost indecomposable.

(ii) If I is projective then U is a simple projective module. In particular, this is the case when $\mathrm{gen}(I) \leq \aleph_0$.

Proof. (i) Assume $I = I_0 \oplus I_1$ and $\mathrm{gen}(I_k) \geq \aleph_0$, $k \leq 1$. By [G1, Lemma 9.7], there exist orthogonal idempotents $e_0, e_1 \in R$ such that $I_k \subset e_k R$, $k \leq 1$. Put $J_0 = e_0 R \oplus I_1$ and $J_1 = e_1 R \oplus I_0$. Then $I \subset J_k$, $k \leq 1$. If $x \in J_0 \cap J_1$, then $x = e_0 r_0 + i_1 = e_1 r_1 + i_0$ for $r_k \in R$, $i_k \in I_k$, $k \leq 1$. So $e_0 r_0 = e_0 x = e_0 i_0 = i_0$, $k \leq 1$, and $x = i_0 + i_1 \in I$. So $J_0 \cap J_1 = I$, in contradiction with the uniformity of U.

(ii) If I is projective then $I = \oplus \sum_{\alpha<\kappa} e_\alpha R$ for some non-zero idempotents $e_\alpha \in R$, $\alpha < \kappa$. By part (i), $\kappa < \aleph_0$, so $\mathrm{gen}(I) < \aleph_0$ and the assertion follows. □

3 ISOLATED POINTS OF ZIEGLER SPECTRA

Let R be a regular ring and T be a complete theory of modules. Denote by A_T the *annihilator* of T, i.e. $A_T = \mathrm{Ann}_R(M)$ for each (some) model M of T, [P1, p.163].

Let $\varphi(v)$ be a pp-formula in one free variable v. By [P1, Proposition 16.17], there is $e \in \mathcal{E}_R$ such that $\varphi(v)$ is equivalent to "$ve = 0$", whence $\varphi(M) = \{m \in M \mid M \models \varphi(m)\} = \{m \in M \mid me = 0\}$ for each module M.

Now, the model-theoretic notions defined in the introduction take a specific form:

LEMMA 3.1. *Let R be a regular ring and T be a complete theory of modules closed under products. Put $\bar{R} = R/A_T$. Then $T = \mathrm{Th}(\bar{R}^{\aleph_0})$ and $\mathcal{I}(T) = \mathrm{Zg}_R \cap$Mod-$\bar{R} = \mathrm{Zg}_{\bar{R}}$.*

Proof. Clearly, any model, M, of T is a faithful right $\bar{R}$-module. In particular, $\bar{R}$ embeds into a product of copies of M. Since $\bar{R}$ is regular, we infer that $T = \mathrm{Th}(\bar{R}^{\aleph_0})$, and the models of T are exactly the faithful right $\bar{R}$-modules. It follows that $\mathcal{I}(T) = \mathrm{Zg}_R \cap$Mod-$\bar{R}$. Since any right ideal of R is generated by idempotents, all injective right $\bar{R}$-modules are injective as (right R-)modules. So $\mathrm{Zg}_R \cap$Mod-$\bar{R} = \mathrm{Zg}_{\bar{R}}$. □

LEMMA 3.2. *Let R be a regular ring, T be a complete theory of modules, and $N \in \mathcal{I}(T)$. Put $\bar{R} = R/A_T$. Then*

(i) *N is an isolated point of $\mathcal{I}(T)$ iff there is $\bar{g} \in \mathcal{E}_{\bar{R}}$ such that $\{N\} = \{M \in \mathcal{I}(T) \mid M\bar{g} \neq 0\}$;*

(ii) *N is isolated by a T-minimal pair iff there is $\bar{g} \in \mathcal{E}_{\bar{R}}$ such that $\{N\} = \{M \in \mathcal{I}(T) \mid M\bar{g} \neq 0\}$ and $\bar{g}\bar{R}$ is an indecomposable right $\bar{R}$-module.*

Proof. Let $\varphi(v)$ and $\psi(v)$ be pp-formulas and $e, f' \in \mathcal{E}_R$ be such that $\varphi(v)$ and $\psi(v)$ is equivalent to "$ve = 0$" and "$vf' = 0$", respectively. Assume that $\varphi \leq_T \psi$. Then $eR + f'R = eR \oplus e'R$ for some $e' \in \mathcal{E}_R$ with $ee' = e'e = 0$, [G1, Proposition 2.13]. By the assumption, $e' \in A_T$, so $\varphi(v)$ is equivalent to "$vf = 0$" modulo T, where $f = e + e'$. Then $fR = f'R \oplus gR$, where $g \in \mathcal{E}_R$ and $f'g = gf' = 0$, [G1, Proposition 2.11]. Clearly, $\psi \leq_T \varphi$ iff $g \in A_T$. If $g \notin A_T$, put $\bar{g} = g + \bar{R}$. Then $(\varphi/\psi)_T = \{M \in \mathcal{I}(T) \mid M\bar{g} \neq 0\}$ and the characterization of isolated points of $\mathcal{I}(T)$ follows. Moreover, if ψ/φ is a T-minimal pair then $\bar{g}\bar{R}$ is an indecomposable right $\bar{R}$-module, and the assertion follows. □

Also the isolation condition can be expressed in a simpler way:

LEMMA 3.3. *Let R be a regular ring. Then the following are equivalent:*

(i) *R satisfies the isolation condition;*

(ii) *If N is an isolated point of $\mathrm{Zg}_{\bar{R}}$, then there is a minimal pair of pp-formulas isolating N, for any factor-ring $\bar{R}$ of R.*

Proof. By Lemmas 3.1 and 3.2(i), the isolated points of $\mathrm{Zg}_{\bar{R}}$ coincide with those of $\mathcal{I}(T)$, where $T = \mathrm{Th}(\bar{R}^{\aleph_0})$. By Lemma 3.2(ii), the points isolated by minimal pairs of pp-formulas of the language of $\bar{R}$-modules coincide with those isolated by T-minimal pairs of pp-formulas of the language of R-modules. □

Now, we show that any isolated point $N \in \mathcal{I}(T)$ is in case (I) of §2, hence a hull of a simple module:

LEMMA 3.4. *Let R be a regular ring, T be a complete theory of modules, and $N \in \mathcal{I}(T)$ be isolated. Then there are $g \in \mathcal{E}_R$ and $S \in$ simp-(R/A_T) such that $N = E(S)$, $Sg \neq 0$, and $\{N\} = \{M \in \mathcal{I}(T) \mid Mg \neq 0\}$.*

Proof. Put $\bar{R} = R/A_T$. Assume $N \in \mathcal{I}(T)$ is isolated, but not a hull of a simple right $\bar{R}$-module. Let $g \in \mathcal{E}_R \setminus A_T$ be such that $\{N\} = \{M \in \mathcal{I}(T) \mid Mg \neq 0\}$. Then $E(S)g = 0$ for each $S \in$ simp-$\bar{R}$, whence $g \in \bigcap_{S \in \text{simp-}\bar{R}} \mathrm{Ann}_R(S) = \mathrm{Rad}(\bar{R}) = A_T$, a contradiction. So $N = E(S)$ for some $S \in$ simp-$\bar{R}$, and similarly, $Sg \neq 0$. □

THEOREM 3.5. *The following are equivalent:*

(i) *There exists a regular ring, R, which does not satisfy the isolation condition;*

(ii) *There is a simple non-artinian regular ring, R', such that $\mathrm{card}(\mathrm{Zg}_{R'}) = 1$.*

Proof. (i) $\Longrightarrow$ (ii): Possibly replacing R by a factor-ring by Lemma 3.3, we can assume that there exists an isolated point $N \in \mathrm{Zg}_R$ such that there is no minimal pair of pp-formulas isolating N. Let $g \in \mathcal{E}_R$ be such that $\{N\} = \{M \in \mathrm{Zg}_R \mid Mg \neq 0\}$. Our premise implies that gR does not have an indecomposable decomposition. Moreover, by Lemma 3.4, there is $S \in$ simp-R such that $N = E(S)$ and $Sg \neq 0$.

Put $R' = gRg$. By [G1, Theorem 1.7], R' is a regular ring. Since gR has no indecomposable decomposition, R' is not artinian. Consider $0 \neq S' = Sg \in$ Mod-R'. If $0 \neq X'$ is a right R'-submodule of S', $0 \neq x' \in X'$ and $s \in S$, then there is $r \in R$ such that $sg = x'rg = x'grg \in X'$. So $X' = S'$ and S' is a simple right R'-module.

Let $N' \in \mathrm{Zg}_{R'}$. Then $N' \simeq E_{R'}(R'/I')$, where I' is a right ideal of R' such that $0 \neq R'/I' \in$ Mod-R' is uniform. Let I be a right ideal of R generated by I'. Then $I = I'R$ and $I' = Ig$. Thus, $I \subset gR$.

We shall prove that gR/I is a uniform module: assume $(e+I)R \cap (f+I)R = 0$ for some $e, f \in \mathcal{E}_R \cap gR$. By Lemma 0.1(i), there exist idempotents $e' = ge'g \in R'$ and $f' = gf'g \in R'$ such that $eR = e'R$ and $fR = f'R$. So $(e'+I)R \cap (f'+I)R = 0$. Then $(e'+I')R' \cap (f'+I')R' = 0$, whence either $e' \in I'$ or $f' \in I'$. So either $eR = e'R \subseteq I$ or $fR = f'R \subseteq I$. This proves that gR/I is uniform.

So $E(gR/I)$ is isomorphic to an element of Zg_R and $E(gR/I)g \neq 0$. This means that $E(gR/I) \simeq E(S)$, whence there is a right ideal J of R such that $I \subset J \subseteq gR$ and $J/I \simeq S$. Then $S' = Sg \simeq (Jg+I)/I \simeq Jg/Ig = Jg/I' \subseteq R'/I'$ (in Mod-R'). This gives $N' = E_{R'}(S')$. Since N' was an arbitrary element of $\mathrm{Zg}_{R'}$, we get $\mathrm{card}(\mathrm{Zg}_{R'}) = \mathrm{card}(\text{simp-}R') = 1$.

Let $0 \neq L' \neq R'$ be a two-sided ideal of R', and P' be a maximal right ideal of R' containing L'. Then $R'/P' \simeq S'$ and $L' \subseteq \mathrm{Ann}_{R'}(S')$. Since $\mathrm{Rad}(R) = 0$, there is a maximal right ideal Q' of R' not containing L'. Then $L' \not\subseteq \mathrm{Ann}_{R'}(R'/Q') = \mathrm{Ann}_{R'}(S')$, a contradiction. This proves that R' is a simple ring.

(ii) $\implies$ (i): We have $\mathrm{Zg}_R = \{E(S)\}$, where $\{S\} =$ simp-R. Trivially, $E(S)$ is an isolated point of Zg_R. Since $\mathrm{Soc}(R) = 0$, Lemma 0.1(ii) implies that gR is not indecomposable for any $0 \neq g \in \mathcal{E}_R$. In particular, there is no minimal pair of pp-formulas isolating $E(S)$, and the isolation condition does not hold. □

Let T be a regular ring. Recall that any ring $R = eTe$ where $0 \neq e \in T$ is an idempotent, is called a *corner* of T. The proof of Theorem 3.5 actually shows that if T does not satisfy the isolation condition, then some factor-ring of T has a corner, R, such that $\mathrm{card}(\mathrm{Zg}_R) = 1$.

COROLLARY 3.6. *Let R be a regular ring. Then R satisfies the isolation condition provided that either*

(i) *all primitive factors of R are artinian (e.g., R is abelian regular);* or

(ii) *R is semiartinian;* or

(iii) $\mathrm{card}(R) < 2^\omega$.

Proof. Assume R is a regular ring which does not satisfy the isolation condition. Let $\bar{R}$ be a factor-ring of R such that there is an isolated point of $\mathrm{Zg}_{\bar{R}}$ which is not isolated by a minimal pair (see Lemma 3.3). Let $R' = gRg$ be the corner of $\bar{R}$ as in the proof of Theorem 3.5.

(i) Clearly, all primitive factors of $\bar{R}$ are artinian, and $R' \simeq \mathrm{End}_R(gR)$ is a simple regular ring with the same property ([G1, Corollary 6.4]). Then R' is artinian, a contradiction.

(ii) Clearly, $\bar{R}$ is semiartinian, and Lemma 0.1(i) implies $R' = gRg$ has a non-zero socle, for gR contains a minimal right ideal. So R' is not simple, a contradiction.

(iii) Clearly, $\mathrm{card}(R') < 2^\omega$. By induction on $n < \omega$, we define complete sets, $E_n = \{g_{\bar{a}} \mid \bar{a} = (a_0, \ldots, a_{n-1}) \in \{0,1\}^n\}$, of orthogonal idempotents of R' as follows: $E_0 = \{g\}$.

Since R' is a simple regular ring, for each $\bar{a} = (a_0, \ldots, a_{n-1}) \in \{0,1\}^n$ there exist non-zero orthogonal idempotents $e, f \in R'$ such that $g_{\bar{a}} = e + f$. Put $\bar{b} = (a_0, \ldots, a_{n-1}, 0)$, $\bar{c} = (a_0, \ldots, a_{n-1}, 1)$, $g_{\bar{b}} = e$ and $g_{\bar{c}} = f$. Since E_n is a complete set of orthogonal idempotents of R', so is $E_{n+1} = \{g_{\bar{d}} \mid \bar{d} = (d_0, \ldots, d_n) \in \{0,1\}^{n+1}\}$. For each $x \in \{0,1\}^\omega$, define a sequence $(\overline{a_i} \mid i < \omega)$ as follows: $\overline{a_0} = (x_0 \oplus 1)$ and $\overline{a_i} = (x_0, \ldots, x_{i-1}, x_i \oplus 1) \in \{0,1\}^{i+1}$ for all $i < \omega$. Put $I_x = \oplus \sum_{i<\omega} g_{\overline{a_i}} R'$. Since $g_{(x_0)} \notin I_x$, there is a maximal right ideal, M_x, of R' containing I_x. By Theorem 3.5, $S' \simeq R'/M_x$. Moreover, take $y \in \{0,1\}^\omega$ such that $y \neq x$. Then $I_x + I_y = R'$, whence $M_x + M_y = R'$. Take $s'_x \in S'$ such that $M_x = \mathrm{Ann}_{R'}(s'_x)$. Then $s'_x \neq s'_y$. This means that $\mathrm{card}(S') \geq 2^\omega$, in contradiction with $\mathrm{card}(R') < 2^\omega$. □

The case (i) and (iii) generalizes the commutative and the countable case, respectively, considered in [P1, p.328] and in [Z, Lemma 8.3]. Also, we can use §1 to provide an alternative proof in the case (ii): if T^* is the largest complete theory of modules, then Lemma 1.2 shows that T^* has continuous part zero, and [P1, Theorem 10.16] implies that R satisfies the isolation condition.

References

[EM] P.C.Eklof and A.H.Mekler, *Almost Free Modules*, North-Holland, New York, 1990.

[G1] K.R.Goodearl, *Von Neumann Regular Rings*, Pitman, London, 1979; *Second Ed.*, Krieger, Melbourne, FL, 1991.

[G2] K.R.Goodearl, *Artinian and noetherian modules over regular rings*, Comm. Algebra **8** (1980), 477-504.

[OR] A.Orsatti and N.Rodinò, *On the endomorphism ring of an infinite dimensional vector space*, Proc. Conf. "Abelian Groups and Modules" (Padova '94), Kluwer, Dortdrecht, 1995.

[P1] M.Prest, *Model Theory and Modules*, Cambridge Univ.Press, Cambridge, 1988.

[P2] M.Prest, *Remarks on elementary duality*, Ann. Pure Appl. Logic **62** (1993), 183-205.

[R1] N.Rodinò, *Cyclic modules over* End(V_D) *whose endomorphism ring is an ultrapower of D* (to appear).

[R2] N.Rodinò, *Simple modules over* End(V_D) *with prescribed annihilator* (to appear).

[Z] M.Ziegler, *Model theory of modules*, Ann. Pure Appl. Logic **26** (1984), 149-213.

Katedra algebry MFF UK, Sokolovská 83, 186 00 Praha 8, Česká Republika

E-mail address: trlifaj@karlin.mff.cuni.cz

On Weakly Injective Continuous Modules

DINH VAN HUYNH, Institute of Mathematics, P.O.Box 631, Hanoi, Vietnam

S. K. JAIN, S. R. LÓPEZ-PERMOUTH, Department of Mathematics, Ohio University, Athens, OH 45701, U.S.A.

Abstract. We study several properties of weakly injective modules. In particular, we show that, while every finitely generated weakly injective continuous module is injective, this implication does not necessarily hold for infinitely generated modules. On the other hand, weakly injective quasi-continuous modules do behave better than arbitrary weakly injective modules in the sense that any direct summand of a weakly injective quasi-continuous module is again weakly injective.

The question of when the domain of weak injectivity of certain modules is closed under direct sums and under essential extensions is also considered. This study yields characterizations of quasi-injective modules and of right SI-rings in terms of weak injectivity.

1. INTRODUCTION

The concept of weak relative injectivity of modules was introduced originally in [JL1] to study semiperfect rings for which every cyclic module is isomorphic to an essential submodule of a projective module. Since then, the study of this concept has been furthered extensively, a convenient reference for background material on the subject is [JL2]. Among other applications, weakly injective modules have served to study semiprime Goldie rings [LRY], [Z], QI rings [JLS], [LR], direct sums of matrix rings over local QF rings [JLOS] and other rings [L].

Weakly injective modules are closed under finite direct sums but not necessarily under direct summands. Similarily, essential extensions of weakly injective modules are again weakly injective but, certainly, essential submodules of weakly injective modules need not be weakly injective. Finally, while the domain of weak injectivity of a module is closed under quotient modules and submodules, it is not closed under direct sums or essential extensions. In this paper we study when the converses of the above results also hold.

It is known that a finitely generated weakly injective continuous module is indeed injective. In this paper we provide examples to show that this implication is not true in general when M is not finitely generated. Then we study rings for which the implication does hold. Rings satisfying the condition that weakly injective continuous modules are injective seem to be abundant and hard to characterize in general. We show examples of such rings which do not satisfy any obvious chain conditions as well as an example of a noetherian ring having non-injective weakly injective continuous modules.

Throughout this paper, all modules are right unital modules unless otherwise stated. For a module M, either notation $\hat{M}$ or $E(M)$ serves to denote the injective hull of M. For any other notation or concept not defined here the reader might refer to [AF] or [F].

2. WEAKLY SELF-INJECTIVE MODULES

Given two modules M and N, we say that M is weakly N-injective if for every homomorphism $\varphi : N \to E(M)$ there exists a submodule $X \subset E(M)$ which is isomorphic to M such that $\varphi(N) \subset X$. If the module M is weakly N-injective for every finitely generated module N, we say that M is weakly-injective. Additionally, in this paper we will be concerned with the following

2.1. Definition. *A module M is said to be weakly self-injective if it is weakly M-injective.*

A module M is weakly N-injective if and only if it is weakly K-injective for every submodule K of N if and only if it is weakly N/K-injective for every submodule K of N. On the other hand, the ring R of upper triangular 2×2 matrices over a field is weakly R-injective but not weakly R^2-injective. Therefore, the domain of weak injectivity of a module is not necessarily closed under (finite) direct sums. The following two propositions pertain when the domain of weak injectivity of a weakly self-injective module is closed under direct sums.

2.2. Proposition. *The following conditions are equivalent for a weakly self-injective module M:*

(1) *for any modules N and K, if M is weakly N-injective and weakly K-injective then M is weakly $N \oplus K$-injective;*
(2) *M is weakly M^n-injective for all $n \in \mathbf{Z}^+$;*
(3) *M is weakly M^2-injective.*

Proof. One needs only to show (3) $\Rightarrow$ (1). Let $\varphi : N \oplus K \to E(M)$. Considering the restrictions of φ to N and K respectively one obtains submodules X_1 and X_2

of $E(M)$, both isomorphic to M such that $\varphi(N) \subset X_1$ and $\varphi(K) \subset X_2$. From the fact that M is weakly M^2-injective one obtains that there exists a submodule $X \subset E(M)$, isomorphic to M such that $X_1 + X_2 \subset X$. It follows that $\varphi(N \oplus K) = \varphi(N) + \varphi(K) \subset X$, proving that M is indeed weakly $N \oplus K$-injective. □

2.3. Proposition. *The following conditions are equivalent for a module M:*

(1) *M is quasi-injective;*
(2) *M is weakly $M^{(\alpha)}$-injective for every cardinal α;*
(3) *M is weakly $M^{(\alpha)}$-injective for $\alpha = |End(\hat{M})|$.*

Proof. Obviously, (1) ⇒ (2) and (2) ⇒ (3). It is only left to show that (3) ⇒ (1). Consider $q(M) = End(\hat{M})(M)$, the quasi-injective hull of M. Write $q(M) = \Sigma_{\varphi \in End(\hat{M})} \varphi(M)$. Then $q(M) = \Phi(M^{(\alpha)})$ where $\Phi : M^{(\alpha)} \to \hat{M}$ is given by $\Phi((m_\varphi)_{\varphi \in End(\hat{M})}) = \Sigma_{\varphi \in End(\hat{M})} \varphi(m_\varphi)$. Since M is weakly $M^{(\alpha)}$-injective, there exists $X \subset \hat{M}$, isomorphic to M such that $q(M) \subset X$. But $X \subset q(M)$ therefore $q(M) = X \cong M$. Hence M is quasi-injective, as claimed. □

As a consequence of the preceeding theorem one gets that

(1) a ring R is such that every module is (weakly) N-injective whenever N is singular if and only if
(2) every singular module is semisimple.

The implicaton (2) ⇒ (1) is trivial. For the injective case of (1) ⇒ (2) notice that given a singular module N and a submodule T of N, since T is N-injective, $T \subset^{\oplus} N$. For the weakly injective case, notice that every singular module N is quasi-injective, by Proposition 2.3, and then proceed as in the injective case.

In the next two propositions we deal with the question of when the domains of weak injectivity of certain modules are closed under essential extensions.

2.4. Proposition. *The following conditions are equivalent for a ring R:*

(1) *for every right module M, the domain of weak injectivity of M is closed under essential extensions;*
(2) *every weakly self-injective module is injective;*
(3) *R is semisimple artinian.*

Proof. Obviously (3) ⇒ (1). Also, if one assumes (1) then every weakly self-injective module M is weakly $E(M)$-injective and hence injective. It only rests to prove that (2) ⇒ (3). Let R be a ring satisfying condition (2). It follows that every completely reducible module must be injective. Hence, R is a right noetherian right V-ring. Assume that R is not semisimple. Then there exists a right ideal $K \subset R$ maximal with respect to the property that $U = R/K$ is not semisimple. It follows that every quotient U/V of U with $V \neq 0$ is semisimple. Now, let X be a nonzero cyclic submodule of U. Then every homomorphism $\varphi : X \to E(X)$ either is a monomorphism or satisfies that $\varphi(X) \subset \text{soc}X$. Hence X is weakly self-injective and so injective by (2). That makes U semisimple, a contradiction. Therefore, R is semisimple artinian. □

Recall that a ring R is called right SI if every singular right R-module is injective [G]. One may use the arguments above to prove the following equivalences.

2.5. Proposition. *The following conditions are equivalent for a ring R:*

(1) *for every singular right module M, the domain of weak injectivity of M is closed under essential extensions;*
(2) *every singular weakly self-injective module is injective;*
(3) *R is a right SI-ring.*

Proof. Clearly (3) $\Rightarrow$ (1) and (1) $\Rightarrow$ (2). If R satisfies condition (2) then every singular semisimple module is injective. We will show that if (2) holds then every singular cyclic module is semisimple and hence injective and thus by [G] R is right SI. In order to get the argument to go through we need to start by showing that under condition (2) every cyclic singular right module is noetherian and then proceed as in the proof of the previous Proposition . Let $M = R/I$ be singular. If M is not noetherian one may consider a strictly ascending chain of submodules of M of the form

$$x_1R/I \subset (x_1R + x_2R)/I \subset \cdots .$$

Let

$$X/I = \bigcup_{k=1}^{\infty}(x_1R + \cdots + x_kR)/I.$$

Let M_1/I be a maximal submodule of x_1R/I. Since x_1R/M_1 is simple and singular, it is injective and therefore $x_1R/M_1 \subset^{\oplus} X/M_1$ and $x_1R/M_1 \subset^{\oplus} (x_1R{+}x_2R)/M_1 = x_1R/M_1 \oplus y_2R/M_1$, for some $y_2 \in (x_1R + x_2R)$. Take M_2/M_1 to be a maximal submodule of y_2R/M_1. Then $(x_1R{+}x_2R{+}x_3R)/M_2 \cong x_1R/M_1 \oplus y_2R/M_2 \oplus y_3R/M_2$ for some $y_3 \in x_1R + x_2R + x_3R$. Continuing this way we can produce for each $i = 2, 3, \ldots$ a submodule M_i maximal in y_iR/M_{i-1}. Let $M^* = \bigcup_{i=1}^{\infty} M_i$. Then X/M^* has a submodule N which is an infinite direct sum of singular simple modules. By (2), N is injective. But then N is a direct summand of the cyclic module M/M^*, a contradiction. Therefore M is noetherian. □

At this point, we would like to mention some related results recently obtained in [HR]:

(1) A ring R is semisimple artinian if and only if every continuous right R-module is injective.
(2) A ring R is right SI if and only if every singular continuous right R-module satisfying the restricted semisimple condition is injective.

Combining the argument in [HR] and that presented here we obtain the following conclusion:

(3) A ring R is right SI if and only if every singular continuous weakly self-injective module satisfying the restricted semisimple condition is injective.

3. WEAK INJECTIVITY AND CONTINUITY

Every weakly injective quasi-injective module is injective. Also, if a finitely generated module M is weakly injective and continuous then M is indeed injective [JL, Examples 1.11]. One could conjecture that weakly injective continuous modules in general are injective. We will show next that this is not necessarily the case even when the ring has very nice structure.

3.1. Proposition. *A right and left PCI domain R which is not a division ring has a weakly injective continuous module which is not injective.*

Proof. Let R be a right and left PCI domain which is not a division ring [C], [CF]. Since R is right and left noetherian, each submodule of $E(R_R)$ is weakly injective [LRY]. Let M be a proper submodule of $E(R_R)$ such that $E(R_R)/M$ is noetherian. Since R is not a division ring, $M \neq 0$. By [HR, Lemma 1], M is continuous. Hence M is a weakly injective continuous non-injective right R-module. □

On the other hand, weakly injective quasi-continuous modules do show some injective-like behaviour. While it is not true in general that direct summands of weakly K-injective modules is weakly K-injective, we observe the following

3.2. Theorem. *Let M be a weakly K-injective quasi-continuous module. Then every direct summand of M is weakly K-injective.*

Proof. Let $M = N \oplus T$ be a weakly K-injective quasi-continuous right R-module. Consider a map $\varphi : K \to E(N)$. Let $K_1 = \varphi(K) \cap N$. Notice that $K_1 \subset' \varphi(K)$. Since N is quasi-continuous, there exists a summand $K^* \subset^{\oplus} N$ with $K_1 \subset' K^*$. So, let $N = K^* \oplus H$. On the other hand, since M is weakly K-injective $\varphi(K) \subset M'$ for some submodule M' of $E(M)$ isomorphic to M. It follows by the quasi-continuity of M that $\varphi(K) \subset' K' \subset^{\oplus} M'$. Let $\varrho : M' \to M$ be an isomorphism. Then $\varrho(K') \subset^{\oplus} M$. Notice that $E(K^*) \cong E(\varrho(K'))$. It follows from the quasi-continuity of M that $K^* \cong \varrho(K') \cong K'$. The identity map $i : \varphi(K) \to \varphi(K)$ extends to an embedding $i^* : K' \to E(\varphi(K)) \subset E(N)$. Since $K_1 \subset' \varphi(K) \subset' i^*(K')$, we get $i^*(K') \cap H = 0$. Thus $\varphi(K) \subset i^*(K') \oplus H \cong K^* \oplus H = N$, proving that N is weakly K-injective. □

3.3. Corollary. *Every direct summand of a weakly injective quasi-continuous module is weakly injective.*

It seems natural next to study rings for which every weakly injective continuous module is injective. Proposition 3.1 illustrates that being noetherian is not sufficient to guarantee for this property to hold. The characterization of such rings appears to be quite complicated. We show next that no reasonable chain condition seems to be necessary either by presenting several examples of rings for which every weakly injective continuous module is injective.

3.4. Example. *Let R be any commutative noetherian ring. Then every weakly injective continuous module is injective.*

Proof. By [R], if R is commutative and noetherian, every continuous module is quasi-injective. Quasi-injective weakly injective modules are always injective. □

3.5. Example. *The ring*

$$R = \{\begin{pmatrix} a & b \\ 0 & a \end{pmatrix} | a \in End_{\mathbf{Z}}(\mathbf{Z}(p^\infty)), b \in \mathbf{Z}(p^\infty)\}$$

is a non-noetherian ring for which every weakly injective continuous right module is injective.

Proof. As seen in Example 24.34 of [F], R is a commutative self-injective uniserial ring with

$$R/Z(p^\infty) \cong End_{\mathbf{Z}}(\mathbf{Z}(p^\infty)),$$

and $End_{\mathbf{Z}}(\mathbf{Z}(p^\infty))$ is isomorphic to the ring of p-adic integers which is commutative and noetherian. Let M be a weakly-injective continuous R-module. Then $M = A \oplus B$ where A has essential socle and $soc(B) = 0$. By Corollary 3.3, A and B are weakly injective. It is easy to see that $\mathbf{Z}(p^\infty) \subset ann_R(B)$ and therefore B is a weakly injective continuous $R/\mathbf{Z}(p^\infty)$-module. By [R], B is a quasi-injective $R/\mathbf{Z}(p^\infty)$-module. It follows that B is an injective $R/\mathbf{Z}(p^\infty)$-module and so B_R is quasi-injective. Since B_R is weakly injective, B is injective. Now, let A_1 be a uniform direct summand of A, say $A = A_1 \oplus A_2$. Then A_1 and A_2 are weakly injective and mutually injective. Since $E(A_1) \cong R_R$, we see that A_1 is isomorphic to an ideal of R. But no proper ideal of R_R may be weakly injective, therefore $A_1 \cong R_R$. Hence, A_1 and A_2 are injective. In conclusion, M is injective, as claimed. □

Rings whose cyclic right modules have finite uniform dimension (called right qfd rings) sometimes play a role similar to that of noetherian rings when studying weakly injective modules [AJL1]. For that reason, while being right noetherian is not necessary for all weakly injective continuous right modules to be injective one might expect that being right qfd is. Our next example rules out this implication.

3.6. Example. *Any commutative semiprimary ring with* $J^2 = 0$ *(where* J *is the Jacobson radical of* R*) has the property that every weakly injective continuous module is injective.*

Proof. If R is a commutative semiprimary ring with $J^2 = 0$ then every continuous module is quasi-injective [R]. Consequently, every weakly injective continuous R-module is injective. □

One could expect that the injective-like behaviour of weakly injective continuous modules might be to the extent that the direct sum of any two such modules would once again be continuous. That this is not the case is the purpose of our next proposition.

3.7. Proposition. *A ring R is such that every weakly injective continuous module is injective if and only if any finite direct sum of weakly injective continuous modules is continuous.*

Proof. Necessity is clear.Conversely, consider a weakly injective continuous module M. If M^2 were continuous then M would be quasi-injective and hence injective. □

Let us refer to a ring R as being right Σ-(weakly) injective if for any set $\mathcal{A}$, the R-module $R_R^{(\mathcal{A})}$ is (weakly) injective. By [F, Proposition 20.3A], a right Σ-injective ring is quasi-Frobenius.

3.8. Remark. *There is a commutative right Σ-weakly injective semiperfect ring which is not semiprimary and not noetherian.*

Proof. Let R be a ring of Example 3.5. Since R is right q.f.d., arbitrary direct sums of weakly injective modules are weakly injective [AJL1]. In particular, R is right Σ-weakly injective. Furthermore, we notice that $R^{(\omega)}$ is not continuous, indeed not even CS, because otherwise R would be quasi-Frobenius by [H, Corollary 2]. □

We conclude with a positive result.

3.9. Proposition. *A left perfect right Σ-weakly injective ring is quasi-Frobenius.*

Proof. Let R be left perfect and right Σ-weakly injective. By [AJL2, Theorem 3.3], R is indeed right self-injective. Consider any cyclic right R-module X. Since R is left perfect, $\text{soc}(X) \subset' X$ and there exists a set $\mathcal{A}$ such that $\text{soc}(X)$ embeds in $R^{(\mathcal{A})}$. Extend the embedding to one of X into $E(R^{(\mathcal{A})})$. Since $R^{(\mathcal{A})}$ is weakly injective we get that X indeed embeds in $R^{(\mathcal{A})}$ hence in R^n, for some $n \in \mathbf{Z}^+$. But then X has finite uniform dimension. Since the cyclic module X was chosen arbitrarily, we have shown that every cyclic module has an essential and finitely generated socle. It follows that R is right artinian (cf. [F, Corollary 19.16B]) and thus quasi-Frobenius. □

Acknowledgements: Dinh Van Huynh gratefully acknowledges the support of the Department of Mathematics, Ohio University, Athens, Ohio. S.R. López-Permouth wishes to thank the Department of Mathematics of Universidad del Valle de Guatemala for their support and hospitality during his sabbatical visit.

References

[AJL1] A.H. Al-Huzali, S.K. Jain and S.R. López-Permouth, *Rings whose cyclics have finite Goldie dimension*, J. Alg. **153** (1992), 37-40.

[AJL2] A.H. Al-Huzali, S.K. Jain and S.R. López-Permouth, *On the weak relative injectivity of rings and modules*, Non-commutative ring theory, Springer-Verlag, Berlin, Heidelberg, New York, 1990, pp. 93-98.

[AF] F.W. Anderson and K.R. Fuller, *Rings and Categories of Modules, 2nd. Ed.*, Springer-Verlag, New York, 1992.

[C] J.H. Cozzens, *Homological properties of the ring of differential polynomials*, Bull. Amer. Math. Soc. **76** (1970), 75-79.

[CF] J.H. Cozzens and C.F. Faith, *Simple Noetherian Rings*, Cambridge University Press, London, New York, 1975.

[F] C.F. Faith, *Algebra II, Ring Theory*, Springer-Verlag, Berlin, Heidelberg, New York, 1976.

[G] K.R. Goodearl, *Singular torsion and the splitting properties*, Memoirs Amer. Math. Soc. **124** (1972).

[H] D.V. Huynh, *A right countably sigma-CS ring with ACC or DCC on projective principal right ideals is left artinian and QF-3*, Trans. Amer. Math. Soc., to appear (1995).

[HR] D.V. Huynh and S.T. Rizvi, *An approach to Boyle's conjecture, to appear*, Proc. Edinburgh Math. Soc..

[JL1] S.K. Jain and S.R. López-Permouth, *Rings whose cyclics are essentially embeddable in projective modules*, J. Algebra **128** (1990), 257-269.

[JL2] S.K. Jain and S.R. López-Permouth, *A Survey on the theory of weakly injective modules*, Computational Algebra, Marcel Dekker, New York, Basel, Hong Kong, 1994, pp. 205-232.

[JLOS] S.K. Jain, S.R. López-Permouth, K. Oshiro and M.A. Saleh, *Weakly projective and weakly injective modules*, Canadian J. Math **46** (1994), 971-981.

[JLS] S.K. Jain, S.R. López-Permouth and S. Singh, *On a class of QI-rings*, Glasgow Math. J. **34** (1992), 75-81.

[L] S.R. López-Permouth, *Rings characterized by their weakly injective modules*, Glasgow Math. J. **34** (1992), 349-353.

[LR] S.R. López-Permouth and S.T. Rizvi, *On certain classes of QI-rings*, Methods in Module Theory, Marcel Dekker, New York, Basel, Hong Kong, 1992, pp. 227-235.

[LRY] S.R. López-Permouth, S.T. Rizvi and M.F. Yousif, *Some characterizations of semiprime Goldie rings*, Glasgow Math. J. **35** (1993), 357-365.

[R] S. Tariq Rizvi, *Commutative rings for which every continuous module is quasi-injective*, Arch. Math. **50** (1988), 435-442.

[Z] Yiqiang Zhou, *Strongly compressible modules and semiprime right Goldie rings*, Comm. Alg. **21** (1993), 687-698.

A $\diamond$-Characterization of Some Small Totally Projective Groups

JANE KIRCHNER WEST [1] Department of Mathematics, Auburn University, Alabama

1 INTRODUCTION

Whether two subgroups of an abelian group are equivalent, i.e., whether there is an automorphism of the group carrying one subgroup onto another, is an interesting structural question which has been investigated on numerous occasions. For it not only refers to the isomorphism of two subgroups, but in a sense how they are positioned in the containing group. Thus it provides structural information about the containing group. But it can also lead to the new classification of various subgroups of groups, in particular, groups which have been classified earlier. See [H1] for an exposition of equivalence theorems in general, and [H2] for an example of their significance to structure theory.

In a classical paper on isotype subgroups of totally projective groups [HM1], one finds this important equivalence theorem.

THEOREM. (Hill and Megibben) Let H and H' be isotype subgroups of the totally projective group G. Then there is an automorphism of G that maps H onto H' if and only if

(1) H and H' have the same Ulm invariants, and

(2) G/H and G/H' are isomorphic as valuated groups when endowed with the coset valuation.

Using the two criteria from the theorem, we develop a certain "test" for the equivalence of subgroups within a given group. As in the theorem, we focus on isotype subgroups.

This work represents part of the author's dissertation research under the direction of Dr. Paul Hill.

DEFINITION. A group G satisfies the **Basic Test for Equivalence of Isotype Subgroups (BTE)** if whenever H and H' are isotype subgroups of G satisfying

(1) H and H' have the same Ulm invariants, and

(2) G/H and G/H' are isomorphic as valuated groups when endowed with the coset valuation, then H and H' are equivalent.

This terminology refers to the fact that if H and H' have the same Ulm invariants, then they contain isomorphic basic subgroups. Of course, in some cases, this also implies that H and H' are isomorphic, but not in general.

The totally projective groups are a very important class of abelian p-groups, as they form the largest class of p-groups which are known to be determined by their Ulm invariants. This class of groups contains the Σ-cyclic groups and is closed under direct sums and summands. It is also known as the class of simply presented groups; an additional characterization of totally projective groups is that they contain a "nice composition series" or an "axiom 3 system" of nice subgroups [F]. Our question is the following: Does the BTE also characterize totally projective groups?

Inherent in some of the work of C. Megibben, there is a hint that the above question cannot be answered in the ordinary ZFC axiomatization of set theory. For on one hand, under Martin's Axiom and the negation of the Continuum Hypothesis, there exist abelian p-groups of cardinality $\aleph_1$ and length at most ω_1 which are not totally projective and yet satisfy the BTE when applied to *countable* isotype subgroups. This is seen in the following equivalence theorem found in [HM2].

THEOREM (MA). (Hill and Megibben) Suppose that the p-group G is a group of cardinality less than the continuum whose countable subgroups have countable closure. Let H and H' be isotype subgroups of G which are countable and have the same Ulm invariants. If $\phi : G/H \to G/H'$ is an isomorphism which preserves the coset valuation, then there exists an automorphism of G that maps H onto H' and induces ϕ.

On the other hand, under $V = L$, a p-group of size $\aleph_1$ without elements of infinite height (also known as "separable" or of length $\leq \omega$), which is not Σ-cyclic, contains nonequivalent isotype subgroups which still satisfy the two criteria of the BTE. This is seen in the proof of the following theorem from [M1].

THEOREM ($V = L$). (Megibben) A Crawley group of cardinality $\aleph_1$ is Σ-cyclic.

The proof uses a specific criterion for Crawley groups, which shows its relevance to the matter of equivalent subgroups.

RICHMAN'S CRITERION. A separable p-group is a Crawley group if and only if its automorphisms act transitively on the dense subsocles of codimension one.

Given a separable p-group of size $\aleph_1$ which is not Σ-cyclic, Megibben uses a version of the $\diamondsuit$-principle, which holds in $V = L$, to construct nonequivalent dense subsocles of codimension one. These would then support pure (and thus isotype, in this setting) subgroups which satisfy the two criteria of the BTE, but cannot be equivalent. Using similar techniques, we are able to generalize this to the setting of a p-group of size $\aleph_1$ and any length, as long as what we call the "non-totally projective length" of the group is countable. This then gives us our characterization, as indicated in the title.

2 PRELIMINARIES

All groups under consideration here are abelian p-groups and are assumed to be reduced. We use the term Σ-cyclic for a direct sum of cyclic groups, saving the abbreviation d.s.c. for a direct sum of countable groups. As noted in the introduction, we are particularly interested in the larger class of groups which contains these, the totally projective groups. See [F] for further information about this class, or an elaboration of any of the terms or notation used here.

The subgroup $p^\alpha G$ of a group G for a given ordinal α plays an important role; it is defined as follows. If $\alpha = \beta + 1$, then $p^\alpha G = p(p^\beta G)$, and if α is a limit ordinal, then $p^\alpha G = \bigcap_{\beta<\alpha} p^\beta G$. The height of an element $x \in G$, denoted $|x|$ or $|x|_G$ is defined as the ordinal α such that $x \in p^\alpha G \backslash p^{\alpha+1} G$. And the length of the group is the least ordinal λ such that $p^\lambda G = 0$. As the group is reduced, some such λ exists. Given a subgroup H of G, the coset valuation, $\|x + H\|$ or $\|x + H\|_G$ is given by $\sup\{|x + h| + 1 : h \in H\}$ where these heights are calculated in G.

A subgroup H of G is nice in G if $p^\alpha(G/H) = (p^\alpha G + H)/H$ for each ordinal α, or equivalently, if every coset contains a proper element. Thus if it is not nice, there is a coset whose valuation is a limit ordinal. A subgroup H is isotype if $p^\alpha G \cap H = p^\alpha H$ for each ordinal α. A subgroup H of G is called "λ-isotype" in G if it is isotype up to λ, i.e., if $p^\alpha G \cap H = p^\alpha H$ for all $\alpha < \lambda$. Note that if λ is a limit, then this is also true for all $\alpha \leq \lambda$. And H is λ-dense in G for a given ordinal λ, if given any x in G there is a sequence of h_α's in H such that $\lim |x + h_\alpha| = \lambda$. In other words, $G = p^\alpha G + H$ for all $\alpha < \lambda$. The latter two properties are easily seen to be inductive properties, i.e., the union of an ascending chain of λ-dense (λ-isotype) subgroups is itself λ-dense (λ-isotype).

Another prominent subgroup of a group G is the socle, $G[p] = \{x \in G : px = 0\}$. This is naturally a vector space over the integers mod p, and is a valuated vector space via the heights of the elements as computed in G.

We assume familiarity with the following set theoretical terms. A cub set in ω_1 is a closed and unbounded subset of ω_1, and a set is stationary in ω_1 if it intersects every cub. Recall that the intersection of a cub with a stationary set is again stationary. A filtration of a group G of cardinality $\aleph_1$ is an increasing sequence of countable subgroups of G, G_α, for $\alpha < \omega_1$, such that whenever α is a limit ordinal, $G_\alpha = \bigcup_{\beta<\alpha} G_\beta$, and $\bigcup_{\alpha<\omega_1} G_\alpha = G$. So if C is a cub set and $\{G_\alpha\}_{\alpha<\omega_1}$ is a filtration of G, then $\{G_\alpha\}_{\alpha\in C}$ is also a filtration of G.

Finally, a filtration $\{G_\alpha\}_{\alpha<\omega_1}$ of a p-group G of cardinality $\aleph_1$ is called a nice composition series for G if G_α is nice in G for all $\alpha < \omega_1$. Although technically this term is used for a filtration of nice subgroups with $|G_{\alpha+1} : G_\alpha| = p$, the fact that $|G_{\alpha+1} : G_\alpha| = \aleph_0$ for a filtration of a group of size $\aleph_1$ easily leads to a filtration of nice subgroups such that the former is true. A group G is totally projective if and only if it has a nice composition series.

3 THE CHARACTERIZATION

As stated in the introduction, the purpose of this paper is to generalize the result from [M1] that under $V = L$, a p-group of length at most ω and cardinality $\aleph_1$ which is not Σ-cyclic contains isotype subgroups with the same Ulm invariants and isomorphic quotients (as valuated groups), but which are not equivalent. This is what gives us our characterization of some small totally projective groups.

MAIN THEOREM ($V = L$). Given a p-group G of cardinality $\aleph_1$ and of any countable length, G is totally projective (a d.s.c.) if and only if G satisfies the Basic Test for Equivalence of isotype subgroups.

Since totally projective groups satisfy the BTE, we must show that a non-totally projective group (of any countable length) does not. We make use of a version of the $\diamondsuit$-principle, which holds in the constructible universe (see [J]). This, of course, is the version most commonly used in group theory.

$\diamondsuit$-PRINCIPLE ($V = L$). Let G be a group of cardinality $\aleph_1$ with filtration $\{G_\alpha\}_{\alpha<\omega_1}$, and let E be a stationary subset of ω_1. Then there is a collection of functions $\{f_\alpha : G_\alpha \to G_\alpha : \alpha \in E\}$ such that for any function $f : G \to G$ the set $\{\alpha \in E : f \restriction G_\alpha = f_\alpha\}$ is stationary in ω_1.

Using this principle, we construct the desired nonequivalent isotype subgroups within a non-totally projective group. The role that the stationary set E from the $\diamondsuit$-principle plays in the non-totally projective group setting is given by the following fact. For completeness, we provide its simple proof.

LEMMA 1. If G is a p-group of cardinality $\aleph_1$ with filtration $\{G_\alpha\}_{\alpha<\omega_1}$, then G is not totally projective if and only if the set $E = \{\alpha : G_\alpha$ is not nice in G$\}$ is stationary in ω_1.

Proof:

Suppose E is not stationary. Let C be a cub such that $C \cap E = \emptyset$. Then $\{G_\alpha\}_{\alpha \in C}$ is a nice composition series for G, and G is totally projective.

If G is totally projective, then G has a nice composition series, say $\{G'_\alpha\}_{\alpha<\omega_1}$. But then $\{G_\alpha\}_{\alpha<\omega_1}$ and $\{G'_\alpha\}_{\alpha<\omega_1}$ must agree on a cub, C, i.e. $\forall \alpha \in C,\ G_\alpha = G'_\alpha$. Then $C \cap E = \emptyset$. □

In order to characterize totally projective groups in the setting of p-groups of size $\aleph_1$ and any countable length, we actually deal with a slightly more general situation. As long as what may be referred to as the "non-totally projective length" of a non-totally projective group G is countable, we can find isotype subgroups which show that G fails to satisfy the BTE. The "non-totally projective length" of a group refers to the smallest ordinal λ such that $G/p^\lambda G$ is not totally projective. Of course, if the length of G is countable, then this ordinal is also. The following fact further describes this "non-totally projective length."

PROPOSITION 2. If λ is the smallest ordinal such that $G/p^\lambda G$ is not totally projective, then λ is a limit ordinal.

Proof:

Suppose that $\lambda = \mu + n$ for some limit μ and $n < \omega,\ n \neq 0$. Then

$$\frac{G}{p^\mu G} = \frac{G/p^\lambda G}{p^\mu (G/p^\lambda G)} \text{ is totally projective.}$$

Since $p^\mu(G/p^\lambda G)$ is a bounded group, this implies that $G/p^\lambda G$ is totally projective. But this contradicts the choice of λ.

□

Under the special hypothesis of a countable "non-totally projective length," we are able to say something more about the stationary set from Lemma 1.

LEMMA 3. If G is a p-group of cardinality $\aleph_1$ which is not totally projective, and $\lambda < \omega_1$ is the least ordinal such that $G/p^\lambda G$ is not totally projective, and $\{G_\alpha\}_{\alpha<\omega_1}$ is a filtration of G, then $E = \{\alpha : \|x + G_\alpha\|_G = \lambda \text{ for some } x \in G\backslash G_\alpha\}$ is stationary in ω_1.

Proof:
First we note that since G is not totally projective, we have that λ is a limit ordinal $\geq \omega$. By Lemma 1, the set

$$\begin{aligned} E_1 &= \{\alpha : G_\alpha \text{ is not nice in } G\} \\ &= \{\alpha : \|x + G_\alpha\|_G = \mu \text{ for some limit } \mu \text{ and } x \in G\backslash G_\alpha\} \end{aligned}$$

is stationary in ω_1. In addition, since $G/p^\lambda G$ is not totally projective, and $\{(G_\alpha+p^\lambda G)/p^\lambda G\}_{\alpha<\omega_1}$ is a filtration thereof, there is a stationary set of α's where there exists $\bar{x} = x + p^\lambda G$ outside of $(G_\alpha + p^\lambda G)/p^\lambda G$ for which the coset valuation is a limit, and necessarily a limit $\leq \lambda$. This implies that the coset valuation of $x + G_\alpha$ is this same limit. Thus we may assume that

$$E_1 = \{\alpha : \|x + G_\alpha\|_G = \mu \text{ for some limit } \mu \leq \lambda \text{ and } x \in G\backslash G_\alpha\}.$$

Write $E_1 = \bigcup_{\mu\leq\lambda} E_\mu$, where

$$E_\mu = \{\alpha : \|x + G_\alpha\|_G = \mu \text{ for some } x \in G\backslash G_\alpha\}.$$

Suppose that E_μ is stationary for some limit $\mu < \lambda$. Then $\forall \alpha \in E_\mu$,

$$\|\bar{x} + \frac{G_\alpha + p^\mu G}{p^\mu G}\|_{\frac{G}{p^\mu G}} = \mu \qquad (\bar{x} = x + p^\mu G)$$

implies that $G/p^\mu G$ is not totally projective by Lemma 1, and is a contradiction. So $\forall \mu < \lambda$, the set E_μ is not stationary. Since the union of fewer than ω_1 nonstationary sets is again nonstationary, we must have that $E = E_\lambda$ is stationary in ω_1.

□

Before we state and prove the theorem which gives us our characterization, we note a few elementary facts. These will be used in the construction of isotype subgroups within a non-totally projective group which satisfy the two criteria of the BTE and yet are not equivalent.

PROPOSITION 4. Let λ be a limit ordinal, and let G be a p-group with length$(G) \geq \lambda$. If H is maximal in G with respect to $H[p] = S$, where S is a λ-dense subsocle of $G[p]$, then H is λ-dense and λ-isotype in G.

Proof:
We show by induction on $\alpha < \lambda$ that $p^\alpha G \cap H \subseteq p^\alpha H$. (The other inclusion always holds.) Suppose it is true $\forall \beta < \alpha$. If α is a limit ordinal, then it is trivial, so assume that $\alpha = \beta + 1$.

Let $h \in p^\alpha G \cap H$. Then $h = pg$ for some $g \in p^\beta G$. If $g \in H$, then $g \in p^\beta G \cap H = p^\beta H$ implies that $h = pg \in p^\alpha H$. So suppose $g \notin H$. Then by the maximality of H, there exists $x \in \langle H, g\rangle$ of order p, and $x \notin H[p]$. Write $x = h_1 - ng$, where $h_1 \in H$, and without loss of generality we may assume that $n = 1$. Then $px = p(h_1 - g) = 0$, so $ph_1 = pg = h$, and $h_1 - g \in G[p]$. Since $H[p]$ is λ-dense in $G[p]$ and $\beta < \lambda$, there exists $h_2 \in H[p]$ such that $(h_1 - g) - h_2 \in p^\beta G$. Then $h_0 = h_1 - h_2 \in p^\beta G \cap H = p^\beta H$, and $ph_0 = p(h_1 - h_2) = ph_1 = h$, so that $h \in p^\alpha H$ as desired.

Now to show that H is λ-dense, we let $g \in G$ and $\alpha < \lambda$, and find $h \in H$ such that $|g - h| > \alpha$. We induct on the order of g. Say $p^n g = 0$ and $p^{n-1} g \neq 0$. Then $p^{n-1} g \in G[p]$, so there exists $x \in H[p]$ such that $|p^{n-1}g - x| > \alpha + n$. So $|x| > n$ and $\exists y \in G$ such that $p^{n-1}y = x$. Then since H is pure in G, $\exists h_1 \in H$ such that $p^{n-1}h_1 = x$. Thus $|p^{n-1}g - p^{n-1}h_1| > \alpha + n > \alpha$. Now we can find $z \in p^\alpha G$ such that $p^{n-1}z = p^{n-1}(g - h_1)$ and we have that $g - h_1 - z$ has order $< n$. So $\exists h_2 \in H$ such that $|g - h_1 - z - h_2| > \alpha$, which implies that $|g - (h_1 + h_2)| > \alpha$. Thus the desired h is $h = h_1 + h_2$.

□

Of course, in order to make use of Proposition 4, we need to establish the existence of such a λ-dense subsocle. In our construction, we also desire to have one of codimension one.

PROPOSITION 5. Let G be a p-group of length $\geq \lambda$, where λ is a countable limit ordinal. Then there exists a λ-dense subsocle S such that $G[p]/S \simeq \mathbf{Z}/p\mathbf{Z}$.

Proof:
First let $\{\alpha_n\}_{n<\omega}$ be a sequence of ordinals such that $\sup\{\alpha_n\} = \lambda$. Then choose $x_0 \in G[p]$ with $\alpha_0 \leq |x_0| < \lambda$, and choose $x_1 \in G[p]$ with $\max\{\alpha_0, |x_0|\} < |x_1| < \lambda$. Given $x_n \in G[p]$, choose $x_{n+1} \in G[p]$ such that $\max\{\alpha_n, |x_n|\} < |x_{n+1}| < \lambda$, obtaining a sequence $\{x_n\}_{n<\omega}$ such that $|x_n| \to \lambda$. Let $R = \langle x_1 - x_n : n < \omega \rangle$, and take $S \subseteq G[p]$ maximal with respect to $x_1 \notin S$ and $R \subseteq S$. Then $G[p] = S + \langle x_1 \rangle$ and S is λ-dense in $G[p]$.

□

Finally, we make one set theoretic observation, which is useful mainly for notational purposes.

PROPOSITION 6. Suppose $\{G_\alpha\}_{\alpha<\omega_1}$ is a filtration of a p-group G of size $\aleph_1$, and that $\{A_\alpha\}_{\alpha \in E}$, where $E \subseteq \omega_1$, is a collection of countable subsets of G. Then there is a cub set $C \subseteq \omega_1$ such that the filtration $\{G_\alpha\}_{\alpha \in C}$ of G has the property that $A_\alpha \subseteq G_{\alpha+1}$ $\forall \alpha \in E \cap C$.

Proof:
Inductively define an increasing function $\psi : \omega_1 \to \omega_1$ as follows.
(1) $\psi(0) = 0$.
(2) If $\psi(\alpha)$ is a limit ordinal, then $\psi(\alpha) = \sup\{\psi(\beta) : \beta < \alpha\}$.
(3) If $\psi(\alpha) \in E$, then $\psi(\alpha + 1) = \min\{\beta > \psi(\alpha) : A_\alpha \subseteq G_\beta\}$.
(4) Otherwise, $\psi(\alpha + 1) = \psi(\alpha) + 1$.
Then $\mathrm{Im}(\psi)$ is a cub, and $C = \mathrm{Im}(\psi)$ has the desired properties.

□

First we state and prove our main result under the conditions for which it holds in the usual ZFC axiomatization of set theory, and then we state the application of the $\diamondsuit$-principle, which gives us our characterization.

THEOREM 7. Let G be a p-group of cardinality $\aleph_1$ which is not totally projective, and let λ be the smallest ordinal such that $G/p^\lambda G$ is not totally projective, and suppose that $\lambda < \omega_1$. Let $\{G_\alpha\}_{\alpha<\omega_1}$ be a λ-isotype filtration of G, and let $E = \{\alpha : G_\alpha$ is not nice in $G\}$. Suppose that $\{f_\alpha : \alpha \in E\}$ is a set of automorphisms $f_\alpha : G_\alpha \to G_\alpha$. Then there exist isotype subgroups H and H' of G which satisfy the two criteria of the BTE, and there exists a stationary subset of E such that no f_α, for α therein, can be extended to an automorphism of G which maps H onto H'.

Proof:

Recall that E itself is stationary by Lemma 1. And by Lemma 3, we may assume that

$$\forall \alpha \in E \; \exists x \in G \backslash G_\alpha \text{ such that } \|x + G_\alpha\| = \lambda.$$

Let S be a λ-dense subsocle of G, say $G[p] = S \oplus \langle z \rangle$. Let $\{s_n\}_{n<\omega} \subseteq S$ be a sequence such that $s_n \to z$ in the λ-topology, i.e., $|s_n - z| \to \lambda$. Without loss of generality, we may assume that $z \in G_0$ and $s_n \in G_0$ for all $n < \omega$. We use Proposition 6 to assume that any two desired elements in $G \backslash G_\alpha$ at stage α are in $G_{\alpha+1}$. This replacement of ω_1 with a cub is what gives rise to the stationary *subset* of E mentioned in the statement of the theorem. However, for practical purposes, this subset is as good as E itself, so we will continue to refer to it as the set E.

Construct H_α and H'_α for $\alpha < \omega_1$ so that (a) $p^\lambda G_\alpha \subseteq H_\alpha, H'_\alpha$, (b) $z \notin H_\alpha, H'_\alpha$, and (c) $H_\alpha, H'_\alpha \subseteq G_\alpha$, as follows.

(1) $H_0 = \langle p^\lambda G_0, s_n : n < \omega \rangle = H'_0$.

(2) If α is a limit ordinal, $H_\alpha = \bigcup_{\beta<\alpha} H_\beta$, $H'_\alpha = \bigcup_{\beta<\alpha} H'_\beta$.

(3) If $\alpha \in E$ and f_α has an extension to some $\phi \in \mathrm{Aut}(G)$, find x_α such that $\|x_\alpha + G_\alpha\| = \lambda$ and the coset $x_\alpha + p^\lambda G$ has minimal order. Let $y_\alpha = \phi(x_\alpha)$, and we may assume that $x_\alpha, y_\alpha \in G_{\alpha+1}$. Put $H_{\alpha+1} = \langle H_\alpha + p^\lambda G_{\alpha+1}, x_\alpha \rangle$ and $H'_{\alpha+1} = \langle H'_\alpha + p^\lambda G_{\alpha+1}, y_\alpha - z \rangle$.

(4) Otherwise, put $H_{\alpha+1} = H_\alpha + p^\lambda G_{\alpha+1}$ and $H'_{\alpha+1} = H'_\alpha + p^\lambda G_{\alpha+1}$.

Clearly all three inductive hypotheses are preserved in cases (1), (2), and (4), as z is a λ-limit of S. It remains to verify that the inductive assumption in (b) is not violated in step (3). First we show that $z \notin \langle H_\alpha + p^\lambda G, x_\alpha \rangle$. Suppose it is. Then $mz = h_\alpha + g_\lambda + p^n x_\alpha$ for some $h_\alpha \in H_\alpha$, $g_\lambda \in p^\lambda G$, $n < \omega$, and $1 \leq m < p$. If $p^n x_\alpha \in p^\lambda G$, then $mz - h_\alpha \in p^\lambda G \cap G_\alpha = p^\lambda G_\alpha \subseteq H_\alpha$ which implies that $mz \in H_\alpha$ so that $z \in H_\alpha$, a contradiction. So n is smaller than the order of the coset $x_\alpha + p^\lambda G$. Now we still have $mz - h_\alpha \in p^n G \cap G_\alpha = p^n G_\alpha$, which implies that $mz - h_\alpha = p^n g_\alpha$ for some $g_\alpha \in G_\alpha$, so $p^n(x_\alpha - g_\alpha) \in p^\lambda G$. But this contradicts the choice of x_α. Therefore $H_{\alpha+1}$ still satisfies the inductive assumption.

Next we show that $z \notin \langle H'_\alpha + p^\lambda G, y_\alpha - z \rangle$. Again, suppose it is. Then $mz = h'_\alpha + g_\lambda + p^n(y_\alpha - z)$ for some $h'_\alpha \in H'_\alpha$, $g_\lambda \in p^\lambda G$, $n < \omega$, and $1 \leq m < p$.

Case 1: $n > 0$.

Then $mz = h'_\alpha + g_\lambda + p^n y_\alpha$ implies that $mz - h'_\alpha = g_\lambda + p^n y_\alpha$ so that $p^n y_\alpha \notin p^\lambda G$. For otherwise $mz - h'_\alpha \in p^\lambda G \cap G_\alpha = p^\lambda G_\alpha \subseteq H'_\alpha$, implying that $z \in H'_\alpha$, a contradiction. Thus $p^n x_\alpha \notin p^\lambda G$, as the fact that $\phi(x_\alpha) = y_\alpha$ implies that the orders of the cosets $x_\alpha + p^\lambda G$ and $y_\alpha + p^\lambda G$ are the same. But we do have that $mz - h'_\alpha \in p^n G_\alpha$ so that $mz - h'_\alpha = g_\lambda + p^n y_\alpha = p^n g'_\alpha$ for some $g'_\alpha \in G_\alpha$, and so $p^n(y_\alpha - g'_\alpha) \in p^\lambda G$. But then $p^n(x_\alpha - g_\alpha) \in p^\lambda G$, where $\phi(g_\alpha) = g'_\alpha$. This again contradicts the choice of x_α.

Case 2: $n = 0$.

Then $mz = h'_\alpha + g_\lambda + y_\alpha - z$ implies that $(m+1)z = h'_\alpha + g_\lambda + y_\alpha$, so that $\|y_\alpha + G_\alpha\| > \lambda$, and hence $\|x_\alpha + G_\alpha\| > \lambda$, a contradiction. Therefore $H'_{\alpha+1}$ also preserves the desired properties. Note that we then have that for each $\alpha \in E$ and $\beta > \alpha$, $y_\alpha \notin H'_\beta$.

Now let $H_{\omega_1} = \bigcup_{\alpha<\omega_1} H_\alpha$, $H'_{\omega_1} = \bigcup_{\alpha<\omega_1} H'_\alpha$, and take H maximal with respect to $H[p] = T$ where $T \supseteq H_{\omega_1}[p]$ and $G[p] = T \oplus \langle z \rangle$. Similarly, take H' maximal with respect to $H'[p] = T'$ where $T' \supseteq H'_{\omega_1}[p]$ and $G[p] = T' \oplus \langle z \rangle$. Note that since $z \notin H'$, we still have that $y_\alpha \notin H'$ for all $\alpha \in E$.

Then by Proposition 4, H and H' are λ-isotype and λ-dense in G. But also, they contain $p^\lambda G$, so they are actually isotype, and they have the same Ulm invariants as G itself. And $G/H \simeq Z(p^\infty) \simeq G/H'$, homogeneous with value λ, so that this is a valuated isomorphism. So H and H' satisfy the criteria of the BTE.

Now suppose, by way of contradiction, that $\pi \in \mathrm{Aut}(G)$ is an extension of some f_α, $\alpha \in E$, for which $\pi(H) = H'$. Then let x_α, y_α, and ϕ be as in case (3). Let $\{t_n\}_{n<\omega} \subseteq G_\alpha$ be a sequence such that $t_n \to x_\alpha$ in the λ-topology. Now $x_\alpha + p^\lambda G \subseteq H$, but then by the continuity of automorphisms,

$$\pi(x_\alpha) + p^\lambda G = \lim_{n\to\infty} \pi(t_n) + p^\lambda G = \lim f_\alpha(t_n) + p^\lambda G$$

$$= \lim_{n\to\infty} \phi(t_n) + p^\lambda G = y_\alpha + p^\lambda G \subseteq H'$$

which is impossible since $p^\lambda G \subseteq H'$ but $y_\alpha \notin H'$.

□

THEOREM 8 ($V = L$). Let G be a p-group of cardinality $\aleph_1$ which is not totally projective, and let λ be the smallest ordinal such that $G/p^\lambda G$ is not totally projective. If $\lambda < \omega_1$, then there exist isotype subgroups H and H' of G such that

(1) H and H' have the same Ulm invariants

(2) $G/H \simeq G/H'$ in the valuated sense

but H and H' are not equivalent in G.

Proof:

Let $E = \{\alpha : \|x + G_\alpha\| = \lambda$ for some $x \in G \backslash G_\alpha\}$ as in Lemma 3, a stationary set. Let $\{f_\alpha : \alpha \in E\}$ be the "prediction maps" as in the $\diamondsuit$-principle. Then construct H and H' as in the proof of the previous theorem. If $\pi \in \mathrm{Aut}(G)$, there is a cub $C \subseteq \omega_1$ such that $\pi(G_\alpha) = G_\alpha$ $\forall \alpha \in C$. Also, the set $S = \{\alpha \in E : \pi \restriction G_\alpha = f_\alpha\}$ is stationary in ω_1, so there is some $\alpha \in C \cap S$. Then for this α, $\pi(G_\alpha) = G_\alpha$ and $\pi \restriction G_\alpha = f_\alpha$ and $\alpha \in E$, so there exists $x_\alpha \in H$, $y_\alpha \notin H'$, but $\pi(x_\alpha) + p^\lambda G = y_\alpha + p^\lambda G \subseteq H'$ which is a contradiction since $p^\lambda G$ is contained in H' as well as H.

□

Thus we do indeed have our "$\diamondsuit$-characterization" of some small totally projective groups. In addition, we can reword the previous theorem to obtain an interesting corollary. Recall that a C_Ω (or C_{ω_1}) group is a p-group such that $G/p^\lambda G$ is totally projective for all $\lambda < \omega_1$. Thus the class of totally projective groups is contained in the class of C_Ω groups.

COROLLARY 9 ($V = L$). Given a p-group G of cardinality $\aleph_1$, if G satisfies the Basic Test for Equivalence of isotype subgroups, then G is a C_Ω group.

We conclude with a couple of related questions. Of course it would be nice to know whether under $V = L$ we have a characterization of totally projective groups, or even d.s.c.'s in general. But the situation of a non-totally projective C_Ω group, even of length ω_1, is surprisingly difficult to deal with. However, it is conjectured that the answer to the first question is no.

QUESTION 1. Under $V = L$, does a C_Ω group of size $\aleph_1$ which is not totally projective satisfy the BTE?

The second question, if answered in the affirmative, would show that the characterization of totally projective groups of countable length and cardinality $\aleph_1$ in terms of the BTE is independent of the ZFC axiomatization of set theory. Recall that it is true when the BTE is restricted to countable isotype subgroups.

QUESTION 2. Under MA_{ω_1}, or possibly some other forcing axiom, does there exist a non-totally projective group of size $\aleph_1$ and countable length which does satisfy the BTE?

REFERENCES

[F] Fuchs, L., *Infinite Abelian groups*, Volumes I & II, Academic Press, New York (1970) and (1973).

[H1] Hill, P., "Equivalence theorems," *Rocky Mountain J. Math.*, **23** (1993), 203-221.

[H2] Hill, P., "On the structure of abelian p-groups," *Trans. Amer. Math. Soc.*, **288** (1985), 505-525.

[HM1] Hill, P. and Megibben, C., "On the theory and classification of abelian p-groups," *Mathematische Zeitschrift* **190** (1985), 17-38.

[HM2] Hill, P. and Megibben, C., "On the structure of primary abelian groups whose countable subgroups have countable closure," *to appear.*

[J] Jensen, R., "The fine structure of the constructible hierarchy," *Ann. Math. Logic* **4** (1972), 229-308.

[M1] Megibben, C., "Crawley's problem on the unique ω-elongation of p-groups is undecidable," *Pacific Journal of Math.*, **107** (1983), 205-212.

[M2] Megibben, C., "ω_1-separable p-groups," Abelian Group Theory, *Proceedings of the Third Conference on Abelian Group Theory at Oberwolfach 1985,* Gordon and Breach Science Publishers (1987), 117-136.

Functors from Locally Completely Decomposable Groups to Mixed Groups

W. WICKLESS
University of Connecticut, Storrs, CT

ABSTRACT. We construct category equivalences from categories of locally completely decomposable groups to categories of mixed groups

In [6] (later slightly modified in [4]) category equivalences were constructed from certain subcategories of the category of locally free torsion-free groups and quasi-homomorphisms to subcategories of the category of self-small mixed groups G such that $G/T(G)$ is finite rank divisible and quasi-homomorphisms. The purpose of this paper is to extend these equivalences from locally free groups to locally completely decomposable groups.

All groups are abelian. All tensor products and all unadorned homomorphisms are taken with respect to Z. For a group H the symbol H_p denotes $Z_p \otimes H$, the usual localization at the integral prime p. As is customary, if T is a torsion group we identify T_p with the $p-$torsion subgroup of T. In contrast, if a sequence of groups is being chosen, one for each prime p, we use the notation $H(p)$. Here there will be no fixed object H being localized. We use QH for the divisible hull of a torsion-free group H. If H_1, H_2 are groups we write $H_1 \doteq H_2$ to indicate that $H_1 \cap H_2$ is of finite index in both H_1 and H_2.

For convenience, we gather together the following well-known facts, which will be referenced repeatedly.

Remarks 0 Call a subgroup F of a group H **full** if H/F is torsion. Let F, F' be finite rank free groups which are each full in some group H. Then, since F, F' are finitely generated, $F \doteq F'$. Thus $F_p = F'_p$ for almost all primes p and $F_p \doteq F'_p$ for the finitely many exceptions. It follows that, for almost all p, $(H/F)_p \cong H_p/F_p = H_p/F'_p \cong (H/F')_p$.

Before we can state and prove our main result we need some definitions.

Definition 1. *Let H be a torsion-free group of finite rank and F a full free subgroup of H. The finite outer type of H ($FOT(H)$) is the type $\tau = [(k_p)]$ where $Z(p^{k_p})$ is (isomorphic to) a maximal finite cyclic summand of the p-group $(H/F)_p$.*

By Remarks 0, the finite outer type τ of H is independent of the choice of the full free subgroup $F \subset H$. The type τ is a **locally free type** in the sense that each entry k_p of a representing height vector for τ is finite. Throughout the paper $\tau = [(k_p)]$ will denote a locally free type.

Definition 2. *A torsion-free finite rank group H is called locally completely decomposable (lcd) if, for every prime p, there is a decomposition*

$$(d^p) \qquad H_p = R(p) \bigoplus D(p)$$

where $R(p)$ is a free Z_p module and $D(p)$ is divisible.

If H is lcd with $FOT(H) = [(k_p)]$ and F is a full free subgroup, then, for almost all primes $p, p^{k_p}R(p) \subset F_p$. Additionally, given F and the decompositions (d^p), we can define a subgroup H^0 of H by the stipulations $H_p^0 = R(p) + F_p$. The finite outer type of H will be the outer type of H^0.

Definition 3. *Let τ be a locally free type. Then $\mathcal{QL}_\tau$ is the category with objects reduced lcd groups of finite outer type less than or equal to τ and maps quasi-homomorphisms.*

It is easy to check that $\mathcal{QL}_\tau$ is an additive category. Moreover if $E : 0 \to K \to H \to L \to 0$ is exact with $H \in \mathcal{QL}_\tau$ then we can choose $E-$ compatible decompositions (d^p) for K, H and L . In view of the comments proceeding Definition 3, we can see that $\mathcal{QL}_\tau$ is closed under torsion-free epimorphic images and pure subgroups.

As defined in [6], the category $\mathcal{QG}$ has objects self-small mixed groups G such that $G/T(G)$ is finite rank divisible. Recall that a group G is **self-small** if $Hom(G, \bigoplus_{i\in I} G_i) \cong \bigoplus_{i\in I} Hom(G, G_i)$ for any indexed set of groups $\{G_i : i \in I\}$ with each $G_i \cong G$. The maps in $\mathcal{QG}$ are quasi-homomorphisms, that is $Hom_{\mathcal{QG}}(G, G') = Q \otimes Hom_Z(G, G')$. A less compact but more revealing description of the objects in $\mathcal{QG}$ is that $G \in \mathcal{QG}$ if and only if G is a reduced mixed group with torsion subgroup $T = \bigoplus T_p$ satisfying:

- each T_p is finite and
- G is (can be embedded as) a pure subgroup of $\prod T_p$ and
- G/T is finite-rank (necessarily divisible) and
- (projection condition) if F is a full free subgroup of G then F projects onto T_p for almost all primes p.

For a locally free type $\tau = [(k_p)]$, $\mathcal{QG}_\tau$ is the full subcategory of $\mathcal{QG}$ with objects satisfying the additional requirement $p^{k_p}T_p = 0$ for almost all p. This requirement is, of course, independent of the choice of $(k_p) \in \tau$. Various aspects of the categories $\mathcal{QG}, \mathcal{QG}_\tau$ are discussed in [2], [1] and [3].

In [6], for each locally free τ, a category equivalence was constructed from the category $\mathcal{QH}_\tau$, with objects torsion-free finite rank groups H such that outer type $H \leq \tau$ and maps quasi-homomorphisms, to $\mathcal{QG}_\tau$. This equivalence is also discussed in [5] and [4].

Let X be a rank one group of type τ and let $H \in \mathcal{QH}_\tau$. Since outer type $H \leq \tau$, H can be embedded in X^n for $n =$ rank H. But τ is locally free so that $X_p \cong Z_p$ for all p. Consequently, each H_p embeds in a free Z_p module. It follows that the groups in $\mathcal{QH}_\tau$ are all lcd; that is $\mathcal{QH}_\tau$ is a full subcategory of $\mathcal{QL}_\tau$. We want to extend the category equivalence of [6] from locally free groups ($\mathcal{QH}_\tau$) to lcd groups ($\mathcal{QL}_\tau$). This will involve extending the mixed group category $\mathcal{QG}_\tau$. We now define a larger category of mixed groups which seems to be appropriate for the job. To facilitate the statement of Definition 4 we establish some notation. Let $T = \bigoplus T_p$ be a torsion group with each T_p finite. For each prime p, let $L(p)$ be a finite rank free Z_p module and put $M(p) = T_p \bigoplus L(p)$.

Definition 4. *The category $\mathcal{QM}$ is the category with objects mixed groups M such that*

- $T = T(M) \subset M \subset_{pure} \prod M(p)$,
- $L(p) \neq 0$ for at least one prime p and
- M/T is of finite rank.
- We also require that M satisfy the following **projection conditions**. Let F be a full free subgroup of M and let, for a fixed prime p, π_p^t [resp. π_p^l] be the natural projection from $\prod M(q)$ onto T_p [resp. $L(p)$]. Then: (a) $\pi_p^t(F_p) = T_p$ for almost all p, (b) $\pi_p^l(F_p) = L(p)$ for almost all p, (c) $\pi_p^l(F_p) \doteq L(p)$ for all p and (d)

$kernel(\pi_p^l) \cap F_p \subset kernel(\pi_p^t) \cap F_p$ for almost all p.

We remark that, in the context of the projection conditions, the projection maps π_p^t and π_p^l are first restricted to the free subgroup $F \subset M \subset \prod M(q)$, then, since T_p and $L(p)$ are p-local groups, extended naturally from F to F_p.

For each $M \in \mathcal{QM}$ and prime p, the modular law provides a direct sum decomposition $M = T_p \bigoplus T_p^*$, with $T_p^* = M \cap [L(p) \bigoplus \prod_{q \neq p} M(q)]$. Let $M' \in \mathcal{QM}$ with torsion subgroup T' and corresponding decompositions $M' = T_p' \bigoplus (T')_p^*$. Set $Hom_c(M, M') = \{\theta \in Hom_Z(M, M') : \theta(T_p^*) \subset (T')_p^* \text{ for almost all } p\}$ (the **coherent homomorphisms**) and $Hom_{\mathcal{QM}}(M, M') = Q \bigotimes (Hom_c(M, M')/[Hom_c(M, M') \cap Hom_Z(M, T')])$.

As in the definition of $\mathcal{QG}_\tau$, the category $\mathcal{QM}_\tau$ is the full subcategory of $\mathcal{QM}$ with objects mixed groups M such that $T(M) = \bigoplus T_p$ satisfies the requirement $p^{k_p}T_p = 0$ for almost all p.

It is routine to check that $\mathcal{QM}$ and $\mathcal{QM}_\tau$ are additive categories. Remarks 0 shows that the projection conditions are independent of the choice of the full free subgroup F.

We now are ready for our main result.

Theorem 5. *For each locally free type τ there is a pair of rank preserving functors (Φ_τ, Γ_τ) providing a category equivalence $\mathcal{QL}_\tau \sim \mathcal{QM}_\tau$.*

Proof. We construct rank preserving functors $\Phi = \Phi_\tau : \mathcal{QL}_\tau \to \mathcal{QM}_\tau$ and $\Gamma = \Gamma_\tau : \mathcal{QM}_\tau \to \mathcal{QL}_\tau$ such that the composites are naturally equivalent to the appropriate identity functors. Let $\tau = [(k_p)]$ and $H \in \mathcal{QL}_\tau$ with full free subgroup F. For each p we have a decomposition (d^p) $H_p = R(p) \bigoplus D(p)$. We know that $p^{k_p}R(p) \subset F_p$ for almost all p. For these p define $T_p = [F_p + D(p)]/p^{k_p}H_p$; otherwise put $T_p = 0$. Since $T_p \subset H_p/\ p^{k_p}H_p$ it follows that $p^{k_p}T_p = 0$ for all p. Additionally $p^{k_p}H_p = p^{k_p}R(p) \bigoplus D(p)$ and $R(p), F_p$ are finite rank free Z_p modules, so that each T_p will be finite. Let $L(p) = F_p/F_p \cap D(p)$. Then $L(p) \cong [F_p + D(p)]/D(p) \subset R(p)$ is a finite rank free Z_p module. Since H is reduced $H_p \neq D(p)$ for at least one p. Hence $L(p) \neq 0$ for at least one p.

Put $M(p) = T_p \bigoplus L(p)$ and define $i : F \to \prod M(p)$ by $i(f) = [i(f)_p]$, where $i(f)_p = f + p^{k_p}H_p \oplus f + F_p \cap D(p)$. Since all groups in $\mathcal{QL}_\tau$ are reduced, if $0 \neq f \in F$, we cannot have $f \in D(p)$ for all p. It follows that i is an embedding. Define $\Phi(H) \subset \prod M(p)$ to be the pure subgroup of $\prod M(p)$ generated by $T = \bigoplus T_p$ and iF. Since T is the torsion subgroup of $\prod M(p)$, taking this pure subgroup makes sense. In the construction of $\Phi(H)$ we have already noted all the parts of Definition 4 except for the projection conditions, and these are easy to verify. Thus $\Phi(H)$ is an object in $\mathcal{QM}_\tau$.

We need to check that, up to isomorphism in $\mathcal{QM}_\tau$, the mixed group $\Phi(H)$ as constructed is independent of the initial choice of the full free subgroup $F \subset H$. Let $C \subset H$ be another full free subgroup and $\Phi(H)^c \subset \prod M(p)^c$ be the object of $\mathcal{QM}_\tau$ constructed by using C instead of F. By Remarks 0, $C \doteq F$, and consequently, $M(p)^c = M(p)$ for almost all p, $M(p)^c \doteq M(p)$ for the remaining p. Choose a positive integer s such that $s(F + C) \subset F \cap C$ and $s[M(p) + M(p)^c] \subset M(p) \cap M(p)^c$ (regarding everything as a subgroup of $\prod[M(p) + M(p)^c]$). Then multiplication by s is a coherent homomorphism carrying $\Phi(H)$ into $\Phi(H)^c$ and vice-versa. It follows that $\Phi(H) \cong \Phi(H)^c$ in the category $\mathcal{QM}_\tau$.

Let $H' \in \mathcal{QL}_\tau$ with full free subgroup F'. To complete the definition of the functor Φ we need to define, for each map $\theta \in Hom_{\mathcal{QL}_\tau}(H, H')$, an appropriate map

$\Phi(\theta) \in Hom_{\mathcal{QM}_\tau}[\Phi(H), \Phi(H')]$. Since θ is a quasi-homomorphism from H to H', we can choose a positive integer s such that $s\theta(H) \subset H'$ and $s\theta(F) \subset F'$. The latter containment holds since F is finitely generated and F' is a full free subgroup of H'. Regard $s\theta$ as a map from iF to $i'F'$. Then, adopting the obvious notation, $s\theta$ naturally induces maps on the associated factor groups: $(s\theta)^t(p) : T_p \to T'_p$, $(s\theta)^l(p) : L(p) \to L'(p)$ and $(s\theta)(p) = (s\theta)^t(p) \oplus (s\theta)^l(p) : M(p) \to M'(p)$. Moreover, if $\pi_p^u, (\pi')_p^u, u = t, l$ are the natural projection maps of Definition 4, restricted to the free subgroups $iF \subset \Phi(H), i'F' \subset \Phi(H')$, then, by construction, $(s\theta)^u(p)\, \pi_p^u = (\pi')_p^u\, (s\theta)$ for $u = t, l$. Using these facts, it is now easy to check that $\prod_p(s\theta)(p) \in Hom[\prod M(p), \prod M'(p)]$ will restrict to a coherent homomorphism from $\Phi(H)$ to $\Phi(H')$. We put

$$\Phi(\theta) = 1/s \bigotimes[\prod_p(s\theta)(p) + Hom_c(M, M') \cap Hom_Z(M, T')].$$

Using the definitions, it is routine to verify that $H \to \Phi(H), \theta \to \Phi(\theta)$ defines a functor from $\mathcal{QL}_\tau$ to $\mathcal{QM}_\tau$.

To construct the inverse functor Γ, let $M \in \mathcal{QM}_\tau$ with $T = \bigoplus T_p \subset M \subset_{pure} \prod M(p), M(p) = T_p \bigoplus L(p)$ where, for each p, T_p is a finite p-group and $L(p)$ is a finite rank free Z_p module. Choose $F \subset M$ a full free subgroup, let π_p^t, π_p^l be the maps from F_p to $T_p, L(p)$ induced by the natural projection maps and denote $N_p^t = \ker \pi_p^t, N_p^l = \ker \pi_p^l$.

We construct a torsion-free finite rank group $H = \Gamma(M)$ with $F \subset H \subset QF$ by defining each localization H_p. Since $M \in \mathcal{QM}_\tau$ and $\tau = [(k_p)]$, we have $p^{k_p}T_p = 0$ for almost all p. Thus, for almost all $p, p^{k_p}F_p \subset N_p^t \subset F_p$. For these p, we have a chain of subgroups of QF: $F \subset F_p \subset p^{-k_p}N_p^t \subset p^{-k_p}F_p$. For these p, define $H_p = [p^{-k_p}N_p^t + QN_p^l] \subset QF$. For the finitely many exceptional p, let $H_p = [F_p + QN_p^l]$. Note that since F is contained in each p-local group H_p, if we define $H = \cap H_p$, then H will be a torsion-free group of rank equal to rank F such that each p-localization is equal to the predefined H_p. Since each H_p is the sum of the free Z_p module $p^{-k_p}N_p^t \subset p^{-k_p}F_p$ and the divisible group QN_p^l, each H_p will be the direct sum of a free and a divisible Z_p-module. Thus H is locally completely decomposable. Additionally, since for almost all $p, p^{-k_p}N_p^t \subset p^{-k_p}F_p$ then $FOT(H) \leq \tau$. Finally, for at least one $p, L(p) \neq 0$. For this p, N_p^l will be a proper pure subgroup of F_p. Hence $H_p \neq D(p)$ for this p and H is reduced. Thus $H \in \mathcal{QL}_\tau$.

It remains to prove that the H constructed above is independent, up to quasi-isomorphism, of the initial choice of F. Let $C \subset M$ be another full free subgroup and let H^c be the torsion-free group constructed by employing C in place of F. In view of Remarks 0, we can regard F and C as subgroups of a common rational vector space V such that, for a finite set of primes P_0, $F_p = C_p$ for $p \notin P_0$ and $F_p \doteq C_p$ for $p \in P_0$. Regarding $H, H^c \subset V$, we have $H_p = H_p^c$ for $p \notin P_0$. For $p \in P_0$, let K_p^l be the kernel of the map from C_p onto $L(p)$ naturally induced by the projection map from $\prod M(q)$

onto $L(p)$. Since $L(p)$ is torsion-free and $C_p \doteq F_p$, then $K_p^l \doteq N_p^l$. Thus, for $p \in P_0$, QK_p^l, the divisible part of H_p^c, coincides with QN_p^l, the divisible part of H_p. Hence, $H^c \doteq H$, as desired.

To complete the definition of Γ, first take $\theta \in Hom_c(M, M')$ and let $\theta\hat{} = \theta + Hom_c(M, M') \cap Hom_Z(M, T')$ be the associated $\mathcal{QM}_\tau$ map from M to M'. Arguing as previously, $k_1\theta(F) \subset F'$ for some positive integer k_1. Thus θ can be regarded as a map from QF to QF'. Put $\Gamma(\theta\hat{}) = \theta \in Hom_Z(QF, QF')$. Recall that, as constructed, $\Gamma(M) \subset QF, \Gamma(M') \subset QF'$. We claim that $\Gamma(\theta\hat{}) = \theta$ is a quasi-homomorphism from $\Gamma(M)$ to $\Gamma(M')$. To prove our claim, first, by the coherence assumption, $\theta(T_p^*) \subset (T')_p^*$ for almost all p. Since, for all p, $N_p^t = F_p \cap T_p^*$, $(N')_p^t = F_p' \cap (T')_p^*$ and $(N')_p^t$ is of finite index in F_p', there is a positive integer k_2 with $k_1k_2\theta(N)_p^t \subset (N')_p^t$ for all p. Next, note that $N_p^l = F_p \cap [T_p \bigoplus \prod_{q \neq p} M(q)] = F_p \cap [T_p \bigoplus p^\omega M], (N')_p^l = F_p' \cap [T_p' \bigoplus p^\omega M']$ and any map from M to M' carries $[T_p \bigoplus p^\omega M]$ into $[T_p' \bigoplus p^\omega M']$. Hence $k_1\theta(N_p^l) \subset (N')_p^l$ for all p. It follows directly from the definitions that $k_1k_2\theta[\Gamma(M)_p] \subset \Gamma(M')_p$ for all p. Hence $\Gamma(\theta\hat{}) = \theta$ is a quasi-homomorphism from $\Gamma(M)$ to $\Gamma(M')$. We remark that if $\theta \in Hom_Z(M, T')$ the first step of our construction gives $k_1\theta(F) = 0$. Thus θ will yield the zero map in $Hom_Z(QF, QF')$. This implies that our definition of $\Gamma(\theta\hat{})$ is independent of the choice of the representative θ for $\theta\hat{}$. Finally, we employ the formula $\Gamma(r/s \bigotimes \theta\hat{}) = r/s \bigotimes \theta$ to extend the domain of Γ to all of $Hom_{\mathcal{QM}_\tau}(M, M')$.

It is easy to check that the correspondence $M \to \Gamma(M), \theta\hat{} \to \Gamma(\theta\hat{})$ is functorial.

Plainly, the functors Φ, Γ are rank preserving, where the rank of a mixed group means the torsion-free rank. To complete the proof of our theorem we need to show that $\Gamma\Phi \sim 1_{\mathcal{QH}_\tau}$ and $\Phi\Gamma \sim 1_{\mathcal{QM}_\tau}$. The equivalence $\Gamma\Phi \sim 1_{\mathcal{QH}_\tau}$ is straightforward. Indeed, if the full free subgroup $iF \subset \Phi(H)$ is used in the construction of $\Gamma\Phi(H)$, then the abelian groups H and $\Gamma\Phi(H)$ are easily seen to be quasi- isomorphic. Any quasi-homomorphism from H to H' will be carried naturally to a quasi-homomorphism from $\Gamma\Phi(H)$ to $\Gamma\Phi(H')$.

The equivalence $\Phi\Gamma \sim 1_{\mathcal{QM}_\tau}$ takes more work to establish and illustrates the need for the many projection conditions in Definition 4. First we show, for each $M \in \mathcal{QM}_\tau$, there exists a $\mathcal{QM}_\tau$ isomorphism $t_M : \Phi\Gamma(M) \to M$. Maintaining our established notation, recall that $\Gamma(M)_p = p^{-k_p}N_p^t + QN_p^l$. Thus, applying the construction of the functor Φ, $\Phi\Gamma(M) \subset \prod M(p)^\#, M(p)^\# = T_p^\# \bigoplus L(p)^\#$, with $T_p^\# = (F_p + QN_p^l)/(N_p^t + QN_p^l), L(p)^\# = F_p/N_p^l$. By parts (b) and (c) of the projection conditions, $L(p)^\# = F_p/N_p^l$ is isomorphic to $L(p)$ for almost all p and embeds into a subgroup of finite index in $L(p)$ for the finitely many remaining p. For each p let $\Psi_p^l : L(p)^\# \to L(p)$ be this isomorphism or embedding.

Consider the natural surjection $F_p \to (F_p + QN_p^l)/(N_p^t + QN_p^l)$. The kernel of this map is $F_p \cap (N_p^t + QN_p^l)$. Since $N_p^t \subset F_p$ the latter intersection is equal to $N_p^t + F_p \cap QN_p^l = N_p^t + N_p^l$ (N_p^l is pure in F_p). By part (d) of the projection conditions the kernel of our surjection is N_p^t for almost all p. Thus, employing projection condition (a), we obtain, for almost all p, an abelian group isomorphism $T_p \cong F_p/N_p^t \cong (F_p + QN_p^l)/(N_p^t + QN_p^l) = T_p^{\#}$. For these p let $\Psi_p^t : T_p^{\#} \to T_p$ be the inverse of that isomorphism. For the finitely many remaining p put $\Psi_p^t = 0$.

Taking direct products of maps, let $\Psi = \prod(\Psi_p^t \oplus \Psi_p^l) : \prod M(p)^{\#} \to \prod M(p)$. We have shown that $\ker \Psi \subset \bigoplus_{p \in P_0} T_p^{\#}$, where P_0 is a finite set of primes, and that $\prod M(p)/image\Psi$ is finite. The map Ψ restricts to a coherent homomorphism from $\Phi\Gamma(M)$ to M with finite kernel and cokernel. It follows that $t_M = 1 \otimes (\Psi + Hom_c[\Phi\Gamma(M), M)] \cap Hom_Z[\Phi\Gamma(M), T])$ will be the desired $\mathcal{QM}_\tau$ isomorphism from $\Phi\Gamma(M)$ to M.

To complete the proof of Theorem 5 we need to show that the isomorphisms $\{t_M : M \in \mathcal{QM}_\tau\}$ are natural. Let $M' \in \mathcal{QM}_\tau$ with full free subgroup F' and torsion subgroup T'. Consider, without loss of generality, a $\mathcal{QM}_\tau$ map of the form $\hat{\theta} = 1 \otimes [\theta + Hom_c(M, M') \cap Hom_Z(M, T')]$, where θ is a coherent homomorphism from M to M' such that $\theta(F) \subset F'$. Naturality for arbitrary maps follows since any $\mathcal{QM}_\tau$ map can be written as $q\hat{\theta}$ with $q \in Q$ and $\hat{\theta}$ a map of this form. For such $\hat{\theta}$, a simple computation shows that $\Phi\Gamma(\hat{\theta}) = t_M \hat{\theta} t_{M'}^{-1}$. This equality establishes the naturality condition and the proof is complete. ◇

In [6], for each non-trivial locally free type τ, inverse category equivalences $f_\tau : \mathcal{QH}_\tau \to \mathcal{QG}_\tau$, $g_\tau : \mathcal{QG}_\tau \to \mathcal{QH}_\tau$ were constructed. (In that construction the objects in $\mathcal{QH}_\tau$ were not allowed to have rank one summands of type τ.) Let $\iota : \mathcal{QH}_\tau \to \mathcal{QL}_\tau$ be inclusion. Our second theorem shows that Φ_τ, Γ_τ can be regarded as extensions of f_τ, g_τ.

Theorem 6. *For each non-trivial locally free type τ, there is an embedding j : $\mathcal{QG}_\tau \to \mathcal{QM}_\tau$ such that $\Phi_\tau \iota = j f_\tau$ and $\Gamma_\tau j = \iota g_\tau$.*

Proof. Let $G \in \mathcal{QG}_\tau$ with $\bigoplus T_p \subset G \subset_{pure} \prod T_p$. Choose a full free subgroup $F \subset G$. For each p, let $\pi_p : \prod T_q \to T_p$ be the natural projection map. Let $d : F \to \prod F_p$ be the diagonal embedding. Define $e : F \to \prod(T_p \bigoplus F_p)$ by $e(f) = \prod \pi_p(f) \oplus d(f)$. Take $j(G)$ to be the pure subgroup of $\prod(T_p \bigoplus F_p)$ generated by $\bigoplus T_p$ and $e(F)$. We leave it to the reader to check that $j(G) \in \mathcal{QM}_\tau$ and that $j(G)$ is independent, up to isomorphism in $\mathcal{QM}_\tau$, of the choice of F.

Suppose $\alpha \in Hom_{\mathcal{QG}_\tau}(G, G')$. Employing the obvious notation and reasoning as in the proof of Theorem 5, we regard α as a quasi-homomorphism F to F', hence from eF to eF'. Using the fact that α is a $\mathcal{QG}_\tau$ map from G to G' and simple properties of the diagonal embedding d, it is not hard to see that an integral multiple of α induces a map from $\prod(T_p \bigoplus F_p)$ to $\prod(T'_p \bigoplus F'_p)$ that restricts to a coherent homomorphism from $j(G)$ to $j(G')$. (For help with the details refer to Proposition 2.1 of [1].) This coherent homomorphism in turn provides, in the obvious way, a $\mathcal{QM}_\tau$ map $j(\alpha)$: $j(G) \to j(G')$.

The fact that $j : \mathcal{QG}_\tau \to \mathcal{QM}_\tau$ is a category embedding and the composition equalities $\Phi_\tau \iota = j f_\tau$, $\Gamma_\tau j = \iota g_\tau$ are routine. $\diamond$

We close with a few remarks.

- Each mixed group category $\mathcal{QM}_\tau$, being equivalent to $\mathcal{QL}_\tau$, enjoys all the well known properties of $\mathcal{QL}_\tau$; for example $\mathcal{QM}_\tau$ is a Krull-Schmidt category with indecomposables of arbitrary rank. Also, an object M of $\mathcal{QM}_\tau$ is indecomposable if and only if its endomorphism ring $E_\tau(M) = Hom_{\mathcal{QM}_\tau}(M, M)$ has no proper idempotents. Note that if M is also an object of $\mathcal{QM}_\sigma$ for some $\sigma \neq \tau$ then the rings $E_\tau(M)$ and $E_\sigma(M)$ are isomorphic, their construction being independent of the particular locally free type under consideration. It follows that M is indecomposable in $\mathcal{QM}_\tau$ if and only if M is indecomposable in $\mathcal{QM}_\sigma$. Using this fact it follows easily that the category $\mathcal{QM}$ is Krull-Schmidt.

- Let X be a rank one group of type τ and let $W_\tau = Hom(\_, X)$ be the corresponding Warfield duality on the category $\mathcal{QH}_\tau$. In [4] it was shown that the composite $f_\tau W_\tau$ is independent of τ; that is if $\tau < \sigma$ and we regard $\mathcal{QH}_\tau$ to be a full subcategory of $\mathcal{QH}_\sigma$, then $f_\tau W_\tau$ is naturally equivalent to the restriction of $f_\sigma W_\sigma$. to $\mathcal{QH}_\tau$. This independence condition was used to construct a duality from $\mathcal{QH}$, the category of all locally free groups and quasi-homomorphisms, to $\mathcal{QG}$. No such result is possible here since Warfield duality is unavailable for locally completely decomposable groups.

- A careful examination of the proof of Theorem 5 shows that the need for coherent homomorphisms in the definition of $\mathcal{QM}_\tau$ arises from the non-uniqueness of the free summand $R(p)$ in the decompositions $H_p = R(p) \oplus D(p)$ for lcd groups H. This non-uniqueness is the major difficulty in dealing with lcd groups as opposed to locally free groups.

References

[1] U. Albrecht, H.P. Goeters and W. Wickless, The flat dimension of mixed abelian groups as E-modules, Rocky Mt. J. Math. 25(2) (1995), 569-90.

[2] S. Glaz and W. Wickless, Regular and principal projective endomorphism rings of mixed abelian groups, Comm. in Algebra 22(4) (1994), 1161-76.

[3] S. Files and W. Wickless, The Baer-Kaplansky Theorem for a class of global mixed abelian groups, to appear in Rocky Mt. J. Math.

[4] A. Fomin and W. Wickless, Categories of mixed and torsion-free abelian groups, to appear in Proceedings of International Conference on Abelian Groups and Modules, Padova.

[5] C. Vinsonhaler and W. Wickless, Realizations of finite dimensional algebras over the rationals, Rocky Mt. J. Math.24(4) (1994), 1553-65.

[6] W. Wickless, A functor from mixed groups to torsion-free groups, Cont. Math. 171 (1995), 407-19.